47th Vibration Institute Annual Training Conference and Expo (VIATC 2024)

AA004950

Covington, Kentucky, USA
7 - 9 August 2024

ISBN: 979-8-3313-2291-5

Printed from e-media with permission by:

Curran Associates, Inc.
57 Morehouse Lane
Red Hook, NY 12571

Some format issues inherent in the e-media version may also appear in this print version.

Vibration Institute
2625 Butterfield Road
Suite 128N
Oak Brook, IL 60523-3415
USA

Phone: (630) 654-2254
Fax: (630) 654-2271

information@vi-institute.org

Additional copies of this publication are available from:

Curran Associates, Inc.
57 Morehouse Lane
Red Hook, NY 12571 USA
Phone: 845-758-0400
Fax: 845-758-2633
Email: curran@proceedings.com
Web: www.proceedings.com

VIATC TRAINING CONFERENCE & EXPO

COVINGTON, KENTUCKY | AUGUST 7-9, 2024

47TH ANNUAL TRAINING CONFERENCE AND EXPO
COVINGTON, KENTUCKY

CONFERENCE PROCEEDINGS

Vibration Institute
www.vi-institute.org
(630) 654-2254

VIATC TRAINING CONFERENCE & EXPO

COVINGTON, KENTUCKY | AUGUST 7-9, 2024

Proceedings of the Vibration Institute Annual Training Conference
VIATC 2024
Covington, KY
August 7 – 9, 2024

Table of Contents

No.	Title	Author	Page
01	Making an Initial Machine Vibration Assessment	George D'Entremont	1
02	Static and Couple Balancing Calculations Using Commonly Available Modern Tools	Steven P. Ciesla, PE and Michael J. Titone, PE	17
03	The Analysis and Comparisons of Different Types of Vibration Rubs	Thomas B. Lantrip	41
04	Unbalance Control of Rotors Using a Smart Bearing: SEMAJIB	Aly El-Shafei, Ph.D.	68
05	Adding Value for Turbomachinery Evaluation with Case Monitoring	Jeff Mauk	80
06	A Mathematical Review and Measured Validation of Dynamic Stiffness of a Simply Supported T-Beam	Tom Calko	98
07	Make Service Companies Your Partner in Reliability	Forrest Pardue	121
08	War Stories of a Vibration Analyst	Monroe Voyles	132
09	Naïve Bayes Classifier Optimization for Bearing Fault Detection	Matthew J. Kenney	174
10	Creating a Virtual Assistant to Vibration Analyst	Mantosh Bhattacharya	188
11	CBM Analysis Using Limited Vibration Data: Classification of Paper Machine Bearing Defects Using ML Models and RUL Prediction	Lloyd Dewey Lee, Jr. "Buddy",	203
12	Dynamic Analysis of Rigid Foundations that Support Dynamic Equipment	Robert J. Sayer, PE	230
13	Foundation Interface Problems as it Relates to Vibration Investigations	Thomas Lantrip	250

14	Analysis and Mitigation of Acoustic-mechanical FCCU Feed Piping and Nozzle Vibration	Scott I. McNeill	273
15	Case Studies on Hydro Turbines with High Vibrations	Bernard Boueri	290
16	Investigation into a Large Boiler Feed Water Pump Thrust Bearing Failure at Bruce Power	James Chin	302
17	Advanced Analytics Support Malfunction Detection on Reciprocating Compressors	Fayyaz Qureshi	313
18	Reciprocating Compressor Crosshead Guide Vibration	Brian Howard	329
19	Belt-Driven Fan Vibration Case Study: The Rotor Resonance that Wasn't … Or Was It?	Neil Tugby	358
20	Standardized Fan Diagnostics According to ISO 13373-5	Aly El-Shafei, Ph.D.	377
21	High-Frequency Vibration Mitigation Through Cylinder Design for Reciprocating Compressors	Tom Stephens, P.E.	400
22	Batteries and IoT - Are they Made for Each Other?	Ed Spence	416
23	Modal Analysis of a Vertical Turbine Pump Using Motion Magnification and Traditional Methods	Andrew Lerche	424
24	Troubleshooting High Vibration on a Steam Turbine Generator Using a New Video Alignment Method and Traditional Vibration Methods	Andrew Lerche	430
25	Online Oil Health Monitoring for Land Fill Gas Engines	Jeremy Sheldon	443
26	Evaluation of Energy Losses Due to Degradation and Defects in Electric Machinery with Electrical Signature Analysis, vibration Analysis, and Motion Amplification	Howard W Penrose, Ph.D., CMRP, CEM	453
27	The Impact of Accelerometer Mounting Techniques on the Accuracy of Vibration Data	Peter Eitnier	469

VIATC TRAINING CONFERENCE & EXPO
COVINGTON, KENTUCKY | AUGUST 7-9, 2024

MAKING AN INITIAL MACHINE VIBRATION ASSESSMENT

George D'Entremont

George D'Entremont recently retired from a 41-year career at PG&E's Diablo Canyon Power Plant. George participated in the initial commissioning of Diablo Canyon; holding positions as a Reactor Operator, Operations and Engineering Instructor, and for the last 26 years as a Predictive Maintenance Engineer. He holds an ISO Level IV Vibration Analyst Certification.

ABSTRACT:
This paper describes strategies and provides a guideline that supports an initial and timely assessment of an elevated or unusual machine vibration condition. A written initial assessment of such a condition at a nuclear power plant is typically expected on the day of discovery. Using vibration data as a baseline, initial assessments are strengthened by incorporating evidence from field inspections, operating history, maintenance history, and corroborating PdM technologies. The intent of these initial assessments, while not necessarily final or perfect, are to serve the immediate needs of the station and provide sufficient insight to move forward.

An Assessment Guideline is provided which may be used as a tool to support a timely initial machine vibration assessment. The information in the guideline is sourced from the training and experience of a tenured vibration analyst at a large nuclear generating station. While the analyst remains responsible for their assessment, under the pressure of time and urgency the guideline prompts a broader and more thorough review of a machine's condition. The intended result of using the Assessment Guideline is a written initial assessment that has a higher probability of remaining valid pending subsequent discoveries and occurrences.

KEY WORDS: Assessment, Guideline, Initial, Machine, PdM, Vibration.

INTRODUCTION:

The origins of this paper come from discussions related to the processes and strategies employed by machine vibration analysts in performing their work. Demographics are such that many vibration analysts are close to retirement. Their replacements are curious to know how they perform their assessments, how they prioritize activities, what they look at, and how they process information. This paper is an attempt by the author to describe his process, using his notes and personal resource material, as a forty-one-year employee at a large nuclear generating facility. The process described here is specifically that which might be employed by an onsite vibration analyst at a nuclear generation facility, but may likely be adopted by analysts in other industrial settings. It also addresses, specifically, those circumstances in which an onsite analyst is asked to provide an initial vibration assessment for a rotating machine, in writing, on the day a machine vibration problem is discovered. The author acknowledges that many such processes and strategies are effective and valid; but attempts, here, to describe his own.

DISCUSSION AND SCOPE:

Over the past 30 years, much has been entrusted to the activities of machine vibration analysts. With the advent of high-speed chips and data storage capabilities, enormous databases of digitized vibration spectra, time waveforms, and other plot formats have become available for evaluation and interpretation. Although efforts are afoot to develop automatic or smart interpretation of this data, this has yet to be fully proven or implemented. Coincident with this, and with the implementation of Reliability Centered Maintenance (RCM) principles, many time-directed maintenance activities have been deleted in favor of condition-based maintenance. The challenge left to condition monitoring practitioners is to distill information in a manner that may be used to make accurate and effective maintenance recommendations, preserving critical functions of the equipment they oversee. A further coincidence is that many of the experienced vibration analysts are retiring, leaving less experienced analysts to interpret vibration information for the same purpose.

Although machine vibration analysis is an extremely useful condition monitoring technology, it commonly requires corroborating data from other sources to provide the most insightful evaluations. These other sources include information from maintenance histories and inspections, operational process data and anomalies, and corroborating PdM technologies. Use of such corroborating data has, to date, been left primarily to the creativity and imagination of the analysts and maintenance professionals doing the work.

An "Initial Assessment Guideline" is submitted as an addendum to this paper that seeks to arm a machine vibration analyst with guidance to review vibration data against corroborating data to make an initial vibration assessment. It is not offered as a series of procedural steps, nor a fault or decision tree that must be proscriptively sequenced through. The guideline is instead offered as an organized accumulation of reminders that a trained analyst likely already knows, allowing him/her to review these reminders in a systematic-like but non-proscriptive manner. The analyst may not be able to complete all elements of the Initial Assessment Guideline in the time available. Under time pressure, however, the author has found that the process of addressing and attempting to complete elements of the guideline provides a structured approach, calms urgency, and commonly yields insight sufficient to write an initial assessment prior to going home that day. The goal of such an assessment, while not necessarily final or perfect, will be to serve the immediate needs of the station and provide sufficient insight to move forward.

INITIAL RESPONSE TO A REPORTED VIBRATION ISSUE:
Machine vibration problems are most commonly identified by trends observed and tracked over time by the vibration analyst, in a manner such that he/she may carefully consider potential causes and appropriate recommendations. The circumstances addressed in this paper, however, are those in which the condition emerges abruptly, requires timely response, and is possibly initially identified by someone other than the analyst, such as by Maintenance or Operations staff. When an analyst is initially notified of such a machine condition, it is best for the analyst to go into a mode of listening and asking questions. Unless the cause of the vibration is obvious and must be urgently addressed, the analyst is best to avoid voicing initial guesses. Guesses that must be retracted later may misdirect an organization's initial response, and chip away at an analyst's otherwise hard-earned good reputation. The analyst should also avoid lingering hallway conversations with people unfamiliar with the issue. Time is limited. A written report is due prior to the analyst going home that day.

Among the first questions an analyst should ask is to identify any source material or initial reports of the condition. The analyst should not rely exclusively on word-of-mouth reports of the condition. Once identified, read the source material. Also identify any personnel that may have observed or have insightful information of the issue. If they are onsite, or otherwise available for interview, make a note for future data-gathering.

Next, perform a review of the machine's existing vibration database, identify and note baseline vibration amplitudes and notable frequencies. Review the type of machine and significant characteristics, such as operating speed, type of bearings, type of couplings, et al.

INITIAL ASSESSMENT GUIDELINE (REFERENCE ADDENDUM)
Once the analyst has completed initial reviews, an "Initial Assessment Guideline" is provided to help the analyst continue their investigation (see ADDENDUM). The information in the guideline is sourced from the training and experiences of a vibration analyst at a large nuclear generating station. This guideline provides no information beyond that which a trained analyst likely already knows or might infer. It doesn't solve a vibration problem for the analyst. The analyst remains responsible for the assessment. But under the pressure of time and urgency the guideline prompts a broader and more thorough review of a machine's condition, considering relevant sources of information beyond just vibration. The intended result is a written initial assessment that has a better chance of remaining valid pending subsequent discoveries and occurrences.

Diagnostic Fundamental: The first page of the "Initial Assessment Guideline" cites the "Forced Vibration Response" equation. It is not intended that the analyst should insert values or try to solve the issue numerically. It, instead, serves to remind the analyst of variables that affect machine vibration; hopefully allowing the analyst's training, experience, and even intuition to interpret the machine's vibration condition.

Confirmatory Vibration Survey and Field Assessment: These activities and inspections are performed locally, at the site of the machine under attention. A portable vibration data analyzer is recommended to perform a confirmatory vibration survey. In addressing additional items under this assessment, the analyst will find that a digital pocket strobe, a spot radiometer, and a camera will be useful.

Operational Assessment: Items under this assessment may be addressed by either personnel interview, and/or review of the station's CMMS (Computerized Maintenance Management System) database. The assessment may result in unanswered questions, prompting future investigations or inspections that may aid the evaluation.

Maintenance Assessment: Items under this assessment may be addressed by either personnel interview, and/or review of the station's CMMS database. The assessment may result in unanswered questions, prompting future investigations or inspections that may aid the evaluation.

Predictive Maintenance (PdM) Assessment: Items under this assessment may be addressed by reviewing the station's vibration monitoring and/or CMMS databases for evidence of previous similar vibration issues or assessments. Also review other available PdM technology databases (e.g. thermography, oil analysis, etc.) for corroborating evidence that might identify anomalies or potential degradation. These reviews will, hopefully, identify corroborating PdM evidence that will contribute to the overall evaluation, and/or prompt future insightful inspections.

Machine Vibration Matrix: This matrix is a nine-page document that arranges "Sources" of machine vibration in vertical columns. These sources are color coded to reflect "conditions" in blue (e.g. Imbalance, misalignment, etc.), "components" in orange (e.g. bearings, couplings, etc.), and "machines" in purple (e.g. fans, motors, etc.). "Symptoms" are arranged in horizontal rows, and include vibration frequency domain, maintenance assessments, et al.

Once assessment data is collected, the analyst returns to his workstation for evaluation. Notes should be taken to summarize findings from the confirmatory vibration survey; identifying notable frequencies responsible for the offending vibration, and whether this vibration is isotropic (evidenced in all directions radial) or anisotropic (dominant in a single radial direction). Findings from the Confirmatory Vibration Survey, combined with insight gained from the Field Assessment, Operational Assessment, Maintenance Assessment, and Predictive Maintenance Assessment, are then tested against the Vibration Matrix.

Use of the Vibration Matrix begins by reviewing the title of each column of potential Sources, and the top two rows of corresponding symptoms (i.e. "Vibration Frequency Domain" and "Vibration Direction/Amplitude"). These top two rows of symptoms are color coded in red, and have emboldened frequencies and vibration directions (*isotropic* or *anisotropic*) for associated sources. Sources of vibration that correspond to symptoms identified in the top two rows of the Matrix are circled by the analyst, sources that do not are struck out. The analyst then follows clues from the Matrix, as deemed appropriate, to narrow potential sources of vibration for the machine condition in question. The result of this process hopefully identifies a cause, or potential causes, of the offending vibration.

Use of the "Initial Assessment Guideline" may parallel use of ISO 13373-3, "Guideline for vibration diagnosis". This ISO standard offers deeper detail and more exhaustive explanations. The "Initial Assessment Guideline", in contrast, offers a more abbreviated approach, with a specific entry point to its matrix by reviewing the top two rows of symptoms for each potential source of vibration. It also has more specific reminders to incorporate maintenance and operation histories, as well as corroborating PdM technologies. The author offers the "Initial Assessment Guideline" as a tool he has found useful to perform timely initial evaluations. Collaborative use of ISO 13373-3 may, however, enhance an analyst's evaluation

COMMON EXPECTIONS OF AN INITIAL VIBRATION ASSESSMENT

The Initial Assessment Guideline does not "create" a written machine assessment, but it does "inform" an assessment. The trained vibration analyst remains responsible for the assessment. By the author's experience, however, once relevant information specified in the Initial Assessment Guideline is collected, and reviewed against diagnostic elements in the Vibration Matrix, potential sources of vibration become evident.

The author's organization commonly requests three things from an initial machine vibration assessment. This committed in writing before the analyst goes home that day.

1. What is likely causing the vibration.
2. A recommendation whether the machine may be left in service?
3. A recommendation identifying what maintenance or inspections should be performed or prepared for?

The Initial Assessment Guideline does not solve a vibration problem, but helps inform a potential cause, or causes, per item #1. This, then, informs good judgment for items #2 and #3.

These are initial assessments. They are similar, in concept, to medical triage; in which conditions are initially assessed, evaluated for potential causes, and prioritized by urgency. These assessments are not necessarily final or perfect, but suitable to move forward. Each initial assessment will, however, be committed to a permanent record, and measured against future findings and occurrences. It is important, then, to consider as much evidence as possible, and apply considered judgement.

SUMMARY

The author has attempted, here, to share a process he has used to perform initial machine vibration assessments. This process was developed, and modified, over a twenty-six-year career as a machine vibration analyst at a large nuclear generating station. It reflects knowledge gained through training and certification as an ISO Level IV Vibration Analyst. Some analysts may, however, disagree with some of the diagnostic elements of this process. Our discipline remains relatively new, is still evolving, and there is much yet to discover. The author welcomes discussion and suggested improvements to this process, hopefully resulting in a product that will aid both the author and other analysts seeking to improve their practices.

ADDENDUM:

Machine Vibration – Initial Assessment Guideline

Machine Vibration – Initial Assessment Guideline

Diagnostic Fundamental (Force Response Eq.):

$$X = \frac{\frac{F_0}{k}}{\sqrt{\left[1-\left(\frac{\omega}{\omega_n}\right)^2\right]^2 + \left[2\zeta\frac{\omega}{\omega_n}\right]^2}}$$

Where:
X = mass motion, inches-peak
F_0 = exciting force, lbf
k = spring stiffness, lbf/in
ω = forcing frequency, rad/s
ω_n = natural frequency, rad/s
ζ = damping ratio = C/C_c

Confirmatory Vibration Survey and Field Assessment (performed locally):

Caution: Observe all site and industry safety rules and protocols! *Approach cautiously. Observe "two-minute rule" prior to approaching machine. If contacting equipment, initiate contact with back of hand or fingernail. Observe "never stand in the bight" rule during machine starts or transients (e.g. close proximity, areas with no escape, radial or axial from rotating elements, close to leakage paths or gasketed flanges, et al.*

Perform a Confirmatory Vibration Survey.
In addition, inspect equipment for the following:

1. Sound anomalies. Try to identify specific source locations of suspected sound anomalies.
2. Radiometer gun temperature anomalies. Reference hot spots for personal safety and diagnostics.
3. **Caution! Observe safety protocols!** "Nibbling" at joints may be sensed by touch. Any relative motion, sensed by touch at either grouted or jointed interfaces, should be noted.
4. Degraded pedestal or grout. Visible evidence of fracture or degradation. Rigid pedestal base vibration >0.03-0.05 ips-p. Machine foot vibration >30% of bearing vibration.
5. Excessive, corroded, or displaced shims.
6. Lubrication anomalies; level, flow, temperature, expelled, misdirected.
7. Inspect visible portions of rotating element; seal leakage, visible runout, shaft shuttling, etc.
8. Strobe tach anomalies. Speed vs historical speed, speed modulation, condition of rotating components, etc.
9. Valve or damper mispositioning.
10. Environmental anomalies (e.g. temperature, moisture, ventilation, corrosive liquids or atmosphere, etc.).
11. If machine bearing vibration appears isotropic, validate with a measurement at 45 deg between orthogonal survey points.

Operational Assessment (performed by personnel interview and/or review of CMMS Database):

1. Anomalies noted during operation, startup, or shutdown.
2. Anomalies noted in operating or process parameters (e.g. load, pressure, temperature, etc.).
3. Different operating condition, lineup, or procedure, "are you running equipment differently now that in the past?"

Maintenance Assessment (performed by personnel interview and/or review of CMMS Database):

1. Details of maintenance history (e.g. recent maintenance outages, repeated failures, etc.).
2. Anomalies noted during recent inspections (unusual as-found conditions, out-of-tolerance conditions, etc.).
3. Anomalies observed during recent maintenance (e.g. difficulty in assembling or aligning, "did it seem to go together right?").
4. New or different procedures, parts, or components used during recent maintenance.

Predictive Maintenance (PdM) Assessment (personnel interview and/or review of PdM Databases):

1. Review the station's Vibration Database for previous similar occurrences and/or assessments.
2. Review the station's PdM Databases for corroborating evidence of anomalies or degradation.

Machine Vibration Matrix – Common Sources and Symptoms of Machine Related Vibration

Machine Vibration – Initial Assessment Guideline

	Unbalance	Resonance	Looseness / Impacting
Vibration Frequency Domain	1X	1X of rotor speed most common. 1X Harmonic. 2XLF or RBPF (motor), Vane Pass (may affect piping or fan wheel; not likely rotor). Any Forcing Frequency. Any Frequency in the presence of impacting or rub.	1X w/ multiple harmonics. "Swoop" pattern across tops of harmonics in spectra. Fractional orders in presence of resonant frequencies or severe condition.
Vibration Direction / Amplitude	Isotropic High Radial. Couple Unbalance - Possible high axial.	Anisotropic - Structural. High in direction of resonance mode. Isotropic - Shaft Critical (Except Split Critical)	Anisotropic Amplitude difference across loose strructural interfaces.
Vibration Time Domain	Sinusoidal	Sinusoidal in the direction of greatest response.	Not sinusoidal. Sharp peaks or truncation
Vibration Phase	Stable. Rotates w/ transducer.	Substantial phase changes w/ speed change.	Unstable. Phase difference across loose structural joints.
Vibration Orbit	Circle / Open Ellipse. Normally forward precession. Possible reverse precession if overhung.	Elongated ellipse along axis of resonant response. Normally forward precession (Split Critical - reverse precession)	
Field/Operation Assessment	Rigid Shaft: Force $\propto$ Speed Squared. Step change - possible material loss. Gradual change - Possible erosion or oxide accumulation on rotating element.	Substantial amplitude changes with speed changes. High balance sensitivity. Sole plate "Hammer test" may ID new grout voids that have allowed *Fn* to approach excitation freqs.	Caution! Observe safety protocols! "Nibbling" at joints may be sensed by touch. Any relative motion, sensed by touch at either grouted or jointed interfaces, is evidence of looseness.
Maintenance Assessment	Newly installed or balanced? Inspect for damaged or corroded rotating element. Axial vibe may be quieted by couple balance correction.	Structural changes w/ recent maint? Recommend quieting resonance before attempting to balance. Use of Bellevelle washer may detune structural resonance.	Undertorqued fasteners, fractured weldments, damaged grout. Visible fretting or denting at loose contact surfaces. Bluing shows insufficient contact.
PdM Assessment	If recently balanced, review applicable documents for acceptable residual balance condition and balancing procedure.	Verified with impact test, modal analysis, runup/coastdown Bode. May rerquire "running" impact test to determine shaft resonant freqs.	NDE evidence. Wear particles in oil if looseness occurrintg within lube boundary. ODS. Motion Amplification.
Notes	Rigid rotors ref ISO 21940-11 and IRD "Balance Rqmts Rigid Rotors". Flexible rotors ref ISO 21940-12. In-situ balance ref ISO 21940-13. Pumps ref API 610-2021, 6.9.3. Use "time synch avg" when balancing near external forces.	May quiet resonance by varying mass, stiffness, damping or speed. Torsional resonance normally undetected?... But generates high radial in gear units.	

Machine Vibration – Initial Assessment Guideline

Machine Vibration Matrix – Common Sources and Symptoms of Machine Related Vibration

	Rub	Softfoot / Distortion	Thermal Variability
Vibration Frequency Domain	**1X, harmonics, non-synch, sub-synch, broad-banded.** Possible "chaotic" symptoms. Multiple possible freqs. May excite non-synch resonant frequencies.	**1X.** **2XLF or RBPF** (Motor). **Vane Pass** (Pump).	**1X.** **1X & Harms w/ Pole Pass S/Bs** (motor).
Vibration Direction / Amplitude	**Any direction.** **Possible long period "oscillation"** from light rub or viscous heating that creates a long period 1X amplitude oscillation... period from a few min to an hour (Morton Effect).	**Any direction.**	**Any direction.** Vibration amplitude change occurs during warm-up following cold start.
Vibration Time Domain	Non-sinusoidal. Truncation.		
Vibration Phase	Unstable. 1X may create circular pattern on polar plot, progressing against shaft rotation over time (Morton Effect).		Phase may vary during warm-up, which may warrant caution if performing balance calc.
Vibration Orbit	Truncation, exterior loops. Full annular rub may result in reverse precession.		
Field/Operation Assessment	Noise or visible indication near area of rub, may be more obvious during coastdown.		Vibration change occurs during warmup to equilibrium conditionns, with change in machine or process fluid temperatures, or change in ventilation or ambient conditions.
Maintenance Assessment	NDE evidence at contact point.	Softfoot check. Pipe strain, piping flanges should match without jacking. Excessive, displaced, or corroded shims.	
PdM Assessment	Thermography - Evidence of high temperature at contact point. Wear particles in oil if rub occurringt within lube boundary.		Thermography - Evidence of temperature anomalies. MCSA - Pole Pass S/Bs possible motor rotor fault.
Notes	Thermal bow may be caused by rub-induced localized rotor heating. Reversing any initiating speed or load change may quiet vibe. Rub may "wipe away" interference ...restoring normal vibration.	Avoid attempt to correct softfoot if pedestal is degraded or soleplate delaminated.	Thermal bow may be caused by localized rotor heating (eg rub, Morton Effect, motor rotor fault, etc),

Machine Vibration – Initial Assessment Guideline

Machine Vibration Matrix – Common Sources and Symptoms of Machine Related Vibration

	External Force	Shaft Misalignment	Coupling Degradation/Condition
Vibration Frequency Domain	**Any Frequency.** Frequency peaks generated by adjacent machinery.	**1X and/or 2X. Possible low-order harmonics** (dependent on coupling type). Evident on both sides of coupling.	**1X, and/or 2X Possible low-order harmonics** (dependent on coupling type). Evident on both sides of coupling. **Fractional Orders** - Severe wear or inadequate lube of geared coupling.
Vibration Direction / Amplitude	**Any direction. Possible long period "beat"** between machine freqs and closely matched external freqs generated by adjacent components.	**Anisotropic.** Changing axial or radial. Presence may *increase or decrease* vibe amplitude.	**Anisotropic** - Degradation, changing axial or radial. Presence may *increase or decrease* vibe amplitude. **Isotropic** - Hub runout, face or rim. Uneven bolt torque. Torsional binding.
Vibration Time Domain		"W" or "M" shape, or double step in sinusoid.	"W" or "M" shape or double step in sinusoid
Vibration Phase		Does not rotate with transducer. Possible approx 180 deg phase shift across coupling... axial or radial.	Degradation - Does not rotate with transducer, possible 180 deg phase shift across coupling... axial or radial.
Vibration Orbit		Pinched, peanut shaped, figure 8. Possible exterior loops.	Pinched, peanut shaped, figure 8. Possible exterior loops.
Field/Operation Assessment	Assess likely external sources and compare frequencies.	May cause adjacent bearings to become loaded or unloaded, corroborated by changing bearing metal temperatures.	Grease expelled inside coupling guard.
Maintenance Assessment	Use "time synch avg" when balancing around external forces.	Newly Installed or aligned? Allowable misalignment ref Piotrowski "Basic Shaft Alignment Workbook".	Observe degraded or worn coupling, hardened or inadequate lubricant. Mis-directed lubricant spray. Geared cpling sleeves don't slide off. Excessive hub runout, rim or face. Uneven coupling bolt torque.
PdM Assessment		Thermography identifies changing bearing housing temp, or increased coupling temp (dependant on coupling type).	Thermography identifies increased coupling temp (dependant on coupling type).
Notes		Ref. Piotrorwski "Shaft Alignment Handbook". May require "Off-line to Running" alignment survey to correct. Avoid attempt to correct shaft alignment if pedestal is degraded or soleplate delaminated.	Torsional Binding - Amplitude or phase change following start/stop cycle, load or speed change. Cannot rely on Vibe Analysis to predict coupling failure.

Machine Vibration – Initial Assessment Guideline

Machine Vibration Matrix – Common Sources and Symptoms of Machine Related Vibration

	Degraded Pedestal	Bent Shaft	Cracked Shaft
Vibration Frequency Domain	1X, or any forcing frerquency w/ possible harmonics.	1X	1X and/or 2X
Vibration Direction / Amplitude	Anisotropic. > 0.03 - 0.05 ips-p @ pedestal base. Machine foot vibe >30% of bearing vibration. Possibly Isotropic?... Location dependent, vertical machine?	Isotropic Possible high axial.	Isotropic Possibly High. Crack may propogate without vibration increase (Dependent on fracture plane).
Vibration Time Domain		Sinusoidal	
Vibration Phase	Pedestal "rocking" indicated by 180 deg phase difference betweeen vert measurements on adjacent machine feet.	Stable. Rotates w/ transducer. Axial approx 180 deg out of phase across element	Phase may change (lag) w/ crack propagation
Vibration Orbit		Circle / Open Ellipse	1X forward precession, w/ possible 2X internal loop.
Field/Operation Assessment	Caution! Observe safety protocols! "Nibbling" at joints may be sensed by touch. Any relative motion, sensed by touch at either grouted or jointed interfaces, is evidence of degradation.	Rigid Shaft: Force ∝ Speed Squared	Progressively decreasing rpm of critical speed on successive rollups may corroborate crack propagation.
Maintenance Assessment	Observed damaged grout. Sole plate "Hammer test" - Hollow or "drum" sound is evidence of grout voids or delaminating sole plate.	Excessive radial runout.	Visual inspection confirms crack.
PdM Assessment	Motion Amplification. Operating Deflection Shape (ODS).		NDE - Evidence of crack
Notes	Simply retorquing embedded bolts is ineffective, must re-establish grout bonding. Consider epoxy fill of grouted base. Avoid attempt to correct alignment or softfoot if pedestal is degraded or soleplate delaminated.	Bent shaft evident at slow-roll speed. "Bowed" shaft may "roll-out" over time, following extended slow-roll, w/ heat, or admission of steam. Possible "hand-baring" process may straighten a bowed shaft.	Often diagnosed initially as unbalance with progressively changing phase as crack propogates. Fracture plane is perpendicular to stress, bending stress may go undetected prior to failure.

Machine Vibration – Initial Assessment Guideline

Machine Vibration Matrix – Common Sources and Symptoms of Machine Related Vibration

	Runout / Eccentric Rotor	Rolling Element Bearing Degradation
Vibration Frequency Domain	1X	**Stage 1 Fatigue Degradation / or Underlubrication** Elevated hi freq accel (>5K Hz) – HFD or Spike Energy. Excites transducer *Fn*. **Stage 2 Fatigue Degradation** Elevated accel (0.5K – 6K Hz) – Defect freqs in demod spectra. Excites Brg *Fn*. **Stage 3 Fatigue Degradation** Defect Freqs in vel spectra (<1K Hz) – harms and/or sidebands. **Stage 4 Fatigue Degradation** Increasing 1X and defect freqs in velocity spectra.
Vibration Direction / Amplitude	**Isotropiic** - If it causes unbalance. **Anisotropic** - If it causes interaction with bushing, wear ring, belt, etc.	**Any Direction**, radial or axial. Evident closest to energy source.
Vibration Time Domain		Consider Replacement (ref accel freq band 10 Hz - 6K Hz): Ball Bearing True-Peak >8g-p Roller Bearing True-Peak >12g-p
Vibration Phase		
Vibration Orbit		
Field/Operation Assessment		Noise may accompany ball-sliding, under-lube, over-lube, early degradation, or false brinelling. Ball-sliding squeal, sometimes intermittant, may be accompanied by elevated housing temperature (increase approx 20 degF?).
Maintenance Assessment	Excessive runout… axial or radial.	Observe damaged ball paths due to fatigue failure, overload, false brinelling, contamination, or fluting. Confirm damage cause via ISO 15243. Ball-sliding - may be due to under-loaded bearing, possibly corrected with wavy-washer.
PdM Assessment		High Fe in oil may confirm bearing damage. High Fe in oil w/o defect freqs, possible Inner Ring/Shaft fretting. Black oil possibly due to poor oil delivery, low oil level, or ball sliding.
Notes	Visually observed runout sometimes interpreted as unbalance, but not necessarily true.	Recommend replacement @ Stage 3 (damage visible at contact surface of rolling elements). Diagnosis of rolling element bearings, ref ISO 13373-3, Annex D. Inadequate/contaminated lubrication - Elevated noise floor, broadband haystacking, or BSFs. Ball-sliding: Haystacking at housing Nat Freqs w/ BPFO sidebands.

Machine Vibration – Initial Assessment Guideline

Machine Vibration Matrix – Common Sources and Symptoms of Machine Related Vibration

	Rolling Elem Bearing Poor Fit / Distortion	Sleeve Bearing Degradation / Condition
Vibration Frequency Domain	**1X** - Cocked bearing. **Defect freqs in Velocity Spectra w/o corroborating high freq warning** - Pinched, Distorted, or pre-loaded bearing.	**1X** - Opening internal clearance, change in shaft centerline position (possibly due to loading or unloading bearing), internal misalignment, shaft scratch, mechanical or electrical runout. **Sub-synch** - Whirl: Approx 0.45X. Whip: Sub-synch locked on shaft resonance. Trapped fluid within hollow shaft or coupling: Approx 0.6X - 0.95X. **Sub-synch on long vert shaft** - Opening internal clearances, pump suction flow obstruction.
Vibration Direction / Amplitude	**Any Direction** Evident closest to energy source. **Axial** - Cocked bearing.	**isotropoic** - Opening internal clearance (opening all radial directions), change in shaft centerline position, low C/L eccentricity ratio, mechanical or electrical runout. **Anisotropic** - Opening internal clearance (dependent on axis of opening), Internal misalignment, improper twist/tilt.
Vibration Time Domain		1X spikes (Prox Probe) - Shaft scratach Random spikes (Prox Probe) - Electro-static discharge.
Vibration Phase		
Vibration Orbit		Whirl/Whip - Forward Precession. Spikes in-line w/ probe (Prox Probe) - Shaft scratch. Randomly directed spikes (Prox Probe) - Electro-static discharge.
Field/Operation Assessment	Elevated housing temps due to pre-load from poor fit or distortion.	Bearing metal or return oil temp change corroborates change in bearing loading or shaft C/L position. Increased return oil flow corroborates opening clearances. Whirl/whip may be initiated due to changes in oil viscosity or pressure.
Maintenance Assessment	Poorly fitted inner ring (<80% contact), may result in bore fretting.	Damaged babbitt or journal. Change in bearing Internal clearance. Change in Lift-Check dimension. Twist/Tilt condition. Journal damage due to abrasion, or electrical discharge/tracking. Failed bearing housing insulation test may explain electrical discharges.
PdM Assessment	Thermography - Elevated housing temps due to pre-load.	Oil analysis may confirm bearing damage. Electric discharge test validates electro-static discharge. Parasitic magnetism test (rotor >2 gause, housing >4 gause) may indicate cause of electric discharge damage.
Notes	Inner ring fretting may cascade quickly to failure; causing heat, expanding inner ring, squeezing bearing clearance, and loss of oil film.	Displacement thresholds: Internal Clearance ref ISO 20816-3, Annex C; Speed ref ISO 20816-3, Annex B; Load ref McHugh "... Turbomachinery". Shaft C/L w/ low eccentricity ratio may yield higher vibration. Reductiion in speed or load may quiet instablilties. Whirl/whip may require design modification for correction, reference Sommerfeld variables (e.g. clearance, viscosity, diameter, speed, etc).

Machine Vibration – Initial Assessment Guideline

Machine Vibration Matrix – Common Sources and Symptoms of Machine Related Vibration

	Belt/Sheave Wear	Centrifugal Pump Condition
Vibration Frequency Domain	1X - Sheave runout (radial or axial). 1X or Belt Frequency - Worn belts or sheaves. 1X and/or 2X - Sheave misalignment, Belt Freq Harmonics - Worn or damaged belts.	1X - Impeller or fluid unbalance. Vane pass - Common... but possibly indicates flow outside Preferred Operating Range, impeller/difuser wear, improper gaps, may excite acoustic resonance. Sub-Synch - Possible rotating stall. On deep-draft pump possible opening bearing clearances or suction flow restriction. Approx 0.40-0.49X - Whirl @ impeller shroud or bearing. Elevated noise floor, broadband haystacks - Insufficient NPSH, cavitation, flow turbulence, vortex sucked into pump suction, reversed impeller rotation (swapped motor leads?).
Vibration Direction / Amplitude	Any Direction	Any Direction
Vibration Time Domain		Non-periodic impacts due to low-flow shuttling.
Vibration Phase		
Vibration Orbit		
Field/Operation Assessment	Belt Flop - Under-tensioned, or unevenly tensioned belts. Worn belts ride low in sheave w/ possible excessive belt dust. Squealing belts may not threaten serviceability.	Rotating stall may occur outside Preferred Operating Range (70-120% BEP). BEP confirmed by reviewing pump curve. Degrading pump curve performance indicates wear. Observed non-periodic axial shaft shuttling at low flow. Noise (rocks) - cavitation due to insufficient NPSH.
Maintenance Assessment	Exposed belt cord. Bulge at belt splices. Cup shaped sheave race. Verify sheave pitch matches belt pitch.	Visually examine for damaged/worn impeller, diffusers, wear rings. Visually examine for flowpath obstruction affecting flowrate. Impeller oxide layer may alter balance condition. Identify improper lift or gap dimensions. Paired ACBB (thrst brg) must be installed back-to-back; ref API 610-2021, para 6.10.1.8; may otherwise affect vibe.
PdM Assessment	Thermography - worn or slipping belts may result in Increased belt temperature.	Monitor pump curve performance. Acoustic resonance corroborated by pressure transducers.
Notes	Belt squeal often mis-diagnoseed as brg squeal. Contrast belt noise from brg noise by spraying water on belt/sheave interface. When repllacinng sheaves, always also replace belts.	Diagnosis ref Olson, "Centrifugal Pump Vibration Diagnostics". Vibe Thresholds ref ISO 10816-7, HI 9.6.4., API 610-2021 6.9.4. Changing wear ring clearance may alter shaft stiffness (Lomakin effect), altering critical speed. Vane pass freq may excite acoustic or piping resonance, not likely to excite rotor resonance.

Machine Vibration – Initial Assessment Guideline

Machine Vibration Matrix – Common Sources and Symptoms of Machine Related Vibration

	Centrifugal Fan Condition	Induction Motor Condition
Vibration Frequency Domain	**1X** - Impeller unbalance. **Vane Pass** - Common. May coincide with fan wheel Fn. **0.6X to 0.75X** - Rotating stall. **0.5 to 2.0 Hz** - Surge. **Non synch** - Vortex Shedding at inlet damper.	**2XLF** - Common (esp 2-pole mtrs), but possible uneven air gap, softfoot, stator frame distortion, top cover fit, loose termination. **2XLF w/ Mult Harms** - Loose stator core, stator winding fault. **1X & Harms w/ Pole Pass SBs** - Common w/ 2-pole motor?.. possible rotor fault or delaminating rotor core. **1X w/ Slip Freq S/Bs** - Eccentric rotor. **RBPF w/ 2XLF SBs** - Common, if multiple S/Bs possible loose rotor bars, reliability likely unaffected. **Stator Slot Pass w/ 1X S/Bs** - Loose stator coils. **6XLF / 12XLF** - VFD related.
Vibration Direction / Amplitude	**Any Direction**	**Any Direction**
Vibration Time Domain		
Vibration Phase	1X Phase change after start/stop cycle - Debris or water in impeller void.	
Vibration Orbit		
Field/Operation Assessment	Stall may occur @ reduced flow. Surge @ unstable operarting point. Surge evident by noise, flow, and amp fluctuation (0.5Hz - 2Hz). Vortex Sheding - Inlet damper throttled (30%-50% open).	Vibe or audible beat between 2XLF and 2X shaft speed is common for 2-pole motor, if changing confirm contributing freqs. Gradaully changing 1X after start - faulted rotor bar heating? Speed modulation at pole-pass freq - faulted rotor bar? "Foghorn" sound at start - Loose stator weldments. 60Hz "buzz" during start w/o turning - improperly terminated.
Maintenance Assessment	Uneven buildup of contaminants, or uneven corrosion or erosion of impeller, may affect balance. Inspect for restriction, reducing flow. Damaged ductiing possibly caused by stall.	Softfoot. Measured uneven airgap (rotating or stationary eccentricity). Damaged stator bars/weldments. Evidence of contact between rotor and stator. Circulating currents corroborated by bearing insulation test, may cause bearing damage.
PdM Assessment	Perf monitoring by tracking fan curve operating point. Dynamic pressure monitoring, and duct vibration, validates stall.	Electrical tests - PI, Surge, RIC, MCSA, Torque Signature Analysis. No reliable test for stator insulation embrittlement.
Notes	Diagnosis ref Sayer "Dynamic Testing of Centrifugal Fan Wheels" and ISO 13373-5. Vibe thresholds ref ISO 14694. 1X Amp change after start/stop cycle - debris or water in impeller void.	Diagnosis ref IEEE Paper No. PCIC-99-20, ISO 13373-9. Vibe Thresholds ref ISO 20816-3, Uncouplled no load test thresholds ref NEMA MG 1-2006, Part 7. Rotor Fault diagnosis load dependent, recommend >80% rated load for accurate results.

Machine Vibration – Initial Assessment Guideline

Machine Vibration Matrix – Common Sources and Symptoms of Machine Related Vibration

	Steam Turbine Condition	Generator Condiiton	Gear Unit Condition
Vibration Frequency Domain	**1X** - Change in balance condition due to material loss from rotating component, change in bearing elevation due to changes in exhaust backpressure or heat sink temperature. **Sub Synch** - Oil whirl at beariing or steam whirl.	**1X** - Real or reactive load change or system disturbance? Possible Rotor Fault... could occur @ load change or system disturbance, causing step chng in amplitude, followed by 20 min additional gradual change. **2XLF** - Common on frame, brg housings, end-windings, and stator core. Acceptance thresholds are vendor specific.	**1X w/ multiple harmonics** - broken tooth (teeth).. **1X and/or GM with 1X S/Bs** - eccentric gear or pitch line runout. **Gear mesh and harms w/ S/Bs**- misaligned or worn gears. **Gear Fn (Intermediate Freqs)**- Underloaded, possible early stage degradation.
Vibration Direction / Amplitude	**Any Direction**	**Any Direction**	**Any Direction** Observed at monitoiring point closest to gear.
Vibration Time Domain			1X Impact - Broken gear tooth
Vibration Phase	Progressively changing phase & amplitude over time - progressing shaft crack?	Phase shift following system disturbance - Possible retaining ring slippage.	
Vibration Orbit			
Field/Operation Assessment	Progressively decreasing rpm of crit spds during rollups may accompany pedestal degradation. Chnge in turb process parameter. Transition between partial arc and full arc steam admission.	Large amplitude vibe increase at field bkr closure - possible rotor fault. H2 seal oil temp reduction may result in Morton Effect.	Audible pulsing at specific speed or load - Torsional Resonance.
Maintenance Assessment	Missing or damaged blade shrouds. Degradation of bearing support structure.	Pole drop test - Rotor Fault.	Broken gear teeth. Excessive or uneven gear wear. Misaligned gears or pitch line runout. Excessive backlash.
PdM Assessment	Any corroborating Turbine Supervisory indicator (e.g. thrust position, diff expansion, shaft centerline, casing expansion, etc).	Flux probe may validate faulted rotor.	Oil Analysis - Wear evidence in oil.
Notes	Diagnosis ref ISO 13373-4. Vibe Thresholds Stm Turb > 40MW ref ISO 20816-2, Gas Turb >3MW ref ISO 20816-4.	Diagnosis >50 MVA ref ISO 13373-10 Vibe Thresholds > 40MW ref ISO 20816-2.	Diagnosis ref Taylor "The Gear Analysis Hndbk", Mitchell "Intro Machine Analysis..." pg 214-232. Vibe thresholds ref ISO 20816-9. Common for gearbox vibe amplitude to vary substantially with load.

VIATC TRAINING CONFERENCE & EXPO

COVINGTON, KENTUCKY | AUGUST 7-9, 2024

STATIC/COUPLE BALANCING USING COMMONLY AVAILABLE MODERN TOOLS

Steven P. Ciesla, PE
Machinery Diagnostic Services Technical Leader, Bently Nevada
Steven.Ciesla@bakerhughes.com

Steven P. Ciesla, P.E. is presently a Machinery Diagnostic Services Technical Leader for Bently Nevada. He re-joined the company in 2013. Prior to that he worked as a predictive maintenance engineer at a nuclear power point where he was responsible for the vibration and lubrication programs. His work history includes being as a field engineer for both Bently Nevada and Westinghouse Process Control. He has over 20 years of hands-on, vibration analysis experience in the power generation industry.

Steve earned a BS in mechanical engineering from California State Polytechnic University, Pomona and an MSE in mechanical engineering from the University of Nevada, Las Vegas. He is a licensed professional engineer in the states of Nevada and California and certified as a Category IV Vibration Analyst through Mobius Institute.

Michael J. Titone, PE
Machinery Diagnostic Services Technical Leader, Bently Nevada
Michael.Titone@bakerhughes.com

Michael J. Titone, P.E. is presently a Machinery Diagnostic Services Technical Leader for Bently Nevada. He joined the company in 2011. In prior job assignments, he worked for a power generation company and a vibration consulting firm in various roles. Since 1989, Michael has been a practicing engineer in the areas of vibration analysis, balancing, condition monitoring, mechanical system design/analysis, and I&C system design.

Michael earned a BS in mechanical engineering from Louisiana Tech University and an MBA from Mississippi College. He also earned 18 hours of graduate credit in mechanical engineering classes. He is a licensed professional engineer in the state of Mississippi and certified as a Category III Vibration Analyst through the Vibration Institute.

ABSTRACT:

Balancing complex machine trains is often performed using expensive commercial software or proprietary software. In both cases, the programs may not be readily available for some users. Regardless of whether the user has access to software, an understanding of the calculation methodology is required to use the computer codes properly. The current paper extends the use of previously published standard balancing calculations using spreadsheets to static and couple modal balancing techniques. The static and couple balancing techniques reduce a two-plane balancing problem into two, single-plane problems. Static and couple balancing techniques are commonly applied to balancing large turbo-machinery rotors as the elevated vibration amplitudes observed may only manifest at a specific vibration mode on a particular rotor. Specifically addressing that mode's vibration can result in acceptable vibration amplitudes in fewer runs than performing true multi-plane balancing activities, thereby returning the unit to service sooner and saving the owner/operator significant cost.

KEY WORDS: Balancing, Multiplane, Static, Couple, Optimization, Excel

INTRODUCTION:

Field balancing of complex machine trains may be accomplished using a variety of techniques including single plane, static-couple, or multiplane. The static-couple technique provides a simple methodology for performing hand calculations when two measurement planes and/or two modes must be considered. The technique is often employed on complex machine trains, especially as the starting point of a balancing campaign. In many cases, based on the authors' experience, addressing a single mode with a single static pair or couple pair weight placement is sufficient to achieve an acceptable balance condition.

For the purposes of this paper, a typical complex machine train is a steam or gas turbine generator with rigid couplings, three to twelve bearings, multiple critical speeds, and multiple operating conditions. The techniques discussed, however, are not exclusive for complex machines. The techniques are equally usable on simpler machines with flexible couplings such as fans.

The measured vibration could be shaft vibration relative to the bearings, absolute bearing vibration, or absolute shaft vibration. For the paper, a well-balanced machine is the goal for a balancing program. Well-balanced for this purpose is: shaft vibration at all bearings below about 2.0 mils pk-pk of 1X vibration at normal operating conditions and/or bearing vibration at or below about 0.5 mils pk-pk of 1X vibration. Note that these vibration levels are stated as examples only. Actual vibration amplitudes for a well-balanced machine should be based on OEM and industry standards. The vibration through critical speeds and off-normal operating conditions should be below alarm levels.

One reason that the static couple (SC) technique often works well to balance complex machinery trains, such as large turbine generators, in the field is that the technique is similar to modal balancing. These machines typically operate in speed ranges where the effects of the first three modes, especially the first two modes, are prominent in the response. Since static weight placements affect mainly the first or third mode response and since couple weight placements affect mainly the second mode response, SC balancing often works well for field balancing these machines.

Another reason for the popularity of SC balancing is that turbine and generator rotors tend to be symmetric and supported by two bearings per rotor with balance planes on the ends of both rotors. This characteristic allows SC balancing to work well. By definition, static and couple weight placements are symmetric. Therefore, the best results from SC balance campaigns are typically obtained on symmetric rotors. However, SC techniques may be applied to non-symmetric rotors such as single flow turbines. In these cases, there can be considerable cross effects from static to couple and from couple to static. Stated differently, static shots can affect the second mode response, and couple shots can affect the first and/or third mode response for non-symmetric rotors. These issues may complicate the process, but they do not rule out the technique.

A previous paper prepared by the authors in 2023 [1] reviewed the calculation techniques and algorithms used in multiplane balancing and how they can be implemented in common programming languages and tools, such as Excel, Excel with VBA, and Python. This paper is an extension of that work, focusing on the implementation of static-couple technique using Excel with the built-in Solver tool to optimize the balance results.

DECIDING WHEN TO USE THE STATIC COUPLE APPROACH - MACHINE TRAIN DESIGN CONSIDERATIONS:

The machine train design for a complex machine such as a large steam turbine generator must be considered when SC balancing is used. There is no standard turbine generator machine train design. The authors have balanced a variety of machines from smaller "three-bearing" machines typically found in paper mills to twelve-bearing designs with six rotors and two bearings per rotor. Some machines have "shared" bearings, where a bearing is used to support two adjacent rotors.

The machine train design also allows for operation at speeds that involve up to three critical speeds. Smaller, stiffer rotors tend to operate above their first critical speed and below their second critical speed. Some low pressure (LP) rotors operate near their second critical speed. Some large, flexible generator rotors operate near their third critical speed.
In addition to the turbine and generator rotors, the couplings and jack shafts can play a significant role in the machine response. In a rigidly coupled machine, these parts act as small rotors.

The locations and numbers of balance planes are also significant. Some rotors have field balance planes at both ends only, some have them only at one end, and some have field accessible midspan balance planes. Also, some rigid couplings include a field balancing plane.

Some factors related to machine train design that support using SC techniques include:

- Symmetric rotor

- Rotor response is highly influenced by a single mode.
- Field accessible balance planes are symmetric.

STATIC-COUPLE CALCULATIONS:

Rotor balancing has been the subject of a great deal of research. Foiles et al. summarized some of this work in their paper from 1998, which listed one hundred and sixty-six references. [2] The section on Modal Balancing references works from K.R. Hopkirk [51]* in 1932, L.P. Grobel [43]* in 1953, and J.R. Lindsey [88]* in 1969 as the basis for static-couple balancing.

* Reference numbers from the Foiles et al. paper

Analysis of the synchronous vibration vectors from the bearings at each end of a rotor can provide insight into which vibration mode is being dominantly excited. To perform this analysis, synchronous vibration vectors from radial vibration probes installed at the same angular location at each end of the rotor are utilized (i.e., the Y probes located at 45° L from top-dead-center from the two bearings supporting a steam turbine LP rotor). The static or in-phase component is determined by adding the two 1X vectors and multiplying the amplitude (length) of the resulting vector by one-half. The couple or out-of-phase components are determined by subtracting the two 1X vectors and multiplying the amplitude (length) of the resulting vector by one-half. Note that the couple calculations will result in two vectors of the same length, but which are 180° out of phase from each other.

Assume that the 1X vibration vectors from the two bearings are denoted as A and B.

$$\boldsymbol{Static_A} = \boldsymbol{Static_B} = \frac{1}{2}(\vec{\boldsymbol{A}} + \vec{\boldsymbol{B}}) \tag{1}$$

$$\boldsymbol{Couple_A} = \frac{1}{2}(\vec{\boldsymbol{A}} - \vec{\boldsymbol{B}}) \tag{2}$$

$$\boldsymbol{Couple_B} = \frac{1}{2}(\vec{\boldsymbol{B}} - \vec{\boldsymbol{A}}) \tag{3}$$

These calculations can be performed graphically, calculated with a vector calculator, or determined with a computer program. Regardless of the technique used to calculate the static and couple components, understanding how the calculations work graphically is helpful for those learning the process.

First, the two original vectors, A and B, are plotted. Next, using a parallel ruler or a pair of triangles, the vectors can be translated so that the tip of A joins the tail of B, and the tip of B joins the tail of A to form a parallelogram. Then the two interior diagonals can be plotted. The gold line represents the sum of the vectors, and the light green line represents the difference of the two original vectors. Figure 1 illustrates the graphical calculation to determine the static vectors. Figure 2 illustrates the graphical calculation to determine the couple vectors, including a depiction of transferring the vectors to the origin to facilitate measuring the amplitudes and angles.

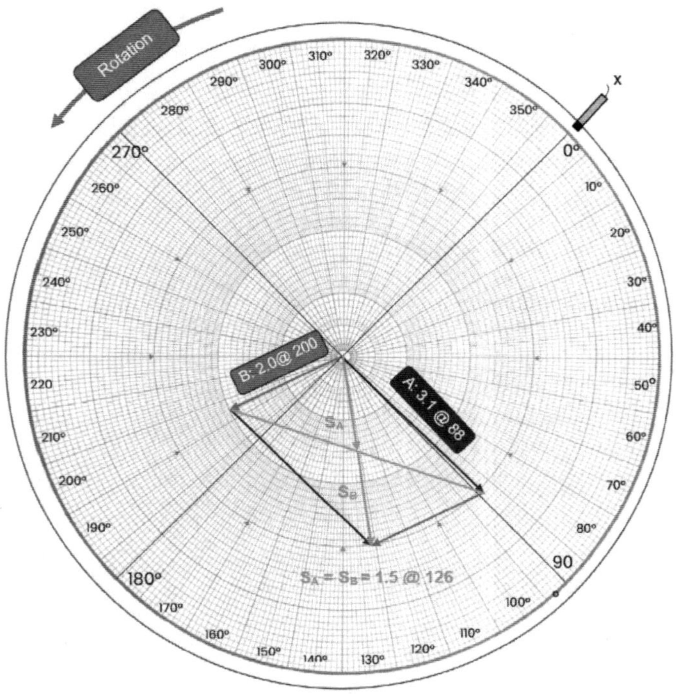

Figure 1, Graphical calculation of static vectors

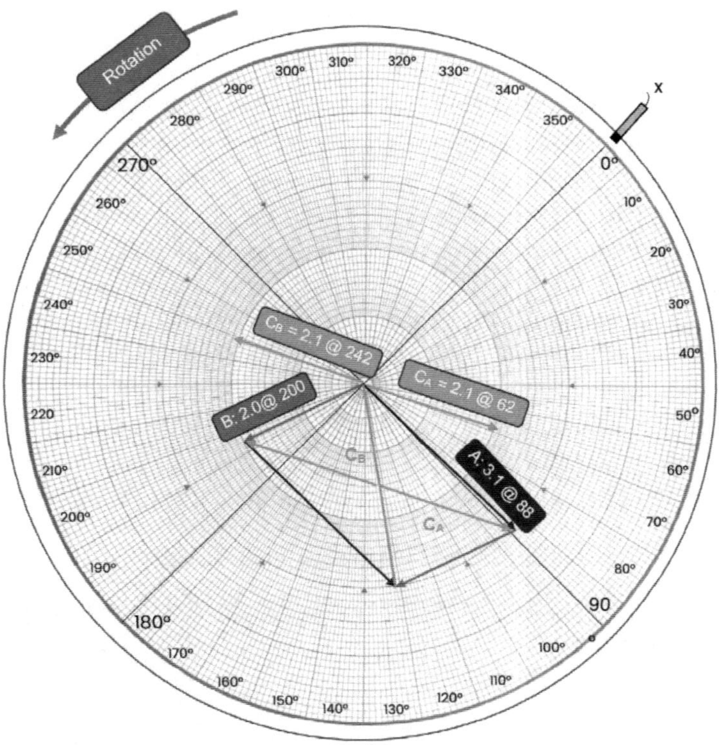

Figure 2, Graphical calculation of couple vectors

In Excel, the vectors are converted from polar coordinates (r,θ) into cartesian coordinates (x,y). The mathematical operations (addition and subtraction) are performed on the cartesian coordinate values independently. Alternatively, the couple components can be determined by taking the vectors and subtracting the static component.

$$\boldsymbol{Couple_A} = \vec{\boldsymbol{A}} - \vec{\boldsymbol{S}} \quad (4)$$

$$\boldsymbol{Couple_B} = \vec{\boldsymbol{B}} - \vec{\boldsymbol{S}} \quad (5)$$

STATIC-COUPLE FIELD BALANCE PROCEDURE:

The balancing process is well-described in a variety of references, including the cited references for this paper. Using the static-couple balancing technique involves the following high level steps:

- Run 0 – Obtain the reference or as-found data {O}
- Calculate the static and couple components for Run 0
- Determine which component to address, static or couple. This decision is discussed below.
- Shot 1 – Install a static pair ($W_{cal,static}$) or a couple pair ($W_{cal,couple}$)calibration weight.
- Run 1 – Operate the machine at conditions similar to those when the reference data was collected. {O+C}
- Calculate the static and couple components with the calibration weight installed.
- Evaluate the effect of the calibration shot {C} and calculate influence coefficients for all measurement locations and conditions. {H}
- Using Run 1 as a reference decide whether to optimize Shot 1 or whether to address the remaining component, static or couple.
- Assuming the first component vibration is satisfactory, install a static or couple pair for the second component, $W_{cal,static}$ or $W_{cal,couple}$, Shot 2
- Run 2 – Operate the machine at conditions similar to those when the reference data was collected.
- Calculate the static and couple components with the calibration weight installed.
- Evaluate the effect of the calibration shot {C} and calculate influence coefficients for all measurement locations and conditions. {H}
- Decide whether to optimize the balance condition or to end the campaign.

The balancing process described above is typically continued until the as-left vibration is acceptable. Figure 3 provides a flow chart illustrating a typical static couple balancing campaign.

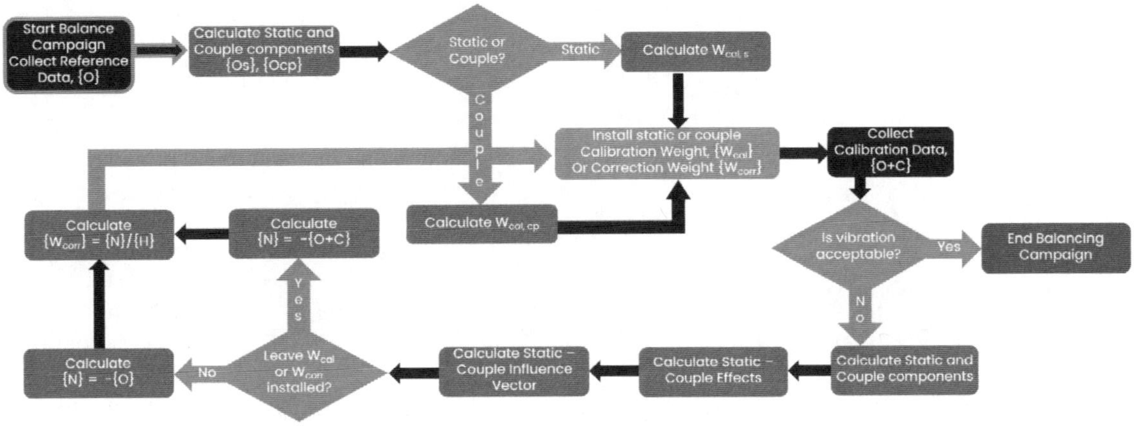

Figure 3, Balancing campaign flow chart

One critical decision point in the balancing process is determining which component, static or couple, to address first. Often, the vibration response can be improved enough to end the balancing campaign by addressing only the static or the couple component. In these cases, making a good decision early in the process can save a significant number of runs, thereby reducing down time and the cost of balancing.

The relative sizes of the static and couple components provide insight into the vibration mode that is most prominent in the response and may guide the balancing engineer on what type of balance correction to implement. If the couple component is dominant, then the rotor is likely running close to the second mode. If the static component is dominant, then the rotor is likely running close to the first mode or the third mode. However, these general statements are not always true. If a rotor is well balanced with respect to the second mode, and significantly imbalanced with respect to the first mode, the first mode response may be dominant even if the balancing speed is closer to the second mode. Figure 4 provides an example of a machine (rotor kit) operating at a speed (7000 rpm) well above the first and second modes while the first mode response remains dominate. Often in these situations, the static couple calculations will show that the static component is dominant. In addition to calculating the static and couple vibration components, it is beneficial to review transient 1X polar plots to see the vibration amplitudes and characteristics associated with each of the modes as was shown in Figure 4.

The following items should be considered when deciding which type of balance correction to implement first, static or couple:

- The mode that is nearest to operating speed is an important consideration. If the machine operates at a speed near the peak of a second mode response, most of the time addressing the second mode with a couple correction is a correct decision.
- The dominant response, static or couple, is a primary consideration.
- The available balance planes should be considered. Some machines do not have midspan access. In some cases, there is only a single balance plane on one end of the rotor.
- The history of the machine should also be a factor. For example, some machines do not respond to static corrections made in the end planes.

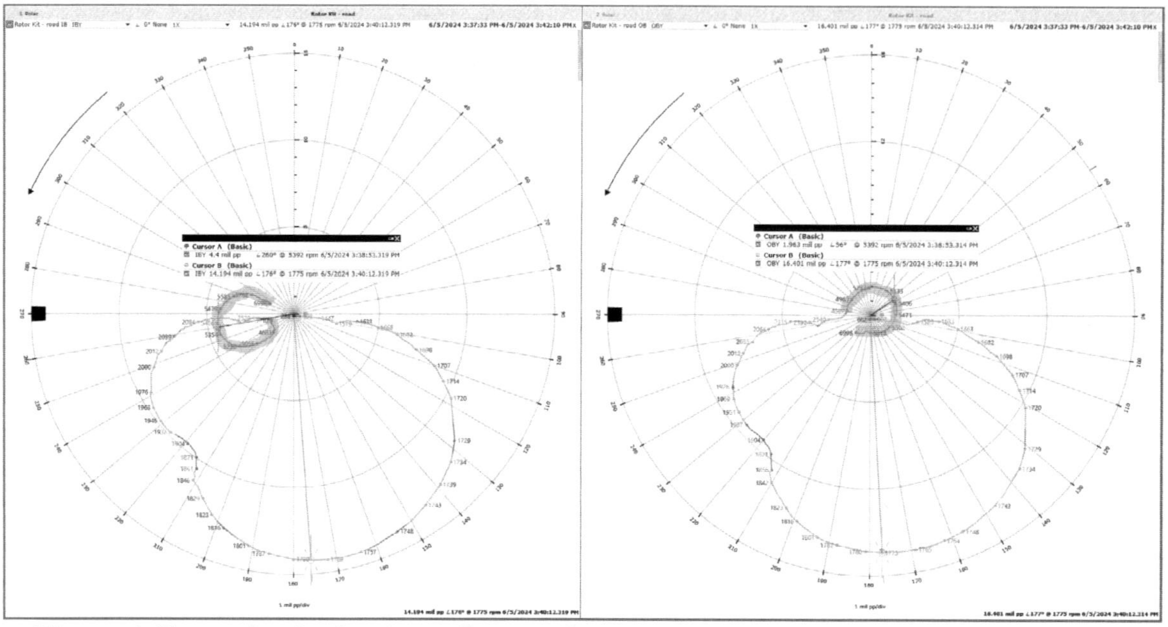

Figure 4, Example of a dominant first mode response highlighted in yellow and a well-balanced second mode response highlighted in blue

USING STATIC COUPLE BALANCING WITH MULTIPLANE BALANCING TECHNIQUES:

The term multiplane (MP) balancing is used to describe many techniques that are used to balance a complex machine train in multiple planes. In the context of this paper, MP balancing is any balance campaign where influence coefficients are calculated based on calibration runs involving a weight change in a single balance plane. Static-couple (SC) balancing is conveniently integrated with MP balancing by using a virtual single plane for static pairs and a virtual single plane for couple pairs. This approach allows using static and couple pairs while using the optimization processes inherent in a MP procedure. The MP approach employed in the field also involves knowledge of the machine modes and incorporates aspects of modal balancing.

In the traditional SC balancing approach, cross effects on other rotors in the machine train are not incorporated into the calculations. Similarly, cross effects from static to couple (couple effects due to static shots) and from couple to static (static effects due to couple shots) are not a part of the calculations. However, the cross effects can be significant. By using a MP technique and incorporating SC balance planes, all cross effects are considered in the optimization calculations.

The spreadsheets developed for this paper incorporate elements of a MP technique by considering the cross effects of a static shot on the couple response and of a couple shot on the static response. The traditional static couple approach would be to calculate an exact solution for the static or couple component using one measurement at each end of the rotor (Y or X probe pairs). The spreadsheets developed for this paper optimize the static or couple correction based on the predicted residual vectors, and the residual calculations includes the cross effects. This technique allows using multiple measurements, which allows consideration of the measurements from all probes (Y and X).

INCORPORATING THE CROSS EFFECTS:

Previous publications by Titone [3] and Yu [4] [5], describe and apply a relationship between static couple and multiplane influence coefficients and describe methodology to account for cross effects between static and couple balance shots. Some aspects of this work were incorporated into the current paper to demonstrate some improvements over the classic static couple balancing process.

Ideally, static weight placements only cause a change in the static response and couple weight placements only cause a change in the couple response. An ideal response is seldom realized in actual field balancing. This limitation of static couple balancing often arises at least in part from a lack of symmetry. Even if a rotor is constructed with almost perfect symmetry, it may be installed a machine train such that it is coupled to rotors with different characteristics on each end. For example, a double flow low pressure rotor may be coupled to a high pressure turbine on one end and a generator on the other end.

The couple response to static pairs and the static response to couple pairs is calculated within the spreadsheets developed for the paper. The implementation of an optimization process to arrive at a solution facilitates using the cross effects.

The spreadsheets developed for the current paper provide the functionality to calculate static and couple influence coefficients from a single end balance shot placed in either end of the rotor. These calculations assume that there are no static cross effects from a couple shot and no couple cross effects from a static shot. This assumption is typically reasonable for symmetric rotors. The calculation procedures for this analysis were discussed in the references cited above and are further described in the following section.

IMPLEMENTING THE PROCEDURE – REQUIREMENTS FOR EXCEL:

The spreadsheet developed to implement static-couple balancing calculations takes advantage of Excel's Solver tool to iteratively determine optimum balance weights using influence vectors obtained from the installation of trial weights in the applicable balance planes (static, couple, or single-ended). The spreadsheet consists of five worksheets (Static, Couple, Static_IV, Couple_IV, and Prediction) with similar structures and calculations. The "Static" and "Couple" worksheets perform the balancing calculations and establish the Solver parameters for static and couple pair balance shots respectively.

The "Static_IV" and "Couple_IV" worksheets are used to calculate static and/or couple influence vectors (IV) from single end-plane balance shots.

Superposition can be used to decompose a single end-plane balance shot as a combination of both static-pair and couple-pair balance shots installed simultaneously. For example, a single end-plane balance shot of 4 oz installed at 0 degrees on Plane A can be represented as a combination of a static-pair of 2 oz at 0 degrees at Plane A and 2 oz at 0 degrees at Plane B and a couple-pair of 2 oz at 0 degrees at Plane A and 2 oz at 180 degrees at Plane B (Figure 5). The "Prediction" worksheet allows the use of the influence vectors determined in the four other worksheets to predict responses and determine balance weights from different conditions (balancing the same machine in the future with different vibration responses or balancing a similar/identical machine).

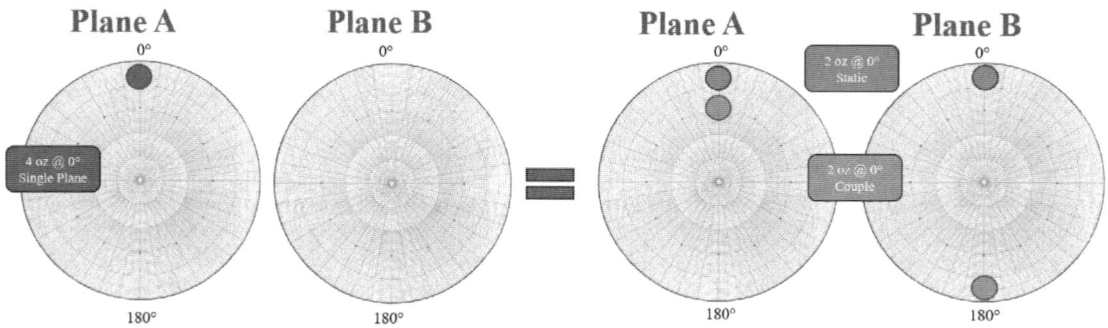

Figure 5, Decomposition of a 4 oz balance weight at 0 degrees at Plane A into a static-pair of 2 oz at 0 degrees at both Plane A and Plane B and a couple-pair of 2 oz at 0 degrees at Plane A and 2 oz at 180 degrees on Plane B

Prior to using the worksheets to balance a rotor, the original vibration data needs to be entered. On each of the worksheets, the cells that require manual input from the user are shown in green (Figure 6). These measurements include slow roll vectors for each probe (if applicable), the original vibration amplitude and phase, the calibration response vector resulting from the installation of the balance weight(s), and the calibration weight (size and angular location). Note that for static and couple pairs, the size and location of the weight installed at one plane (designated as Plane A) dictates the size and location of the weight installed at the other plane (designated as Plane B). The spreadsheet was developed using Plane A to track the balance weight size and location.

For each pair of measurements (Y Probes or X Probes), the worksheet will perform the associated balance calculations. These calculations are similar to those determined in a single plane balance, such as Original Vector, Calibration Vector, Null Vector, and Influence Vector but they are based upon the static and couple components. In addition to calculating an exact-point correction weight to eliminate the static or couple component, the "Static" and "Couple" worksheets also calculate the cross-effect influence vectors H_{CS} and H_{SC} which represent the couple effect from a static-pair and the static effect from a couple-pair respectively.

	Variable	Amplitude	Phase	x	y
Minimize RMS Residual	3.793	1.04	113		
Minimize Maximum (Mini-Max) Amplitude	4.138	0.64	115		
Correction Weight	Wcor	0.64	115	-0.272	0.582
Compensated Predicted Static Result 1Y/2Y	PS1'	2.82	120	-1.430	2.433
Compensated Predicted Static Result 1X/2X	PS2'	3.32	260	-0.553	-3.270
Compensated Predicted Couple Result 1Y/2Y	PC1'	1.92	180	-1.920	-0.008
Compensated Predicted Couple Result 1X/2X	PC2'	2.47	350	2.440	-0.418
Compensated Predicted Result 1Y	PR1'	4.14	144	-3.349	2.430
Compensated Predicted Result 1X	PR2'	4.14	297	1.887	-3.683
Compensated Predicted Result 2Y	PR3'	2.48	79	0.490	2.435
Compensated Predicted Result 2X	PR4'	4.14	224	-2.993	-2.858

Static Y Probes

Measurement #	1Y	1X Amplitude (mil pp)	1X Phase (degrees)	x	y
Slow Roll Vector	SR1	0.00	313	0.000	0.000
Original Response Vector	O1	5.67	135	-4.009	4.009
Compensated Original Response Vector	O1'	5.67	135	-4.009	4.009
Calibration Response Vector	O1+C1	5.01	144	-4.053	2.945
Compensated Calibration Response Vector	O1'+C1	5.01	144	-4.053	2.945
Measurement #	**2Y**	**1X Amplitude (mil pp)**	**1X Phase (degrees)**	**x**	**y**
Slow Roll Vector	SR2	0.00	245	0.000	0.000
Original Response Vector	O2	3.76	61	1.823	3.289
Compensated Original Response Vector	O2'	3.76	61	1.823	3.289
Calibration Response Vector	O2+C2	3.27	75	0.846	3.159
Compensated Calibration Response Vector	O2'+C2	3.27	75	0.846	3.159
Static Vector	S	3.81	107	-1.093	3.649
Static Null Vector	Nstatic	3.81	287	1.093	-3.649
Couple 1 Vector	CP1	2.94	173	-2.916	0.360
Couple 2 Vector	CP2	2.94	353	2.916	-0.360
Couple 1 Null Vector	Ncp1	2.94	353	2.916	-0.360
Couple 2 Null Vector	Ncp2	2.94	173	-2.916	0.360
Static Calibration Response Vector	S(O'+C)	3.45	118	-1.603	3.052
Static Calibration Vector	SC	0.79	229	-0.510	-0.597
Couple 1 Calibration Response Vector	CP(CP1+C)	2.45	182	-2.450	-0.107
Couple 2 Calibration Response Vector	CP(CP2+C)	2.45	2	2.450	0.107
Couple 1 Calibration Vector	CP1_C	0.66	315	0.466	-0.467
Couple 2 Calibration Vector	CP2_C	0.66	135	-0.466	0.467
Static Calibration Weight	Wcal,static	0.40	90	0.000	0.400
Static Influence Vector	Hss	1.96	139	-1.493	1.276
Static Correction Weight	Wcor,static	1.94	147	-1.630	1.051
Couple Influence Vector	Hcs	1.65	225	-1.168	-1.166

Static X Probes

Measurement #	1X	1X Amplitude (mil pp)	1X Phase (degrees)	x	y
Slow Roll Vector	SR1	0.00	313	0.000	0.000
Original Response Vector	O1	4.55	311	2.985	-3.434
Compensated Original Response Vector	O1'	4.55	311	2.985	-3.434
Calibration Response Vector	O1+C1	4.01	305	2.300	-3.285
Compensated Calibration Response Vector	O1'+C1	4.01	305	2.300	-3.285
Measurement #	**2X**	**1X Amplitude (mil pp)**	**1X Phase (degrees)**	**x**	**y**
Slow Roll Vector	SR2	0.00	245	0.000	0.000
Original Response Vector	O2	3.05	235	-1.749	-2.498
Compensated Original Response Vector	O2'	3.05	235	-1.749	-2.498
Calibration Response Vector	O2+C2	3.48	223	-2.545	-2.373
Compensated Calibration Response Vector	O2'+C2	3.48	223	-2.545	-2.373
Static Vector	S	3.03	282	0.618	-2.966
Static Null Vector	Nstatic	3.03	102	-0.618	2.966
Couple 1 Vector	CP1	2.41	349	2.367	-0.468
Couple 2 Vector	CP2	2.41	169	-2.367	0.468
Couple 1 Null Vector	Ncp1	2.41	169	-2.367	0.468
Couple 2 Null Vector	Ncp2	2.41	349	2.367	-0.468
Static Calibration Response Vector	S(O'+C)	2.83	268	-0.123	-2.829
Static Calibration Vector	SC	0.75	170	-0.740	0.137
Couple 1 Calibration Response Vector	CP(CP1+C)	2.47	349	2.423	-0.456
Couple 2 Calibration Response Vector	CP(CP2+C)	2.47	169	-2.423	0.456
Couple 1 Calibration Vector	CP1_C	0.06	12	0.055	0.012
Couple 2 Calibration Vector	CP2_C	0.06	192	-0.055	-0.012
Static Calibration Weight	Wcal,static	0.40	90	0.000	0.400
Static Influence Vector	Hss	1.88	80	0.343	1.851
Static Correction Weight	Wcor,static	1.61	22	1.490	0.610
Couple Influence Vector	Hcs	0.14	282	0.030	-0.138

Figure 6, A screenshot of the Static worksheet showing the required data in green cells and resulting balance calculations and predicted results used by Excel's Solver tool to determine appropriate balance shots

It is natural to describe 1X vibration in polar coordinates (r, θ) as it inherently matches the amplitude and directional nature of vectors. However, in order to perform the balancing calculations in Excel it is necessary to convert the vectors from polar coordinates to cartesian coordinates (x, y). The (x, y) components could also be represented in the Real and Imaginary planes as a complex number (x + yi) shown in Figure 7.

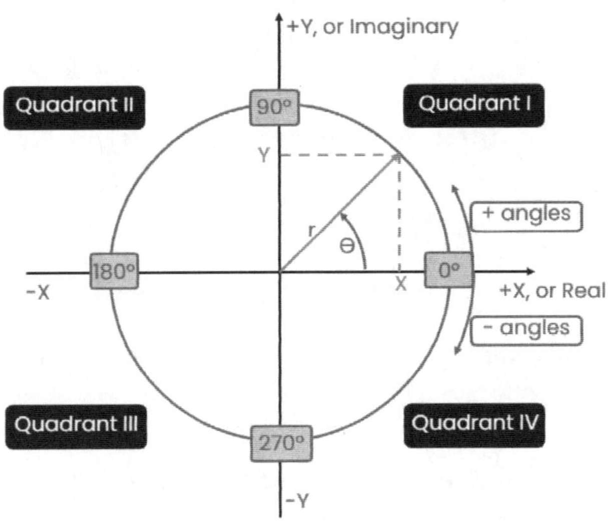

Figure 7, Polar, x-y, and real-imaginary coordinates

The following equations convert the vector from polar coordinates to rectangular components.

$$x = r * \cos(\theta) \quad (6)$$
$$y = r * \sin(\theta) \quad (7)$$

For trigonometric calculations in Excel, the angles must be expressed in radians rather than degrees as shown in Equations 8 and 9.

$$X = R * COS(RADIANS(\theta)) \quad (8)$$
$$Y = R * SIN(RADIANS(\theta)) \quad (9)$$

Conversely, given the x- and y- components of a vector, one can calculate the polar components in Excel as follows.

$$R = SQRT(X\char`^2 + Y\char`^2) \quad (10)$$
$$\theta = MOD(DEGREES(ATAN2(X, Y)), 360) \quad (11)$$

The phase calculation requires additional explanation. The ATAN2 function within Excel returns the angle in radians between -π and +π for the specified x- and y- coordinates. Note that -π radians is equivalent to -180 degrees and +π radians is equivalent to +180 degrees. Additional background information about the ATAN2 function and the use of radians as an angle unit is readily available from online sources.

The DEGREES function converts the angle measurement from radians to degrees. Since the ATAN2 function will produce angles between -180 and +180, the modulo or modulus function, MOD(), within Excel is used to convert the angle from 0 to 360. The MOD() function returns the remainder after a number is divided by a divisor (360 in this application).

In an Excel spreadsheet, vector calculations are implemented by performing calculations on the x- and y- components separately. Adding and subtracting is accomplished by adding or subtracting the x- and y- components. Previously, the static and couple components were defined as:

$$Static_A = Static_B = \frac{1}{2}(\vec{A} + \vec{B}) \quad (12)$$

$$Couple_A = \frac{1}{2}(\vec{A} - \vec{B}) \quad (13)$$

$$Couple_B = \frac{1}{2}(\vec{B} - \vec{A}) \quad (14)$$

In Excel, the x- and y- Static components would be calculated independently as:

$$Static_x = 0.5 * (A_x + B_x) \quad (15)$$

$$Static_y = 0.5 * (A_y + B_y) \quad (16)$$

Additionally, the x- and y- Couple components would be calculated independently as:

$$\boldsymbol{Couple_{Ax} = 0.5 * (A_x - B_x)} \tag{17}$$

$$\boldsymbol{Couple_{Ay} = 0.5 * (A_y - B_y)} \tag{18}$$

$$\boldsymbol{Couple_{Bx} = 0.5 * (B_x - A_x)} \tag{19}$$

$$\boldsymbol{Couple_{By} = 0.5 * (B_y - A_y)} \tag{20}$$

Vector multiplication and division are necessary for calculating predicted static/couple components and influence vectors / correction weights respectively. The influence vector is calculated as follows:

$$\vec{H} = \frac{\vec{C}}{\overrightarrow{W_{cal}}} \tag{21}$$

All three variables in the equation are composed of -x and -y components. Referring back to the Real and Imaginary coordinates (Figure 7), the influence vector equation can be represented as:

$$H_x + iH_y = \frac{C_x + iC_y}{W_{calx} + iW_{caly}} \tag{22}$$

In order to perform the complex division, we multiply the expression by 1, represented as a fraction with the complex-conjugate of the denominator in the numerator and denominator. The complex-conjugate has the same Real and Imaginary values as the original complex number, just that the sign of the Imaginary component is reversed. The multiplication is performed using the distributive property and the expression is simplified by combining like terms and remembering that $i^2 = -1$. The intermediate steps and final results are expressions for the -x and -y components of the influence vector.

$$H_x + iH_y = \left(\frac{C_x + iC_y}{W_{calx} + iW_{caly}}\right)\left(\frac{W_{calx} - iW_{caly}}{W_{calx} - iW_{caly}}\right) \tag{23}$$

$$H_x + iH_y = \left(\frac{C_xW_{calx} - iC_xW_{caly} + iC_yW_{calx} - i^2C_yW_{caly}}{W_{calx}^2 - iW_{calx}W_{caly} + iW_{calx}W_{caly} + W_{caly}^2}\right) \tag{24}$$

$$H_x + iH_y = \left(\frac{C_xW_{calx} - i^2C_yW_{caly} + i(-C_xW_{caly} + C_yW_{calx})}{W_{calx}^2 + W_{caly}^2}\right) \tag{25}$$

$$H_x + iH_y = \left(\frac{C_xW_{calx} - (-1)C_yW_{caly} + i(-C_xW_{caly} + C_yW_{calx})}{W_{calx}^2 + W_{caly}^2}\right) \tag{26}$$

$$H_x + iH_y = \frac{C_xW_{calx} + C_yW_{caly}}{W_{calx}^2 + W_{caly}^2} + i\frac{-C_xW_{caly} + C_yW_{calx}}{W_{calx}^2 + W_{caly}^2} \tag{27}$$

$$H_x = \frac{C_xW_{calx} + C_yW_{caly}}{W_{calx}^2 + W_{caly}^2} \tag{28}$$

$$H_y = \frac{-C_xW_{caly} + C_yW_{calx}}{W_{calx}^2 + W_{caly}^2} \tag{29}$$

A screenshot of the Static worksheet is shown in Figure 8. The user will input data into the green fields for the Slow Roll Vectors (SR), Original Response Vectors (O), Calibration Response Vectors(O+C), and Static Calibration Weight ($W_{cal,\ static}$). The worksheet will calculate the static and couple components as well as the Calibration Vectors (C), Influence Vectors (H_{ss} and H_{cs}) and Static Correction Weight ($W_{cor,static}$). In this example, the Static Correction Weight is calculated as 21.67 oz at 179 degrees for the Y probes and 25.55 oz at 185 degrees for the X probes. Note that the angular reference of the static-pair is relative to the Keyphasor target (notch) and not the Y or X probes when the Keyphasor target is under the Keyphasor probe.

Minimize RMS Residual	4.854	1.04	113		
Minimize Maximum (Mini-Max) Amplitude	5.319	0.64	115		
	Variable	Amplitude	Phase	x	y
Correction Weight	Wcor	0.64	115	-0.272	0.582
Compensated Predicted Static Result 1Y/2Y	PS1'	4.44	19	4.183	1.481
Compensated Predicted Static Result 1X/2X	PS2'	5.20	241	-2.522	-4.544
Compensated Predicted Couple Result 1Y/2Y	PC1'	0.24	305	0.137	-0.196
Compensated Predicted Couple Result 1X/2X	PC2'	0.61	323	0.486	-0.370
Compensated Predicted Result 1Y	PR1'	4.51	17	4.321	1.285
Compensated Predicted Result 1X	PR2'	5.32	247	-2.036	-4.913
Compensated Predicted Result 2Y	PR3'	4.38	23	4.046	1.677
Compensated Predicted Result 2X	PR4'	5.14	234	-3.008	-4.174

Static Y Probes

Measurement #	1Y	1X Amplitude (mil pp)	1X Phase (degrees)	x	y
Slow Roll Vector	SR1	0.00	313	0.000	0.000
Original Response Vector	O1	4.58	15	4.424	1.185
Compensated Original Response Vector	O1'	4.58	15	4.424	1.185
Calibration Response Vector	O1+C1	3.01	49	1.975	2.272
Compensated Calibration Response Vector	O1'+C1	3.01	49	1.975	2.272

Measurement #	2Y	1X Amplitude (mil pp)	1X Phase (degrees)	x	y
Slow Roll Vector	SR2	0.00	245	0.000	0.000
Original Response Vector	O2	4.42	21	4.126	1.584
Compensated Original Response Vector	O2'	4.42	21	4.126	1.584
Calibration Response Vector	O2+C2	3.44	52	2.118	2.711
Compensated Calibration Response Vector	O2'+C2	3.44	52	2.118	2.711

Static Vector	S	4.49	18	4.275	1.385
Static Null Vector	Nstatic	4.49	198	-4.275	-1.385
Couple 1 Vector	CP1	0.25	307	0.149	-0.199
Couple 2 Vector	CP2	0.25	127	-0.149	0.199
Couple 1 Null Vector	Ncp1	0.25	127	-0.149	0.199
Couple 2 Null Vector	Ncp2	0.25	307	0.149	-0.199
Static Calibration Response Vector	S(O'+C)	3.22	51	2.046	2.491
Static Calibration Vector	SC	2.49	154	-2.229	1.107
Couple 1 Calibration Response Vector	CP(CP1+C)	0.23	252	-0.072	-0.220
Couple 2 Calibration Response Vector	CP(CP2+C)	0.23	72	0.072	0.220
Couple 1 Calibration Vector	CP1_C	0.22	185	-0.220	-0.020
Couple 2 Calibration Vector	CP2_C	0.22	5	0.220	0.020
Static Calibration Weight	Wcal,static	12.00	135	-8.485	8.485
[illegible]	[illegible]	[illegible]	[illegible]	0.197	0.066
Static Correction Weight	Wcor,static	21.67	179	-21.669	0.246
[illegible]	[illegible]	[illegible]	[illegible]	0.012	0.014

Static X Probes

Measurement #	1X	1X Amplitude (mil pp)	1X Phase (degrees)	x	y
Slow Roll Vector	SR1	0.00	313	0.000	0.000
Original Response Vector	O1	5.36	246	-2.180	-4.897
Compensated Original Response Vector	O1'	5.36	246	-2.180	-4.897
Calibration Response Vector	O1+C1	4.30	276	0.449	-4.276
Compensated Calibration Response Vector	O1'+C1	4.30	276	0.449	-4.276

Measurement #	2X	1X Amplitude (mil pp)	1X Phase (degrees)	x	y
Slow Roll Vector	SR2	0.00	245	0.000	0.000
Original Response Vector	O2	5.19	233	-3.123	-4.145
Compensated Original Response Vector	O2'	5.19	233	-3.123	-4.145
Calibration Response Vector	O2+C2	4.02	257	-0.904	-3.917
Compensated Calibration Response Vector	O2'+C2	4.02	257	-0.904	-3.917

Static Vector	S	5.24	240	-2.652	-4.521
Static Null Vector	Nstatic	5.24	60	2.652	4.521
Couple 1 Vector	CP1	0.60	321	0.472	-0.376
Couple 2 Vector	CP2	0.60	141	-0.472	0.376
Couple 1 Null Vector	Ncp1	0.60	141	-0.472	0.376
Couple 2 Null Vector	Ncp2	0.60	321	0.472	-0.376
Static Calibration Response Vector	S(O'+C)	4.10	267	-0.227	-4.097
Static Calibration Vector	SC	2.46	10	2.424	0.424
Couple 1 Calibration Response Vector	CP(CP1+C)	0.70	345	0.677	-0.180
Couple 2 Calibration Response Vector	CP(CP2+C)	0.70	165	-0.677	0.180
Couple 1 Calibration Vector	CP1_C	0.28	44	0.205	0.196
Couple 2 Calibration Vector	CP2_C	0.28	224	-0.205	-0.196
Static Calibration Weight	Wcal,static	12.00	135	-8.485	8.485
[illegible]	[illegible]	[illegible]	[illegible]	-0.118	-0.168
Static Correction Weight	Wcor,static	25.55	185	-25.469	-2.087
[illegible]	[illegible]	[illegible]	[illegible]	-0.001	-0.024

Figure 8, A screenshot of the Static Worksheet showing the data entry fields in green and subsequent calculations

Predicted static and couple results from the Y and X probe pairs can be calculated from the derived influence vectors and the proposed correction weight. The predicted results (PS or PC) using the static-pair correction weight are determined from the standard single plane balancing calculation shown below.

$$\overrightarrow{PS} = \vec{S} + \overrightarrow{H_{SS}} W_{cor,static} \tag{30}$$

$$\overrightarrow{PC1} = \overrightarrow{CP1} + \overrightarrow{H_{CS}} W_{cor,static} \tag{31}$$

The predicted results (PS or PC) using the couple-pair correction weight are determined from the standard single plane balancing calculation shown below.

$$\overrightarrow{PS} = \vec{S} + \overrightarrow{H_{SC}} W_{cor,couple} \tag{32}$$

$$\overrightarrow{PC1} = \overrightarrow{CP1} + \overrightarrow{H_{CC}} W_{cor,couple} \tag{33}$$

The calculations are implemented in the worksheets using the previously described methods for vector algebra utilizing the x- and y- components. Note that CP2 is not calculated because it will be identical in amplitude to CP1 and 180 degrees opposite in phase.

Using the data from the previously shown static worksheet, if the Static Correction Weight from the Y probes is utilized, 21.67 oz at 179 degrees, then the predicted static result for the Y probes will be zero (Figure 9). Conversely, if the Static Correction Weight from the X probes is utilized, 25.55 oz at 185 degrees, then the predicted result for the X probes will be zero (Figure 10).

	A	B	C	D	E	F	G
1	Minimize RMS Residual	0.825	1.04	113			
2	Minimize Maximum (Mini-Max) Amplitude	1.183	0.64	115			
3		Variable	Amplitude	Phase		x	y
4	Correction Weight	Wcor	21.67	179		-21.669	0.246
5	Compensated Predicted Static Result 1Y/2Y	PS1'	0.00	0		0.000	0.000
6	Compensated Predicted Static Result 1X/2X	PS2'	0.91	266		-0.056	-0.913
7	Compensated Predicted Couple Result 1Y/2Y	PC1'	0.52	258		-0.110	-0.504
8	Compensated Predicted Couple Result 1X/2X	PC2'	0.51	16		0.489	0.136
9	Compensated Predicted Result 1Y	PR1'	0.52	258		-0.110	-0.504
10	Compensated Predicted Result 1X	PR2'	0.89	299		0.433	-0.776
11	Compensated Predicted Result 2Y	PR3'	0.52	78		0.110	0.504
12	Compensated Predicted Result 2X	PR4'	1.18	243		-0.545	-1.049

Figure 9, The predicted static result for the Y probes is zero when the static correction weight for the Y probes is used

	A	B	C	D	E	F	G
1	Minimize RMS Residual	0.853	1.04	113			
2	Minimize Maximum (Mini-Max) Amplitude	1.479	0.64	115			
3		Variable	Amplitude	Phase		x	y
4	Correction Weight	Wcor	25.55	185		-25.469	-2.087
5	Compensated Predicted Static Result 1Y/2Y	PS1'	0.92	230		-0.593	-0.710
6	Compensated Predicted Static Result 1X/2X	PS2'	0.00	0		0.000	0.000
7	Compensated Predicted Couple Result 1Y/2Y	PC1'	0.60	258		-0.122	-0.585
8	Compensated Predicted Couple Result 1X/2X	PC2'	0.49	28		0.436	0.228
9	Compensated Predicted Result 1Y	PR1'	1.48	241		-0.714	-1.295
10	Compensated Predicted Result 1X	PR2'	0.49	28		0.436	0.228
11	Compensated Predicted Result 2Y	PR3'	0.49	195		-0.471	-0.125
12	Compensated Predicted Result 2X	PR4'	0.49	208		-0.436	-0.228

Figure 10, The predicted static result for the X probes is zero when the static correction weight for the X probes is used

In addition to calculating the predicted static and couple response, it is possible to calculate the predicted response for individual probes. The equations for the static and couple components from two vectors (A and B) are repeated below.

$$\boldsymbol{S} = \frac{1}{2}(\boldsymbol{A} + \boldsymbol{B}) \tag{34}$$

$$\boldsymbol{CP_1} = \frac{1}{2}(\boldsymbol{A} - \boldsymbol{B}) \tag{35}$$

Rearrange the static equation (34) to solve for B.

$$2S = (A + B) \tag{36}$$

$$B = 2S - A \tag{37}$$

Rearrange the couple equation (35) to solve for A.

$$2CP_1 = A - B \tag{38}$$

$$A = 2CP_1 + B \tag{39}$$

Substitute the B equation (37) into the A equation (39) and simplify.

$$A = 2CP_1 + (2S - A) \tag{40}$$

$$A = 2CP_1 + 2S - A \tag{41}$$

$$2A = 2CP_1 + 2S \tag{42}$$

$$A = CP_1 + S \tag{43}$$

Substitute the A equation (39) into the B equation (37) and simplify.

$$B = 2S - A \tag{44}$$

$$B = 2S - (CP_1 + S) \tag{45}$$

$$B = S - CP_1 \tag{46}$$

Based upon the algebra performed above, the predicted results for individual probes are calculated from the static and couple components. These calculations are implemented into the worksheets and are shown below the predicted static and couple components (Figure 11).

	A	B	C	D	E	F	G
1	Minimize RMS Residual	0.853	1.04	113			
2	Minimize Maximum (Mini-Max) Amplitude	1.479	0.64	115			
3		Variable	Amplitude	Phase		x	y
4	Correction Weight	Wcor	25.55	185		-25.469	-2.087
5	Compensated Predicted Static Result 1Y/2Y	PS1'	0.92	230		-0.593	-0.710
6	Compensated Predicted Static Result 1X/2X	PS2'	0.00	0		0.000	0.000
7	Compensated Predicted Couple Result 1Y/2Y	PC1'	0.60	258		-0.122	-0.585
8	Compensated Predicted Couple Result 1X/2X	PC2'	0.49	28		0.436	0.228
9	Compensated Predicted Result 1Y	PR1'	1.48	241		-0.714	-1.295
10	Compensated Predicted Result 1X	PR2'	0.49	28		0.436	0.228
11	Compensated Predicted Result 2Y	PR3'	0.49	195		-0.471	-0.125
12	Compensated Predicted Result 2X	PR4'	0.49	208		-0.436	-0.228

Figure 11, Location of individual probe predicted results in the balancing worksheets

The balancing engineer is now tasked with determining what solution is best or practical based upon the size of available/standard balance weights and angular locations to place them. The solution to this optimization problem can be approached using Excel's built in Solver tool. "Solver is a Microsoft Excel add-in program that can be used for what-if analysis. Use Solver to find an optimal (maximum or minimum) value for a formula in one cell — called the objective cell — subject to constraints, or limits, on the values of other formula cells on a worksheet. Solver works with a group of cells, called decision variables or simply variable cells that are used in computing the formulas in the objective and constraint cells. Solver adjusts the values in the decision variable cells to satisfy the limits on constraint cells and produce the result you want for the objective cell." [6]

Link to Solver description: https://support.microsoft.com/en-us/office/define-and-solve-a-problem-by-using-solver-5d1a388f-079d-43ac-a7eb-f63e45925040#:~:text=Solver%20is%20a%20Microsoft%20Excel,formula%20cells%20on%20a%20worksheet

The Solver can be installed in Excel by following the steps listed below: [6]

1. In Excel 2010 and later, go to File > Options
2. Click Add-Ins, and then in the Manage box, select Excel Add-ins.
3. Click Go.
4. In the Add-Ins available box, select the Solver Add-in check box, and then click OK.
   Notes:
   - If the Solver Add-in is not listed in the Add-Ins available box, click Browse to locate the add-in.
   - If you get prompted that the Solver Add-in is not currently installed on your computer, click Yes to install it.
5. After you load the Solver Add-in, the Solver command is available in the Analysis group on the Data tab.

Link to Solver installation steps: https://support.microsoft.com/en-us/office/load-the-solver-add-in-in-excel-612926fc-d53b-46b4-872c-e24772f078ca

To utilize the Solver tool, an Objective cell must be selected to optimize (minimize, in this application). The least squares technique [7] minimizes the RMS residual. The balancing engineer needs to select whether to determine the RMS residual from the static-couple predicted results or from the individual probe predicted results. To the authors, it seems intuitive to utilize the individual probe predicted results as those values are what are seen and trended by the machine owner and represents the vibration amplitudes of the machine in question. The generalized form of the RMS residual is presented in Equation 47 followed by the specific calculation for a static solution using 4 probes in Equation 48.

$$\boldsymbol{RMS} = \sqrt{\frac{\sum_1^n \boldsymbol{PR_n}^2}{\boldsymbol{n}}} \qquad (47)$$

$$RMS = \sqrt{\frac{PR_1^2+PR_2^2+PR_3^2+PR_4^2}{4}} \tag{48}$$

Alternatively, the Objective cell can be selected as the maximum of all of the individual probe predicted results. This parameter is designated as mini-max i.e., minimize the maximum value in the generalized form, Equation 49 and applicable to 4 probes in Equation 50.

$$mini - max = \max(PR_n) \tag{49}$$

$$mini - max = \max(PR_1, PR_2, PR_3, PR_4) \tag{50}$$

The step-by-step use of the Solver tool in Excel for multi-plane balancing is described in the authors' 2023 paper [1]. These instructions are directly applicable to the worksheets presented in this paper. Although not necessary, the authors have found that adding upper and lower constraints to the variable cells in the Solver Parameters window and selecting the "Require Bounds on Variables" from the Options > GRG Nonlinear tab results in faster results (Figure 12). Selecting the constraints for the phase angle is straight-forward, greater than 0 and less than 360. Selecting the appropriate constraint for the correction weight requires the use of engineering judgement. Another application of the Solver tool is to only use a single variable cell for either the correction weight or phase angle if the balancing engineer has restrictions on those parameters. For example, if the Solver computes a correction weight of 22.0 oz at 170 degrees, but balance weights are fixed at 5.0 oz each, then the Solver can be run to find the optimum angle for a correction weight of 20.0 oz and 25.0 oz and the results compared. Conversely, if the balance plane has threaded holes at specific angular locations or a desired angular location is unavailable (balance weight already installed or filling slot prevents the use of that angle), then a specific angle can be used, and Solver can determine the optimum size of the correction weight.

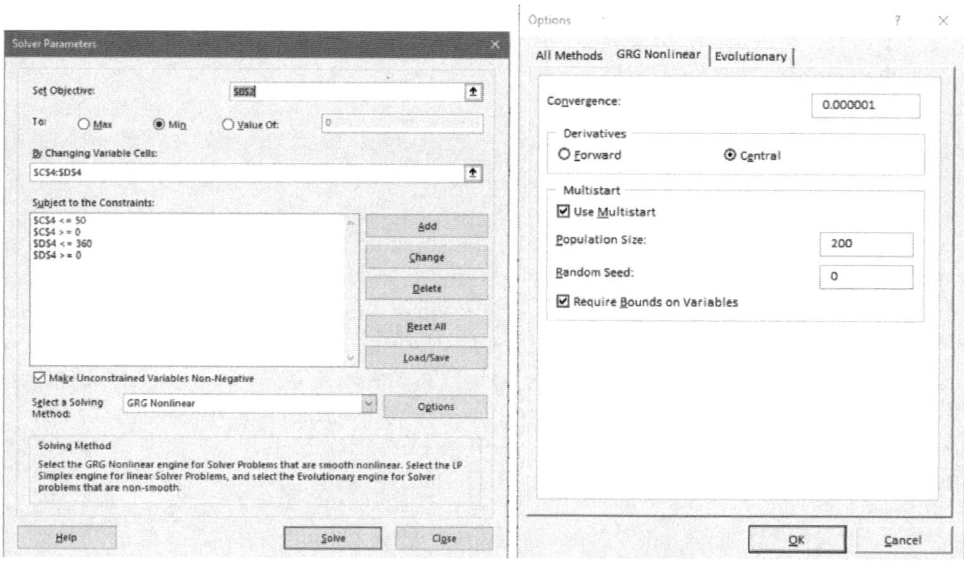

Figure 12, The Solver Parameters window (left) showing added constraints for the size and location of the correction weight and the GRG Options tab (right) showing the selection of the "Require Bounds on Variables" requirement

Using Solver to minimize the RMS Residual, results in an RMS Residual value of 0.700 and a maximum amplitude of 0.945 mil pp (Figure 13). Using Solver to minimize the Mini-Max, results in an RMS Residual value of 0.759 and a maximum amplitude of 0.837 mil pp (Figure 14). Note the trade-off between the two methods, the RMS Residual method results in the smallest RMS Residual value and a slightly higher maximum amplitude while the Mini-Max method results in a slightly higher RMS Residual value and a smaller maximum amplitude. The correction weights for both methods are nearly identical, 23.41 oz at 182 degrees (RMS Residual) and 23.71 oz at 179 degrees (Mini-Max).

	A	B	C	D	E	F	G
1	Minimize RMS Residual	0.700	23.41	182			
2	Minimize Maximum (Mini-Max) Amplitude	0.945	23.71	179			
3		Variable	Amplitude	Phase		x	y
4	Correction Weight	Wcor	23.41	182		-23.396	-0.817
5	Compensated Predicted Static Result 1Y/2Y	PS1'	0.42	230		-0.269	-0.323
6	Compensated Predicted Static Result 1X/2X	PS2'	0.50	266		-0.031	-0.498
7	Compensated Predicted Couple Result 1Y/2Y	PC1'	0.55	258		-0.115	-0.541
8	Compensated Predicted Couple Result 1X/2X	PC2'	0.50	21		0.465	0.178
9	Compensated Predicted Result 1Y	PR1'	0.95	246		-0.384	-0.864
10	Compensated Predicted Result 1X	PR2'	0.54	324		0.434	-0.320
11	Compensated Predicted Result 2Y	PR3'	0.27	125		-0.153	0.217
12	Compensated Predicted Result 2X	PR4'	0.84	234		-0.496	-0.676

Figure 13, Predicted 1X vibration vectors, RMS Residual, and Mini-Max amplitude using Solver to minimize the RMS Residual Value

	A	B	C	D	E	F	G
1	Minimize RMS Residual	0.759	23.41	182			
2	Minimize Maximum (Mini-Max) Amplitude	0.837	23.71	179			
3		Variable	Amplitude	Phase		x	y
4	Correction Weight	Wcor	23.71	179		-23.706	0.414
5	Compensated Predicted Static Result 1Y/2Y	PS1'	0.42	194		-0.411	-0.102
6	Compensated Predicted Static Result 1X/2X	PS2'	0.63	290		0.212	-0.591
7	Compensated Predicted Couple Result 1Y/2Y	PC1'	0.55	256		-0.137	-0.530
8	Compensated Predicted Couple Result 1X/2X	PC2'	0.53	20		0.494	0.185
9	Compensated Predicted Result 1Y	PR1'	0.84	229		-0.548	-0.632
10	Compensated Predicted Result 1X	PR2'	0.81	330		0.706	-0.406
11	Compensated Predicted Result 2Y	PR3'	0.51	123		-0.275	0.429
12	Compensated Predicted Result 2X	PR4'	0.82	250		-0.282	-0.775

Figure 14, Predicted 1X vibration vectors, RMS Residual, and Mini-Max amplitude using Solver to minimize the Mini-Max Value

VALIDATION WITH ROTOR KIT DATA:

A rotor kit was purposely unbalanced and tested with static-pair, couple-pair, and singled-ended calibration weights to validate the proper operation of the balancing worksheets developed and described in this paper.

TEST CASES WITH FIELD DATA:

Case #1: Couple Pair on Steam Turbine HP/IP Rotor

The HP/IP rotor was found to have elevated direct and 1X vibration at the full speed no load (FSNL) condition with a maximum 1X amplitude of 5.6 mil pp at the 2X probe. The initial field balancing used the rotor's mid-span plane to reduce the static component. Even though the static components remained larger than the couple components, a couple-pair calibration weight was installed to further reduce the synchronous vibration amplitudes. The Couple worksheet shows the vibration measurements and associated static-couple calculations for the Y and X probe pairs (Figure 15). Note that significant differences between the couple correction weights calculated from the Y probes (13.21 oz at 255 degrees) and the X probes (5.49 oz at 278 degrees).
Using the Solver tool to minimize the RMS residual for the predicted results, results in a correction weight of 5.08 oz at 276 degrees, with an RMS residual of 0.950 and a mini-max amplitude of 2.160 mil pp as shown in Figure 16. Using the Solver tool to minimize the mini-max results in a correction weight of 5.74 oz at 280 degrees, with an RMS residual of 0.954 and a mini-max amplitude of 2.126 mil pp as shown in Figure 17. These results of both methods are extremely close. Note how minimizing the RMS Residual or the mini-max affects the other parameter. On other machines or balance jobs, the discrepancy between the two parameters can be much more significant.

Couple Y Probes

Measurement #	1Y	1X Amplitude (mil pp)	1X Phase (degrees)	x	y
Slow Roll Vector	SR1	0.00	313	0.000	0.000
Original Response Vector	O1	1.11	338	1.029	-0.416
Compensated Original Response Vector	O1'	1.11	338	1.029	-0.416
Calibration Response Vector	O1+C1	1.28	353	1.270	-0.156
Compensated Calibration Response Vector	O1'+C1	1.28	353	1.270	-0.156

Measurement #	2Y	1X Amplitude (mil pp)	1X Phase (degrees)	x	y
Slow Roll Vector	SR2	0.00	42	0.000	0.000
Original Response Vector	O2	1.90	15	1.835	0.492
Compensated Original Response Vector	O2'	1.90	15	1.835	0.492
Calibration Response Vector	O2+C2	2.24	2	2.239	0.078
Compensated Calibration Response Vector	O2'+C2	2.24	2	2.239	0.078

Static Vector	S	1.43	2	1.432	0.038
Static Null Vector	Nstatic	1.43	182	-1.432	-0.038
Couple 1 Vector	CP1	0.61	228	-0.403	-0.454
Couple 2 Vector	CP2	0.61	48	0.403	0.454
Couple 1 Null Vector	Ncp1	0.61	48	0.403	0.454
Couple 2 Null Vector	Ncp2	0.61	228	-0.403	-0.454
Static Calibration Response Vector	S(O'+C)	1.75	359	1.755	-0.039
Static Calibration Vector	SC	0.33	347	0.322	-0.077
Couple1 Calibration Response Vector	CP(CP1+C)	0.50	194	-0.484	-0.117
Couple2 Calibration Response Vector	CP(CP2+C)	0.50	14	0.484	0.117
Couple 1 Calibration Vector	CP1_C	0.35	104	-0.081	0.337
Couple 2 Calibration Vector	CP2_C	0.35	284	0.081	-0.337
Couple1 Calibration Weight	Wcal,cpl1	7.54	310	4.847	-5.776
Couple Influence Vector	Hcc	0.05	154	-0.041	0.020
Couple Correction Weight	Wcor,cpl1	13.21	255	-3.452	-12.755
Static Infuence Vector	Hsc	0.04	37	0.035	0.026

Couple X Probes

Measurement #	1X	1X Amplitude (mil pp)	1X Phase (degrees)	x	y
Slow Roll Vector	SR1	0.00	313	0.000	0.000
Original Response Vector	O1	1.80	169	-1.767	0.343
Compensated Original Response Vector	O1'	1.80	169	-1.767	0.343
Calibration Response Vector	O1+C1	1.60	259	-0.305	-1.571
Compensated Calibration Response Vector	O1'+C1	1.60	259	-0.305	-1.571

Measurement #	2X	1X Amplitude (mil pp)	1X Phase (degrees)	x	y
Slow Roll Vector	SR2	0.00	245	0.000	0.000
Original Response Vector	O2	3.38	243	-1.534	-3.012
Compensated Original Response Vector	O2'	3.38	243	-1.534	-3.012
Calibration Response Vector	O2+C2	3.01	208	-2.771	-1.176
Compensated Calibration Response Vector	O2'+C2	3.01	203	-2.771	-1.176

Static Vector	S	2.12	219	-1.651	-1.334
Static Null Vector	Nstatic	2.12	39	1.651	1.334
Couple 1 Vector	CP1	1.68	94	-0.116	1.678
Couple 2 Vector	CP2	1.68	274	0.116	-1.678
Couple 1 Null Vector	N1	1.68	274	0.116	-1.678
Couple 2 Null Vector	N2	1.68	94	-0.116	1.678
Static Calibration Response Vector	S(O'+C)	2.06	222	-1.538	-1.373
Static Calibration Vector	SC	0.12	341	0.113	-0.039
Couple1 Calibration Response Vector	CP(CP1+C)	1.25	351	1.233	-0.197
Couple2 Calibration Response Vector	CP(CP2+C)	1.25	171	-1.233	0.197
Couple 1 Calibration Vector	CP1_C	2.31	306	1.349	-1.875
Couple 2 Calibration Vector	CP2_C	2.31	126	-1.349	1.875
Couple1 Calibration Weight	Wcal,cpl1	7.54	310	4.847	-5.776
Couple Influence Vector	Hcc	0.31	356	0.305	-0.023
Couple Correction Weight	Wcor,cpl1	5.49	278	0.786	-5.433
Static Infuence Vector	Hsc	0.02	31	0.014	0.008

Figure 15: A screenshot of the Couple worksheet showing the vibration measurements and associated static-couple vibration calculations

From EQ 47

From EQ 49

		B	C	D	E	F	G
1	Minimize RMS Residual	0.950	5.08	276			
2	Minimize Maximum (Mini-Max) Amplitude	2.160	5.74	280			
3		Variable	Amplitude	Phase		x	y
4	Correction Weight	Wcor1	5.08	276		0.531	-5.052
5	Compensated Predicted Couple Result 1Y/2Y	PC1'	0.40	216		-0.321	-0.235
6	Compensated Predicted Couple Result 1X/2X	PC2"	0.14	119		-0.069	0.122
7	Compensated Predicted Static Result 1Y/2Y	PS1'	1.59	355		1.583	-0.126
8	Compensated Predicted Static Result 1X/2X	PS2'	2.13	221		-1.603	-1.398
9	Compensated Predicted Result 1Y	PR1'	1.31	344		1.262	-0.362
10	Compensated Predicted Result 1X	PR2'	1.91	3		1.905	0.109
11	Compensated Predicted Result 2Y	PR3'	2.10	217		-1.672	-1.276
12	Compensated Predicted Result 2X	PR4'	2.16	225		-1.533	-1.521

Figure 16: Screenshot from the Couple worksheet showing the predicted results from minimizing the RMS Residual

	A	B	C	D	E	F	G
1	Minimize RMS Residual	0.954	5.08	276			
2	Minimize Maximum (Mini-Max) Amplitude	2.126	5.74	280			
3		Variable	Amplitude	Phase		x	y
4	Correction Weight	Wcor1	5.74	280		0.997	-5.653
5	Compensated Predicted Couple Result 1Y/2Y	PC1'	0.38	211		-0.328	-0.201
6	Compensated Predicted Couple Result 1X/2X	PC2"	0.09	310		0.059	-0.072
7	Compensated Predicted Static Result 1Y/2Y	PS1'	1.62	355		1.615	-0.135
8	Compensated Predicted Static Result 1X/2X	PS2'	2.12	221		-1.591	-1.403
9	Compensated Predicted Result 1Y	PR1'	1.33	345		1.287	-0.336
10	Compensated Predicted Result 1X	PR2'	1.94	2		1.944	0.066
11	Compensated Predicted Result 2Y	PR3'	2.13	224		-1.532	-1.475
12	Compensated Predicted Result 2X	PR4'	2.12	219		-1.651	-1.331

Figure 17: Screenshot from the Couple worksheet showing the predicted results from minimizing the mini-max parameter

Case #2: Static Pair on Steam Turbine Generator Rotor

The generator rotor was reported to be experiencing elevated direct and 1X vibration amplitudes following a repowering project. The generator rotor is supported between bearings #4 and #5. At baseload conditions (~167 MW), the compensated 1X vibration amplitudes from both X probes exceeded 3 mil pp. The Static worksheet shows the vibration measurements and associated static-couple calculations for the Y and X probe pairs (Figure 18). Both probe pairs indicate a dominant static component 2.50 mil pp at 18 degrees for the Y probes and 3.25 mil pp at 60 degrees for the X probes. The associated couple components are 0.15 mil pp and 0.38 mil pp respectively. A static-pair calibration weight of 12.0 oz was installed at 135 degrees against rotation from the Keyphasor trigger at the bearing #4 and #5 end planes.

The exact solutions to eliminate the static components using the Y and X probe pairs indicate 20.48 oz at 169 degrees and 23.27 oz at 171 degrees respectively (Figure 18). Using the Solver tool to minimize the RMS residual for the predicted results, results in a correction weight of 22.03 oz at 170 degrees, with an RMS residual of 0.285 and a mini-max amplitude of 0.352 mil pp (Figure 19). Using the Solver tool to minimize the mini-max results in a correction weight of 22.51 oz at 170 degrees, with an RMS residual of 0.292 and a mini-max amplitude of 0.308 mil pp. These results of both methods are extremely close in the size and location of the static-pair balance weights. In this instance, individual balance weights are 4 oz each, so additional optimization could be performed solving for the balance weight angle associated with weights of 20 oz (5 weights) or 24 oz (6 weights) at each weight plane.

Static Y Probes

Measurement #	4Y	1X Amplitude (mil pp)	1X Phase (degrees)	x	y
Slow Roll Vector	SR1	0.00	0	0.000	0.000
Original Response Vector	O1	2.58	15	2.493	0.668
Compensated Original Response Vector	O1'	2.58	15	2.493	0.668
Calibration Response Vector	O1+C1	1.61	49	1.058	1.217
Compensated Calibration Response Vector	O1'+C1	1.61	49	1.058	1.217
Measurement #	5Y	1X Amplitude (mil pp)	1X Phase (degrees)	x	y
Slow Roll Vector	SR2	0.00	0	0.000	0.000
Original Response Vector	O2	2.42	21	2.262	0.868
Compensated Original Response Vector	O2'	2.42	21	2.262	0.868
Calibration Response Vector	O2+C2	1.44	52	0.884	1.132
Compensated Calibration Response Vector	O2'+C2	1.44	52	0.884	1.132
Static Vector	S	2.50	18	2.378	0.768
Static Null Vector	Nstatic	2.50	198	-2.378	-0.768
Couple 1 Vector	CP1	0.15	319	0.115	-0.100
Couple 2 Vector	CP2	0.15	139	-0.115	0.100
Couple 1 Null Vector	Ncp1	0.15	139	-0.115	0.100
Couple 2 Null Vector	Ncp2	0.15	319	0.115	-0.100
Static Calibration Response Vector	S(O'+C)	1.52	50	0.971	1.174
Static Calibration Vector	SC	1.46	164	-1.407	0.406
Couple 1 Calibration Response Vector	CP(CP1+C)	0.10	26	0.087	0.043
Couple 2 Calibration Response Vector	CP(CP2+C)	0.10	206	-0.087	-0.043
Couple 1 Calibration Vector	CP1_C	0.15	101	-0.029	0.148
Couple 2 Calibration Vector	CP2_C	0.15	281	0.029	-0.148
Static Calibration Weight	Wcal,static	12.00	135	-8.485	8.485
				0.107	0.059
Static Correction Weight	Wcor,static	20.48	169	-20.102	3.907
				0.010	-0.007

Static X Probes

Measurement #	4X	1X Amplitude (mil pp)	1X Phase (degrees)	x	y
Slow Roll Vector	SR1	0.00	0	0.000	0.000
Original Response Vector	O1	3.36	246	-1.365	-3.065
Compensated Original Response Vector	O1'	3.36	246	-1.365	-3.065
Calibration Response Vector	O1+C1	2.30	276	0.241	-2.291
Compensated Calibration Response Vector	O1'+C1	2.30	276	0.241	-2.291
Measurement #	5X	1X Amplitude (mil pp)	1X Phase (degrees)	x	y
Slow Roll Vector	SR2	0.00	0	0.000	0.000
Original Response Vector	O2	3.19	233	-1.917	-2.544
Compensated Original Response Vector	O2'	3.19	233	-1.917	-2.544
Calibration Response Vector	O2+C2	2.02	257	-0.455	-1.969
Compensated Calibration Response Vector	O2'+C2	2.02	257	-0.455	-1.969
Static Vector	S	3.25	240	-1.641	-2.805
Static Null Vector	Nstatic	3.25	60	1.641	2.805
Couple 1 Vector	CP1	0.38	317	0.276	-0.260
Couple 2 Vector	CP2	0.38	137	-0.276	0.260
Couple 1 Null Vector	Ncp1	0.38	137	-0.276	0.260
Couple 2 Null Vector	Ncp2	0.38	317	0.276	-0.260
Static Calibration Response Vector	S(O'+C)	2.13	267	-0.107	-2.130
Static Calibration Vector	SC	1.68	24	1.534	0.674
Couple 1 Calibration Response Vector	CP(CP1+C)	0.38	335	0.348	-0.161
Couple 2 Calibration Response Vector	CP(CP2+C)	0.38	155	-0.348	0.161
Couple 1 Calibration Vector	CP1_C	0.12	54	0.071	0.099
Couple 2 Calibration Vector	CP2_C	0.12	234	-0.071	-0.099
Static Calibration Weight	Wcal,static	12.00	135	-8.485	8.485
				-0.051	-0.130
Static Correction Weight	Wcor,static	23.27	171	-22.978	3.665
				0.002	-0.010

Figure 18: A screenshot of the Static worksheet showing the vibration measurements and associated static-couple vibration calculations

	A	B	C	D	E	F	G
1	Minimize RMS Residual	0.285	22.03	170			
2	Minimize Maximum (Mini-Max) Amplitude	0.352	22.51	170			
3		Variable	Amplitude	Phase		x	y
4	Correction Weight	Wcor	22.03	170		-21.695	3.825
5	Compensated Predicted Static Result 1Y/2Y	PS1'	0.19	212		-0.165	-0.103
6	Compensated Predicted Static Result 1X/2X	PS2'	0.18	256		-0.044	-0.175
7	Compensated Predicted Couple Result 1Y/2Y	PC1'	0.11	133		-0.078	0.084
8	Compensated Predicted Couple Result 1X/2X	PC2'	0.28	353		0.279	-0.036
9	Compensated Predicted Result 1Y	PR1'	0.24	184		-0.243	-0.019
10	Compensated Predicted Result 1X	PR2'	0.32	318		0.235	-0.211
11	Compensated Predicted Result 2Y	PR3'	0.21	245		-0.087	-0.187
12	Compensated Predicted Result 2X	PR4'	0.35	203		-0.323	-0.139

Figure 19: A screenshot of the Static worksheet showing the predicted results from minimizing the RMS Residual

	A	B	C	D	E	F	G
1	Minimize RMS Residual	0.292	22.03	170			
2	Minimize Maximum (Mini-Max) Amplitude	0.308	22.51	170			
3		Variable	Amplitude	Phase		x	y
4	Correction Weight	Wcor	22.51	170		-22.168	3.909
5	Compensated Predicted Static Result 1Y/2Y	PS1'	0.25	209		-0.221	-0.122
6	Compensated Predicted Static Result 1X/2X	PS2'	0.12	265		-0.009	-0.118
7	Compensated Predicted Couple Result 1Y/2Y	PC1'	0.12	133		-0.082	0.088
8	Compensated Predicted Couple Result 1X/2X	PC2'	0.28	354		0.279	-0.031
9	Compensated Predicted Result 1Y	PR1'	0.30	186		-0.303	-0.034
10	Compensated Predicted Result 1X	PR2'	0.31	331		0.270	-0.149
11	Compensated Predicted Result 2Y	PR3'	0.25	237		-0.139	-0.210
12	Compensated Predicted Result 2X	PR4'	0.30	197		-0.289	-0.087

Figure 20: A screenshot of the Static worksheet showing the predicted results from minimizing the mini-max parameter

CONCLUSIONS:

The work performed for this paper was an extension to the work performed for the previous paper and presented at VI ATC 2023. [1] The current paper demonstrated how to perform S/C calculations with off-the-shelf tools. Additionally, the work provided an improvement over the traditional static-couple methodology of using an exact solution. By using the optimization techniques shown, the user can incorporate additional measurements and accommodate constraints such as fixed weight sizes and locations.

NOMENCLATURE:

A = vector quantity used to demonstrate calculations
B = vector quantity used to demonstrate calculations
C = vibration caused by calibration weight or weight pair
CP = couple vibration component
CP1 = couple vibration component end 1
H = influence vector
H_{cs} = couple influence vector due to a static balance weight pair
H_{cc} = couple influence vector due to a couple balance weight pair
H_{sc} = static influence vector due to a couple balance weight pair
H_{ss} = static influence vector due to a static balance weight pair
Mini-Max = maximum value from a group of values
MP = multiplane
N = null vector
N_s = static null vector
N_{cp} = couple null vector
O = original vibration
O + C = vectorial combination of original and calibration response vibration
PS = predicted static response vector
PC1 = predicted couple response vector end 1
PR = predicted result
RMS = root-mean square
S = static vector
SC = static couple
SR = slow roll vector
$W_{cal,static}$ = static calibration weight (pair)
$W_{cal,couple}$ = couple calibration weight (pair)

REFERENCES:

[1] M. J. Titone and S. P. Ciesla, "Multiplane Balancing Using Commonly Available Modern Tools," in *Vibration Institute Annual Training Conference*, Niagara Falls, 2023.

[2] W. C. Foiles, P. E. Allaire and E. J. Gunter, "Review: Rotor balancing," *Shock and Vibration 5,* pp. 325-336, 1998.

[3] M. J. Titone, "Application of Analytical Relationship Between Static Couple and Multiplane Influence Coefficients - Two Field Balancing Examples," in *Vibration Institute Annual Training Conference*, Lexington, 2019.

[4] J. J. Yu, "On Two-Plane Balancing of Symmetric Rotors," in *Proceedings of ASME Turbo Expo 2012*, 2012.

[5] J. J. Yu, "Relationship of Influence Coefficients Between Static-Couple and Multiplane Methods on Two-Plane Balancing," *ASME Journal of Engineering for Gas Turbines and Power,* vol. 131, pp. 012508-10, 2009.

[6] "Microsoft Support," [Online]. Available: support.microsoft.com.

[7] T. P. Goodman, "A Least-Squares Method for Computing Balance Corrections," *Journal of Engineering for Industry,* pp. 273-279, August 1964.

VIATC TRAINING CONFERENCE & EXPO

COVINGTON, KENTUCKY | AUGUST 7-9, 2024

THE ANALYSIS AND COMPARISONS OF DIFFERENT TYPES OF VIBRATION RUBS

Thomas B. Lantrip
Consulting Engineer – Southern Company Services
tblantri@southernco.com
tblantri@icloud.com

Tom Lantrip is a consulting engineer with Southern Company Generation and works in the Performance and Reliability Department - Balancing and Vibration Analysis group. Tom has worked at Southern Company for 41 years, primarily in the field of balancing and vibration analysis, with an emphasis on turbines, fans, and pump components. He has a background in structural dynamics and in structural and seismic design, vibration analysis, and balancing of rotating machinery. Tom graduated from the University of Alabama with a B.S. in Civil Engineering, attended the University of California at Berkeley, and has an M.S. in Civil Engineering from the University of Alabama at Birmingham. Tom is a member of ASCE and is a professional engineer in the state of Alabama and Florida and holds a Category IV certification from the Institute.

ABSTRACT:

Vibration "rubs" are a common issue on machines that contain close clearances, typically found on medium to high-speed machines. Traditional "rub" case histories indicate certain behaviors and characteristics. The data provided here includes updated observations of many characteristics of different type of "rub" events and explanations as to the nuances of the spectra, orbits, time waveforms and polar plots.

KEY WORDS: Amplitude, Phase Angle, Filtered Frequency, Time Waveform, Spectra, Rubs, Clearances, Seals,

INTRODUCTION:

Shaft relative probes are typically used for "rub" indications, however additional plots from bearing housings, provide information as well. Some "rubs" are considered minor and not damaging, however some of these "rubs" can result in considerable damage and consequences. Techniques and various plot formats can help in the diagnosis to rule in/out other causes. A couple different machine-type cases are presented.

- High-Speed Motor/Compressor Train after Rebuild
- Turbine Rotor(s) after Outages
- Fan Rub after Repairs

Vibration analysis on critical components and problematic machines will often take on additional effort in the form of continuous monitoring techniques. This analysis technique has more advantages as compared to traditional analysis methods when collecting discrete "route-based" vibration measurements. The continuous monitoring methods allows the analyst to quantify the vibration character and conditions during various conditions including time, loading environment, ambient temperature, production level, processes and starting and stopping conditions. Often the vibration community collects steady-state vibration conditions for a short period of time; performs averaging of a discrete point and location and moves on to the next data point. In these cases, the analyst is not aware of the vibration characteristics after the data point is collected including the changes that may be occurring. The presumption normally is that the vibration conditions are steady, stable and the content of the collected data is reflected in the single snapshot.

Occasionally, when starting and stopping a machine or adjusting load/process we encounter changes in our vibration that manifest themselves in diverse ways, primarily in higher vibration levels than normal and/or secondarily in varying levels including decreasing vibration levels. When we set our alarm levels, we normally set an upper bound limit to capture the crossing of the alarm line, and not the lower bound limit. It would be very unusual to set a lower bound limit however change limits can be valuable insight into developing or acute problems.

To expand on the primary condition where we have an increasing vibration condition that exhibits a higher vibration level than the alarm or an upward moving trend, we would typically examine the spectrum plot. Evaluation of the spectral content would follow, and normally we would expect the primary frequency to include the fundamental operating speed frequency. We classically refer to this fundamental frequency as the 1X filtered frequency, operating speed frequency, or running speed frequency.

Based on countless situations on motors, fans, and turbines the historical evidence suggests the fundamental characteristics of rubs of the fundamental frequency of interest will occur at the operating speed. This is seen countless times on machinery in the power generation industry. Many situations when rubs are encountered the following characteristics are observed.

- Unsteady vibration over an extended period (from 1 minute to 1 hour).
- Fundamental frequency at operating speed frequency.
- Phase angle of operating speed frequency changing. (when data is available)

- Multiple bearings are changing, not typically just one bearing location.
- If outfitted with multiple sensors, i.e. (X and Y probe, Horizontal and Vertical), both sensors will see changes.

As specified earlier, if the analyst does not collect vibration data continuously or often enough to see the differences in the longer period of interest it may not be apparent of the changes.

MOTOR/COMPRESSOR RUB CASE HISTORY:

Below is a case history of a high-speed motor compressor train, where a vibration character exhibited increasing levels over a short period of time after starting up the unit to initiate production from the compressor.

The vibration characteristics indicated a low amplitude on initial full speed operation at 1800 rpm on a synchronous motor.

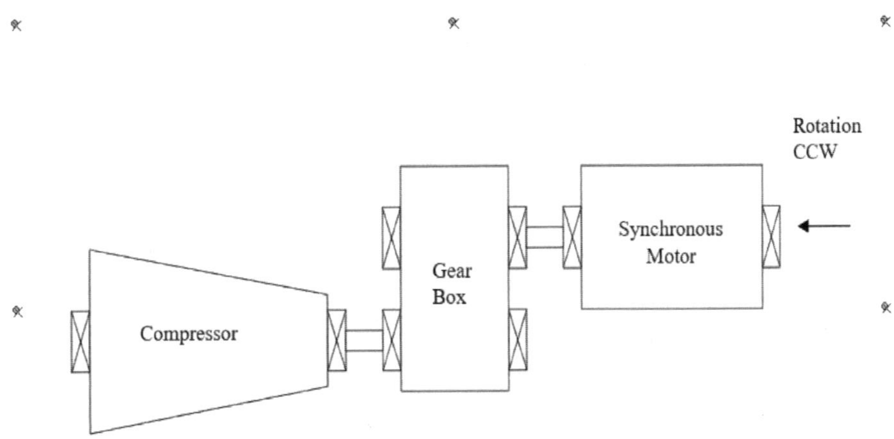

Figure 1 : Compressor Train and Motor Layout for 1800 rpm Synchronous Motor Vibration

Figure 2 provides a typical spectrum of condition when rubbing can occur. The primary frequency is coincident with the operating speed. In this example, the frequency is occurring at 30 hz with an amplitude of 0.327 mils p-p.

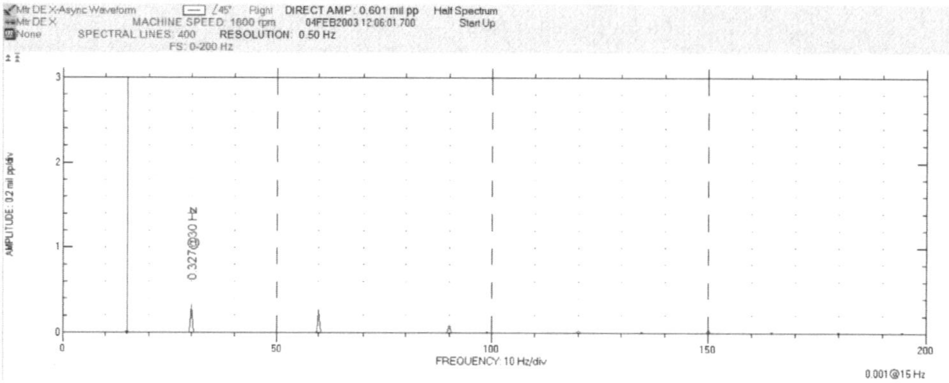

Figure 2 : Spectrum Plot for Motor Drive End (DE) indicating low 1X Amplitude Initially (30 hz running speed)

Figure 3 provides the orbit and time waveform for the initial condition of the Figure 2 spectrum. The time waveform has some unusual spikes, but this is believed to be due to some scratches or other surface conditions on the shaft from a proximity probe shaft measurement.

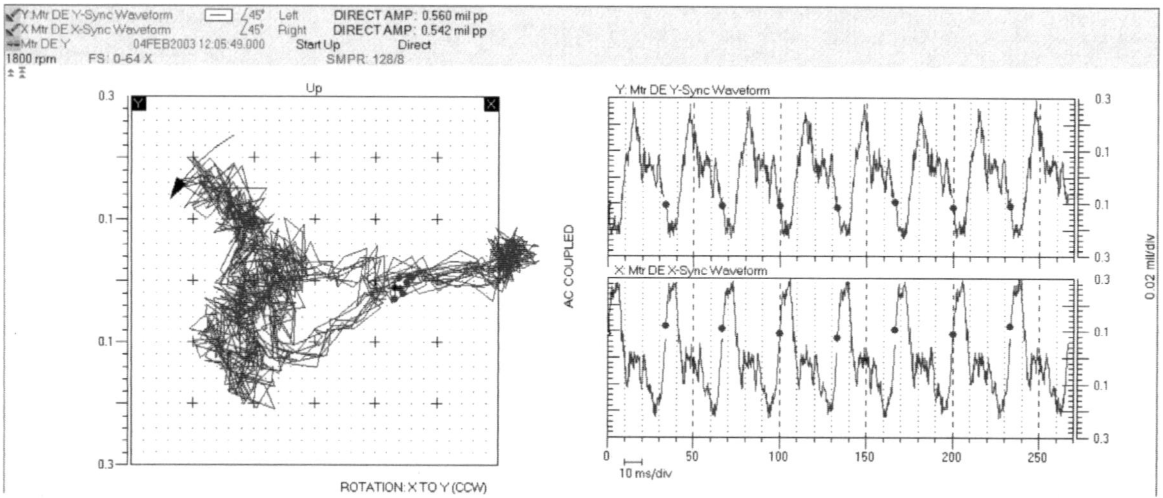

Figure 3 : Orbit and Waveform Plot for Motor Drive End (DE) indicating extremely low Overall Amplitude Initially (30 hz running speed)

During the operation of the motor, some minutes later, the vibration at the operating speed is much higher at 2.78 mils p-p as shown in Figure 4. The primary amplitude is fundamentally still at the operating frequency, but an increase at the harmonic multiples of two times, three times, etc. are indicated as well.

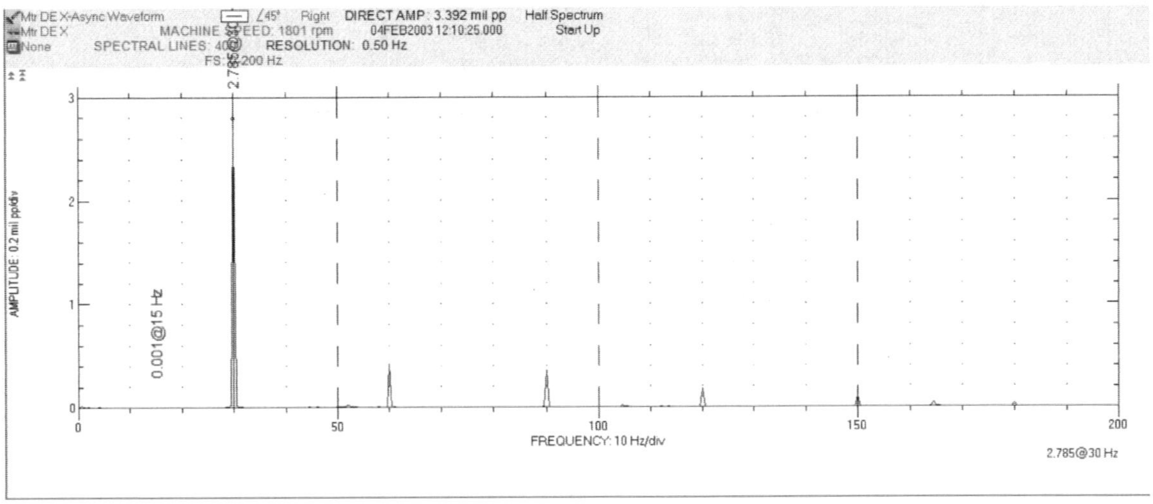

Figure 4 : Spectrum Plot for Motor Drive End (DE) indicating much higher 1X Amplitudes after 4 minutes operation (30 hz running speed)

The corresponding orbit corresponding to this level of vibration is provided in Figure 5 orbit and waveform. Notice the shape of the orbit and the apparent "flattened shape" of the orbit on the top. The time waveform indicates primarily one times shape with narrow shape instead of a more classic sine wave shape. Figure 6 provides the shaft centerline (shaft position using two orthogonal probes) with the corresponding orbits during the transition from the pre-rub condition with lower levels of vibration to the post-rub condition with a similar orbit to that shown in Figure 5.

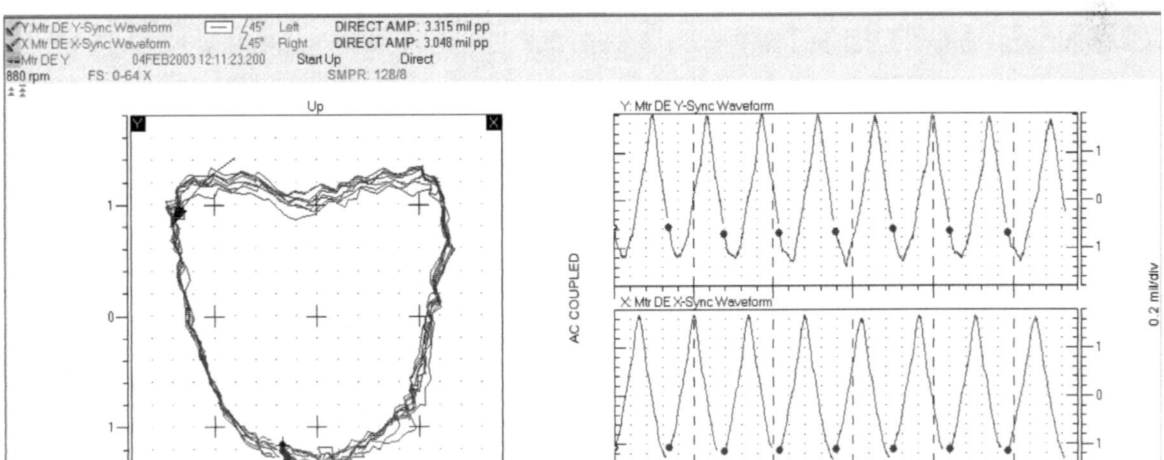

Figure 5 : Orbit and Waveform Plot for Motor Drive End (DE) indicating much higher Overall Amplitudes after 4 minutes operation (30 hz running speed)

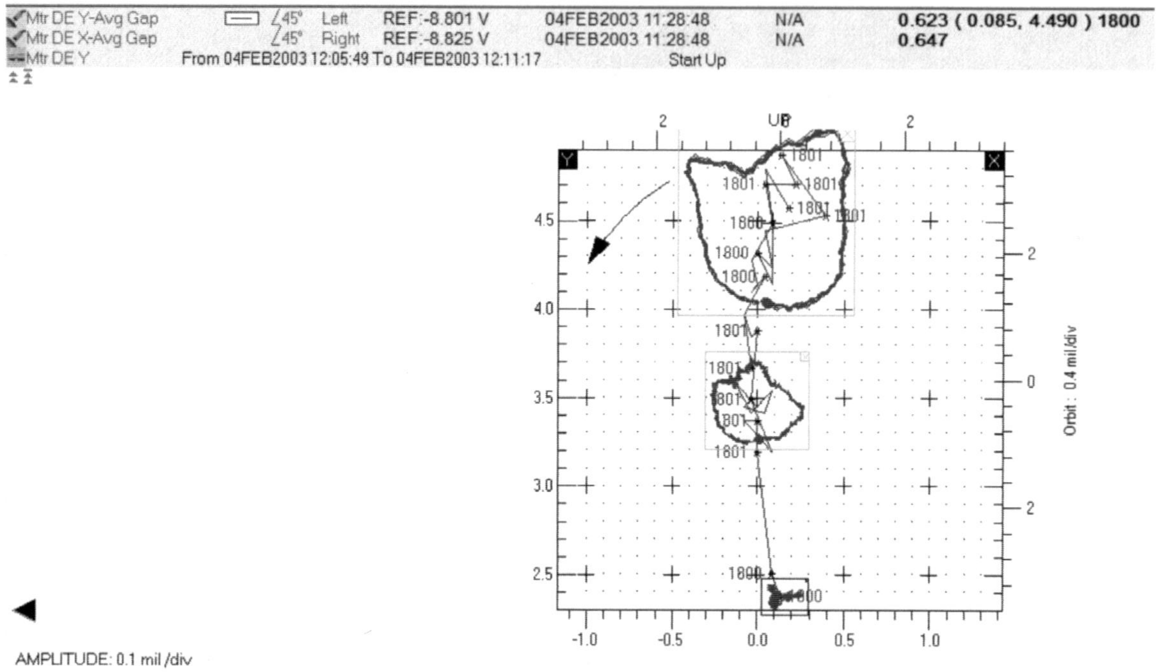

Figure 6 : Shaft Centerline Plot Motor Drive End (DE) indicating shaft position and Orbit During Transition Low to Higher Amplitudes after 4 minutes operation (30 hz op speed)

Figure 7 provides the trend plot of the drive end (DE) for the Y probe with one time amplitude (red line), and one times phase (red line in upper plot) and the overall amplitude (blue line) all plotted versus time. The time scale is over a 5-minute period, and the corresponding vibration levels increase from less than 1 mils to 3+ mils in this period. In addition, review of the 1X phase (red line upper plot) indicates a change to the phase during this period from 200 to 235 degrees through the 5 minutes time.

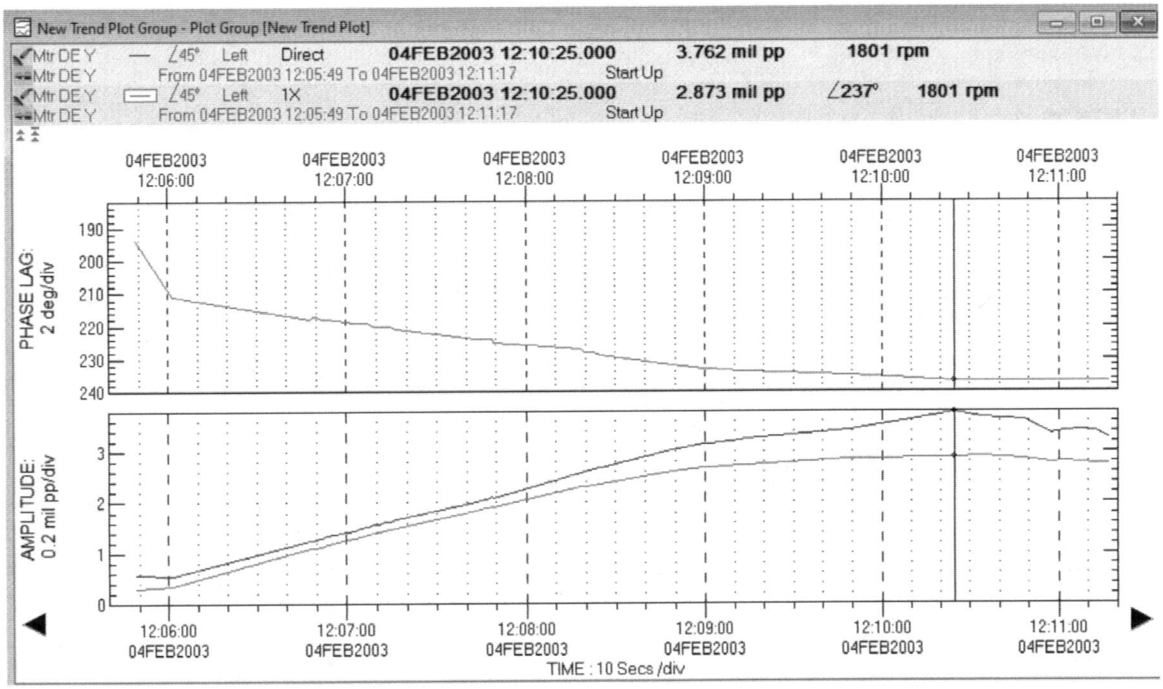

Figure 7 : Trend Plot for Motor Drive End (DE) indicating increasing higher 1X Amplitudes, Phase Change and Overall Amplitude after 4 minutes operation (30 hz)

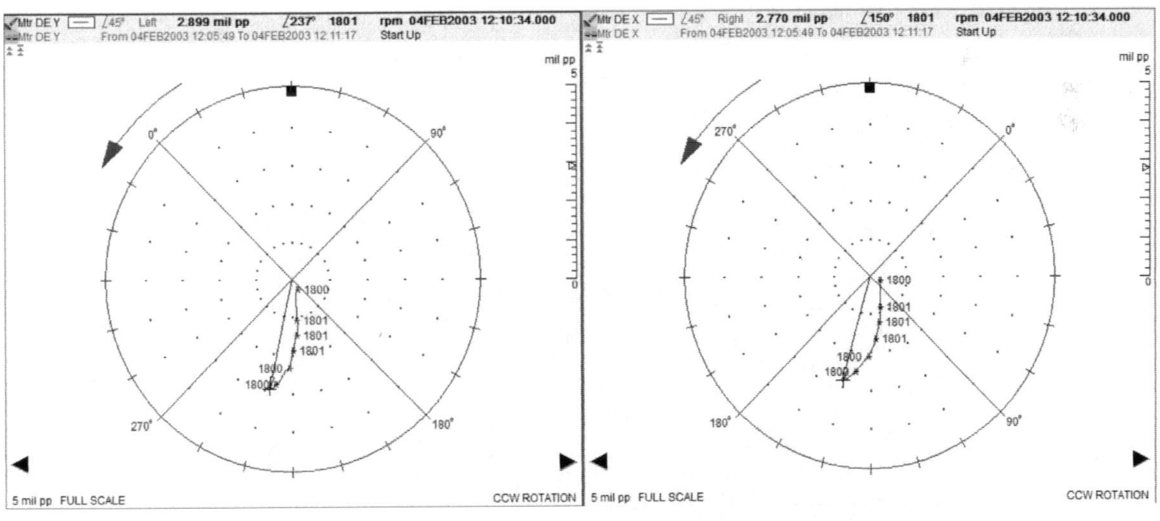

Figure 8 : Polar Plot for Motor Drive End (DE) indicating much higher 1X Amplitudes and Phase Migration after 4 minutes operation (30 hz running speed)

Figure 8 provides the polar plot of the drive end (DE) for the Y and X probe with one time amplitude, and one times phase represented in a polar format. The polar form allows the presentation of the amplitude and phase changes collectively in an efficient plot to monitor the conditions. The polar plot provides a convenient manner to easily view changes to the amplitude and phase vs time. This is a valuable plot to recognize the real-time condition of the apparent rub.

After an inspection of the subject motor, based on the evidence of the vibration above, the motor was found to have an internal component that was resting against the shaft element.

Figure 9 : Photograph of Drive End Shaft Air Seal and Rotor Rub Marks

Figure 10 : Photograph of Drive End Shaft Air Seal Heat Affected Zone and "Burn" marks on Micarta Material

The air dust seals were rubbing on the rotating shaft thus creating the rub. This rub was creating a "hot" spot on the shaft which led to a thermal bow which created a "thermal" unbalance which was different amplitudes and phase from the original unbalance condition. Once the air dust seals were moved out of the way, the vibration levels returned to their previous condition of less than 1 mils p-p.

STEAM TURBINE RUB EXAMPLE:

Figure 11 represents a vibration trend of the amplitude and the phase angle as a function of speed. This plot i to as a bode' plot and is helpful to identify the natural frequencies of the measurement observed. The trend lines indicate the 1X amplitude (red line) of the operating speed frequency order with the corresponding 1X phase (red line) as well as the overall amplitude (blue line) as a function of speed from 0 to 3600 rpm.

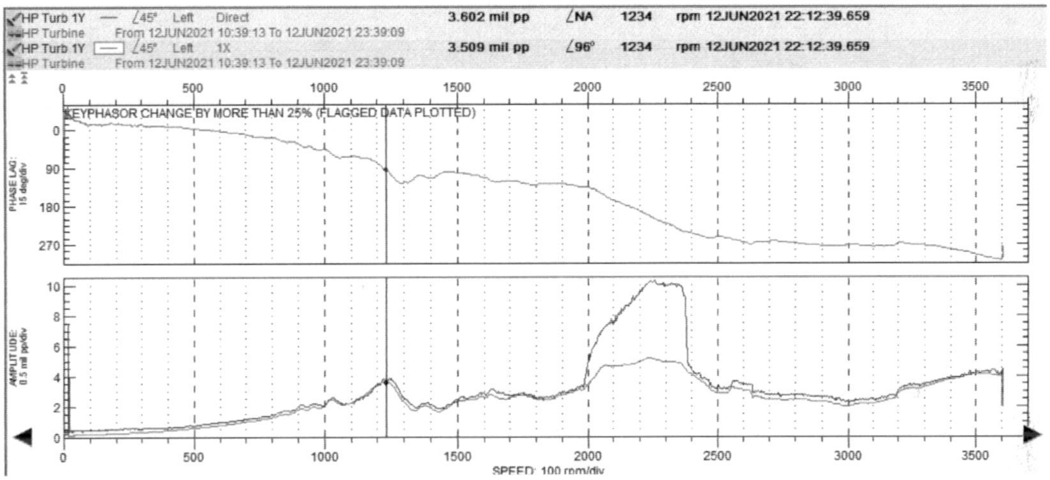

Figure 11 : High Vibration Excursion During Transient Speed Coast-down on Steam Turbine from 3600 rpm (Critical Speed at 1234 rpm)

Figure 12 provides the appropriate locations on the High Pressure (HP) turbine rotor at the bearings on each end of the rotor as denoted as Bearings 1 and 2. Each bearing has an X and Y proximity-probe pair, and these four plots below represent the various bode' plots of the respective location.

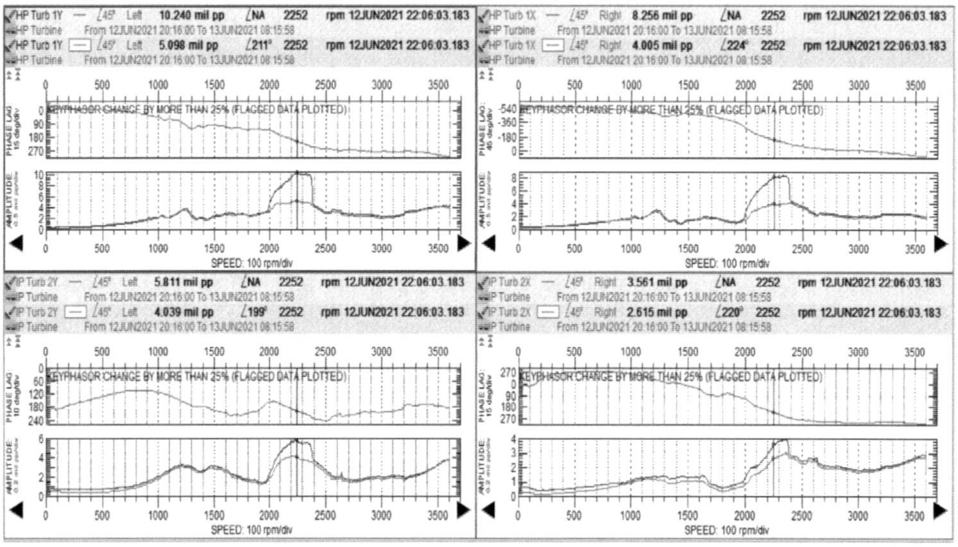

Figure 12 : High Vibration Excursion During Transient Speed Coast-down on Steam Turbine from 3600 rpm (Bearing 1 and 2 Y and X probe)

Review of the plots 11 and 12 suggest that the amplitude of the vibration peaks at the highest level in the range of speed between 2000 to 2400 rpm at close to ten mils which is consistent as a second alarm or trip level of vibration. The plot indicates the 1X amplitude and phase is peaking (red line), but additional levels of vibration are present in the overall plot (blue line) in this speed range. The amplitude of the overall and the 1X closely mimic one another in every other speed range, however at this 2000 to 2400 rpm range, the levels diverge.

Figure 13 provides the shaft centerline and corresponding orbits at the various speeds. The orbit indicates primarily a one times shape (oval typically) until the speed range of 2000 to 2400 rpm occurs.

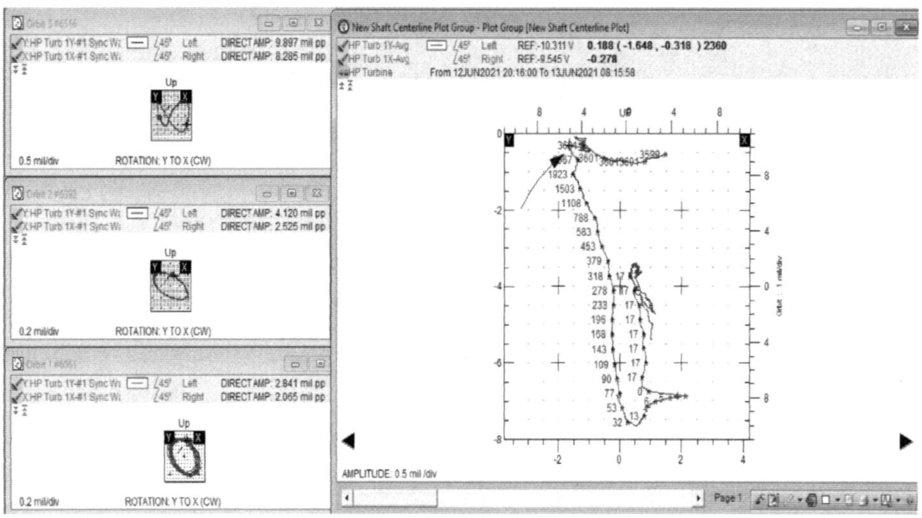

Figure 13 : Shaft Centerline of Bearing 1 during Transient Coast-down from 3600 rpm

In this speed range the spectrum and the orbit suggest the occurrence of an 1/2 X component (1/2 speed component) at 2380 rpm captured in Figure 14 and 15. This spectrum indicates at 2380 rpm that the spectrum is not dominated by the 1X frequency but at the ½ X frequency with harmonics at intervals of ½ speed. Figure 15 is provided to show the harmonics ½ order harmonics (displayed incorrectly as one times harmonics at the speed of 2340 rpm).

Figure 16 provides the orbit of the amplitudes for Bearing 1 during onset of the apparent rub. In addition, the time waveform provides the cycle of vibration amplitudes versus time which indicates two speed crossings from the tachometer signal for one cycle of vibration. The exact repeating of the tachometer (dots) crossing indicates true ½ order amplitudes [1].

Another plot (Figure 17) below provides an aid to demonstrate the source of the frequency and harmonics including the ½ order and harmonics in the range of speed where the amplitude peaks around 2000 to 2400 rpm. This plot is known as a waterfall plot, which presents amplitude in spectral form, vs speed and time.

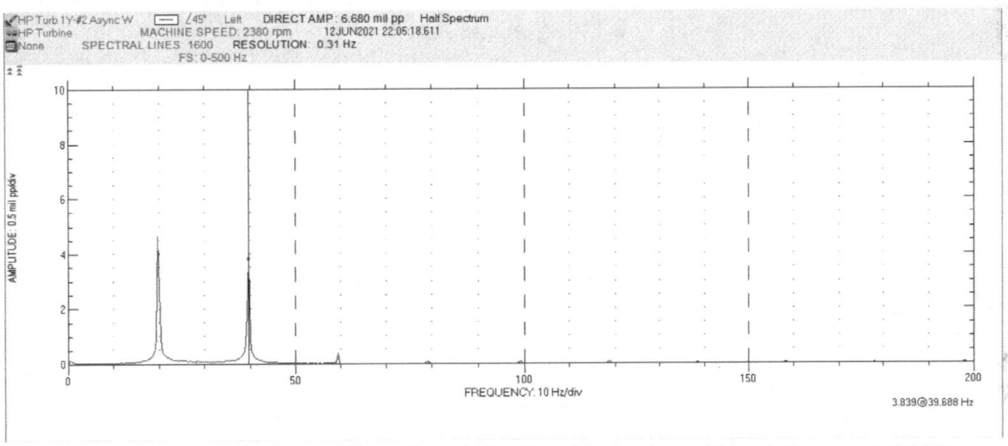

Figure 14 : Spectral Plot of Bearing 1 during Transient Coast-down at 2380 rpm

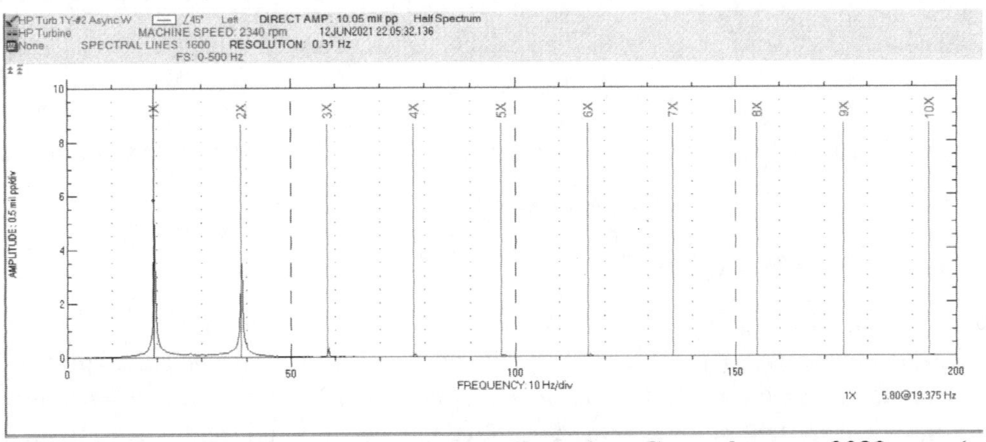

Figure 15 : Spectral Plot of Bearing 1 during Transient Coast-down at 2380 rpm (with ½ order harmonics)

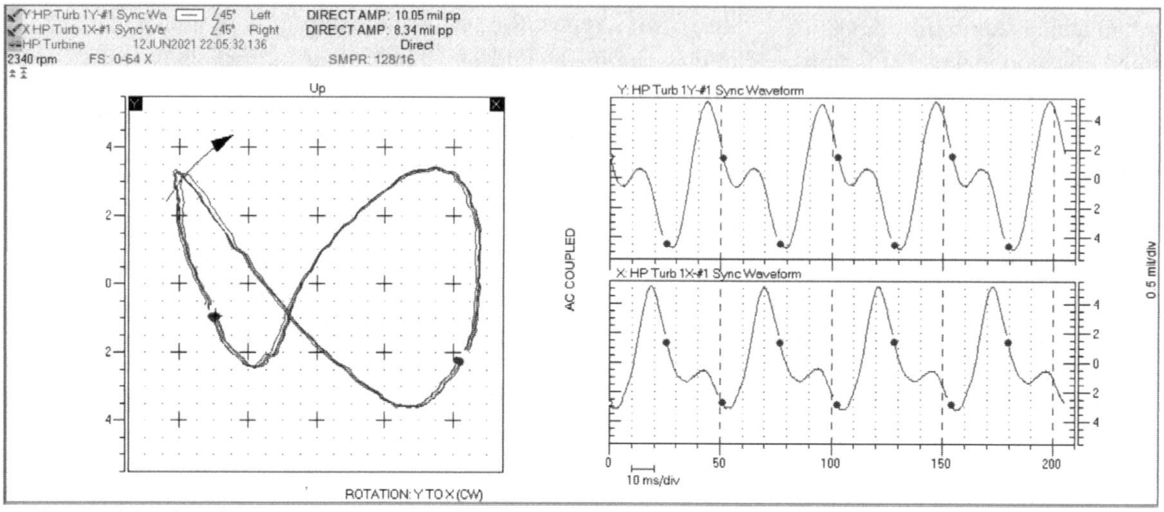

Figure 16 : Orbit and Time Waveform Plot of Bearing 1 during Transient Coast-down at 2340 rpm (39 hz)

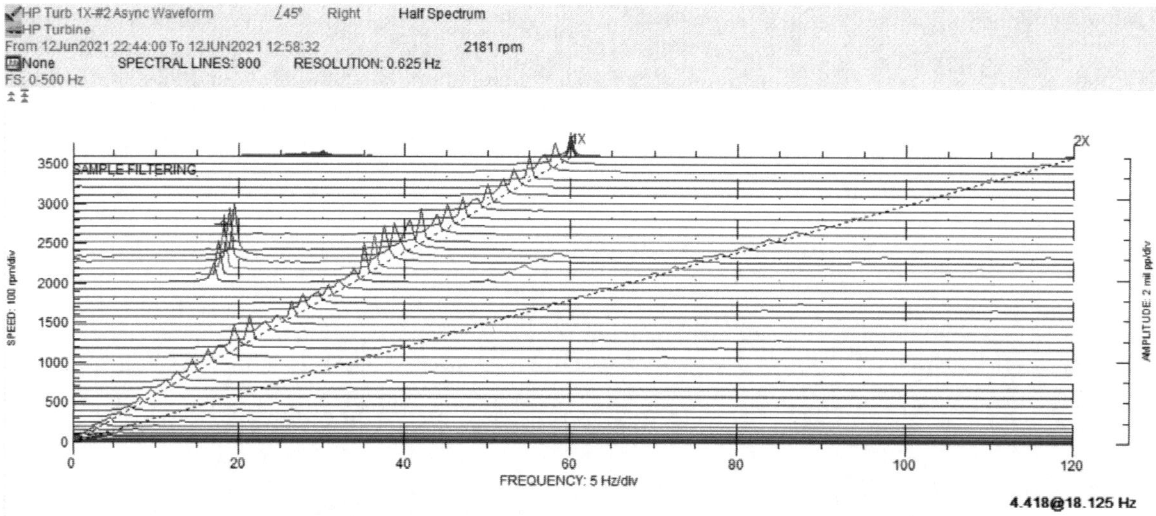

Figure 17 : Waterfall Plot of Spectral Content during Coast-down at 2181 rpm range

(1/2 order and 1.5X order)

STEAM TURBINE RUB EXAMPLE:

Another example of a rub occurs on steam turbines during commissioning or after a major refurbishment maintenance outage. Figure 18 provides a trend of a vibration signal during a steady-state operation condition. The figure depicts the 1X amplitude and one times phase with the corresponding overall vibration amplitude. Notice the 1X amplitude and overall trend are remarkably similar levels. The phase plot above shows the corresponding phase during this time. The amplitude indicates an increase over time until the unit is cycled down.

Figure 19 provides the corresponding spectrum during the time when the amplitudes were highest. Note, the primary frequency is occurring at the operating speed of thirty cycles (1800 rpm). This suggests that no ½ harmonic is present in the spectrum during the onset of the apparent rub condition.

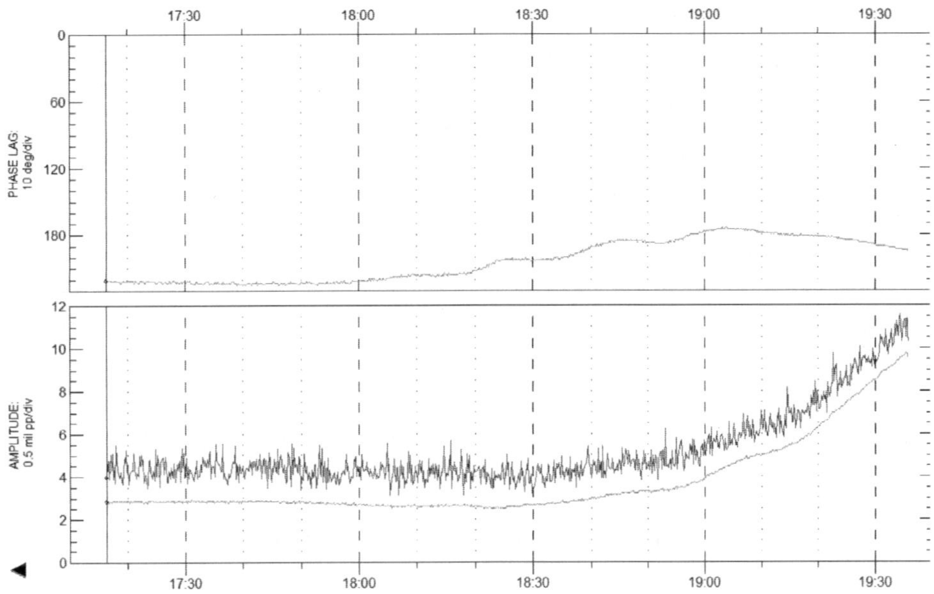

Figure 18 : Vibration Trend over 2 + hours on 4 pole turbine-generator

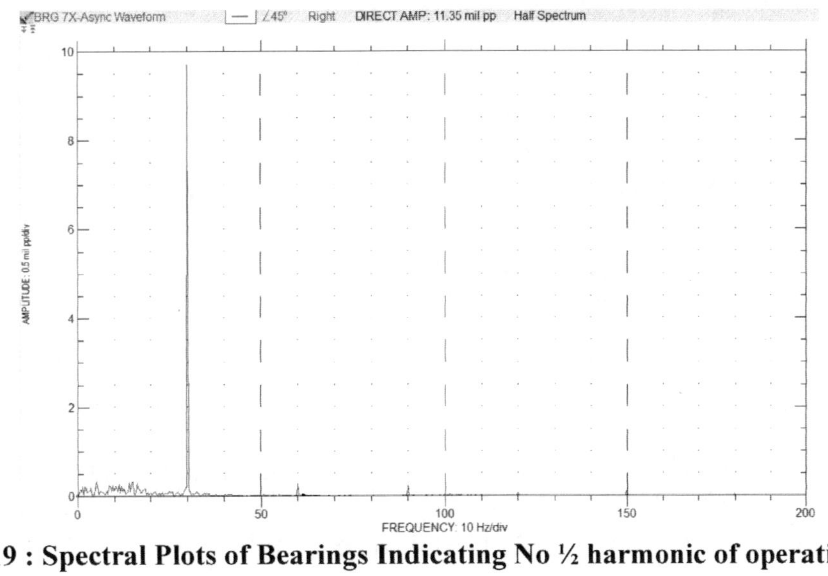

Figure 19 : Spectral Plots of Bearings Indicating No ½ harmonic of operating speed

Figure 20 provides the orbit and polar plot during the same onset of the rub. Notice the variation in amplitude in the polar plot and corresponding phase changes that are occurring. The phase location does not return to the original location on the polar plot, which may suggest the rub is still in progress.

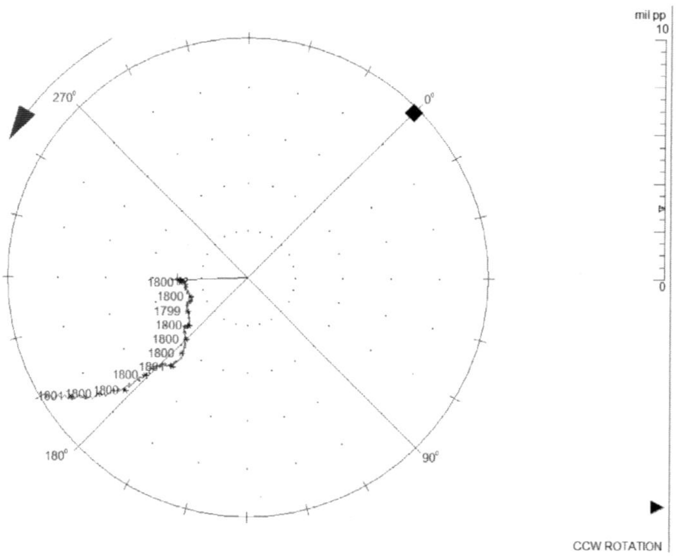

Figure 20 : Orbit and Polar Plot

Figure 21 provides the orbit time waveform, notice the flat spots on the time waveform.

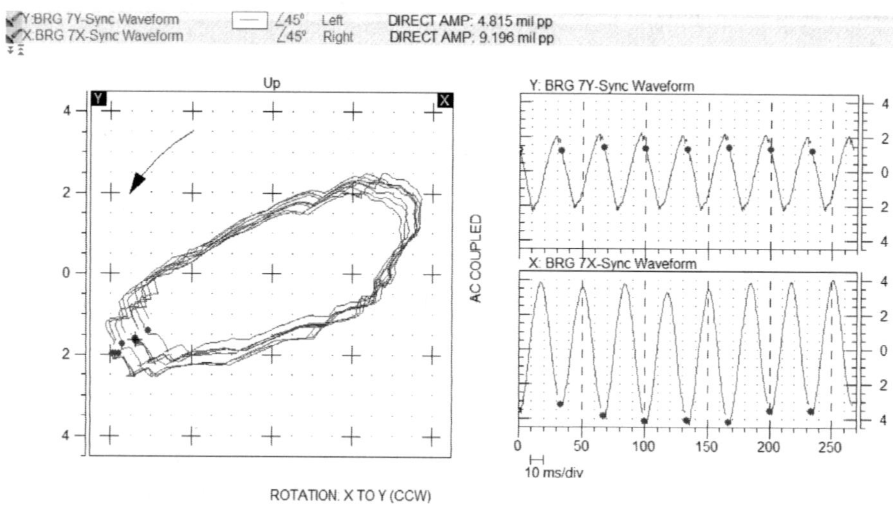

Figure 21 : Orbit During Rub

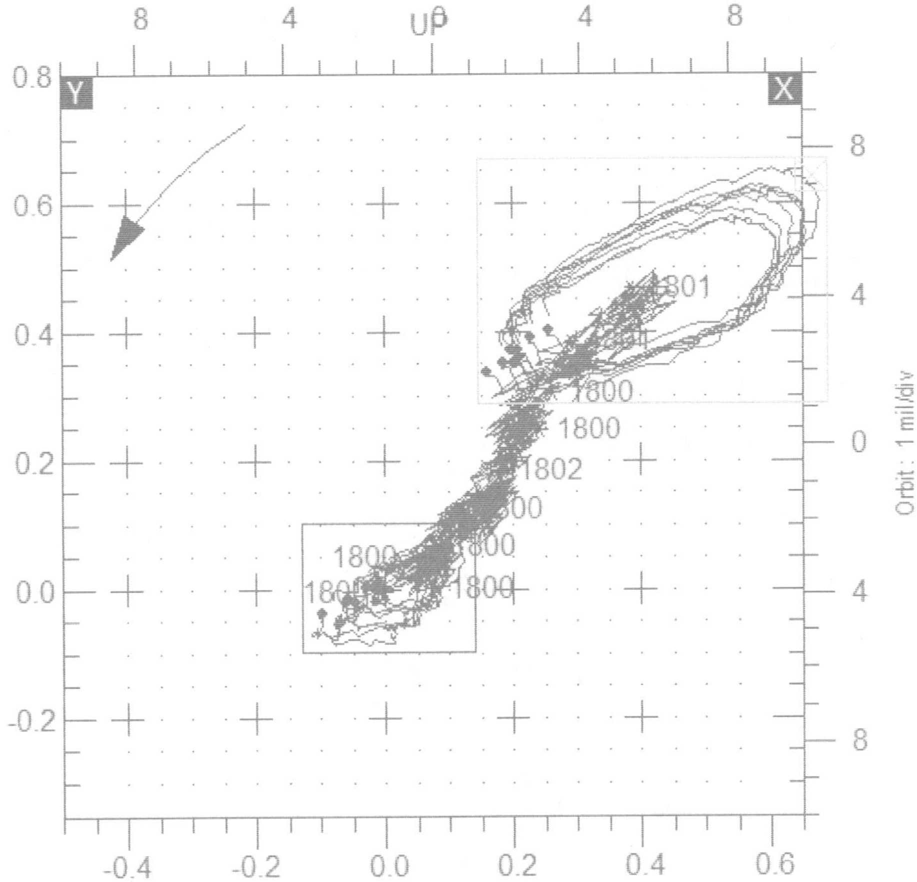

Figure 22 : Shaft Centerline and Orbit During Transient Level of Rub

Figure 22 provides the shaft centerline position during this time and the corresponding orbit shape as a function of time.

1X FREQUENCY RUBS EXPLAINED AND DECOMPOSED:

- When new equipment is commissioned (packing, bearings, and seals are installed, the clearances between the rotor and stationary packing is intentionally or non-intentionally remarkably close to improve performance.
- The result of the rotor and the packing touching during rotation is what we call "rubbing."
- Occasionally, this "rubbing" creates a hot spot on the shaft surface.
- This hot spot leads to a rotor bow, which typically leads to a change in the vibration.

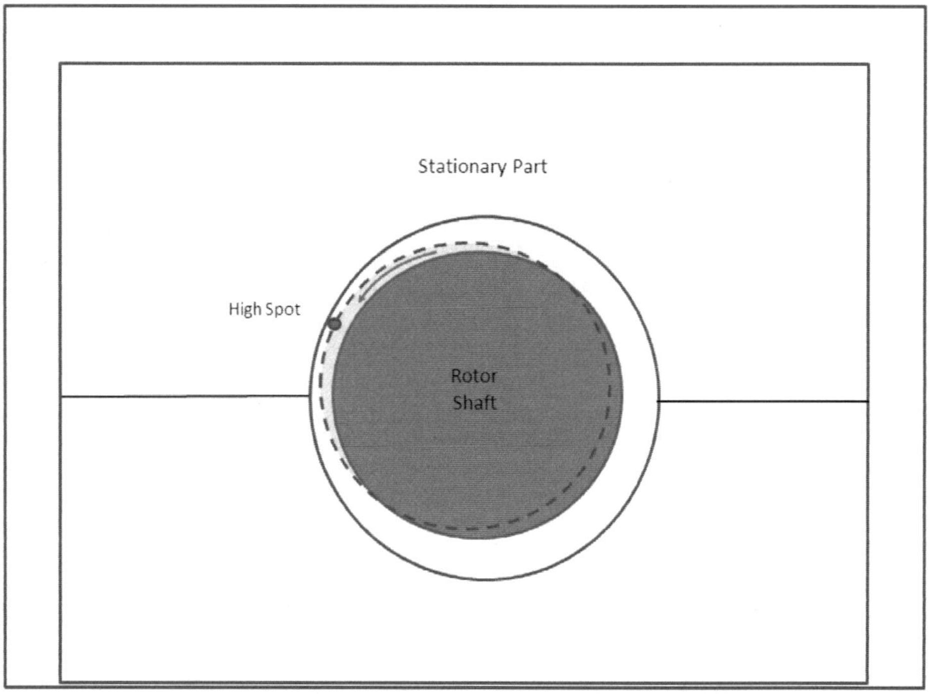

Figure 23 : Rotor within Stationary Clearance

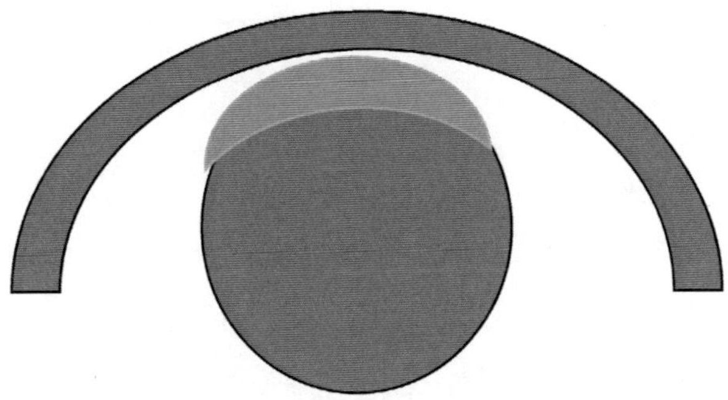

Figure 24 : Rub and "Hotspot"

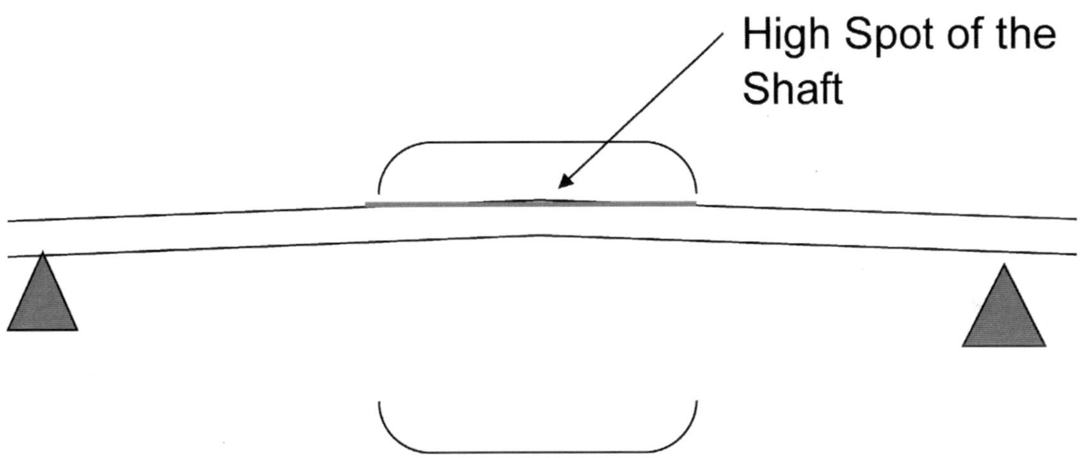

Figure 25 : Rub and "Hotspot" along Rotor Length

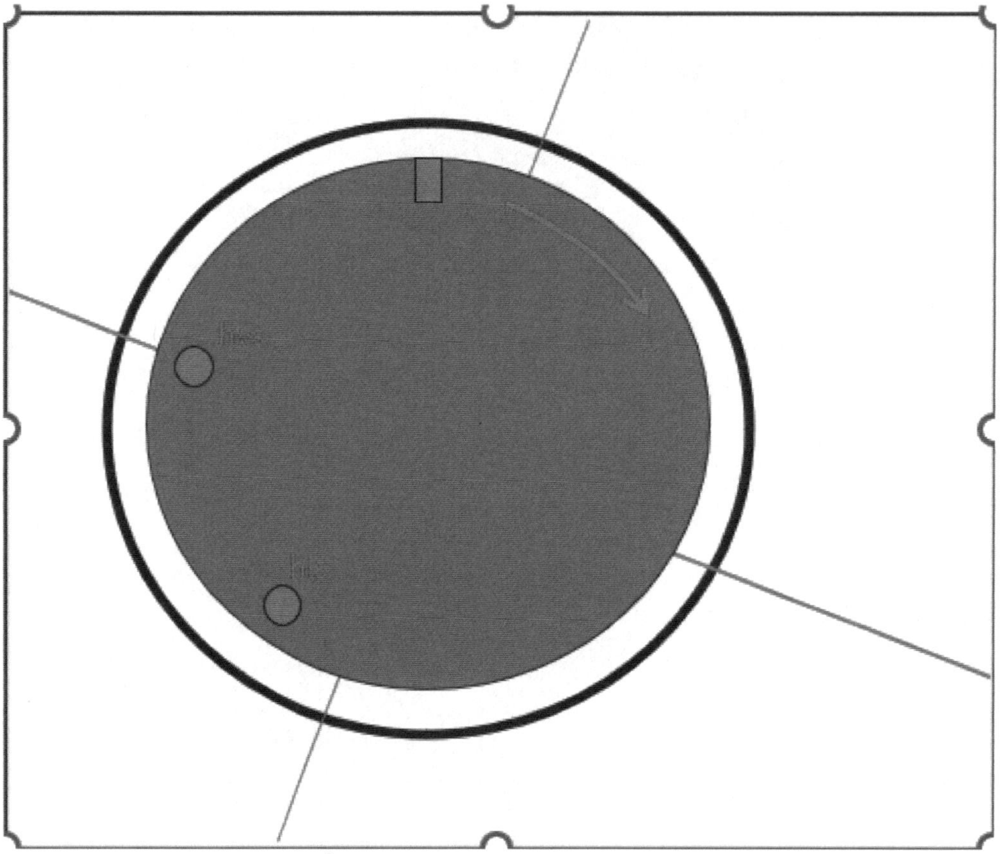

Figure 26 : Unbalance Heavy Spot (HVS) and High Spot Response (HIS)

Typical condition of unbalance with and heavy spot and high spot description as depicted in Figure 26. From rotor dynamics and balancing theory (Ref. Jackson, Gunter, Eshleman) the heavy spot leads the high spot by an angle depending on the location and relationship to the natural frequencies and critical speeds.

As the heavy spot (HVS) in displacement is located on the rotor, the high spot (HIS) lags the heavy spot. So as the (HIS) is close to the rub location as shown in Figure 27, the (HIS) starts to touch the stationary and a high-temperature zone is developed as shown in the previous description shown in Figure 24.

The high-temperature zone expands the rotor fibers in that location which results in the rotor bowing and moving in that location. The hot zone fibers push towards the stationary location.

This bowing creates a new (HVS') and the resulting (HIS) is a combination of the original (HIS) and (HIS') which causes the (HIS) to move from its original location. This creates the mechanism of a cycling movement of the (HIS) phase angle which is what we observe during the phase migration typically seen in one time amplitude excursion for rubs.

Figure 29 provides the bode' plot for the startup, steady-state condition and coast-down. This indicates the offset in the phase angle and the increased rub amplitude on coast-down. This is often the signature for a cold post-outage sequence resulting in a steady-state rub and inevitable coast-down forced by the operator taking action to shut down the machine in question.

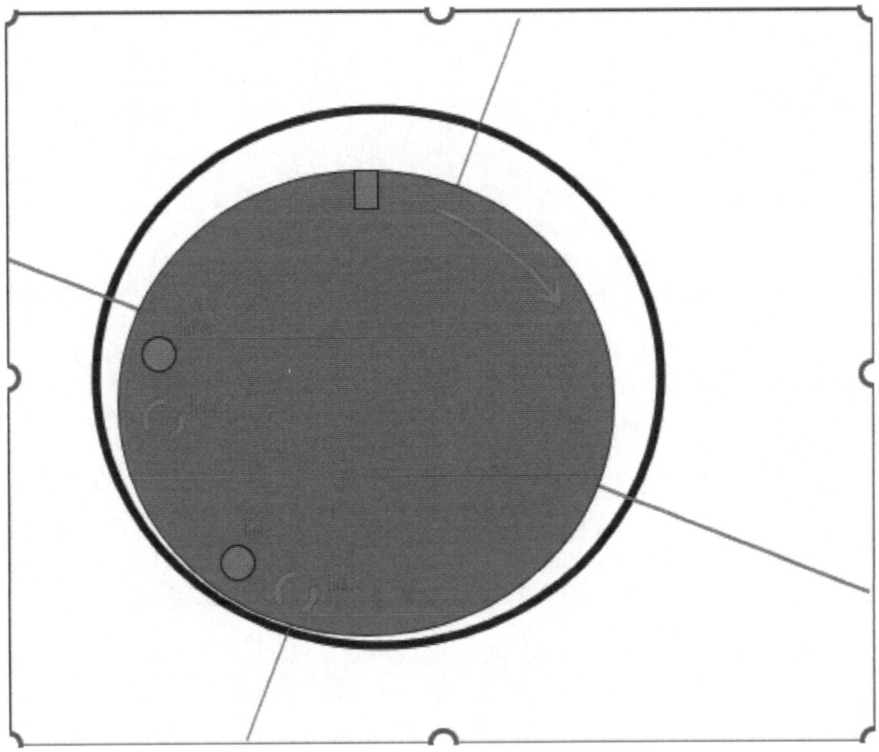

Figure 27 : Modified Unbalance Heavy Spot (HVS') and High Spot Response (HIS')

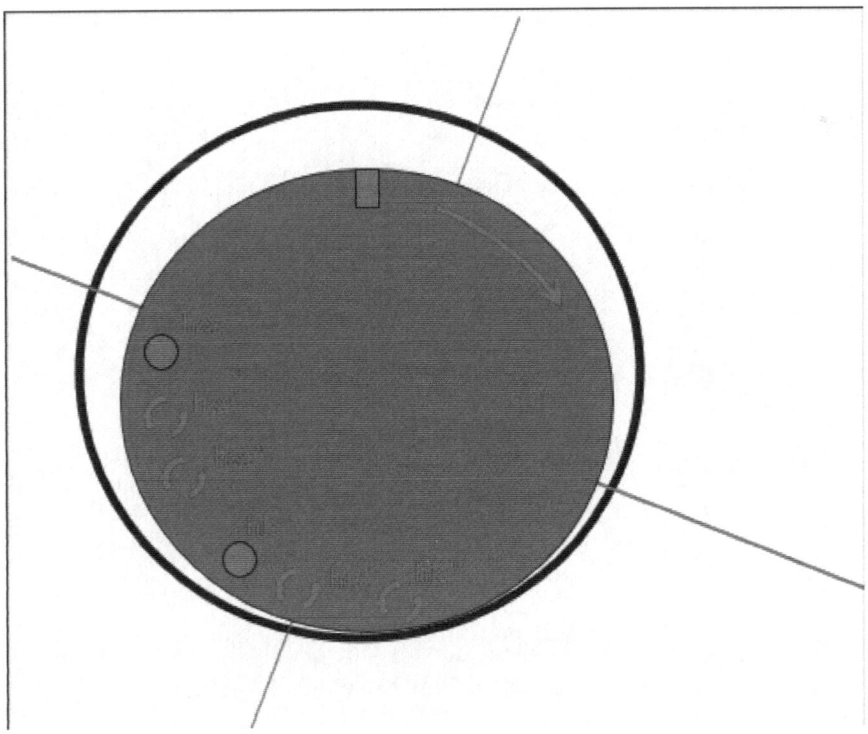

Figure 28 : Progressive Modified Unbalance Heavy Spot (HVS") and High Spot Response (HIS")

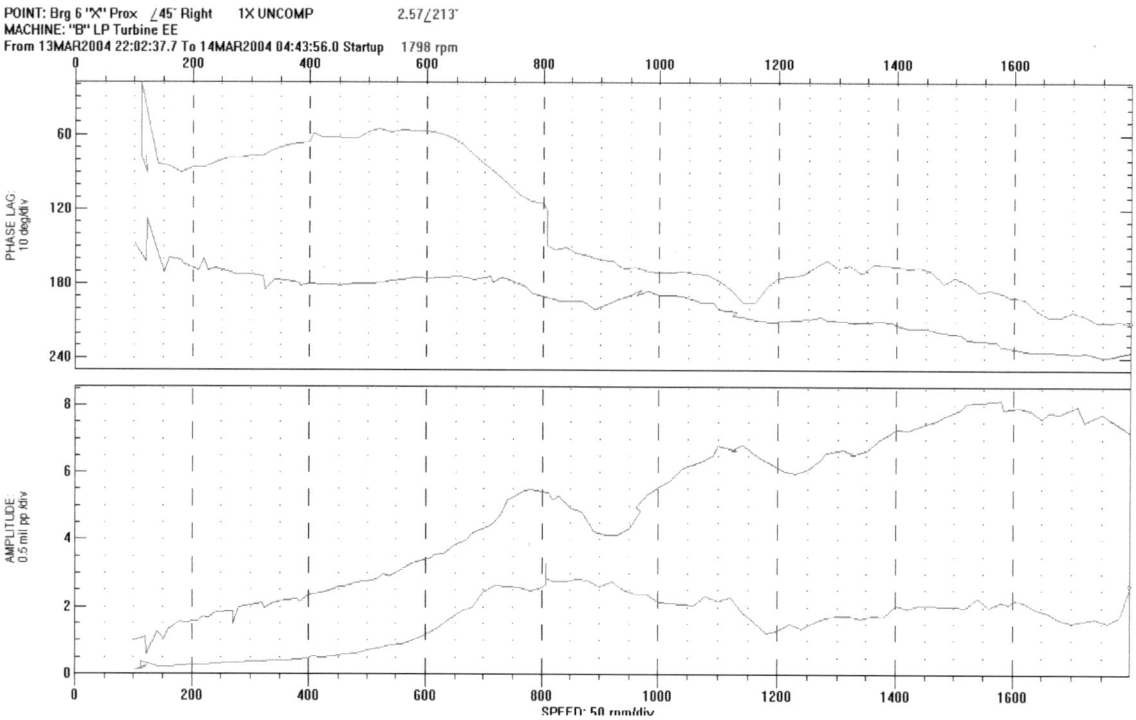

Figure 29 : Typical Startup and Coast-Down Plot (Bode) Exhibiting "Rub"

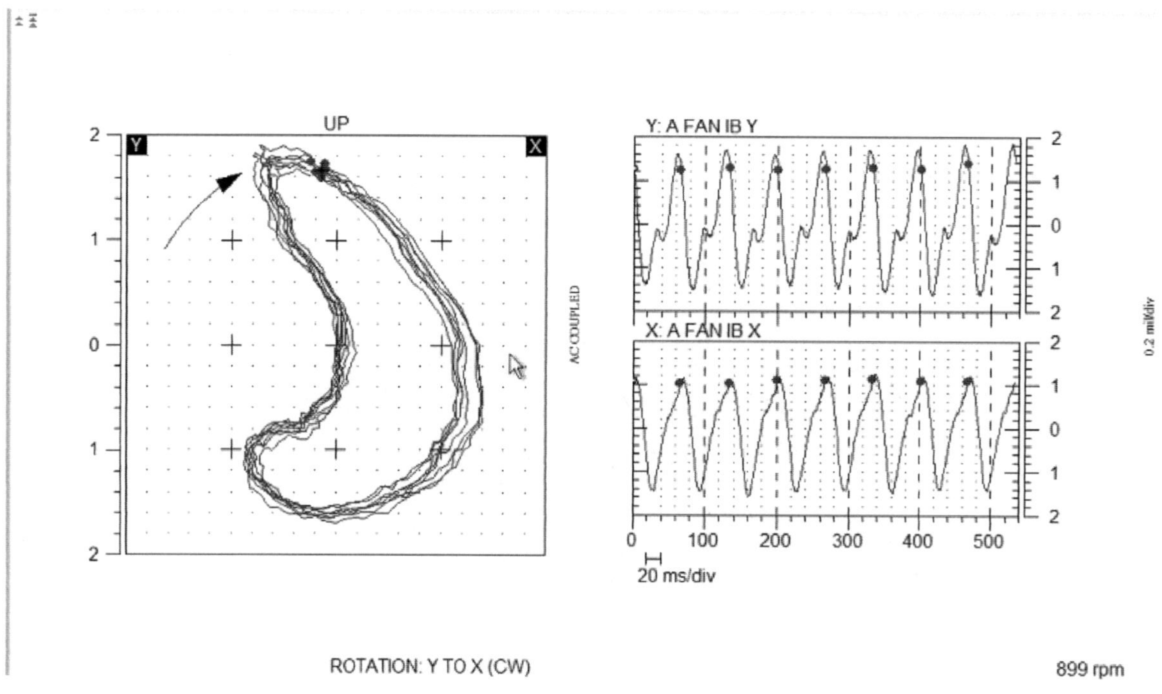

Figure 30 : Rub on Fan Bearings Typical

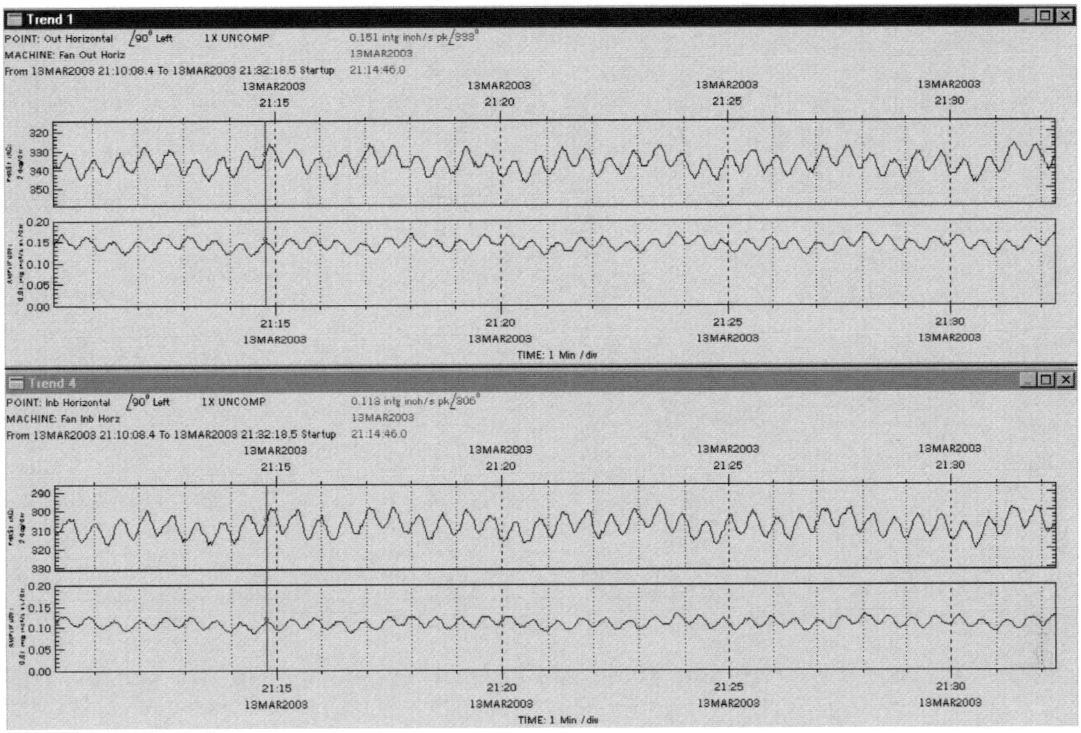

Figure 31 : Trend of Vibration on Horizontal Sensors on Inboard-Outboard End of Fan

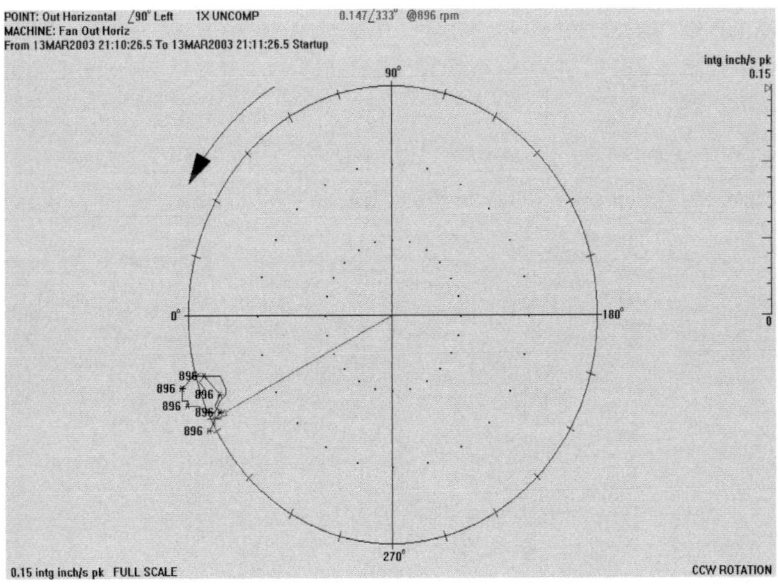

Figure 32 : Polar Plot on Outboard Horizontal Bearing of Fan

Figure 33-38 provide the various trends, polar plots and orbits that occurred over a multi-day period to where the amplitudes and phase angles were constantly changing. The amplitudes and phase locations in the polar plot are cycling past the original (HIS) to the opposite quadrants and the levels would recover to low values. These plots seem to indicate the phase angles never seem to return to any stability over time. The unit was eventually cycled off-line for outage with damage in the rotor depicted upon inspection at a bearing oil-seal.

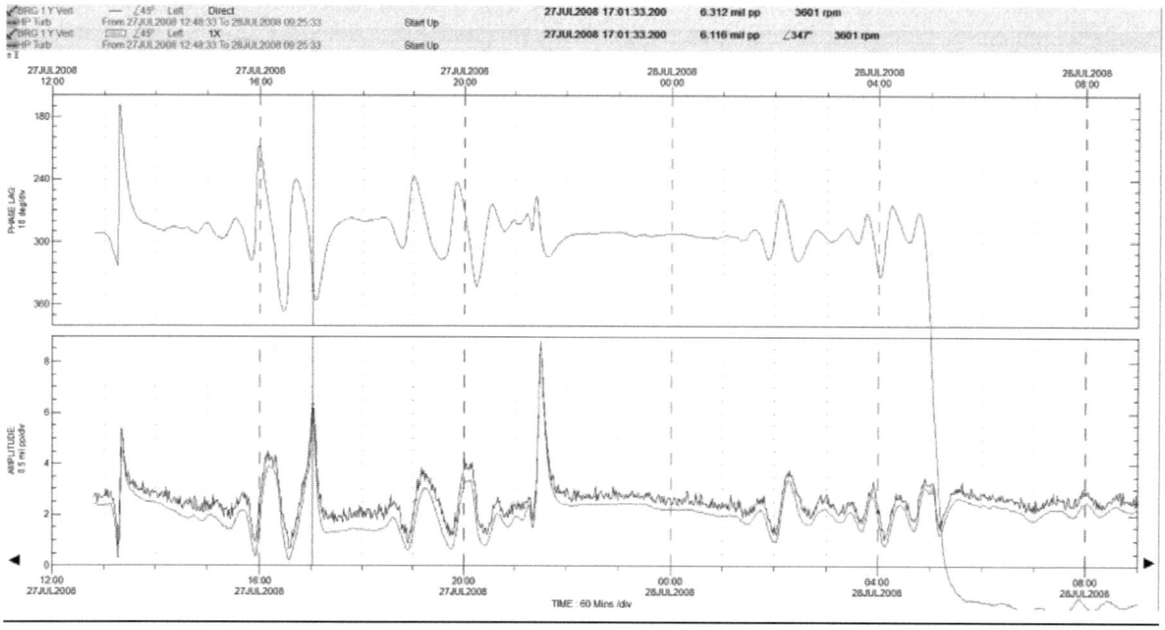

Figure 33 : 1Y Probe Vibration Amplitude and Phase Trend during Rub on Steam Turbine with Cycling Levels (Below Trip Limits)

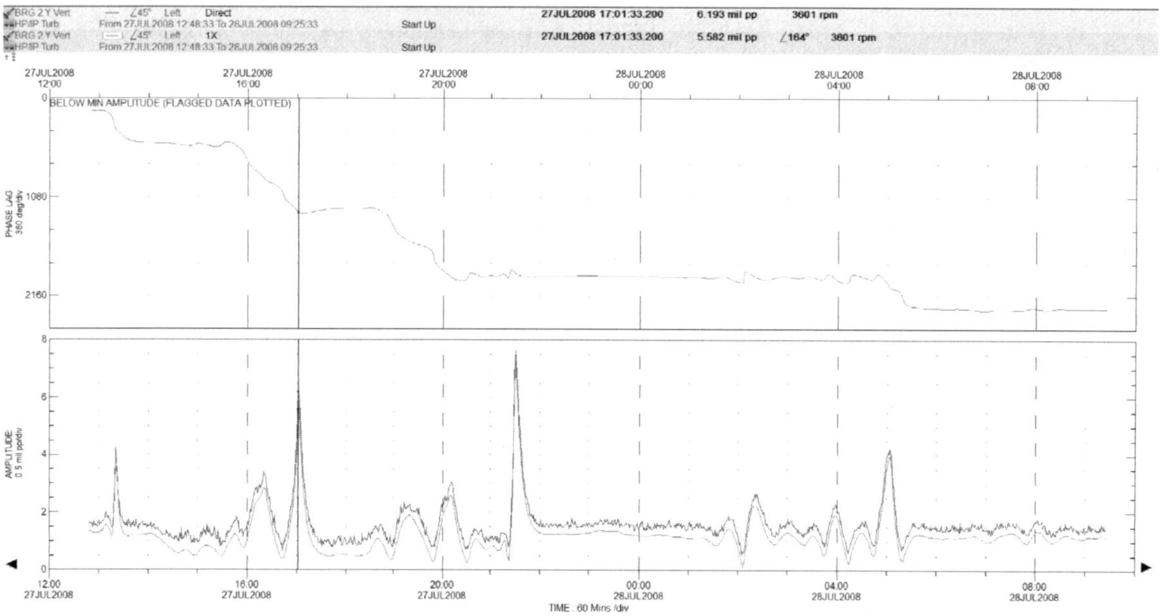

Figure 34 : 2Y Probe Vibration Amplitude and Phase Trend during Rub on Steam Turbine with Cycling Levels (Below Trip Limits)

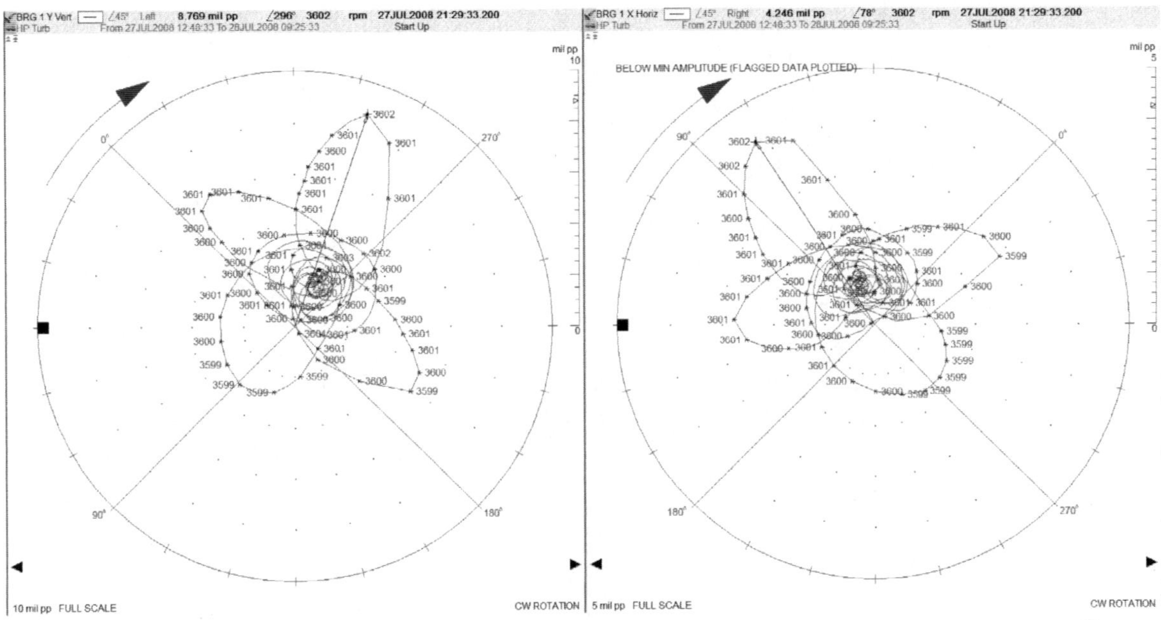

Figure 35 : 1Y Probe Vibration Amplitude and Phase Trend during Rub on Steam Turbine with Cycling Levels (Below Trip Limits)

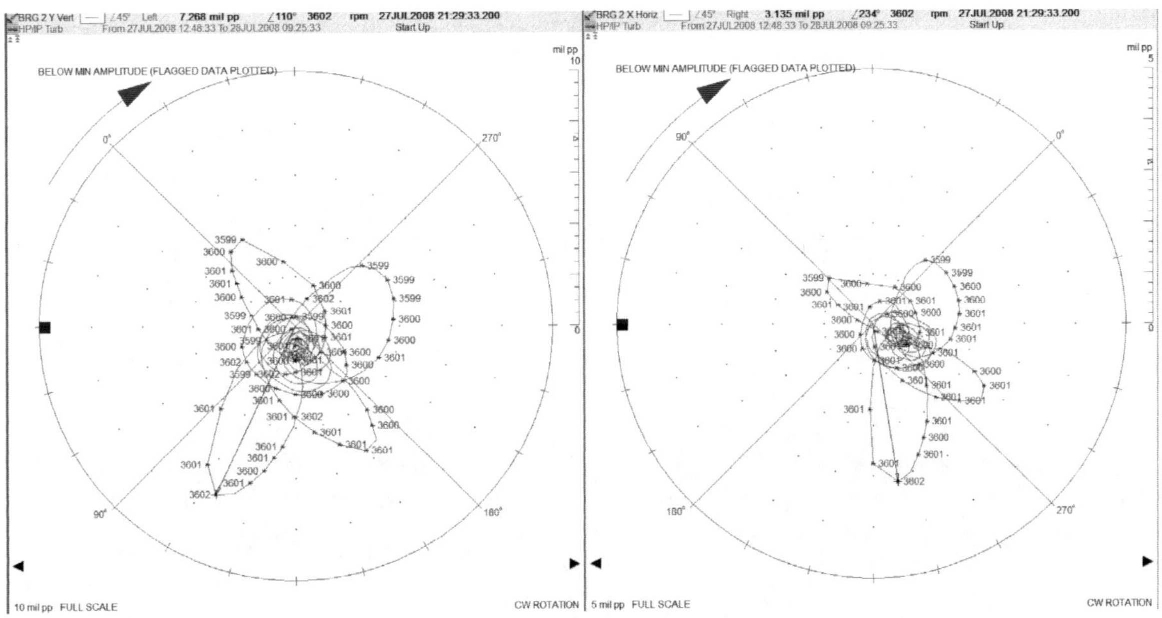

Figure 36 : 2Y Probe Vibration Amplitude and Phase Trend during Rub on Steam Turbine with Cycling Levels (Below Trip Limits)

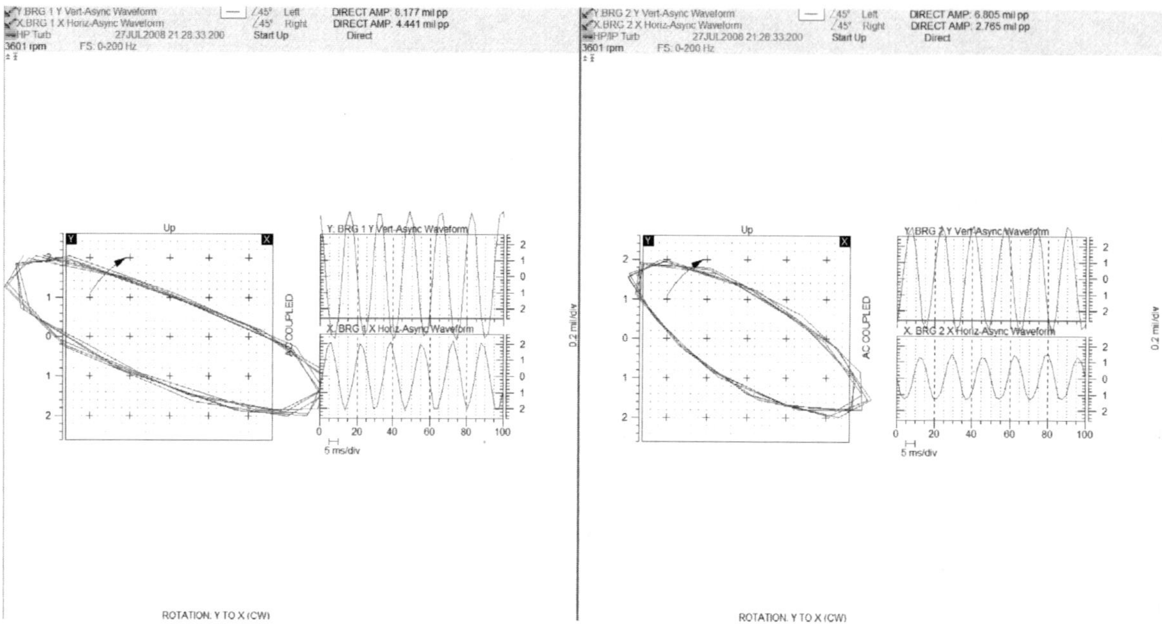

Figure 37 : 1Y and 2Y Orbit Plots during Rub on Steam Turbine with Cycling Levels (Below Trip Limits)

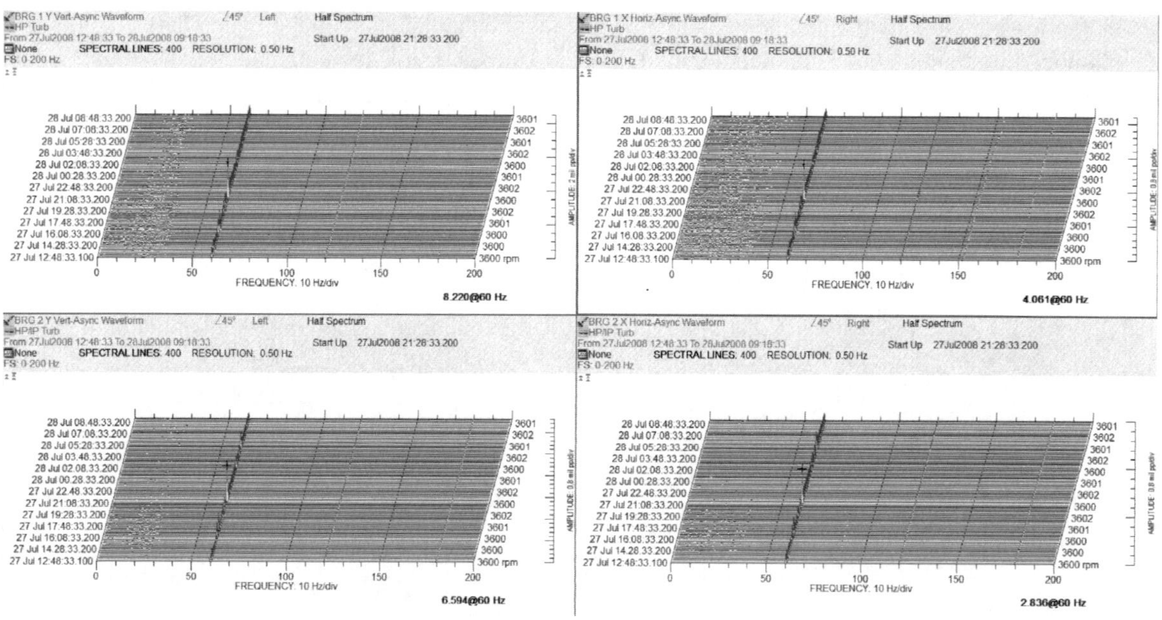

Figure 38 : 1Y and 2Y Orbit Spectral Maps during Rub on Steam Turbine with Cycling Levels (Below Trip Limits)

CONCLUSIONS - TYPICAL SYMPTOMS AND INDICATIONS

- Normally, vibration levels (when not rubbing) at a set speed do not fluctuate by more than 30 % (+/-).
- During startups and coast-downs, when an extremely dramatic difference occurs on coast-down good chance of rub.
- When the phase angle starts to "roll or wander," good chance of a rub.
- Typically, a rub frequency occurs at one time operating speed, but occasionally, the classic symptom is (½) X frequency also. (Rare occurrence)
- Typically, the rub clears or rubs free. This is a consequence of one of the following mechanisms including:

1. Rotor position moves away from the stationary component rubbing.
2. Stationary component moves out of the way of the rotating component.
3. Either of these could be a function of time (short duration or longer duration)
4. Operator takes the machine off-line for inspection or remediation.

Rub in Turbine Rotor

Figure 39 : Extreme Rub Damage on Steam Turbine

Figure 40 : Extreme Rub Damage on Steam Turbine (Rotor Cleaned)

ACKNOWLEDGEMENTS:

I would like to thank the Southern Company for the many years of support and opportunities to learn and practice machinery diagnostics.

REFERENCES:

[1] Bently, Donald E. and Hatch, Charles T. "Fundamentals of Rotating Machinery Diagnostics (Chapter 21 – Rubs and Looseness)" Textbook – 2002

VIATC TRAINING CONFERENCE & EXPO
COVINGTON, KENTUCKY | AUGUST 7-9, 2024

UNBALANCE CONTROL OF A SMART ELECTRO-MAGNETIC ACTUATOR JOURNAL INTEGRAL BEARING (SEMAJIB)

Aly El-Shafei
RITEC, Cairo, Egypt
Email address elshafei@ritec-eg.com

Dr. Aly El-Shafei is a Certified Vibration Analyst, Category IV. He holds a Ph.D. from the Massachusetts Institute of Technology (MIT). In addition to serving as Managing Director at RITEC, Dr. El-Shafei is also a Professor of Vibration Engineering at Cairo University, Egypt and is the Vice President of the Vibration Institute, USA. Prof. El-Shafei chairs the ISO working group ISO/TC108/SC2/WG10 responsible for developing international standards on machinery vibration diagnostics ISO 13373. Dr. El-Shafei is known as a seasoned diagnostics engineer and has solved many chronic problems worldwide. He is also known worldwide for his work on Squeeze Film Dampers, and for developing new technology for the control of rotating machinery vibration. Prof. El-Shafei is the recipient of the distinguished Grand African Innovation Award in 2017. Dr. El-Shafei has received 5 US patents and has published more than 80 papers on rotordynamics and machinery diagnostics.

ABSTRACT:

This paper introduces the concept of using a smart integrated bearing as a unique load carrying element, which is able to effectively carry the load by a journal bearing integrated with an electro-magnetic actuator (flooded in oil) that is able to control the rotor behaviour. The electro-magnetic actuator is used to control both the oil whip instability in the journal bearing as well as single plane and two-plane unbalance in the rotor. The theory is presented, and numerical and experimental results are presented that illustrate the success of the SEMAJIB and algorithms developed to control rotor vibrations.

KEYWORDS: SEMAJIB, Smart Bearing, Unbalance Control, Electro-Magnetic Actuator, Journal Bearing.

INTRODUCTION:

This paper summarizes the development of the Smart Electro-Magnetic Actuator Journal Integrated Bearing SEMAJIB [1-2] that was first introduced by the author over a decade ago. The SEMAJIB combines a fluid film bearing (FFB) with an electro-magnetic actuator (EMA) in one integrated device and capitalizes on the advantages of Fluid Film Bearings (FFBs) and Active Magnetic Bearings (AMBs). FFBs, in particular Journal Bearings (JBs), are superior load carrying elements due to their larger load carrying capacity and their ability to introduce passive damping to the rotor system. AMBs, on the other hand, have a certain feature which is their capabilities as controlling elements. AMBs can provide variable and controllable stiffness and damping and additionally can provide unbalance control and many other control features.

This paper presents the SEMAJIB [1-2], a smart high performance integrated bearing that combines a In all cases, the fluid film bearing shall carry the load, whereas the electro-magnetic actuator can be used as a pure controller or both as a controller and a load carrying element. In the latter case the electro-magnetic actuator can be considered as an active magnetic bearing (AMB).

The integration of AMBs and JBs in one device has clear advantages. The SEMAJIB has all of the advantages of JBs and AMBs, and avoids all of the shortcomings of AMBs and JBs. The SEMAJIB is a superior load carrying element due to its larger load carrying capacity and its ability to introduce passive damping to the rotor system. Moreover, it is free from whip instabilities and has capabilities as a controller. A SEMAJIB can provide variable and controllable stiffness and damping and additionally can provide unbalance control and many other control features. Most importantly, there is no need for a backup bearing as the rotor is carried on the JB in all cases.

The author and his students have developed and published many papers describing the development of the SEMAJIB. In [3] controlling oil whip using the SEMAJIB through multiple algorithms is discussed. It is shown that damping control is an effective method for controlling the SEMAJIB [3]. Instability control and unbalance control using the SEMAJIB are introduced in [4]. An important contribution which shows that oil does not adversely affect the performance of AMBs is presented in [5]. In fact, it is shown that oil in an AMB actually provides some minor improvements in AMB performance. The testing of a rotor on one SEMAJIB and one rolling element bearing using PID control is introduced in [6], while fuzzy logic control to the SEMAJIB is introduced in [7], and $H\infty$ control to the SEMAJIB is introduced in [8] and includes discussion of load sharing between the AMB and the JB. Testing of a rotor on two SEMAJIB bearings and the ability to transgress the oil whip instability for the first mode and the second mode by applying PD control is discussed in [9]. In fact, this work [9] is a clear indication of the success of the SEMAJIB. It shows that the SEMAJIB can carry a high load rotor at high speeds and with ability to control multiple instabilities.

In all of the above experiments an off-the-shelf programmable controller is used. The control algorithms discussed in the previous paragraph were all implemented experimentally, and were quite successful. The choice of the control algorithm is a matter of choice for each application. In many cases, it is important to instruct the magnetic bearing not to interfere with the load carrying by the JB. In fact the $H\infty$ controller [8] actually tends to carry some of the load on the

AMB, while the PD controller tends to act as a controller only. The problem is that the AMB likes to center the rotor, while the JB tends to move the rotor center downwards and sideways. These two competing devices need a controller designed to carry the load on the JB and keep all the AMB power for control. Only in special circumstances (like relocating resonances) should the AMB be allowed to carry the load.

In this paper, presents new data on unbalance control. In the design of the SEMAJIB, the AMB is designed to be able to carry the static load. However, the load is carried by the JB, so actually the AMB is quite a sizable controller. It thus can sustain large unbalance forces (even at resonance). Moreover, the concept of a smart bearing is emphasized, where the bearing can correct malfunctions in a machine, thus increasing availability and reliability of machines. The paper starts by presenting the results of the instability control, followed by the theory and results of the Unbalance Control. The use of the SEMAJIB as a smart bearing is then discussed.

EXPERIMENTAL RESULTS: INSTABILITY CONTROL:

A test rig was developed at Cairo University to test the SEMAJIB. It consisted of 50 mm diameter, 1.7 m long shaft, with 4 discs and mounted on a SEMAJIB [6] (see Figures 1 and 2).

Figure 1 Test Rig used in testing the SEMAJIB [6]

Figure 2 SEMAJIB used in testing [6]

An elaborate testing scheme is described in [6] for the test brig and for the SEMAJIB. The main purpose of the testing was to determine the capability of the AMB to control the JB instability. Figure 3 shows the waterfall diagram showing the run-up of the test rig. The 1x component increases at the critical speed and decreases thereafter, while the oil whip component appears when the oil whirl component reaches the critical speed and locks at the whip frequency. A PD controller was then used with the AMB within the SEMAJIB, and the results are shown in the waterfall of Figure 4, where the oil whip instability was eliminated. Figure 5 shows the spectrum at the operating speed with and without control. It is clear that the PD controller was able to control the oil whip instability and provide a stable bearing.

It is thus concluded that the SEMAJIB is a stable bearing that can transgress critical speeds and is an instability free bearing, due to the controller capabilities of the AMB.

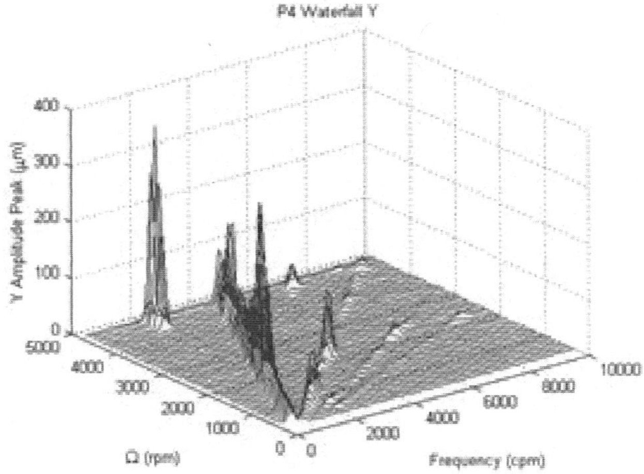

Figure 3 Waterfall diagram showing oil whip instability [6]

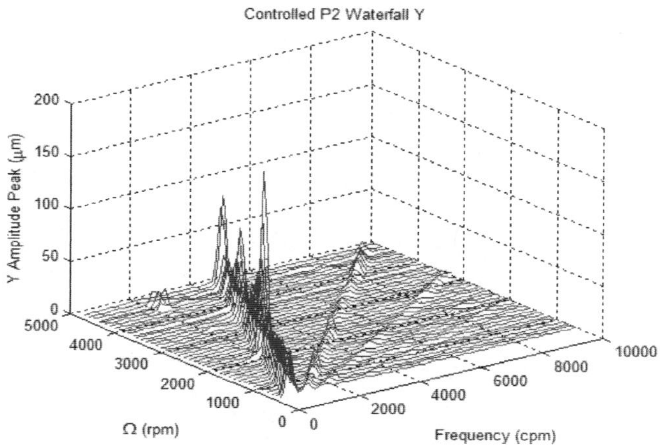

Figure 4 Waterfall diagram with SEMAJIB showing the elimination of the oil whip instability [6]

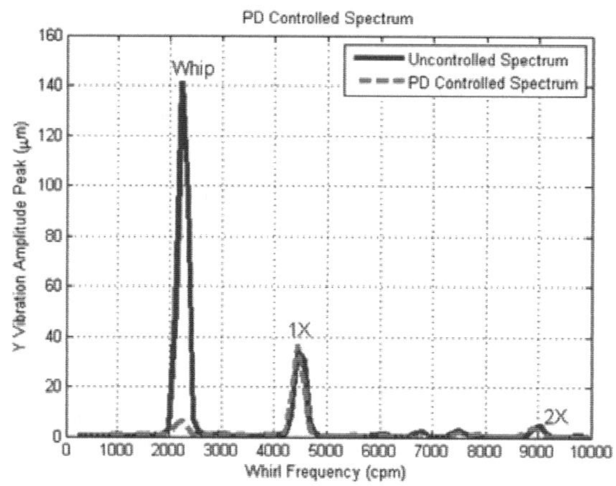

Figure 5 Spectrum showing the elimination of oil whip instability [6]

It was suggested that actually high-speed rotating machinery may actually encounter higher mode instabilities. Three controllers were developed PD control, Fuzzy Logic (FL) control and H∞ control. All three controllers were applied experimentally [7-9] and were quite successful. Figure 6 shows the application of the 3 controllers on the test rig in eliminating the first mode instability, while Figure 6shows the application of the 3 controllers to eliminate second mode instability [9].

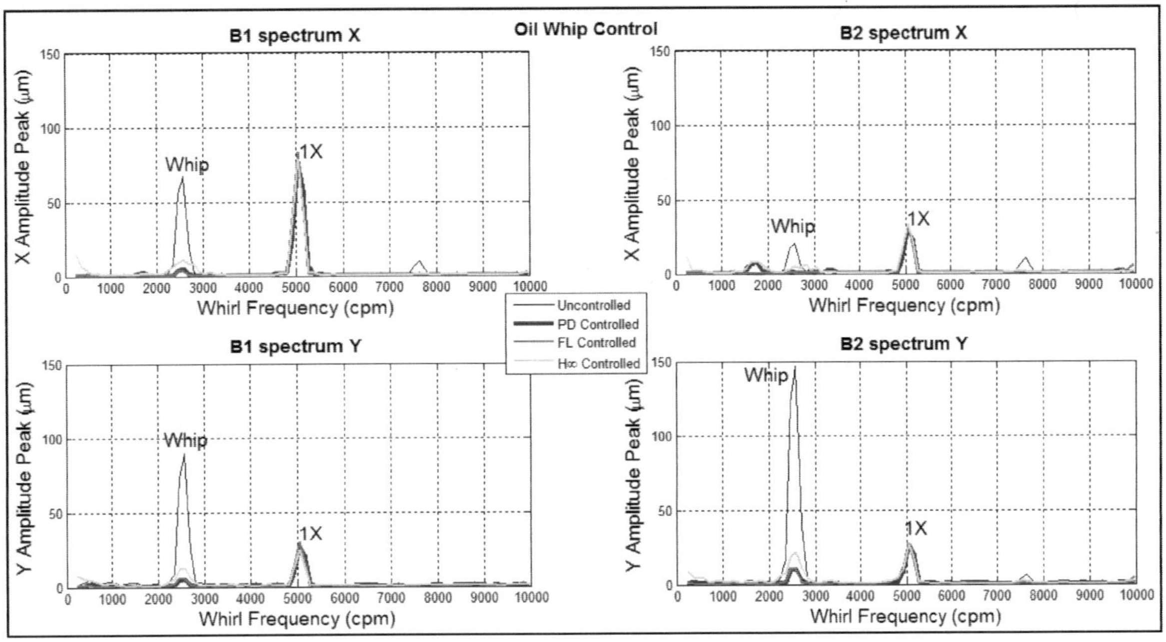

Figure 6 Comparison of controllers for first mode oil whip control [9]

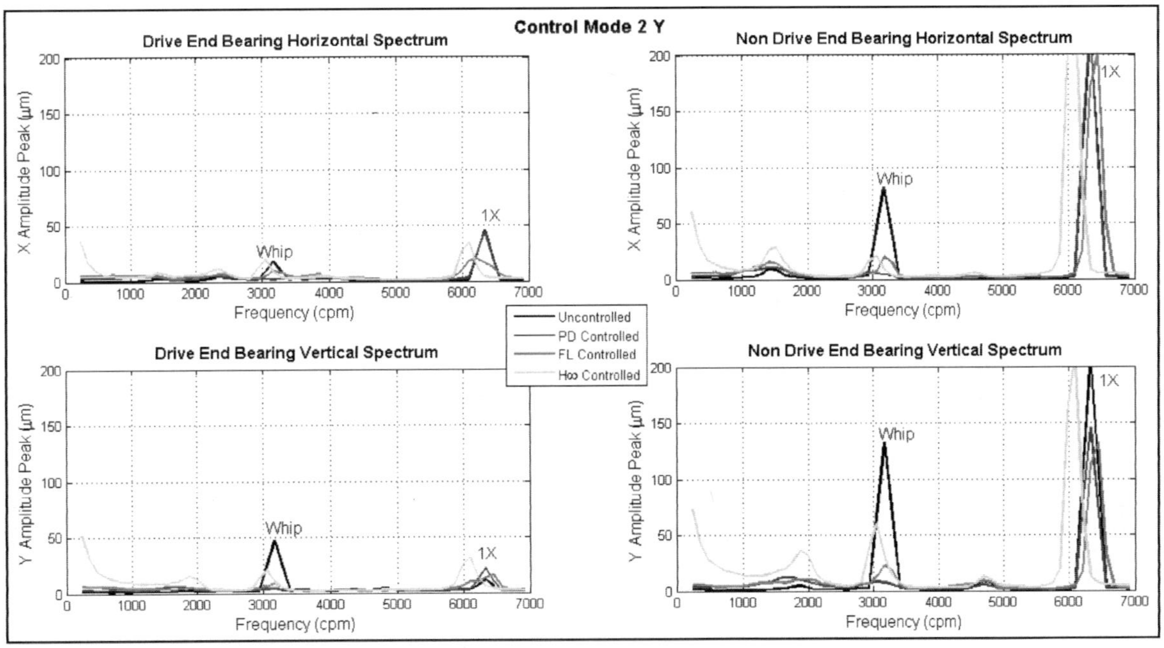

Figure 7 Comparison of controllers for second mode oil whip control [9]

Table 1 shows the average reduction by the 3 controllers. It is is clear that the H∞ controller [8] is the lesser of the three in controlling the instability, since it tends to provide stiffness. This is evident in Figure 7 where the H∞ controller was able to move the critical speed by about 600 rpm.

Table 1 Average Reduction by SEMAJIB Controller

% Reduction	PD	FL	H∞
First Mode Instability	94	93	82
Second Mode Instability	90	86	56

UNBALANCE CONTROL:

The mathematical model for a rotor supported on journal bearing and subjected to external forces and magnetic forces is:

$$\boldsymbol{M}\{\ddot{x}\} + \omega\boldsymbol{G}\{\dot{x}\} + \boldsymbol{K}\{x\} = -\boldsymbol{C}_b\{\dot{x}\} - \boldsymbol{K}_b\{x\} + \{f_{ext}\} + \{f_{bal}\} \qquad (1)$$

Where x is the displacement vector, M is the mass matrix of the rotor, G is the gyroscopic effect matrix, K is the stiffness matrix of the rotor, Cb and Kb are the damping matrix and stiffness matrix from journal bearing, fext is the unbalance vector that is present on the rotor and fbal is the balancing force vector to be applied by the magnetic bearings.

Assuming synchronous forced response in steady state at frequency ω:

$$\{x\} = \{X\}e^{i\omega t} \quad , \quad \{f_{ext}\} = \{F_{ext}\}e^{i\omega t} \quad , \quad \{f_{bal}\} = \{F_{bal}\}e^{i\omega t} \qquad (2)$$

Substitute from (2) into (1):

$$[-\boldsymbol{M}\omega^2 + i\omega^2\boldsymbol{G} + \boldsymbol{K} + i\omega\boldsymbol{C}_b + \boldsymbol{K}_b]\{X\} = \{F_{ext}\} + \{F_{bal}\} \qquad (3)$$

$$\{X\} = [-\boldsymbol{M}\omega^2 + i\omega^2\boldsymbol{G} + \boldsymbol{K} + i\omega\boldsymbol{C}_b + \boldsymbol{K}_b]^{-1}[\{F_{ext}\} + \{F_{bal}\}] \qquad (4)$$

$$\{X\} = \{X_o\} + \boldsymbol{H}\,\{F_{bal}\} \qquad (5)$$

Where Xo is the original synchronous vibrations without balancing and H is the influence coefficient matrix which relates the applied synchronous forces to the displacements at the measuring points. H matrix is square when the number of measuring points is equal to the number of control directions (This is 4x4 in our case: a vertical and a horizontal direction in two different planes)

H is unknown and is estimated by measuring the input and output of several trials and employing a least square estimator as follows:

Construct a sequence of trial forces and corresponding synchronous vibrations.

$$[\{X_1\}\{X_2\}\,\{X_3\}\,....\{X_n\}] = [\{X_o\}\,\{X_o\}\,\{X_o\}\,....\{X_o\}] + \boldsymbol{H}[\{F_{bal1}\}\{F_{bal2}\}\{F_{bal3}\}\,....\{F_{bal\boldsymbol{n}}\}] \qquad (6)$$

or:

$$\boldsymbol{X} = \boldsymbol{X}_o + \boldsymbol{H}\boldsymbol{F}_{bal} \tag{7}$$

Where X, Xo and Fbal are matrices in which each trial occupies a column.

$$\boldsymbol{H}\boldsymbol{F}_{bal} = \boldsymbol{X} - \boldsymbol{X}_o \tag{8}$$

$$\boldsymbol{H} = (\boldsymbol{X} - \boldsymbol{X}_o)\boldsymbol{F}_{bal}^T(\boldsymbol{F}_{bal}\boldsymbol{F}_{bal}^T)^{-1} \tag{9}$$

It is desired to obtain minimum vibrations at vector {X}, this is achieved by setting {X} to zero in equation (7), the control law is obtained as:

$$F_{opt} = -\boldsymbol{H}^{-1}X_o \tag{10}$$

The same control law may be obtained in a more formal approach as follows [10]:

Set a performance index J such that:

$$J = X^T X = (X_o^T + F_{bal}^T\boldsymbol{H}^T)(X_o + \boldsymbol{H}F_{bal}) = X_o^T X_o + X_o^T\boldsymbol{H}F_{bal} + F_{bal}^T\boldsymbol{H}^T X_o + F_{bal}^T\boldsymbol{H}^T\boldsymbol{H}F_{bal}$$
(11)

Find FBal that minimizes J and set it equal to zero:

$$\frac{\partial J}{\partial F_{bal}^T} = \boldsymbol{H}^T X_o + \boldsymbol{H}^T\boldsymbol{H}F_{bal} = 0 \tag{12}$$

$$\{F_{opt}\} = -\boldsymbol{H}^{-1}\{X_o\} \tag{13}$$

It should be noted that a complex representation is used to perform the calculations. Each synchronous displacement or force, which has amplitude and phase, is transformed to complex representation. Balancing was carried out in two planes. The measurements and the forces were applied in the horizontal and vertical directions of each plane. Hence, the dimensions are:

X:4x1

Xo:4x1

H:4x4

Fbal:4x1

This means that a minimum of four trials is required to estimate the H matrix. A moving batch scheme is used in which the oldest trials are removed and the most recent ones are retained and used for estimation.

The experimental results for the unbalance control using the SEMAJIB is shown in Figures 8 to 11.

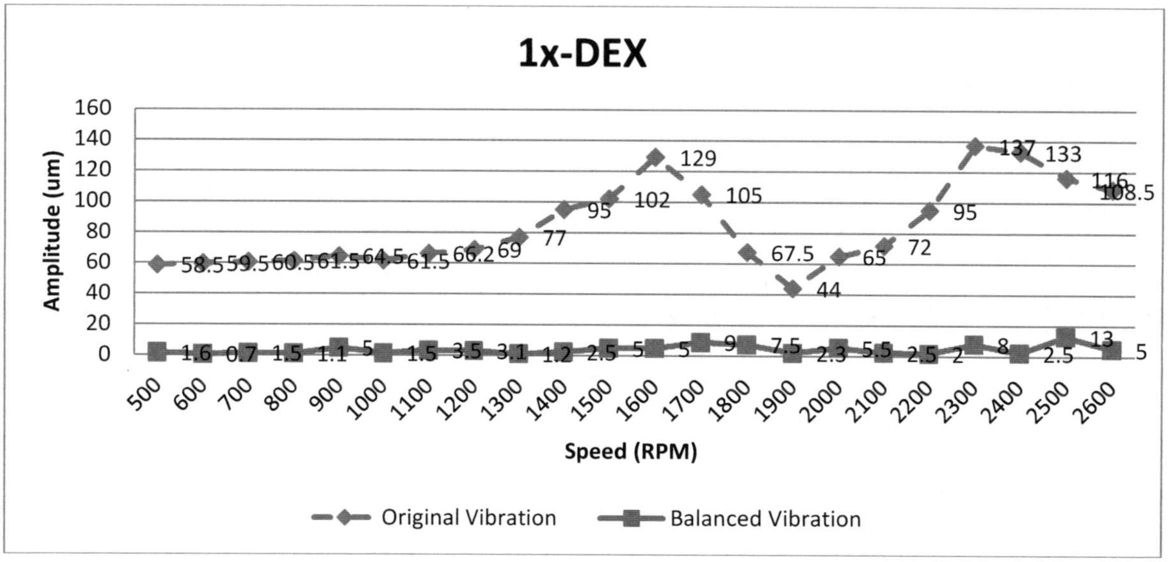

Figure 8 Unbalance control on the drive-end bearing - horizontal

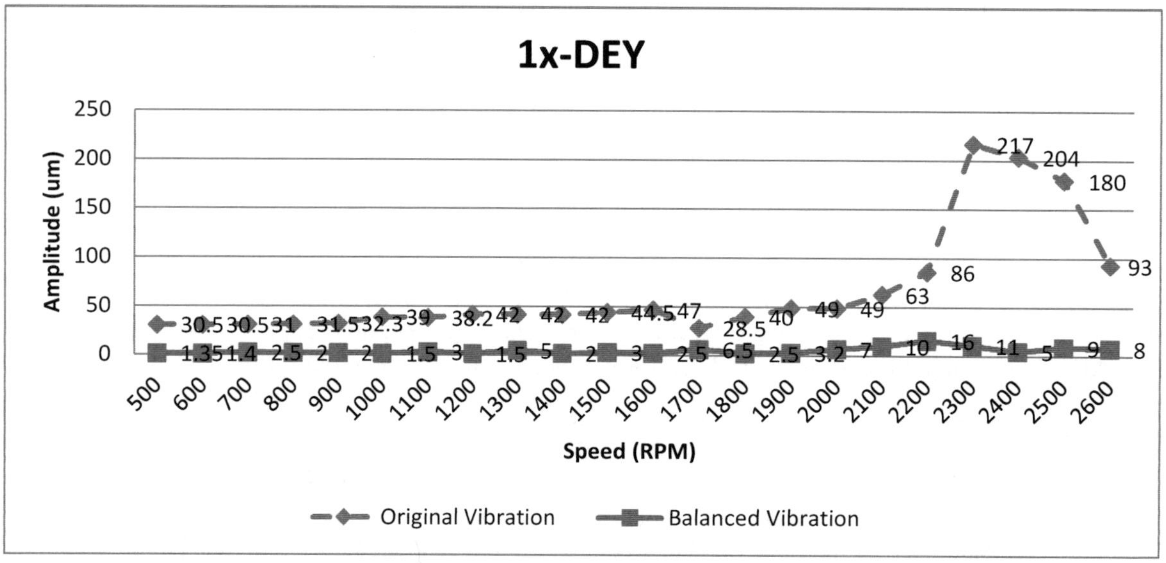

Figure 9 Unbalance control on the drive-end bearing - vertical

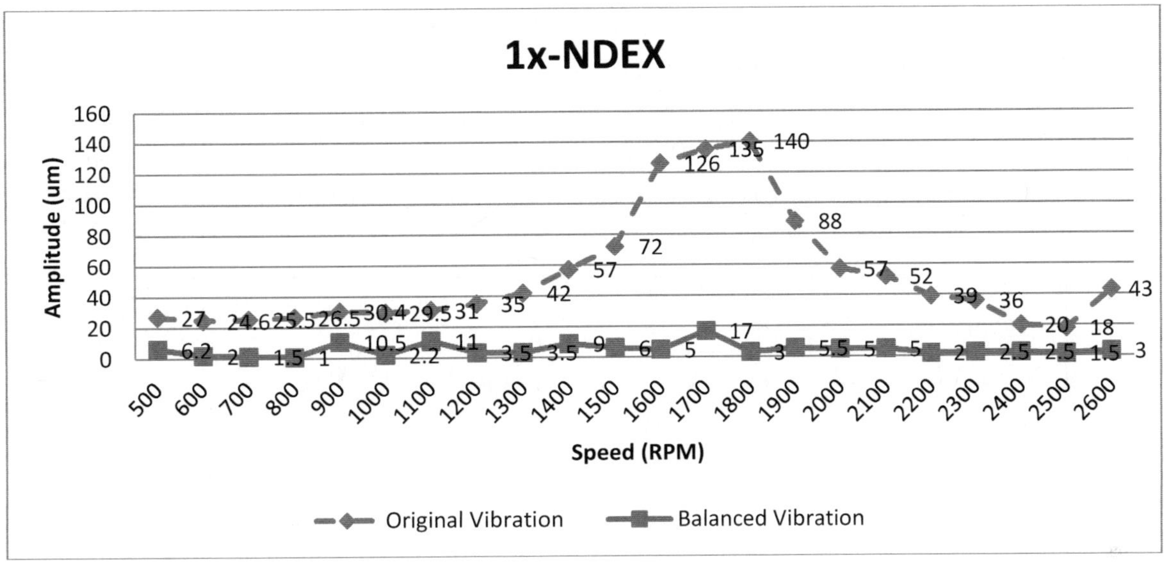

Figure 10 Unbalance control on the non-drive-end bearing - horizontal

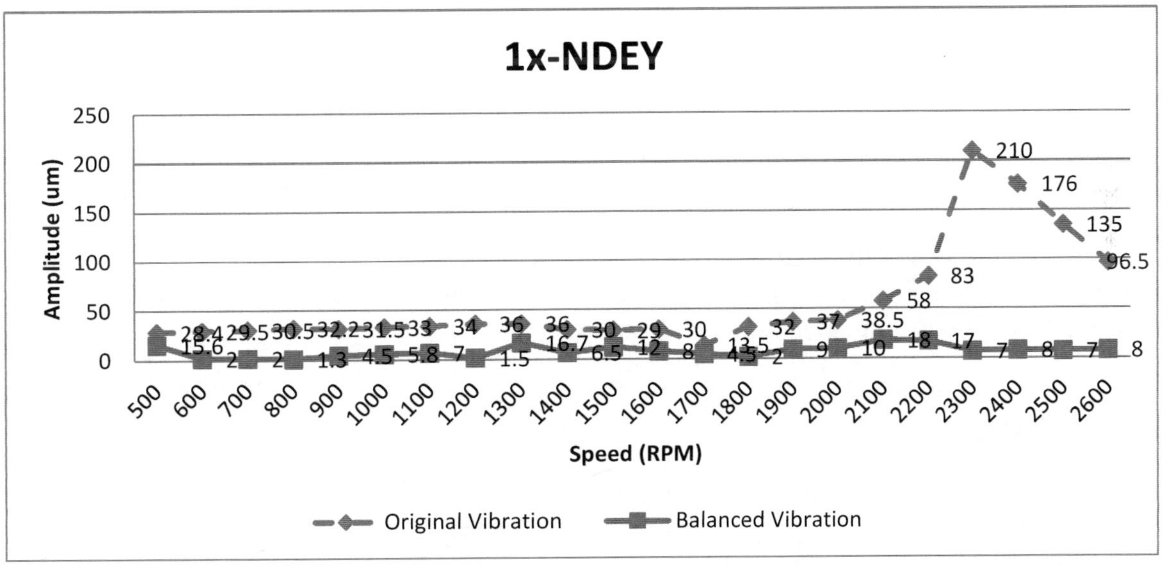

Figure 11 Unbalance control on the non-drive-end bearing - vertical

It is clear that the unbalance was quite effective for this rotor achieving more than 90% reduction at resonance. This is quite remarkable in particular that the rotor transversed multiple critical speeds.

Smart Bearing Concept:

Fig. 12 shows a block diagram of the SEMAJIB system [11]. The rotor is subjected to external forces Fext, however the rotor states x and x' affect the JB which in turn provide a bearing force Fb that is added to the magnetic bearing force Fm. The feedback states x and x' are electronically directed to a programmable controller that provides a current through a power amplifier to the AMB thus producing the magnetic force Fm.

This is inherent in an AMB system, however in an AMB there is limited control action, since most of the AMB work is to carry the static load. For a SEMAJIB, the static load is carried by the JB, leaving the AMB to act as an EMA. This allows the SEMAJIB to really act as a smart bearing.

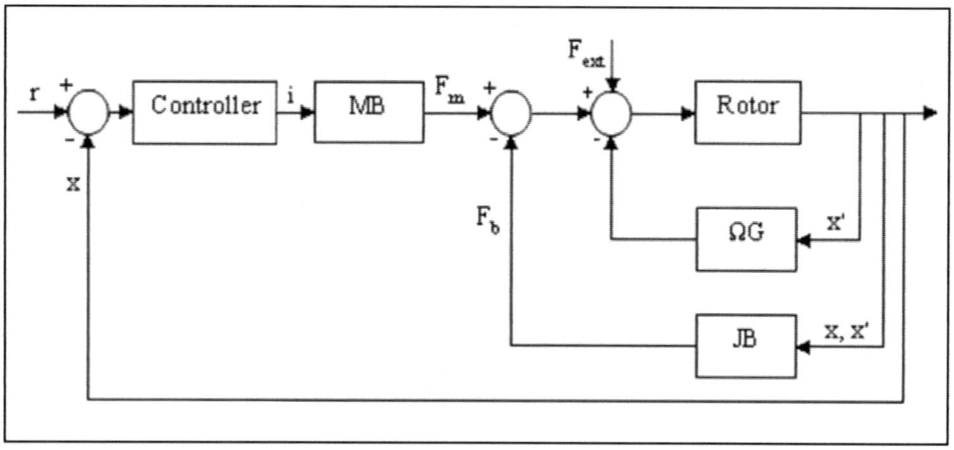

Figure 12 SEMAJIB Feedback System

In fact, if Fext in Figure 12 is considered as an unbalance, then the SEMAJIB acts to control the unbalance, which has been shown earlier in this paper. If however Fext is a misalignment then the SEMAJIB acts to control misalignment, if it is looseness then the SEMAJIB acts to control looseness. This is possible because of the unique spectral maps of each of these faults, and the AMB within the SEMAJIB that acts as an EMA with abundant capacity. In fact, it is envisioned that the smart bearing concept can be used to extend the current condition-based maintenance (CBM) strategy to include smart condition based corrective maintenance (SCBCM).

CONCLUSION:

This paper demonstrated the unbalance control capabilities of the SEMAJIB. The theory was presented, and the experimental results show an extremely effective vibration control achieving more than 90% reduction at resonance.

The concept of a smart machine was briefly reviewed. The application of smart technology should improve the operation of rotating machinery. The net result is the improved availability and reliability of machines.

REFERENCES:

[1] El-Shafei, A. Methods of Controlling the Instability in Fluid Film Bearings, US Patent 7,836,601, 2010.

[2] El-Shafei, A. Integrated Journal Bearing, U.S. Patent No. 10,612,592 B2,, 2020

[3] El-Shafei, A., and Dimitri, A.S. Controlling Journal Bearing Instability Using Active Magnetic Bearing, ASME Trans., Journal of Engineering for Gas Turbine and Power, Vol. 132 Jan., No. 1, 2010.

[4] Dimitri, A.S., and El-Shafei, A. Instability Control and Unbalance Compensation of Flexible Rotors Supported on Journal Bearings Using Magnetic Bearings, Proceedings of the 8th IFToMM International Conference on Rotordynamics September 12-15, KIST, Seoul, Korea, 2010.

[5] El-Hakim, M., Dimitri, A.S, Saqr, T., Mahfoud, J., Adly, A.A., and El-Shafei, A. Numerical and experimental identification of a combined Journal-Magnetic bearing: Smart Integrated Bearing, Proceedings of the 10th International Conference on Vibrations in Rotating Machinery, IMechE, London UK, p. 399-407, 11-13 Sep, 2012.

[6] El-Shafei, A., Dimitri, A.S., Saqr, T., and El-Hakim, M. Test Rig Characterization and Dynamic Testing of a Smart Electro-Magnetic Actuator Journal Integrated Bearing, Proc.9th IFToMM Int.Conf. on Rotordynamics, Sep. 22-25, Milan, Italy, Mechanisms and Machine Science Vol. 21, Springer, 2014.

[7] Dimitri, A.S., Mahfoud, J. and El-Shafei, A. Oil Whip Elimination using Fuzzy Controller, J. Eng. Gas Turbines Power, Vol. 138, No.6, 2015.

[8] Dimitri, A.S., El-Shafei, A., Adly, A.A., Mahfoud, J. Magnetic Actuator Control of Oil Whip Instability in Bearings, IEEE Transactions on Magnetics, Vol. 51, No. 11, 2015.

[9] El-Shafei, A., Dimitri, A.S., and Mahfoud, J. PD Control of a Smart Electro-Magnetic Actuator Journal Integrated Bearing (SEMAJIB), Proceedings the 11th IMechE International Conference on Vibrations in Rotating Machinery, Manchester, UK, September 2016, paper C1030, pp. 239-250, 2016.

[10] C. Knospe, R. W. Hope, S. J. Fedigan, and R. D. Williams Experiments in the Control of Unbalance Response using Magnetic Bearings, vol. 5, 1995.

[11] El-Shafei, A. (2018) SEMAJIB: A Versatile High Performance Smart Bearing, Proceedings of the 10th International Conference in Rotordynamics - IFToMM. Mech. Mach. Sci., Springer, Cham, 2019, pp. 372–385. https://doi.org/10.1007/978-3-319-99262-4_27.

VIATC TRAINING CONFERENCE & EXPO

COVINGTON, KENTUCKY | AUGUST 7-9, 2024

ADDING VALUE FOR TURBOMACHINERY EVALUATION WITH CASE MONITORING

Jeffrey M Mauk has worked as a Machinery Equipment Monitoring Technician at BP Whiting Refinery in Indiana for the past 17 years, with principal duties focused on centrifigal compressors, turbine generators and large critical process rotating equipment. He completed a maintence machinist aprrenticeship certification with the US Department of Labor, and he has over 40 years expierence working in heavy process industry from Steel, Chemical, gas and oil, and manufacturing. He has worked in predictive maintenance for 25 years and holds a Cat III certification with the Vibration Institute.

ABSTRACT:

Non-Contacting Displacement transducers are valuable instrumentation for relative displacement, voltage gap, and overall conditions with several plotting formats for monitoring and diagnosing Turbomachinery. However, highly valuable case data can often be ignored from changing conditions. Measuring velocity and acceleration of the case and comparing it with the relative proximity probes, provides more information for better diagnosis in some cases. Acceleration measurements for impacts and gear mesh frequencies are necessary for proper monitoring and deterioration of gear components. Quarterly case monitoring of turbomachinery is a simple cost-effective method for an analyst's engagement in a hands-on approach for adding value to the equipment monitoring program.

WHY COLLECT CASE DATA:

This discussion comes up from time to time regarding the need for collecting casing vibration data. Do I really need to spend time walking around a large piece of process machinery with a high-speed light rotor or a compressor when most of the signal is attenuated from the large case? What value is really added to the monitoring program? We've invested thousands of dollars, in proximity probes, supervisory protection, API 670. Why do you want to waste time taking case data? I have seen some examples through the years, that not only **added value** but also were instrumental in the diagnosing of the fault. Sharing these examples hopefully will provide evidence as I make my case for case data.

CASE STUDY ONE, OBVIOUS FAULTS:

What may appear obvious to us, may not always be the case to others. Asking what faults exist and what conditions to monitor and analyze to best evaluate the changes is necessary. It requires knowledge to make the correct setup to prove you are right and confident to stand by your conviction.

A five-stage segmented blower, that's driven by a 3000 HP 2 pole motor, provides combustion air for the SRU (Sulfur Recovery Unit), supplying 1600k scf/h. All three blower systems have separate lube oil supply skids and are fully equipped with proximity probes, with an entire suite with axial trip protection. The units have online software platforms for analyzing trending, orbits, phase, etc. And, on a quarterly basis, case data is collected and evaluated for percent change, overall setpoints, shape of waveform and spectrum analysis due to large thermal growth on the NDE of the blower.

After a rebuild of one of the blowers was placed back into service. The blower had higher vibration measures at startup, higher than when we took it out of service for key up, rebuild. This is never a good starting point; you always plan for the same or lower vibration levels. Discussion ensued as to the possible cause of the increased magnitude at the 4x probe, which was 1x vibration. Engineering staff whom I have the most affectionate amiable respect and admiration shared a difference of opinion of the possible faults. There were strong suggestions of misalignment of the coupling due to thermo-growth and struggles with maintenance staff setting up alignment targets. This situation formed an opinion of misalignment as the most probable cause. Orbit shape and phase are excellent diagnostic plots and tools for determining imbalance and pre-load or misalignment conditions. Upon studying both orbit and shape at both the DE and NDE of the blower and the motor, no indication of misalignment response existed as shown in Figure 1. Not fully convinced of a misalignment, or more accurately stated, there was no evidence of misalignment more information was needed to determine what fault existed. Piezo Velocity Sensors were case mounted using a virtual analyzer collecting simultaneous data while incorporating voltage pulse from the proximity probe speed and trigger for phase relationship analysis. The proximity probes don't pick up the 120 Hz response created by electrical force. To Properly identify and separate the frequencies of misalignment from motor electrical faults, correct evaluation and data sampling requirements are necessary.

What are the main tools we use to identify misalignment? What are the differences between those symptoms and the symptoms electrical faults?

"Misalignment occurs when the centerlines of two shafts or two bearings are not collinear. In fact, when the center lines of two shafts or two bearings are properly aligned, they will form one collinear center line without any angular or offset distortion." Jack B. Catlin VI Cat 3 Pg. 289.

- 1x events
- 180 Degree Phase shift events
- High axial events
- Clipped time waveform

- High 2x events
- Pre-loaded waveform

Based upon the data, the orbit isn't preloaded. There is no 180 degrees phase shift between the axial sensors across the bearings. The 1st and 2nd orders were at 90 degrees, (proper compensation for sensor position adjusted). The spectrum and waveform (Figure 2) from the piezo velocity sensor at bearing 3 have very low severity measures in both the peak waveform and the RMS spectrum. Both are within Normal Unrestricted Operation for new and repaired equipment. (Table 5.2 Vibration Guidelines for Condition Evaluation VI Cat 2 Study Guide pg. 111.)

We can rule out misalignment.

Electrical malfunctions in the stator and armature can be categorized in 6 major faults: (Table 4.5 VI Cat 2 Training Manual page 92)

- Air Gap
- Broken Rotor Bars
- Eccentric Rotor
- Stator Flexibility
- Off Magnetic Center
- Stator Shorts

The first step is finding speed and not guessing from a spectrum. Use a trigger proximity probe or photo eye. You must know the RPM.

Then perform proper calculations for setup of Fmax, widowing, and number of samples for recording the data. You don't want to be standing there collecting more data than you need. Separating the 2X component or 2nd order from the 120 Hz line frequency (2XLF) is paramount to know the exact nature of the fault. Often mechanical problems cause 2X running speed vibration and electrical problems cause 2XLF peaks. The equation for calculating the number of lines of resolution (NL) is as follows:

NL (number of lines) = 2 x Fmax x WF / Resolution.

Where:

- NL = number of lines of resolution
- Fmax = maximum frequency
- WF = window factor, 1.5 for a Hanning window
- Resolution = difference between the two frequencies to be resolved.

Lowest resolvable frequency, LRF= N/Fmax isn't resolution. You ***MUST*** know the difference between the two frequencies.

The resolution required to separate two time running speed (2X) from two times line frequency (2XLF) is calculated by the following:

- Two times running speed = 2X = 2*3598 RPM = 7178 RPM
- Converting to Hertz 7178 RPM/60 = 119.63 Hz
- Calculate the difference between 2XLF and 2X = 120 Hz - 119.63 = 0.366 Hz.

Knowing that a resolution of 0.366 Hz is required, the number of lines (NL) required with an Fmax of 200 Hz and a Hanning window can be calculated by the following steps:

- NL = 2 * 200 Hz (Fmax) * 1.5 (Hanning)/.366 = 1639 lines
- Therefore, use 3200 lines of resolution to separate 2X from 2X LF.

The spectrum in Figure 3 clearly shows the 2nd order of running speed at 119.63 Hz. This is verifiable evidence that the highest peak in the spectrum isn't the 2nd order. An additional spectrum (Figure 4) shows the actual peak at 120.06 Hz. Therefore, the fault is electrical in nature.

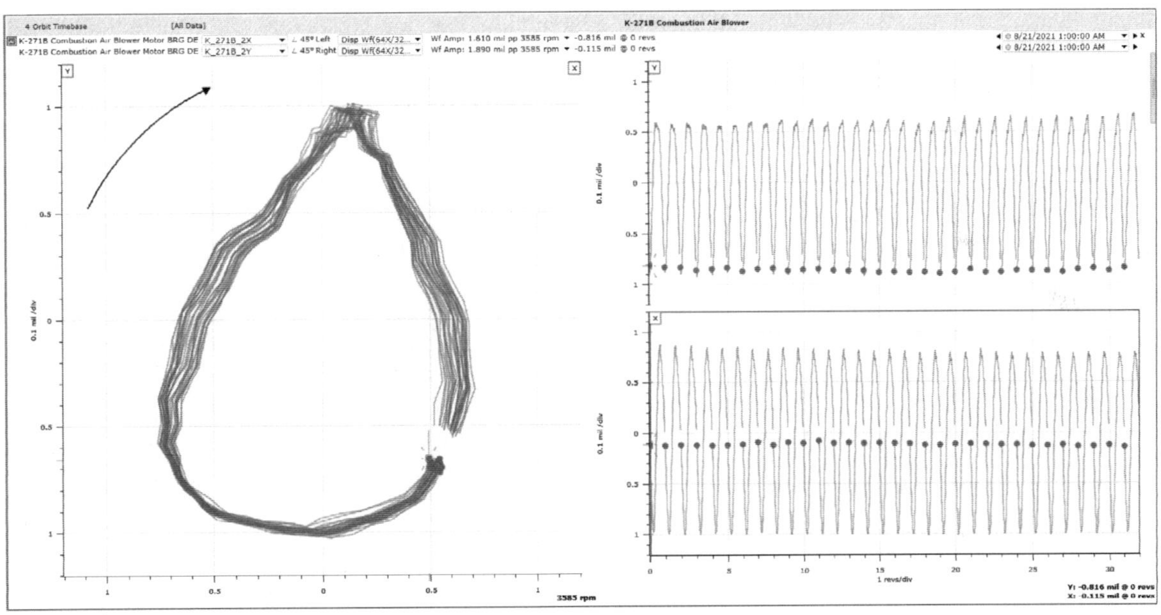

Figure 1

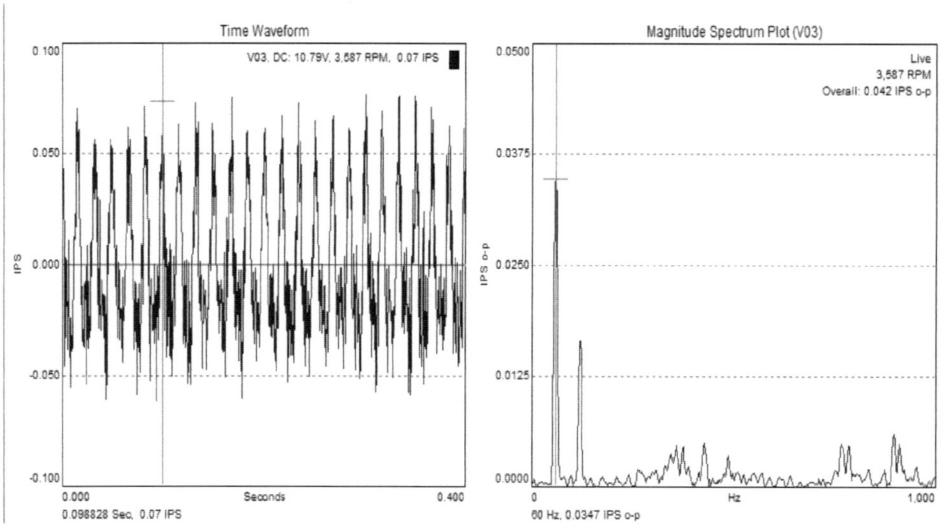

Figure 2

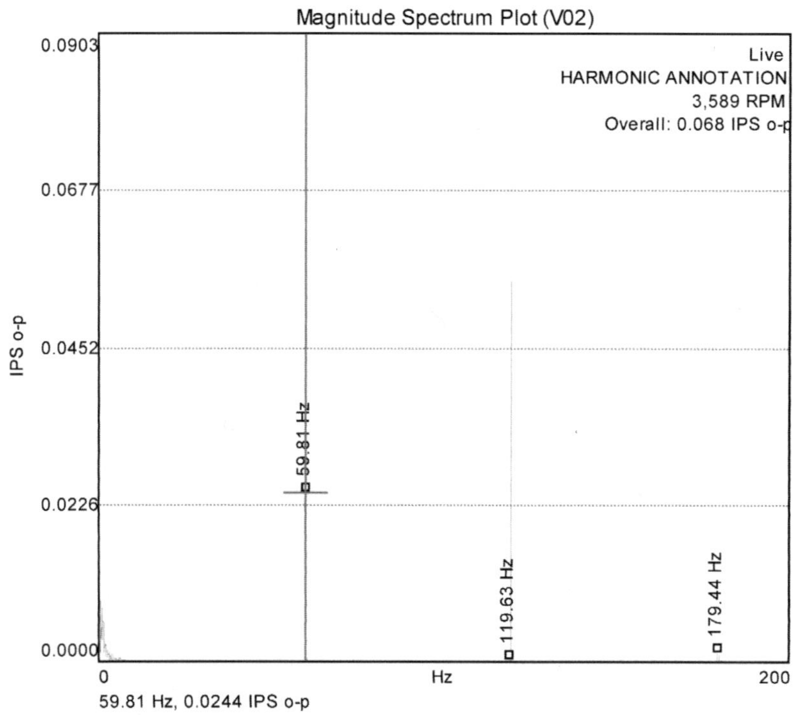

Figure 3: 3200N 200 Hz Hanning WF

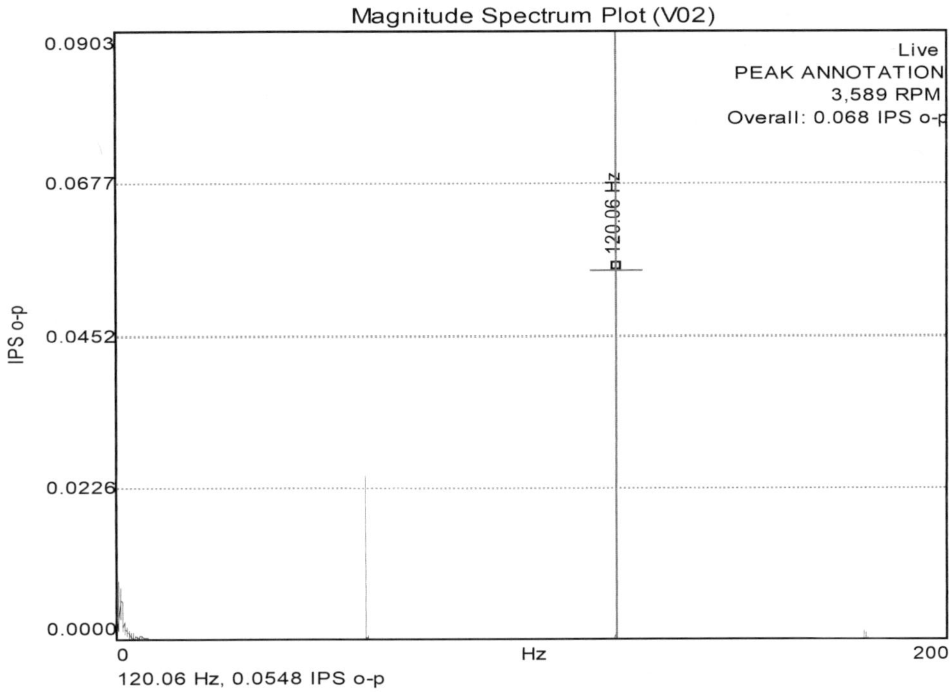

Figure 4: 6400N 200 Hz Flat Top WF

CASE STUDY ONE CONCLUSION:

No misalignment exists, based on the vibration data. There is some minor imbalance at the NDE of blower rotor. The radial vibration to bearing clearance ratio (R/C) is about 0.4. These conditions should be monitored for changes. A 2XLF peak in the spectrum due to some air-gap variation is very common in 2-pole motors. The current case data (0.183 ips dpk or 0.13 RMS) is at the surveillance level and should be monitored for change. Severity values are established by industry standards for evaluation of overall conditions. Knowing you're collecting the data correctly and instrumentation is tested and verified allows informed decisions.

CASE STUDY TWO INTERMEDIATE:

A turbine driven horizontally split centrifugal compressor for the alkylation unit. The compressor is part of the refrigeration system for the reactors, pulling hot overhead vapors from the reactors, compressing, and condensing those vapors, and returning them to the reactors. This cycle results in a net cooling effect, which keeps the reactors at desirable operation temperatures and pressure.

The unit was taken out for TAR (turn around) both rotors were pulled, and recommissioned rotors were installed back into the existing cases. During normal quarterly case monitoring activities, data was collected almost by sheer luck. (I do not believe in luck?) There was an increase of overall vibration levels at the 2H location (Figure 5). Software alarm setpoints within the online monitoring system were set at 0.5 mil pp and had not triggered to indicate a change.

Upon viewing the vibration, amplitude had increased also at the 2x and 2y, proximity probes. The orbit shapes prior to change are shown in Figure 6 and Figure 7.

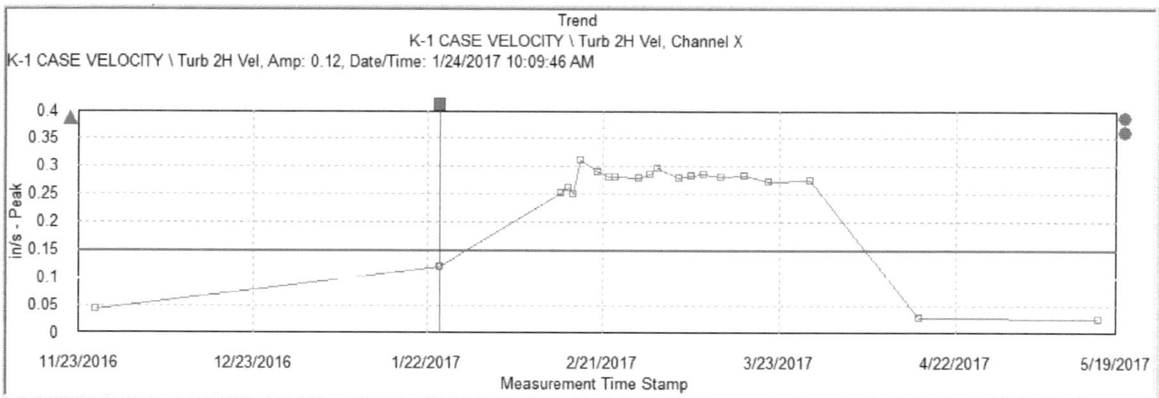

Figure 5

At the first step change on 1-24-17, the vibration measured had increased by 100%. Bearing 2 DE of the turbine 1x proximity probe measured 0.4 mils pp and increased to 0.8 mils pp. Bearing 2 1y proximity probe measured 0.6 mils pp and it also had increased to 1.2 mils pp. The increase of vibration was consistent at both the probes and the case with no attenuation in the response. Monitoring continued at this point with an awareness that something had changed. The process of evaluating the machine began.

Spectral data from Bearing 1 showed the 1X component had increased while the 2X component remained close to previous levels. While Bearing 2 spectral data prior to the change had harmonic content response indicating smooth steady state condition with a typical look of a spectrum at low vibrations. There was no pre-load indication or unbalance dominating the spectrum. Orbits prior to change showed low levels with minimum amplitude to drive a 1X phase. The relative phase between the 1x and 1y probes was 170 degrees, with a very flat response dominated by the 1X frequency. First thoughts were pre-load, rub, or unbalance. Vibration measurements remained at these levels for 2 weeks. Then another step change occurred.

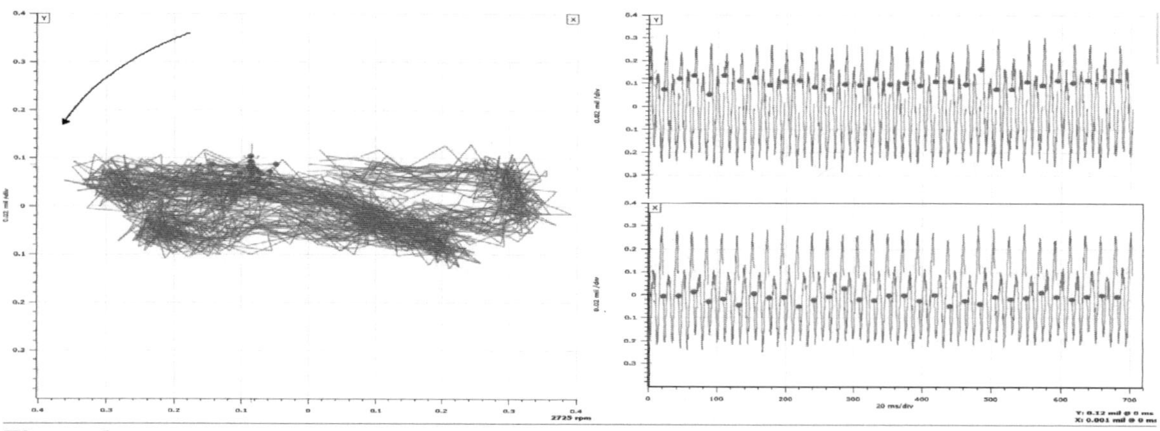

Figure 6

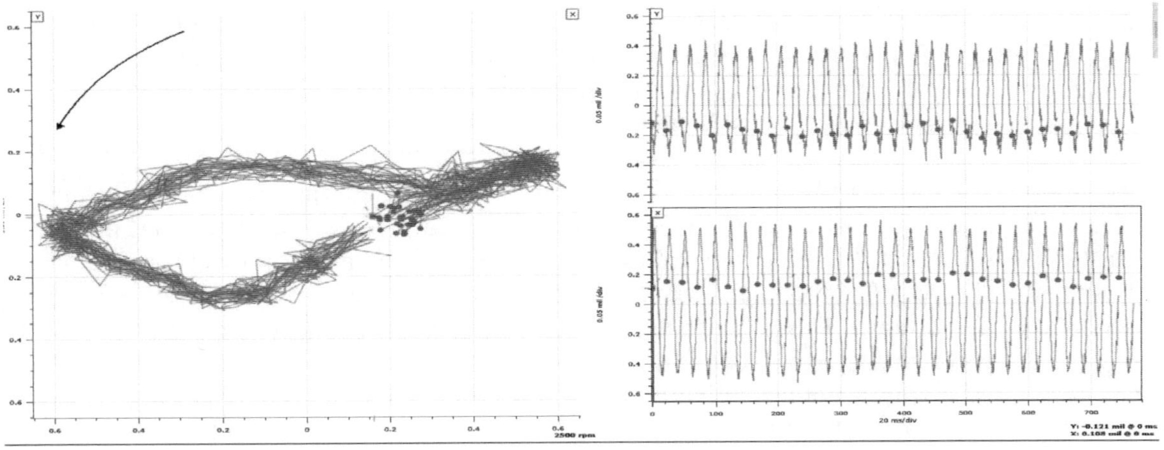

Figure 7

The next step change saw an increase at 2y. (Figure 8) The amplitude exceeded the software setpoint of 1.5 mils pp. It was shortly after these changes that I made others aware of the changes in the compressor vibration and sought further help from the engineering staff. I also began collecting case data at an increased frequency. The case data was remarkably high compared to a converted value of mils pp to ips velocity. Concerns were expressed over lube oil piping, Trip and Throttle valve, and ancillary systems vibrating. Historical information included a test run that had data from a rub in the labyrinth seal.

Some Common 1x faults include the following:

- Mass Unbalance
- Eccentricity
- Shaft Bow
- Thermal Bow.

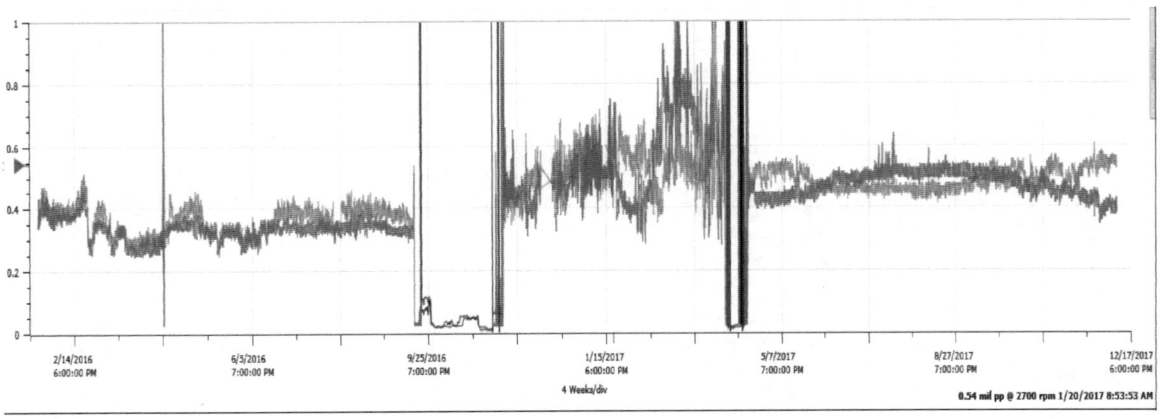

Figure 8

More information was required to make a better decision. A virtual analyzer was configured for with piezo-electric velocity sensors for measuring the case data. The sensors were placed at the same locations as the proximity probe housing, and the Keyphasor pulse was used from the proximity brought directly into the virtual analyzer phase marker reference channel. This would give a direct relationship for the high spot for comparison to the proximity probes. The evaluation would be required to consider the lag from velocity to mils.

The analysis of the case data combined with proximity probe data, confirmed to this analyst that the fault was unbalance. Corporate Engineering was contacted to review the data. This is where the historical rub information was cascaded to plant personal. Their position was that the fault was a rub, and they would not consider that another possibility existed. This was well understood by all. To say I did not question my decision would be an understatement, but I felt strongly that I was right. An Engineer within our group, has much experience with high level analysis as a lead trouble shooting engineer at a previous firm. He was very helpful in confirming what I believed to be true, and I remember his statement that rotors don't always react and fall within the book listed guidelines.

Interestingly the phase shift at the probes and the case sensors (90 Deg apart from each other) were both were at around the 90 Degree's =/- 20 deg. The time waveforms from the direct case measurements are 180 out of phase with perfect sine waves from the piezo velocity sensors. I highly recommend having set of these sensors with integration at the sensor for frequencies in the velocity data sampling range of 10 to 1000 Hz. You get a much nicer signal verses integrating acceleration to velocity in the software. Figure 9 provides a plot with the time waveforms. This information, including shape, phase, and 1X vibration, was convincing evidence that unbalance was the issue. Sometimes the obvious is staring us in the face, and we do not want to believe what the instrumentation is telling us.

There are a few moments in an analyst's career when many of those much smarter than you share a different opinion. When they pulled the top half of the split case of the turbine housing, and the 1st and 2nd stage were missing blades, it felt good. I was standing next to someone who shared a different opinion of the fault who looked at me and gave me a fist bump acknowledging I was right.

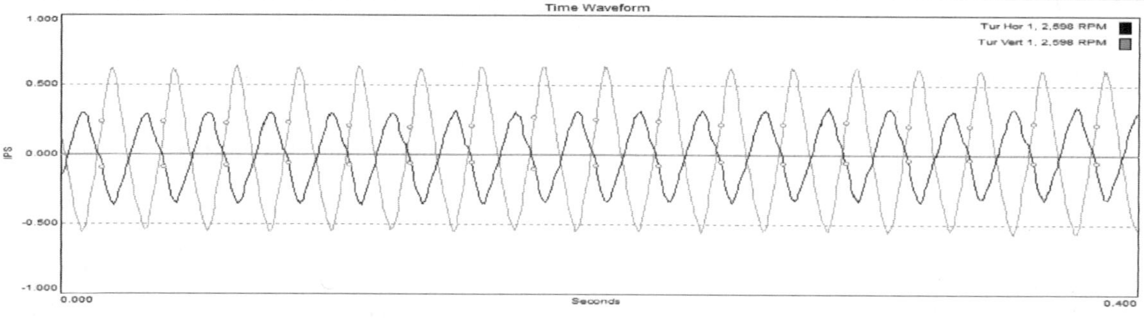

Figure 9

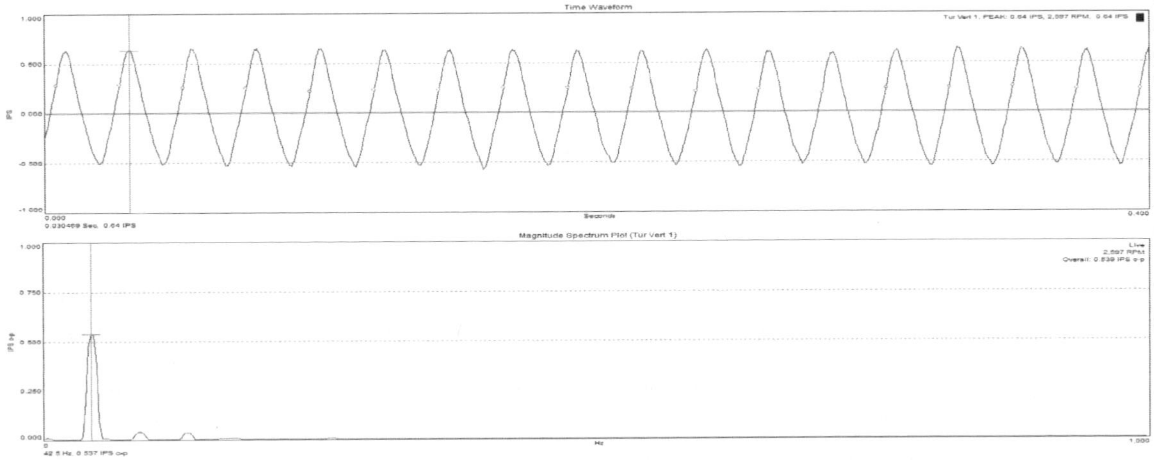

Figure 10

Figure 11

CASE STUDY TWO CONCLUSION:

Rotor unbalance was caused by the loss of blades as shown in Figure 11. The case was moving closely in phase with the shaft, as shown by the piezo velocity sensors phase matching the proximity probes phase angles. This provided confirmation with two agreeing data points that the problem was in fact unbalance. The virtual analyzer provided valuable information in real-time multi-channel comparison referenced to the online diagnostic software. The higher absolute case magnitude with lower relative shaft vibration is not something normally seen. Blade and shroud failure were attributed to generally poor workmanship. Metallurgical analysis confirmed blade failures were due to HCF (high cycle fatigue). Shroud analysis found inadequate peening and improper shroud construction.

Not all diagnostic problems fit into the table of how it should look. I have heard others say that this work is not an exact science. The blades came off on the first and second rotor discs at the NDE end. The probes that saw the highest amplitude change were at the DE of the rotor. The case data was higher at the NDE where the mass unbalance from loss of blades occurred.

CASE STUDY THREE ADVANCED:

A vertically split barrel type, dual staged centrifugal compressor with between stage intercooling. Its drive train consists of an 18,0000 HP synchronous 4 pole motor (1800 RPM), driving a dual helical gear and pinion gearbox with a 3.023:1 gear ratio and an output speed of 5442 RPM. During a turn-around (TAR), the compressor barrel was replaced, and the motor was inspected. No work was conducted on the gearbox, other than removal of an inspection cover and general inspection of gears. No conditions showed change. The unit was placed back in service.

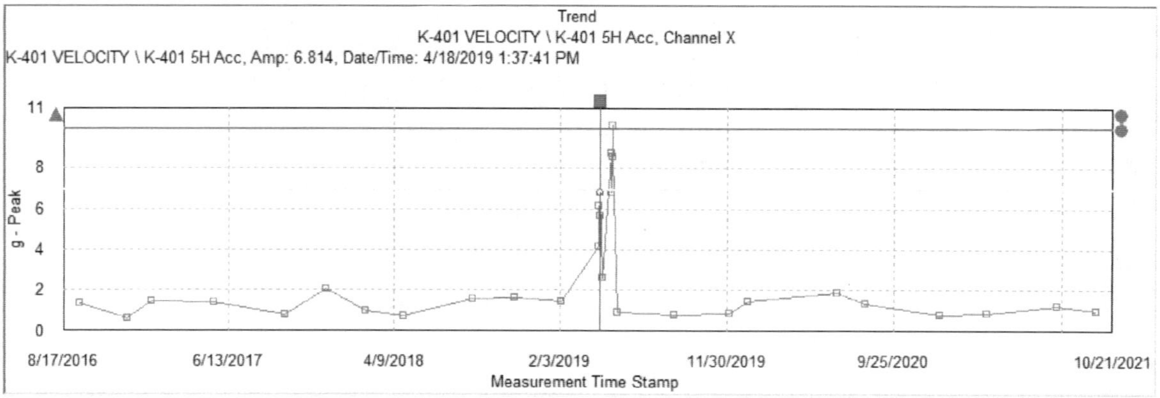

Figure 12

There are two piezo-electric accelerometer sensors stud mounted in the horizontal locations, one on the low-speed gear and the other at the high-speed pinion. The gearbox was well within normal vibration measures and below a software alarm for vibration technician to notice the first change. After a few months the case vibration began to increase as shown in the trend plot provided in Figure 12. The increase triggered further investigation into the step change.

Each analyst approaches the system at which they begin the process of elimination to narrow down the fault condition from their own perspective. My thought process involves plastic soldiers I played with as a child. Beginning with the obvious, I use the data to knock down the soldiers until only a few are standing. It is usually the guy laying on the ground, pointing his rifle directly at me. When the culprit is fully exposed it's always, well I knew it all the time. For the analysis our first identification is frequency, like turning the dial on the car radio, where is the signal coming from? What mechanical, electrical process forces can cause that signal. And so, the journey begins.

The gearbox case sensors were the first to show a change, so naturally. Looking at the proximity probes makes perfect sense to begin the process of evaluation. The orbits were examined, and there was no increase of magnitude. Figure 13 provides the orbit at the time the case magnitude was beginning to increase. Figure 14Fig, 12 provides the orbit about one month later, and there are some slight changes. However, if the case vibration hadn't increased there was nothing in the obit that would lead you to question the condition at the time. The axial probes measuring linear movement of the compressor rotor coupled to the high speed (HS) pinion showed no rotor movement. There are also axial probes on the low speed (LS) motor, and these probes showed no shift in axial position either. Figure 15 and Figure 16 provide trend plots from the axial measurements. All relative vibration at the compressor remained the same too. The gathered data lead to a conclusion that the best tool was not relative proximity probes for the analysis of this gearbox fault.

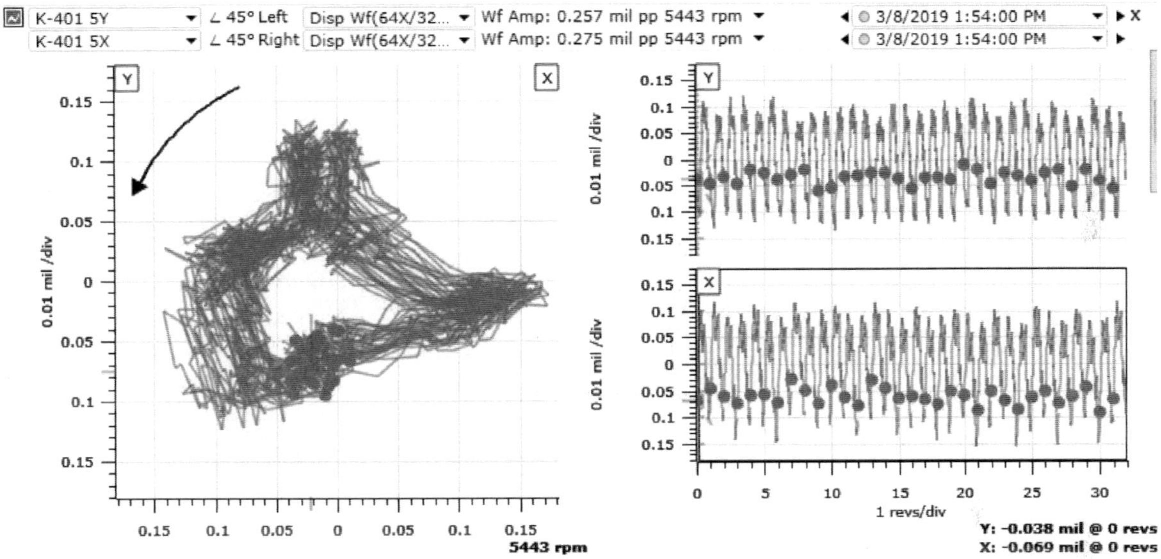

Figure 13

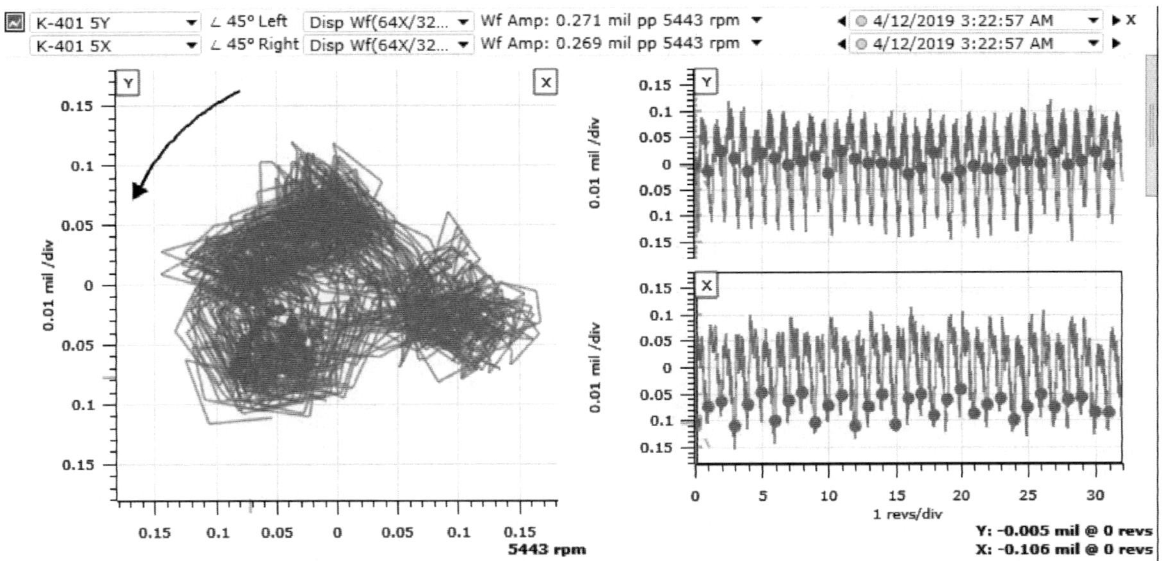

Figure 14

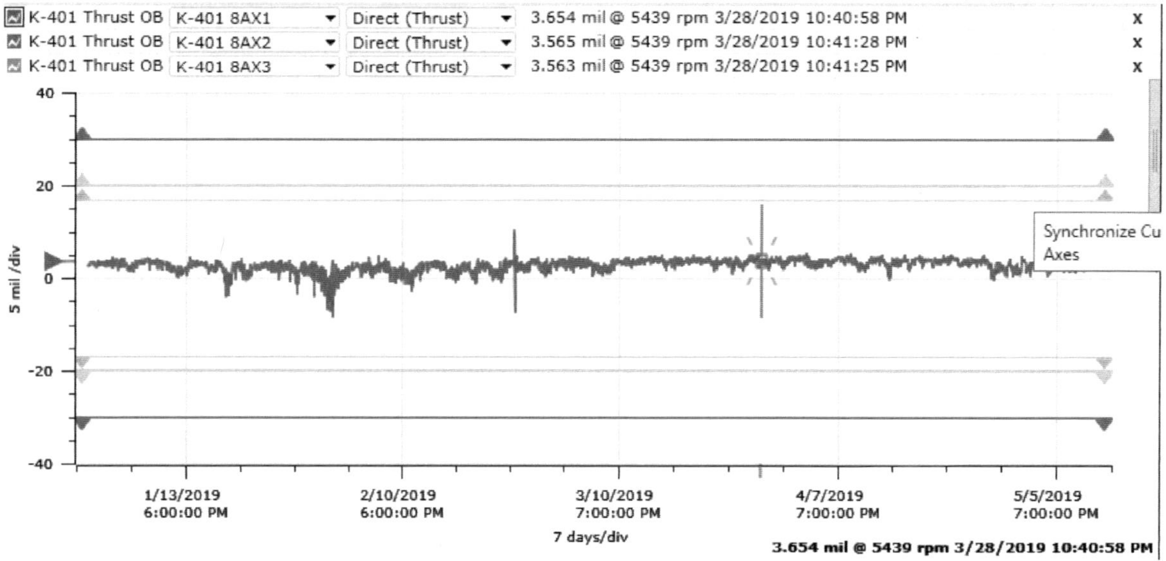

Figure 15

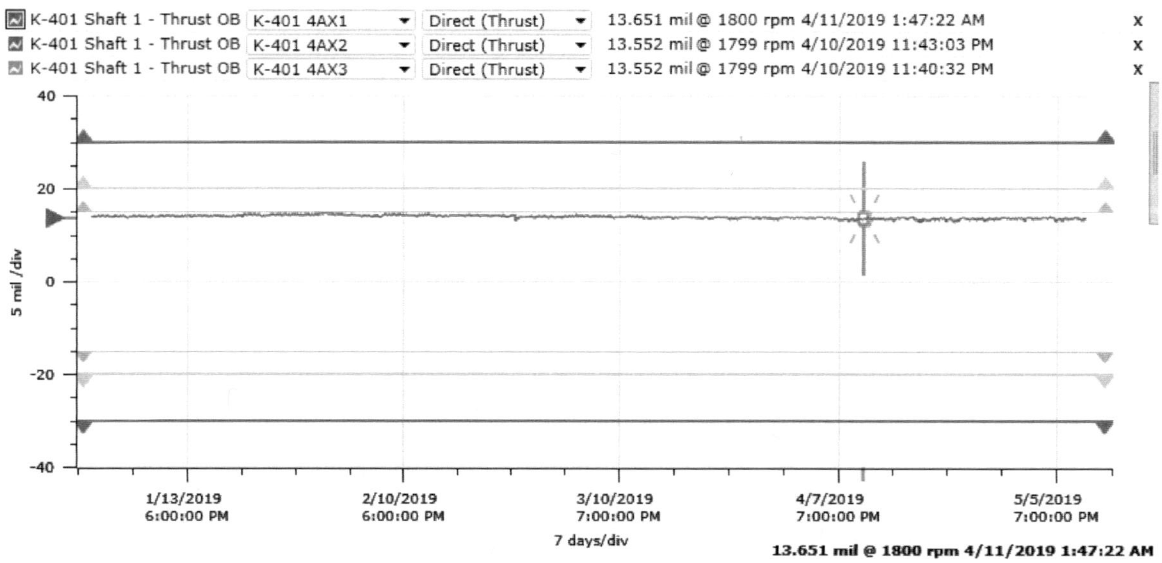

Figure 16

A multi-channel virtual analyzer was configured with data sampling at 10 locations and deployed as shown in Figure 17. The motor and compressor had no increased magnitude. There was no reason to collect virtual data at those locations.

General | Channels | Phasemarkers | External Inputs | Processing | Criteria | Diagnosis | Outputs

Select Channel:

#	Active	Name	Type	mV/EU	Units	Offset	Coupling	Angle	PM
1	Yes	GB H3	Acceleration	100	G	0	IEPE DC	90R	PM1
2	Yes	GB V3	Acceleration	100	G	0	IEPE DC	0L	PM1
3	Yes	GB H4	Acceleration	100	G	0	IEPE DC	90R	PM1
4	Yes	GB V4	Acceleration	100	G	0	IEPE DC	0L	PM1
5	Yes	GB A4	Acceleration	100	G	0	IEPE DC	180L	PM1
6	Yes	GB H5	Acceleration	100	G	0	IEPE DC	90L	PM1
7	Yes	GB V5	Acceleration	100	G	0	IEPE DC	0L	PM1
8	Yes	GB H6	Acceleration	100	G	0	IEPE DC	90L	PM1
9	Yes	GB V6	Acceleration	100	G	0	IEPE DC	0L	PM1
10	Yes	GB A6	Acceleration	100	G	0	IEPE DC	0L	PM1
11	No	GB A6	Acceleration	100	G	0	IEPE DC	180L	PM1
12	No		Acceleration	100	G	0	IEPE DC	0L	PM1
13	No		Acceleration	100	G	0	IEPE DC	0L	PM1
14	No		Acceleration	100	G	0	IEPE DC	0L	PM1
15	No		Acceleration	100	G	0	IEPE DC	0L	PM1
16	No		Acceleration	100	G	0	IEPE DC	0L	PM1

Edit... Export...

Figure 17

A voltage pulse was taken from the motor speed. Since the motor was a synchronous motor, the speed was exactly 1800 RPM. The pulse of the HS rotor speed could not be used in the analyzer due to only one filtering instrument to isolate a possible ground fault. We did not want any power blips that could possibly cause an issue at the supervisory protection system. Using a machine table as shown in Figure 18 allows a quick view of overalls and sensor conditions. A bias voltage should be used to confirm you have a good connection and to confirm that the sensor is powered and operating correctly.

Machine Table

Machine	Component	Point	Bias	OA	1x	2x	1xP	Speed
GB	GB_LS	GB H3	13.01 V	0.14 IPS o-p	0.01 IPS o-p	0.00 IPS o-p	264 °	1,800 RPM
GB	GB_LS	GB V3	12.88 V	0.37 IPS o-p	0.00 IPS o-p	0.00 IPS o-p	104 °	1,800 RPM
GB	GB_LS	GB H4	13.03 V	0.12 IPS o-p	0.01 IPS o-p	0.00 IPS o-p	270 °	1,800 RPM
GB	GB_LS	GB V4	10.76 V	0.02 IPS o-p	0.01 IPS o-p	0.00 IPS o-p	113 °	1,800 RPM
GB	GB_LS	GB A4	13.12 V	0.12 IPS o-p	0.01 IPS o-p	0.00 IPS o-p	156 °	1,800 RPM
GB	GB_LS	GB H3 Gs	13.01 V	7.55 G o-p	0.00 G o-p	0.00 G o-p	175 °	1,800 RPM
GB	GB_LS	GB V3 Gs	12.88 V	19.38 G o-p	0.00 G o-p	0.00 G o-p	10 °	1,800 RPM
GB	GB_LS	GB H4 Gs	13.03 V	6.05 G o-p	0.00 G o-p	0.00 G o-p	180 °	1,800 RPM
GB	GB_LS	GB V4 Gs	10.76 V	0.64 G o-p	0.00 G o-p	0.00 G o-p	22 °	1,800 RPM
GB	GB_LS	GB A4 Gs	13.12 V	5.60 G o-p	0.00 G o-p	0.00 G o-p	68 °	1,800 RPM
GB	GB_HS	GB H5	12.96 V	0.19 IPS o-p	0.01 IPS o-p	0.00 IPS o-p	87 °	1,800 RPM
GB	GB_HS	GB V5	12.50 V	0.39 IPS o-p	0.00 IPS o-p	0.00 IPS o-p	168 °	1,800 RPM
			-	-	-	-	-	-
GB	GB_HS	GB H6	12.84 V	0.23 IPS o-p	0.01 IPS o-p	0.00 IPS o-p	93 °	1,800 RPM
GB	GB_HS	GB V6	13.00 V	0.26 IPS o-p	0.00 IPS o-p	0.00 IPS o-p	318 °	1,800 RPM
GB	GB_HS	GB A6	12.63 V	0.17 IPS o-p	0.00 IPS o-p	0.00 IPS o-p	146 °	1,800 RPM
GB	GB_HS	GB H5 Gs	12.96 V	9.43 G o-p	0.01 G o-p	0.00 G o-p	356 °	1,800 RPM
GB	GB_HS	GB V5 Gs	12.50 V	12.37 G o-p	0.00 G o-p	0.00 G o-p	90 °	1,800 RPM
			-	-	-	-	-	-
GB	GB_HS	GB H6 Gs	12.84 V	11.24 G o-p	0.01 G o-p	0.00 G o-p	3 °	1,800 RPM
GB	GB_HS	GB V6 Gs	12.63 V	6.97 G o-p	0.00 G o-p	0.00 G o-p	57 °	1,800 RPM

Figure 18

Viewing the table helps direct the analyst to the highest magnitude, which is GB_LS. Not that looking at all the spectrum and waveforms isn't important, but looking for the highest severity is the best place to start. The first peak that appears in the spectrum is 3900 Hz at 12.8 G rms as shown in Figure 19. Zooming into the spectrum to determine the side-banding will determine the most probable gear with the problem. Zoom Spectrum at GB V3 has side-banding at the HS pinion, but the highest amplitude of the fault is at the LS bull gear.

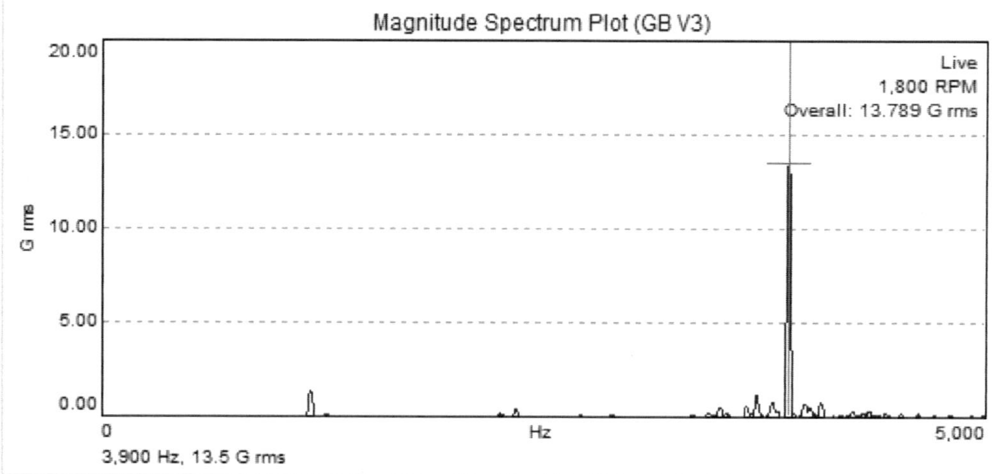

Figure 19

Some common gear faults include:

- Cracked or broken teeth
- Gear defects from manufacturing, assembly, and installation.
- Pitting, scouring.

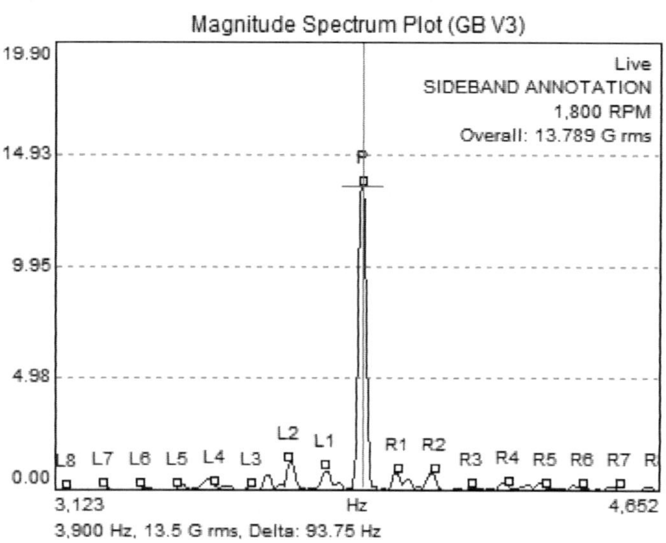

Sid...	Hz	G rms
P	3,900	13.4...
L1	3,8...	0.878
R1	3,9...	0.772
L2	3,7...	1.195
R2	4,0...	0.707
L3	3,6...	0.050
R3	4,1...	0.121
L4	3,525	0.192
R4	4,275	0.194
L5	3,4...	0.100
R5	4,3...	0.091
L6	3,3...	0.056
R6	4,4...	0.101
L7	3,2...	0.042
R7	4,5...	0.081
L8	3,150	0.033

Figure 20: Flattop Window 800N/5KHz

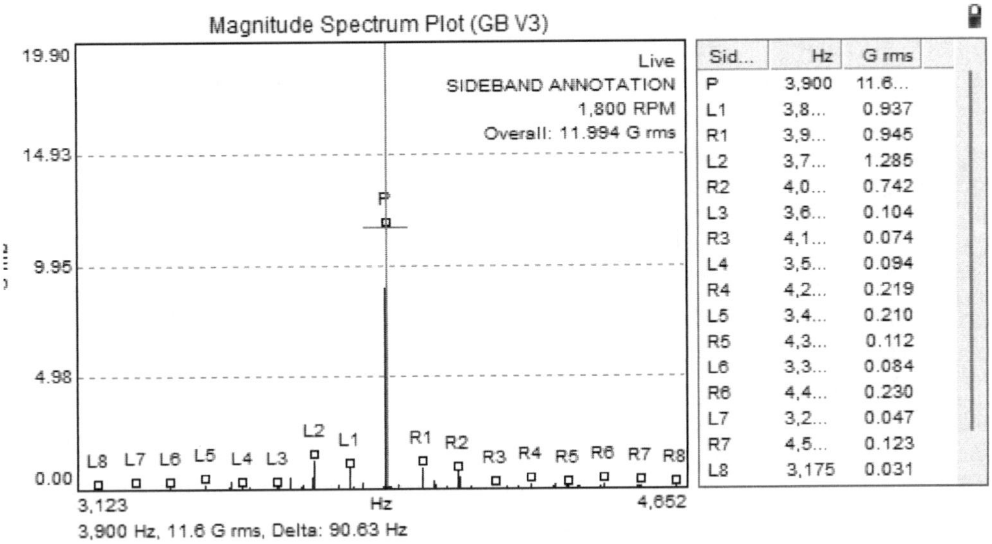

Figure 21: Hanning Window 6400N/5KHz

The time waveform is consistent with the spectral peak. Using two cursors over ten cycles on the time waveform plot provided in Figure 22, the calculated frequency is 387.88 Hz for the ten cycles or 3878.8 Hz for an average cycle. Using 10 peaks gives you a good average as the waveform is the unaltered response of the force measured with the sensor. That is close enough in the time waveform plot.

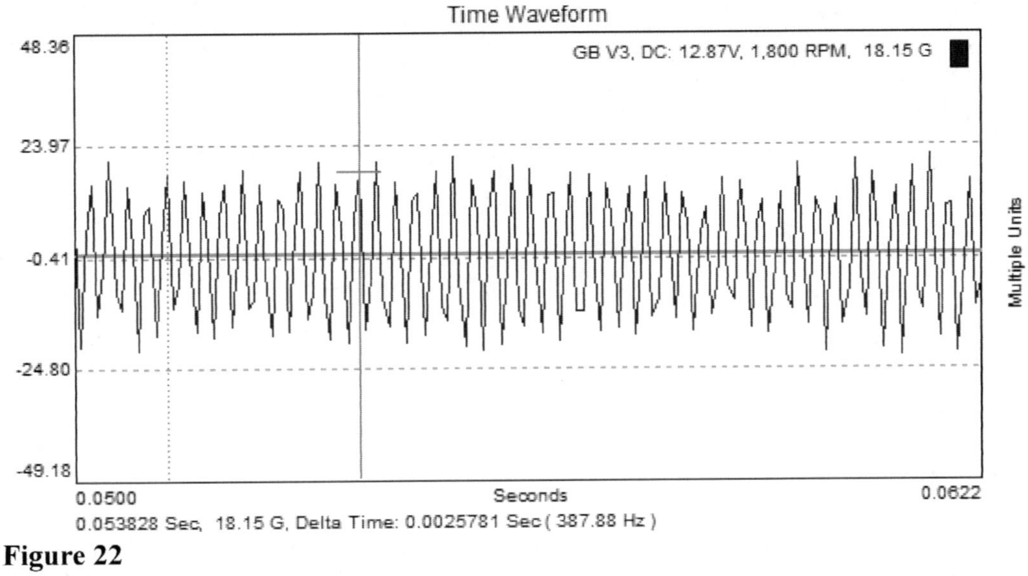

Figure 22

The LS gear has 130 teeth, and the HS gear has 43 teeth. The frequencies of interest for the gearbox include the gear mesh frequency, which is calculated and verified as follows:

- 1800 RPM / 60 sec/min = 30 Hz
- For the LS gear, GMF = 30 Hz x 130 teeth = 3900 Hz

- For the HS gear, GMF = 3900 Hz by definition.
- 3900 Hz / 43 = 90.7 Hz
- 90.7 Hz * 60 seconds/minute = 5441 RPM, and that matches the HS shaft speed.

The delta from the side bands was about 93 Hz. The analyzer was set at a flattop window to obtain a better amplitude within the spectral bin. There are no other mechanical forces, but a higher resolution spectrum collected to confirm the side-band frequency.

Some interesting points can be observed from the spectral plots and from the time waveform plots. The flattop window indicated a pk measurement of 13.5 g's rms at 3900 Hz, while the data processed with a Hanning window indicated a somewhat lower value at 11.6 g's rms. (Figure 20 and Figure 21). However, the highest pk coming from the time-waveform was 25.4 g's pk as shown in Figure 23. The spectrum is where we identify frequency with faults, but our best evaluation of severity comes from the unaltered time waveform. **Yes, you can make good decisions from spectral data but you are missing valuable information by not viewing the waveform.**

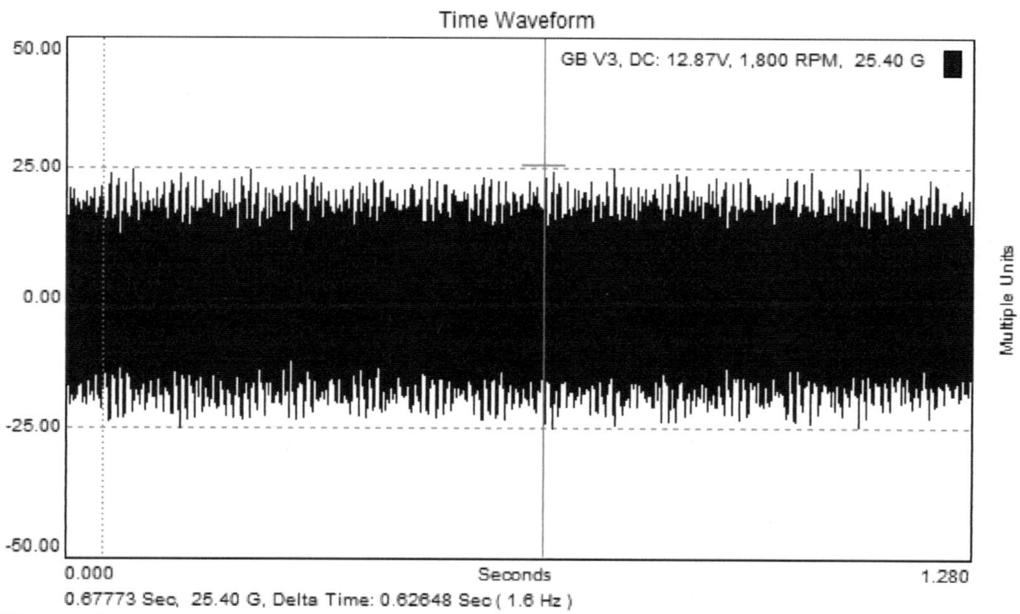

Figure 23

CASE STUDY THREE CONCLUSION:

The HS pinion was destroyed with several broken teeth, the LS gear also suffered damage. The diagnosis lead in the right direction that the HS pinion was bad, but a full failure analysis found the root cause of the failed gear set. The gearbox wasn't worked on during the over haul. During the alignment, a 0.030 shim was left out of the coupling shims, which loaded the gear. This was thought to be the main cause of the failure but further testing was done on the metallury of the gears itself. When the unit was taken down the procedure wasn't clear for operations isolation of the compressor. A cleaning solution was used to flush out high H2S (hydrogen sulfied gas). The dry gas seals skid was shutdown and the vapors of the cleaning solution worked its way thoughout the lube oil system, this compromised the metallurgy of the gears along with the loading the gears abnormally causing a premature failure. It was the case monitoring of the gearbox that measured the change in vibration and the case analysis that provided the precise diagnostic answer to the failure.

VIATC TRAINING CONFERENCE & EXPO
COVINGTON, KENTUCKY | AUGUST 7-9, 2024

A MATHEMATICAL REVIEW AND MEASURED VALIDATION OF DYNAMIC STIFFNESS OF A SIMPLY SUPPORTED T-BEAM

Thomas Calko, PE
E^2G | The Equity Engineering Group, Inc
Shaker Heights, Ohio

Thomas Calko, PE, is a consultant at E^2G in the REVD Team primarily working in the petroleum and petrochemical industries. He performs advanced vibration data analysis and root cause evaluations to solve structural dynamics problems on structures and rotating equipment. He is a Category IV vibration analyst, a certified welding engineer, and a certified welding inspector. Prior to working at E^2G, he worked in maintenance engineering at nuclear power plant as a welding engineer and vibration engineer. He holds engineering degrees from Youngstown State University (BSME) and the University of Pittsburgh (MSME).

ABSTRACT:

This paper provides a theoretical development of dynamic stiffness using both continuous systems and single degree of freedom methods. The mathematical development is presented for a simply supported T-beam that concludes with the calculation of the theoretical dynamic stiffness. This mathematical development provides some insights into the dynamics that are included in the structural dynamic stiffness. Dynamic stiffness testing was also completed on the same theoretical simply supported T-beam and the measured dynamic stiffness is compared to the theoretical. Some recommendations are provided when performing modal testing to improve accuracy of the measured dynamic stiffness.

KEY WORDS: dynamic stiffness, modal testing, FRF

INTRODUCTION:

When most engineers consider stiffness, they tend to think about static stiffness. The static stiffness of a cantilever beam is $K_{static} = 3EI/L^3$ and the static stiffness of a simply supported beam is $K_{static} = 48EI/L^3$. These are simple, linear equations that are derived from strength of materials which quantify the amount of deflection that occurs when a static force is applied. However, this structural characteristic is far more interesting when the dynamics are considered, because the stiffness of a structure is not static – it is dynamic. The structural stiffness is a function of the frequency of applied dynamic loading.

For example, a 12 inch long cantilever beam with a solid round cross section of 1 inch outer diameter weighs 2.67 lbf and has a static stiffness of 5,028 lbf/inch. When a 1,000 lbf load is applied to the end of the cantilever, the beam deflects 0.199 inches. However, when a 1,000 lbf dynamic load is applied at 283 Hz (near resonance), that same cantilever beam has a stiffness of 100 lbf/in and deflects 10 inches!

This paper will review the fundamentals of dynamic stiffness by presenting the following topics:

- Theoretical calculation of dynamic stiffness using continuous systems methods
- Theoretical calculation of dynamic stiffness using single degree of freedom methods
- Dynamic stiffness test results of a simply supported beam
- Case study of a rotor dynamics analysis utilizing dynamic stiffness

The main objectives of this paper are to (1) highlight the key characteristics of dynamic stiffness and (2) reinforce the importance of determining the dynamic stiffness using test methods when troubleshooting structural dynamics problems. Dynamic stiffness data can be very useful when finite element models are utilized in these efforts and the boundary conditions can be modeled using dynamic stiffness from test data. This approach can save significant time by simplifying the FEA model since additional structure does not need to be modeled. It also improves accuracy by utilizing boundary conditions that match the actual stiffness and mass distribution of the structure. And this data can be measured in a simple, accurate, and cost-effective manner. Reference 1 provides additional technical basis for utilizing dynamic stiffness in rotor a dynamic analysis and includes technical guidance in those evaluations.

BASIC ARRANGEMENT USED IN THIS PAPER:

Throughout this paper the same basic arrangement is utilized: a simply supported T-section as shown in Figure 1 with properties as summarized in Table 1. Figure 2 provides the first three calculated natural frequencies and mode shapes of this arrangement using the methods for continuous systems (Reference 2).

THEORETICAL DEVELOPMENT OF DYNAMIC STIFFNESS: CONTINUOUS SYSTEMS:

This section will present a hand calculation method to determine the dynamic stiffness from the basic equation of motion for the arrangement shown in Figure 1. This method demonstrates the structural dynamics principles that are used to determine the dynamic stiffness. Although this method utilizes a very simple arrangement, a more complex arrangement could be studied using FEA, which is beyond the scope of this paper.

The theoretical dynamic stiffness is calculated using a forced response analysis for the mid-point of the beam. The displacement of the beam is calculated for a range of frequency inputs from 0.1 Hz to 150 Hz and the dynamic stiffness is calculated as shown.

$$Stiffness_{@Frequency} = \text{Force}_{\text{@Frequency}}/Displacement_{@Beam\ Mid-Point} \tag{1}$$

The development of the theoretical displacement for this beam is summarized below and is fully detailed in Reference 3. Figure 3 shows a free body diagram of beam and an equilibrium element in the beam that is undergoing free vibrating in the transverse direction, where element dx is has mass and acceleration as defined below.

$$Mass = \rho A dx \qquad \textit{where } \rho = \textit{density}, \quad A = \textit{cross sectional area} \tag{2}$$

$$Acceleration = \partial^2 y/\partial t^2 \tag{3}$$

The summation of forces in the vertical direction result in the following equation.

$$V - V - \frac{\partial V}{\partial x}\partial x - \rho A dx \frac{\partial^2 y}{\partial t^2} = 0 \tag{4}$$

The summation of moments at the element center result in the following equation; these terms sum to approximately 0 since an infinitely small element cannot support a moment.

$$-Vdx + \frac{\partial M}{\partial x} \approx 0 \tag{5}$$

Substituting V from Equation (5) into Equation (4) results in

$$\frac{\partial^2 M}{\partial x^2} dx = -\rho A dx \frac{\partial^2 y}{\partial t^2} \tag{6}$$

A moment can also be defined as the flexural rigidity (EI) times the curvature of the beam.

$$M = EI \frac{\partial^2 y}{\partial x^2} \tag{7}$$

This definition of moment is substituted into Equation (6)

$$\frac{\partial^2}{\partial x^2}\left(EI \frac{\partial^2 y}{\partial x^2}\right) dx = -\rho A dx \frac{\partial^2 y}{\partial t^2} \tag{8}$$

The general partial differential equation describing the dynamic equilibrium of a freely vibrating beam of constant cross section relates the flexural rigidity of the beam with the inertia force.

$$EI \frac{\partial^4 y}{\partial x^4} dx = -\rho A dx \frac{\partial^2 y}{\partial t^2} \tag{9}$$

This method of continuous structures is used to determine the theoretical dynamic stiffness for the simply supported T-section in Figure 1 and is compared to test data of this same arrangement later in this paper. Therefore, the following boundary conditions are applied to determine the solution of this beam.

- Simply supported beam configuration
- The forced response of the beam is driven by a force applied at the middle of the beam
- The deflection is determined at the middle of the beam
- The steady state vibration response is of interest; the initial transient response is not included in this analysis
- Damping effects are not included in this analysis

$$y = \frac{2PL^3 sin(ft)}{EI\,\pi^4}\left[\frac{\sin(\pi/2)}{1^4-\alpha^2} - \frac{\sin(3\pi/2)}{3^4-\alpha^2} + \frac{\sin(5\pi/2)}{5^4-\alpha^2} - \cdots\right] \tag{10}$$

where P = *Amplitude of the dynamic load*
f = *frequency (rad/sec)*
t = *time*
L = *Length of the beam*
E = *Modulus of Elasticity*
I = *Moment of Inertia*
α = *Frequency ratio* = ω/ω_n

This equation provides the displacement of a point at the mid-span at a specified frequency over time. Since the maximum displacement is of interest and the response is sinusoidal, this equation can be adjusted to calculate the maximum displacement at a specified frequency. For a sine wave, the maximum amplitude occurs at 1/4 of the period, T. Or expressed in terms of frequency:

$$t_{MaxAmplitude} = \frac{1}{4\,\omega} \qquad \text{where } \omega = \text{frequency (Hz)} \tag{11}$$

The equation for maximum displacement becomes:

$$y_{MaxAmplitude} = \frac{2L^3}{EI\,\pi^4} P sin(f/4\omega)\left[\frac{\sin(\pi/2)}{1^4-\alpha^2} - \frac{\sin(3\pi/2)}{3^4-\alpha^2} + \frac{\sin(5\pi/2)}{5^4-\alpha^2} - \cdots\right] \tag{12}$$

This equation is composed of three factors: the static stiffness of the beam (1st term), the dynamic force amplitude and frequency (2nd term), and the amplification factor at modes 1,3,5, etc. (3rd term) Equation (12) is utilized with the beam properties from Table 1 to calculate the maximum displacement at the beam mid-span at a range of frequencies from 0.1 Hz to over 200 Hz. The absolute value of the calculated displacements are used in the dynamic stiffness calculations to determine positive stiffness values. The dynamic stiffness at each frequency is then calculated using the applied force and the maximum displacement from this time waveform response as shown in Equation (1).

The following conclusions and observations are made regarding the maximum displacement per unit force or compliance.

(1) Fundamental Amplitude Response. Using a logarithmic scale, Figure 4 shows the absolute value of amplitude per unit force or compliance generally decreases as

frequency increases until proximate to another resonance. Amplitudes increase slowly as the resonance approaches, then increase sharply in the immediate vicinity of the resonance. This sharp amplitude response is due to the amplification factor for each mode as shown in Equation (12).

(2) Resonances. Natural frequencies are shown in Figure 4 as sharp increases in displacement per unit force. Note that Mode 2 does not have a response due to the node being at the mid-span (see Figure 2), which is the measurement point for this analysis.

(3) Amplitudes Decrease as Frequency Increases. The Mode 1 compliance curve in Figure 6 shows after the resonance, the amplitude continually decreases as frequency increases. This decreasing trend in vibration amplitudes may be related to the reason why vibration limits for structural members is higher at higher frequencies, or why damage accumulation is smaller at higher frequencies.

(4) Force-Response 180 Degree Phase Change at Resonance – Using a linear scale and calculated non-absolute value displacements, Figure 5 shows that the sign of the displacement changes from positive to negative at the resonance. This is due to the 180-degree phase change that occurs at a natural frequency.

(5) Amplitude Contributions of Consecutive Modes Decreases – In Equation (12) it is observed that each mode has a factor in the denominator that increases as the mode increases. This directly results in the conclusion that the amplitudes of consecutive modal contributions decreases as the mode increases (Reference 3).

The following conclusions and observations are made regarding the dynamic stiffness.

(6) Resonances Are Local Stiffness Minima. Since the dynamic stiffness is the inverse of the compliance, or amplitude per unit force, then the dynamic stiffness around the resonances will decrease as the inverse of the amplification factor. As shown in Figure 7, the dynamic stiffness at the excited natural frequencies approach 0 lbf/in. The theoretical stiffness at a resonance is 0 lbf/in; the reason the curve in Figure 7 is not 0 is due to the limited number of decimal places used in these calculations and the exact natural frequency.

(7) Static Stiffness. The dynamic stiffness at low frequencies approaches the condition of static loading and the static stiffness can be determined from this method. Using this method, the static stiffness is determined to be 94.5 lbf/in as shown in Figure 8. From Table 1, the calculated 1 DOF static stiffness was determined to be 93.9 lbf/in, which is within 1% of the dynamically calculated static stiffness.

(8) Dynamic Stiffness is Greater than Static Stiffness Except at Resonances. Figure 7 shows that the dynamic stiffness generally is larger than the static stiffness – except near resonances, where the stiffness theoretically decrease to 0 lbf/in.

(9) Dynamic Stiffness is a Nonlinear Function of Frequency Squared. Stiffness increases as frequency increases. From the perspective of mode shapes for increasing mode numbers, this makes sense because the number of nodes in the mode shape increase with mode number (as shown in Figure 2). Since these nodes act as pins, the beam length between pins is shorter and maximum deflections are smaller for a given load, which is an effective increase in stiffness. The dynamic stiffness of Mode 1 and Mode 3 in Figure 9 and Figure 10 show that stiffness increases as a function of the frequency squared.

(10) Peak Stiffness Frequencies Result from Modal Interactions. Figure 7 shows two resonance locations and one location that has sharp increase in stiffness. This peak

stiffness location occurs at 90 Hz and is due to the interaction between Mode 1 and Mode 3. As shown in Equation (12), each mode as an alternating sign; this can be seen in the mode shapes in Figure 2. Figure 11 and Figure 12 show as frequency increases towards 90 Hz, the Mode 1 amplitude is negative and is decreasing and the Mode 3 amplitude is positive and slowly increasing. This results in a net decreasing Mode 1,3 amplitude and at 90 Hz the amplitude crosses 0 in/lbf; at this frequency the dynamic stiffness increases sharply and approaches infinity. Therefore, the peak stiffness frequencies are caused by the inverse amplitude interaction of consecutive modes.

THEORETICAL DEVELOPMENT OF DYNAMIC STIFFNESS: SINGLE DEGREE OF FREEDOM (SDOF) SYSTEMS:

Using a single degree of freedom model and the equations of motion, a very useful relationship can be developed for dynamic stiffness. This approach also provides insight into the effects of damping that were not included in the continuous systems model. The SDOF development is presented in most mechanical vibration textbooks in the section on forced response of damped SDOF systems, such as Reference 4 and Reference 5.

The general equation for dynamic equilibrium for a harmonic force is given below:

$$m\ddot{x} + c\dot{x} + kx = Fe^{i\omega t} \quad \textbf{(13)}$$

where m = *equivalent mass of structure*
$\ddot{x}$ = *acceleration*
c = *damping factor*
$\dot{x}$ = *velocity*
k = *static stiffness*
x = *displacement*
F = *complex amplitude of dynamic force*
ω = *frequency*
t = *time*

The solution to this differential equation is:

$$x(t) = Xe^{i\omega t} \quad \textbf{(14)}$$

where X = *complex amplitude of displacement*

By performing the appropriate derivatives on Equation (14) and substituting into Equation (13), the solution becomes:

$$-m\omega^2 Xe^{i\omega t} + i\omega c\, Xe^{i\omega t} + kXe^{i\omega t} = Fe^{i\omega t} \quad (15)$$

By cancelling the $e^{i\omega t}$ terms and rearranging, the stiffness is represented as:

$$Complex\ Stiffness = \frac{F}{X} = (k - m\omega^2) + i(\omega c) \quad (16)$$

Since magnitude and phase are included in the complex stiffness, the amplitude can be determined by taking the square root of the sum of the squares.

$$Dynamic\ Stiffness = \sqrt{(k - m\omega^2)^2 + (\omega c)^2} \quad \textbf{(17)}$$

For the beam summarized in Table 1 that was evaluated using the continuous systems method, the dynamic stiffness is plotted using Equation (17) and is show in Figure 13. The following conclusions and observations are made regarding the dynamic stiffness.

(11) The SDOF Method is Equivalent to the Continuous Method for 1st Mode – As shown in Figure 13, the SDOF method outlined in this section provides a calculated dynamic stiffness that is very close to that calculated using the continuous systems method. However, the SDOF method is limited to the first mode, since it represents a single spring-mass system and does not include the other modal effects as shown in Figure 13.

(12) Primarily a Function of K_{static} and Inertia – As shown in Equation (17), the amplitude of the dynamic stiffness is a function of the static stiffness and the inertia of the structure. From Equation (15) and Equation (13), it is seen that the $m\omega^2$ term is developed from the $m\ddot{x}$ or inertia term in the equilibrium equation. Equation (17) also clearly shows that the stiffness increases as the frequency squared, which was identified in conclusion (9) above.

(13) Damping Provides Minor Increase in Stiffness – Figure 13 shows the dynamic stiffness curve with 1% and 25% damping for the simply supported T-section. This curve shows a very small increase in stiffness as damping increases. In general for this arrangement, the damping is within 2% of the calculated stiffness using the continuous systems method, except near the 1st natural frequency, where the difference is much larger since the continuous method does not account for any damping.

OVERVIEW OF PERFORMING DYNAMIC STIFFNESS TESTING:

Performing a simple 2 channel impact test can provide the dynamic stiffness curve. Sometimes it is only possible to calculate or approximate the dynamic stiffness – you are unable to visit the site or the component is in the design stage. Other times it is just easier, faster, and more accurate to measure it. With a properly set up data acquisition system (DAQ), it can take a only few minutes to perform the impact test and display the measured dynamic stiffness. Additionally, performing a measurement and using the FRFs (Frequency Response Functions) incorporates the complicated stiffness and mass distribution of the actual structure. This section highlights some of the key steps to perform an accurate test and measure the dynamic stiffness.

Dynamic Stiffness FRF

The primary FRF of interest when performing a dynamic stiffness test is the inverse of the compliance FRF.

$$Dynamic\ Stiffness\ FRF = \frac{Frequency\ Spectrum_{Force}}{Freqeuncy\ Spectrum_{Response\ in\ Displacement}} \tag{18}$$

Sensor Calibration

Since an accurate amplitude of the dynamic stiffness FRF is critical, it is important to use the actual sensitivity for the hammer and response sensor, such as an accelerometer; the use of nominal sensitivities is discouraged for this type of testing. Both the hammer and the accelerometer should be calibrated or have the sensitivity checked prior to use. Since accelerometer calibration is a standard portion of a vibration program, only the force hammer sensitivity check will be summarized below, as this type of calibration is typically less common.

Force Hammer Sensitivity Check

To check the sensitivity of a force hammer, the following basic steps should be performed.

- Setup a mass pendulum test as shown in Figure 14. To perform this check, a mass of known weight is mounted as a pendulum which allows motion of the mass in one direction without lateral or spinning motion.
- Using a DAQ or vibration meter, perform several impact tests at the range of forces that will be utilized for the field testing. This ensures the sensitivity is accurate at the force levels that are actually used for the dynamic stiffness measurement. Use the calibrated sensitivity for the force hammer and the accelerometer.
- Record the maximum measured force applied by the impact hammer and maximum acceleration measured by the accelerometer; multiply the acceleration by the mass of the test pendulum. This data can be recorded in a spreadsheet.
- Plot the measured force by the measured mass x acceleration for all impact tests performed against a 1:1 line.
    - By Newton's 2nd Law, F = Mass x Acceleration, and an accurate hammer sensitivity should result in the measured force being equivalent to the measured mass x acceleration. Therefore, these measured data points should plot on a 1:1 line. Figure 15 shows the force hammer sensitivity check used for the simply supported T-section and the most accurate hammer sensitivity is 1.08 lbf/G.
- If the data doesn't fall within tolerance to the 1:1 plot, modify the hammer sensitivity and re-perform the test.

Performing Dynamic Stiffness Test

When the stiffness of a structure at a specific location is of interest, the dynamic sensitivity should be measured at that location using the following steps basic steps. Note that when performing the dynamic stiffness test, all sequential modes may not be excited due to the location of the impact on the structure as shown in Figure 2 if the stiffness at the mid-span of the beam is of interest. However, that is not problematic since those modes are not active at this location.

- Select the location where the dynamic stiffness needs to be measured and install the accelerometer close to the impact location. Placing the accelerometer on the opposite side of the hammer location can be a convenient location that reduces the risk of overloading the accelerometer from the impact.
    - Typically, 100 mV/G accelerometers provide acceptable range of frequency responses, so these are often used. However, if the dynamic stiffness at lower frequencies is of primary interest, a 500 mV/G accelerometer may be a better option. However as shown in Figure 8, at low frequencies the stiffness approaches the static stiffness.
- Setup the DAQ or vibration meter to double integrate the accelerometer to displacement in inches.
- Select a high pass filter setting appropriate to the interest in low frequency dynamic stiffness. Typically a 2 Hz high pass filter setting is acceptable. For dynamic stiffness measurements, double integration of an acceleration signal tends to produce frequency content less than 2 Hz that may not be useful or accurate.
- Complete the impact testing by performing several impacts and using the average; typically four impacts is sufficient. The waveform and spectrum of the impact should

look similar to Figure 16. The time waveform shows a single impact and the force spectrum shows a smooth curve with amplitude decreases as frequency increases. The impact at very low frequences (<10 Hz) may include significant noise which and may require additional averages or careful interpretation to utilize properly.
  - There are many additional considerations for properly performing impact or modal testing; additional training through the Vibration Institute or references, such as Reference 6 are recommended.

Discussion on Test Results
After the impact tests are completed, the dynamic stiffness transfer function can be calculated or viewed in the vibration analysis software. To provide a direct comparison to the theoretically calculated dynamic stiffness, dynamic stiffness testing was performed on the same simply supported T-section (Figure 1). The measured results are provided directly from testing and the theoretical solution is based on the continuous systems method using a unit load input to calculate the displacement amplitude and dynamic stiffness. This method provides a direct comparison of the measured vs theoretical solution.

- Table 3 shows the calculated vs measured natural frequencies are within 10%.
- Table 4 summarizes the static stiffness values determined using direct static displacement measurements, static stiffness calculations, dynamic stiffness calculations, and dynamic stiffness measurements. These values are all within 10% of the calculated static stiffness using beam theory.
- Figure 19 shows the calculated vs measured dynamic frequency curves, and these curves show very good agreement across the frequency range.
  - The natural frequency of the 1st and 3rd modes are indicated by the sharp decrease in stiffness and are well aligned.
  - The minimum amplitude corresponding to the dynamic stiffness at the natural frequency is also well aligned.
  - The maximum stiffness, which corresponds to the compliance 0-crossing frequency, is also well aligned to the theoretical curve. The test data shows an increase in stiffness around 72 Hz; the source of this is uncertain and does not have practical significance to the dynamic stiffness.

CASE STUDY:

The effects of using different dynamic stiffness values was utilized on a recently analyzed rotor-bearing system. The machine is essentially a vertically-oriented, overhung fan that is belt-driven from a VFD motor. The structural steel framing supports the rotor through the rolling element bearings as shown in Figure 20.

Modal testing was completed on this machine and the dynamic stiffness of this complex structure was measured and is shown in Figure 21. The first structural natural frequency was measured at 14.4 Hz. From this data, the static stiffness was determined to be approximately 70,000 lbf/in, as indicated in Figure 21. Additionally, a finite element analysis (FEA) model was built and the static stiffness was calculated to be 75,000 lbf/in, which is within 10% of the measured static stiffness. Using the measured natural frequency and the FEA calculated static stiffness, the modal mass of the structure was then calculated to be 9.17 lbf*sec^2/in.

Using these calculations, the stiffness was calculated using the SDOF dynamic stiffness (Eq 17) and is plotted in Figure 21. Also shown in this plot is a discretized dynamic stiffness based on the measured FRF values. Variations of these two approaches were utilized in the rotor dynamic analysis to demonstrate the value and limitations of these approaches.

Figure 22 and Figure 23 show the critical speed maps utilizing a range of dynamic stiffness assumptions. The calculated modal mass of the support was adjusted to achieve better alignment of the rotor and structural natural frequencies with the test data. The following conclusions are drawn from Figure 22 and Figure 23.

- The results using the SDOF calculated dynamic stiffness are very similar to those results using the 0-250 Hz measured dynamic stiffness. In this case, this simplified calculation provides results that are similar to the measured FRF test data.
- It is important to use a large enough frequency range in the FRF data. Although the natural frequencies of interest are less than 20 Hz, at higher bearing stiffnesses, utilizing dynamic stiffness values up to 60 Hz provide lower calculated natural frequencies than using 0-125 Hz or 0-250 Hz FRF data.
- If the modal response is relatively simple, even for a complex support structure, the SDOF dynamic stiffness can provide accurate results. However if the support system includes other modal interactions, then this approach should not be used and can result in an inaccurate model.

CONCLUSION:

This paper presented the development of dynamic stiffness from the simple case of a simply supported T-beam. Following this theoretical development, a test case was performed on the same simply supported T-beam to validate the theoretical basis presented. This theoretical/test approach provided some insight into the mechanics behind the dynamic stiffness curve and modal interactions with respect to stiffness.

Lastly, a case study was briefly presented which reinforces the concept that utilization of measured dynamic stiffness data is a good method to incorporate the dynamics of a support structure into a rotor dynamics model and improve accuracy.

REFERENCES:

1. API-684, API Standard Paragraphs, Rotordynamic Tutorial: Lateral Critical Speeds, Unbalance Response, Stability, Train Torsionals, and Rotor Balancing; 1st Edition; section 2.4
2. Blevins, RD; Formulas for Natural Frequency and Mode Shape; 2001 Edition; pg 107-113
3. Timoshenko, S; Vibration Problems in Engineering; 4th Edition; pg 380-385; 415-416; 423; 435-439
4. Rao, S; Mechanical Vibrations; 4th Edition; pg 235-237
5. Edwins, D.J.; Modal Testing: Theory, Practice and Application; 2nd Edition; pg 28-32
6. Avitable, Peter; Modal Testing: A Practitioner's Guide; 1st Edition

FIGURES

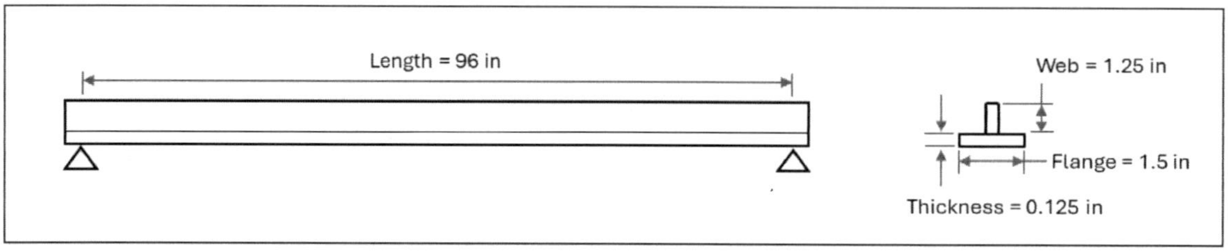

Figure 1. Simply supported T-section used for dynamic stiffness testing and analysis.

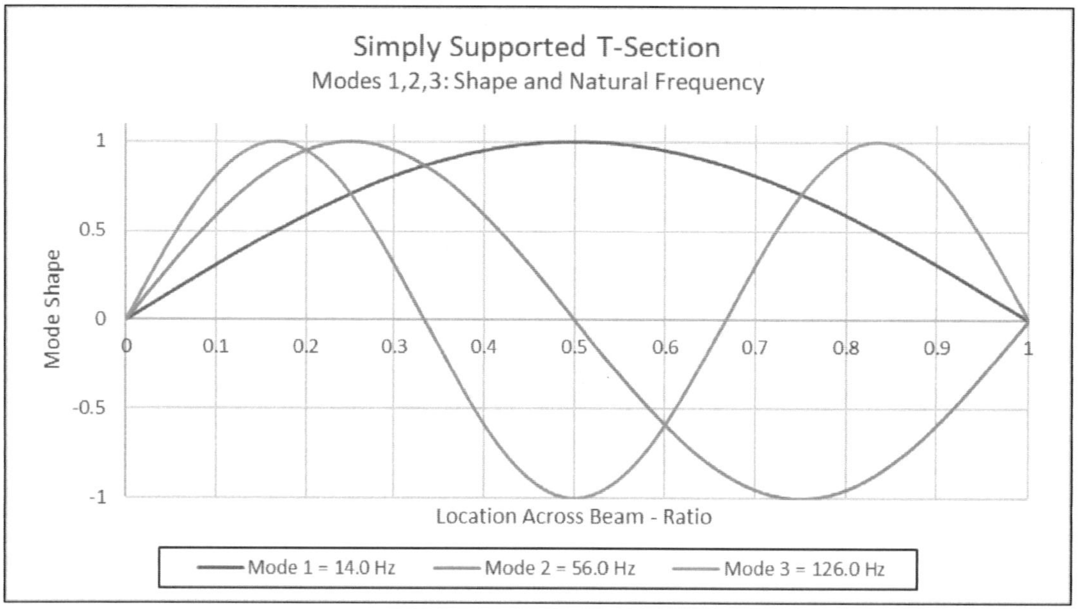

Figure 2. Mode shapes and natural frequencies for modes 1-3 for simply supported T-section.

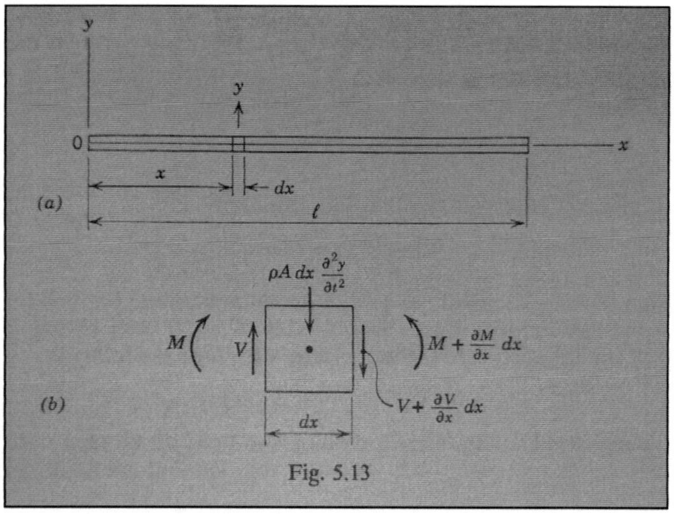

Figure 3. Free body diagram of element in beam with bending load applied.

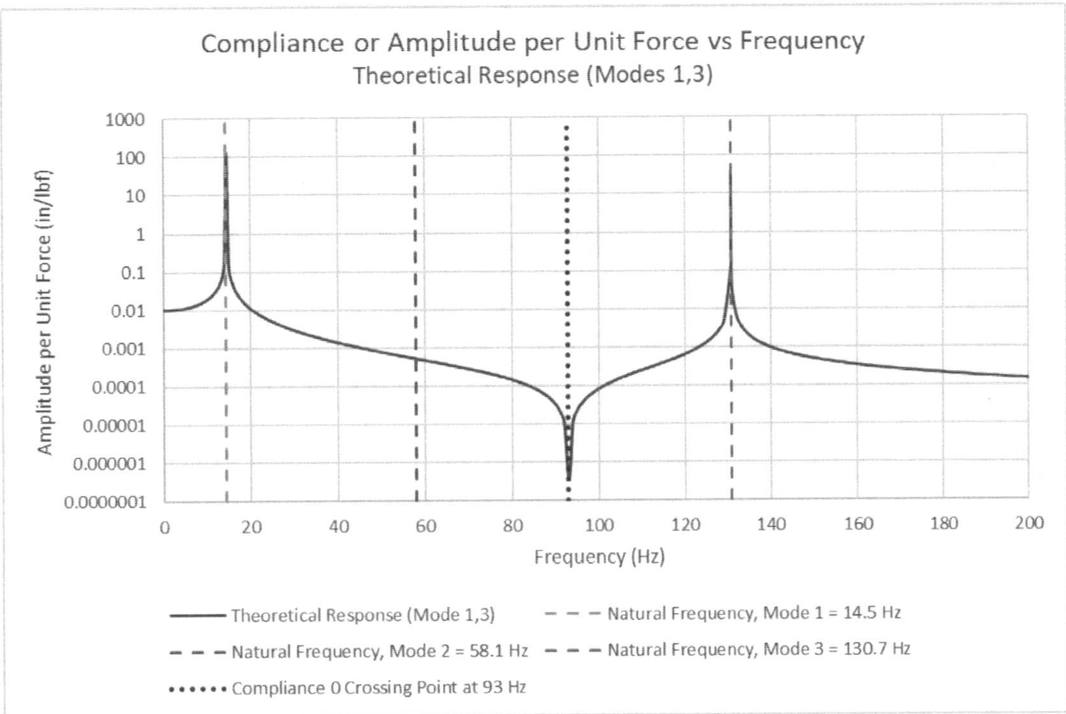

Figure 4. Theoretical compliance or amplitude per unit force with resonances shown.

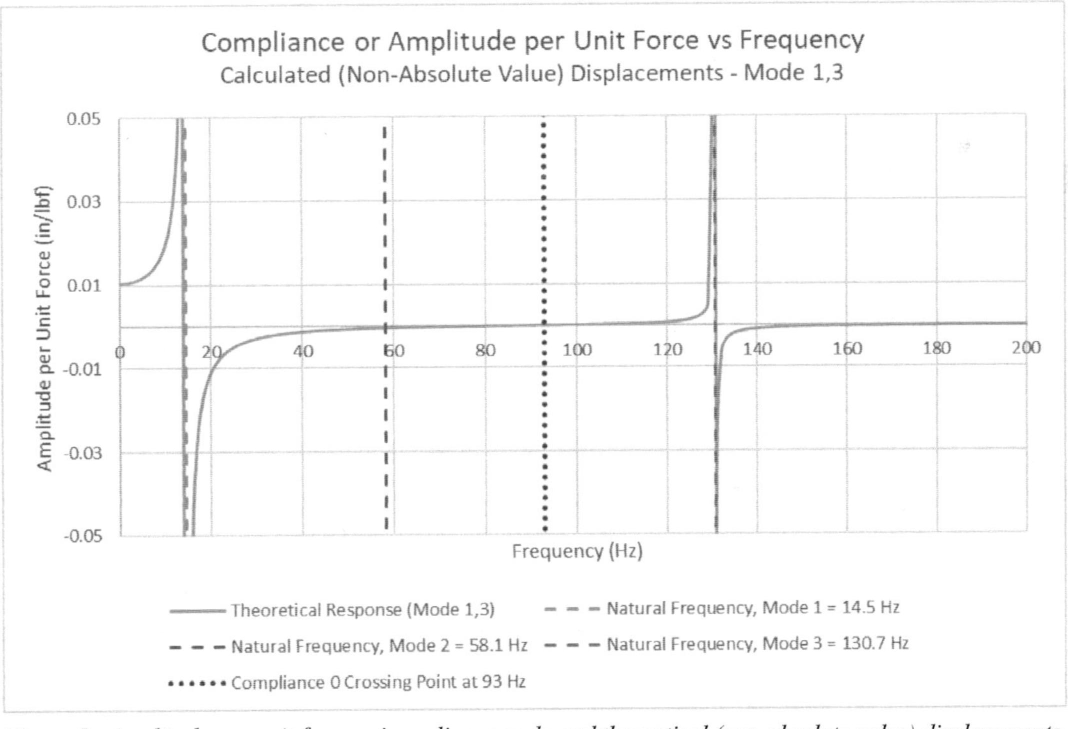

Figure 5. Amplitude per unit force using a linear scale and theoretical (non-absolute value) displacements.

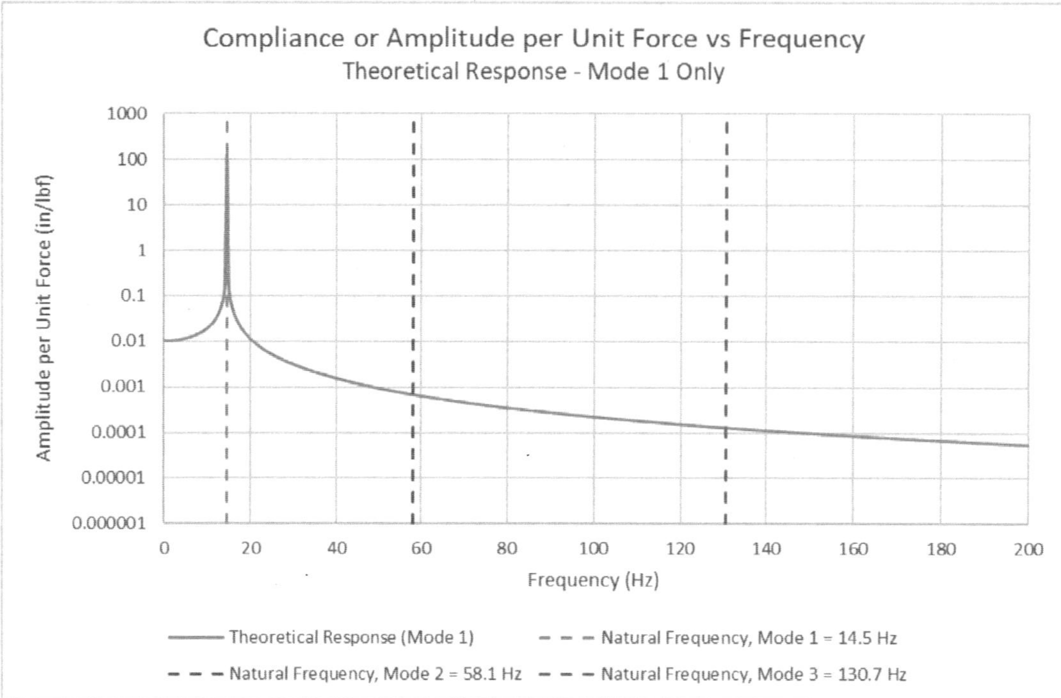

Figure 6. Theoretical mode 1 compliance curve.

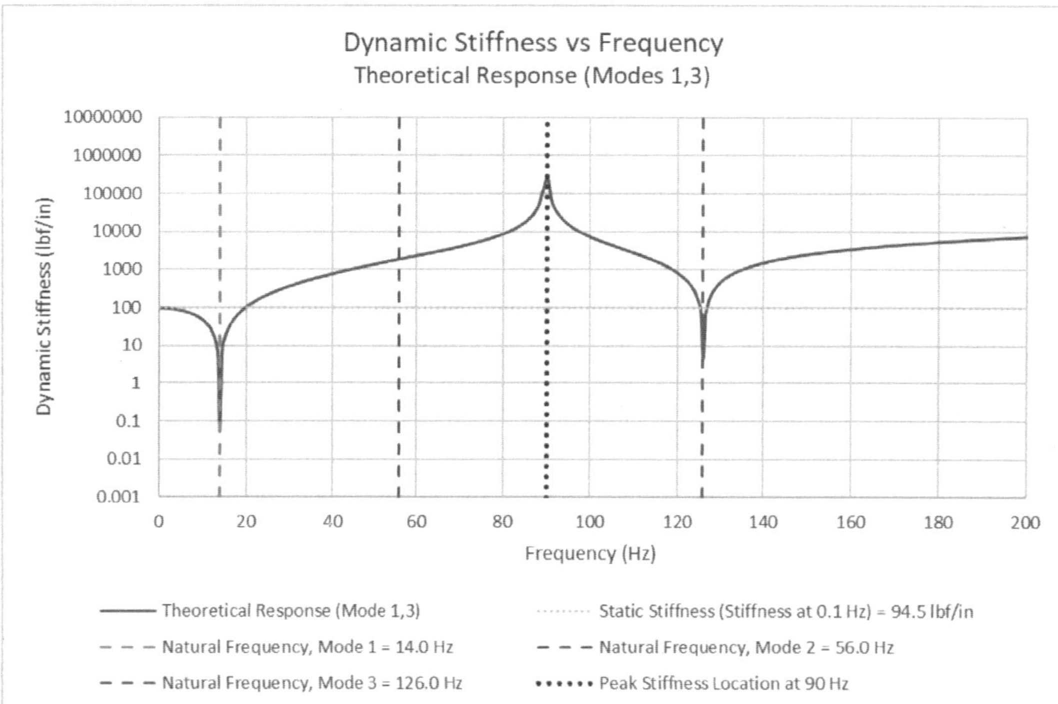

Figure 7. Theoretical dynamic stiffness with resonances shown.

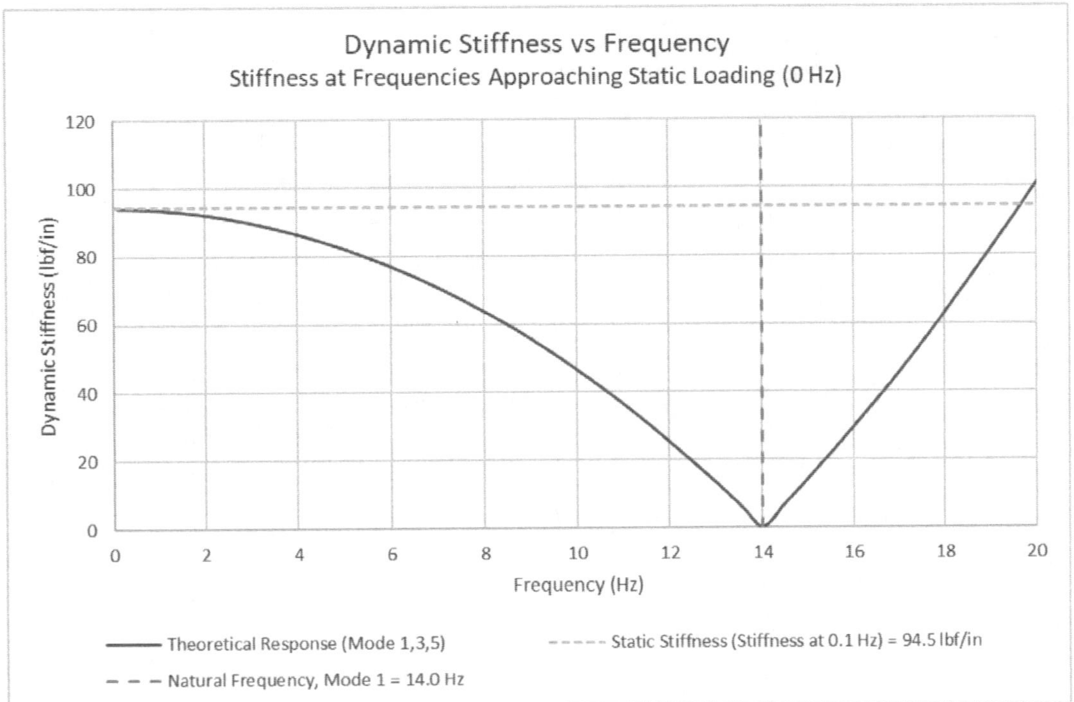

Figure 8. Theoretical dynamic stiffness at low frequencies showing the static stiffness is 94.5 lbf/in.

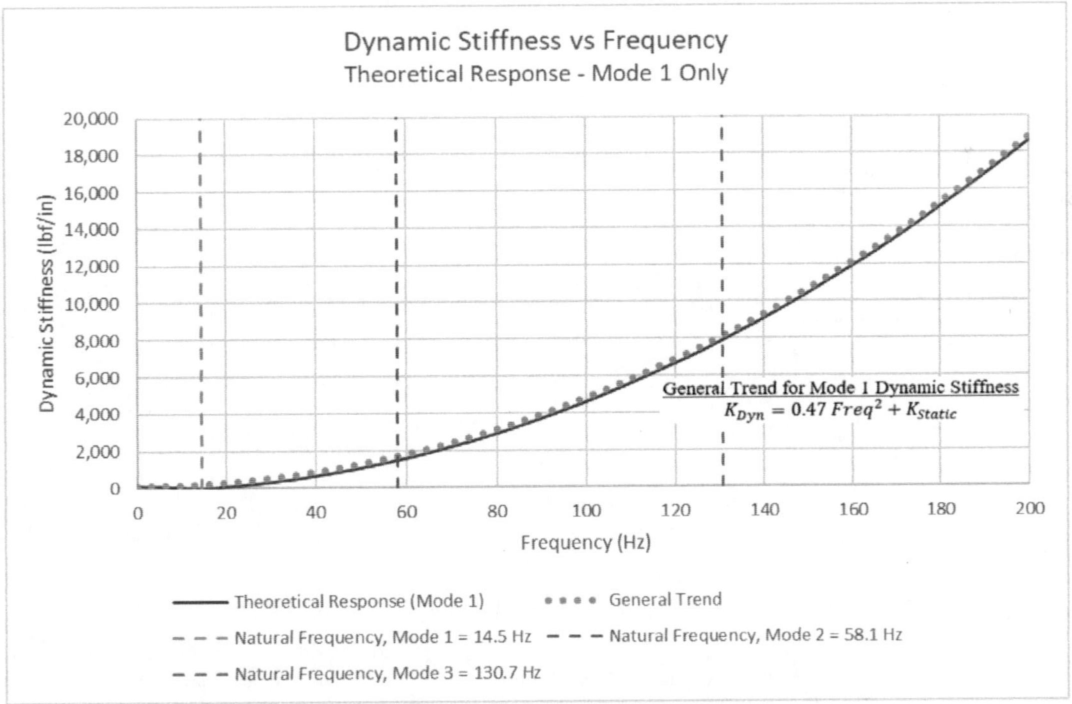

Figure 9. Dynamic stiffness of mode 1 showing relationship with frequency squared.

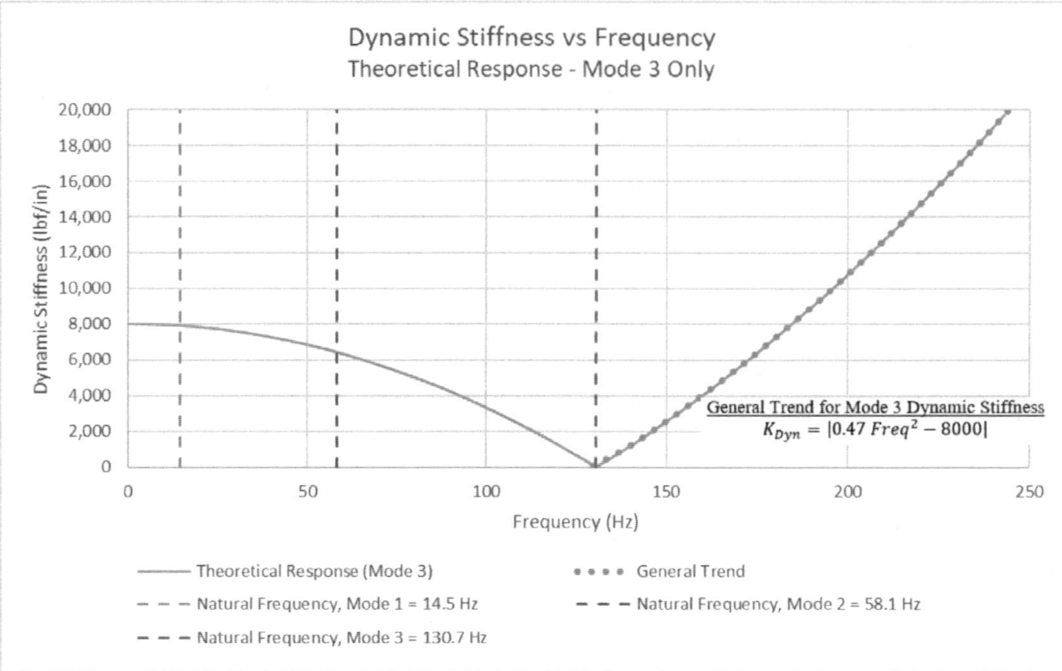

Figure 10. Dynamic stiffness of mode 3 showing relationship with frequency squared.

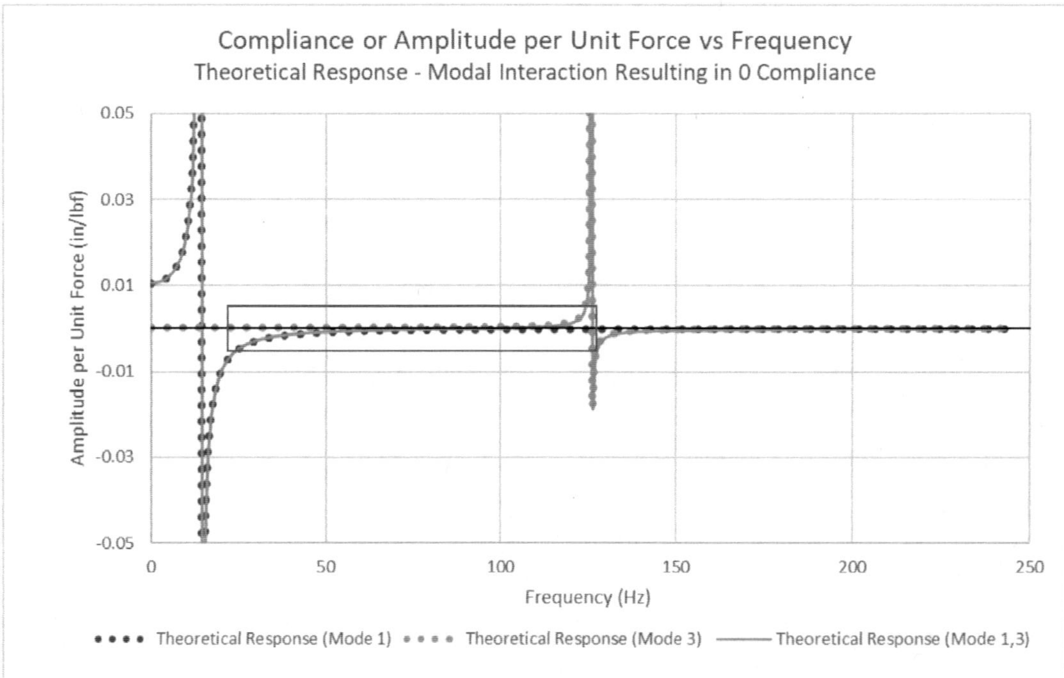

Figure 11. Compliance plot showing mode 1 and mode 3 interaction and area of interest for 0-compliance.

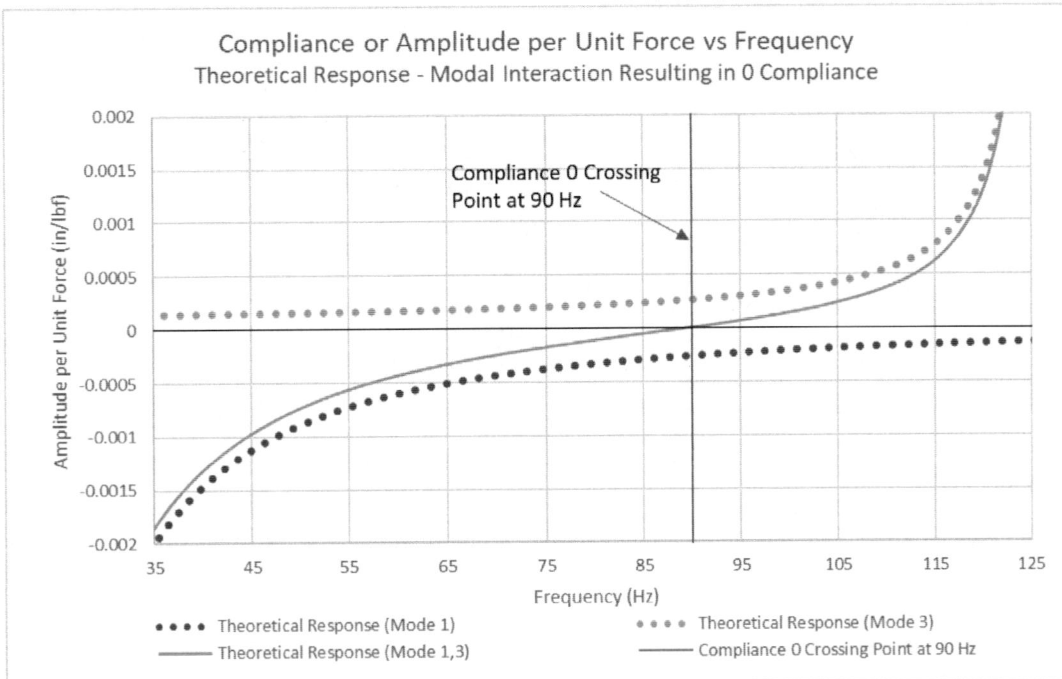

Figure 12. Zoomed in from Figure 11 - Compliance plot showing mode 1 and mode 3 interaction and area of interest for 0-compliance.

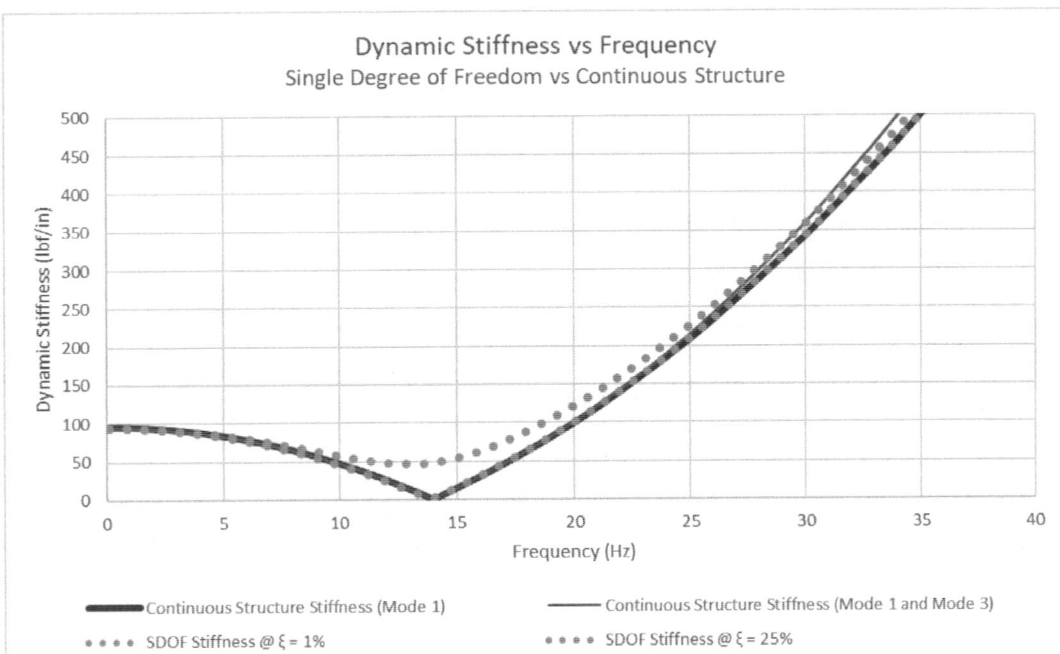

Figure 13. Dynamic stiffness comparison using continuous SDOF methods and effects of damping.

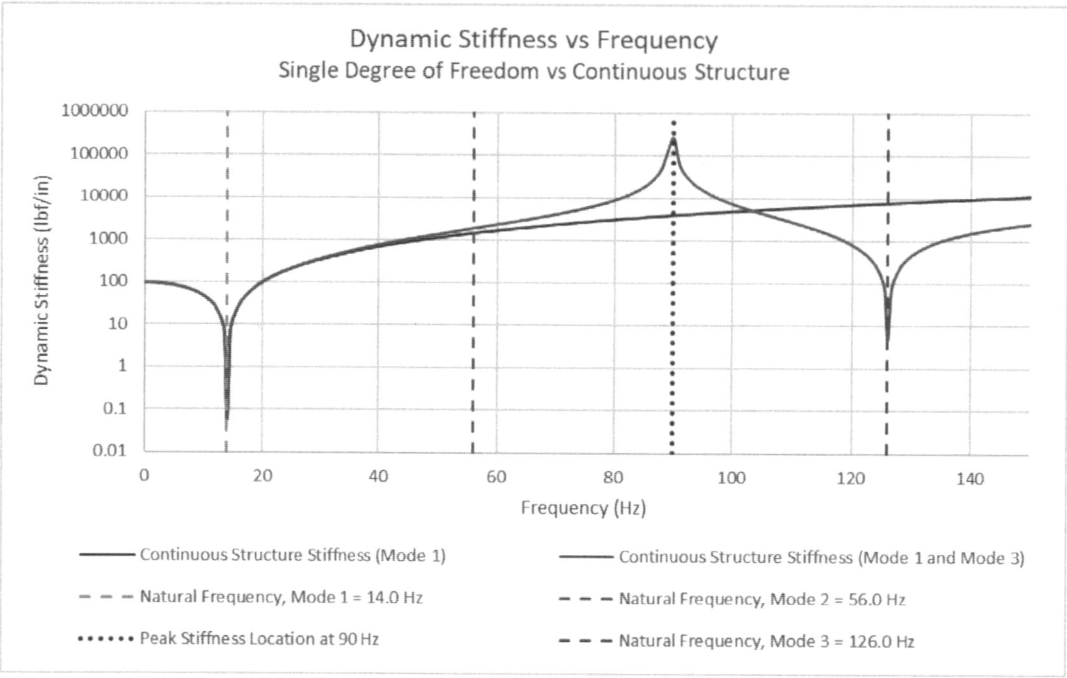

Figure 14. Dynamic stiffness comparison using continuous SDOF methods.

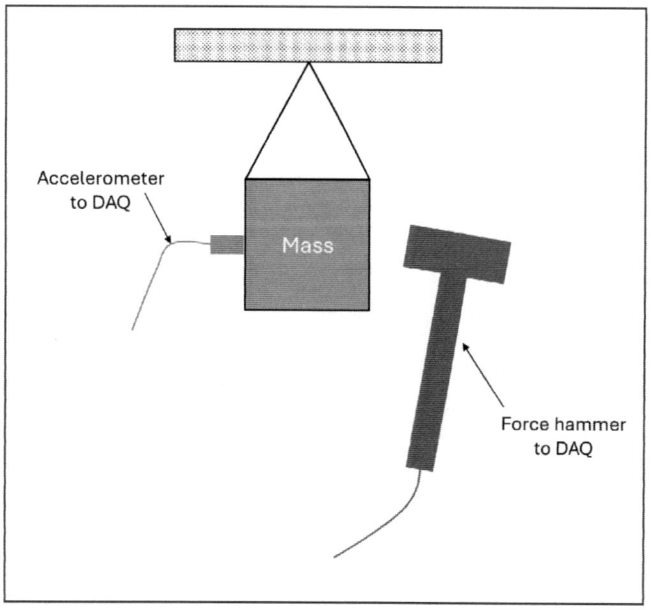

Figure 15. Test setup to check force hammer sensitivity.

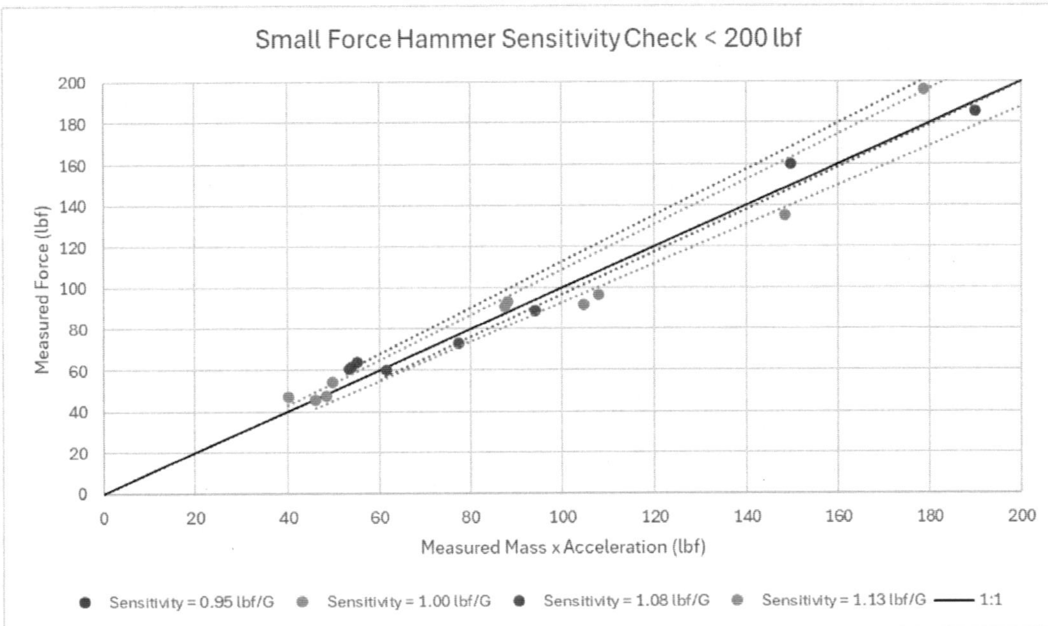

Figure 16. Force hammer sensitivity check for modal testing; the sensitivity of 1.08 lbf/G is most accurate.

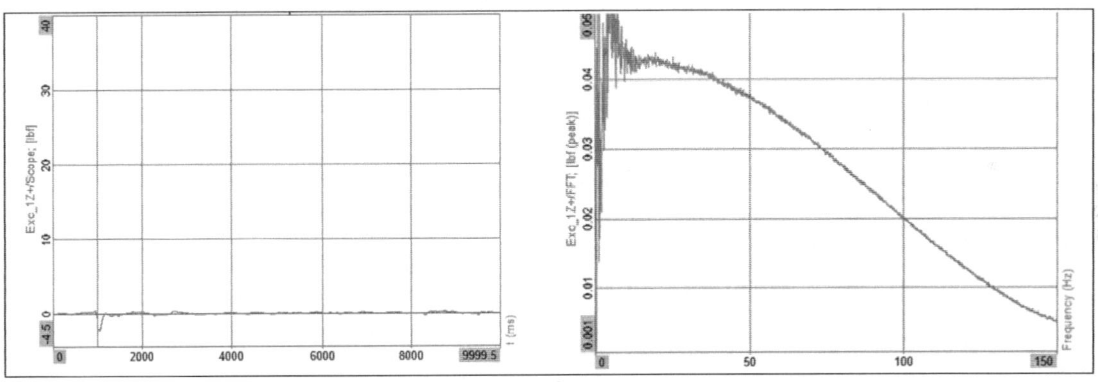

Figure 17. Time waveform and spectrum for the impact force.

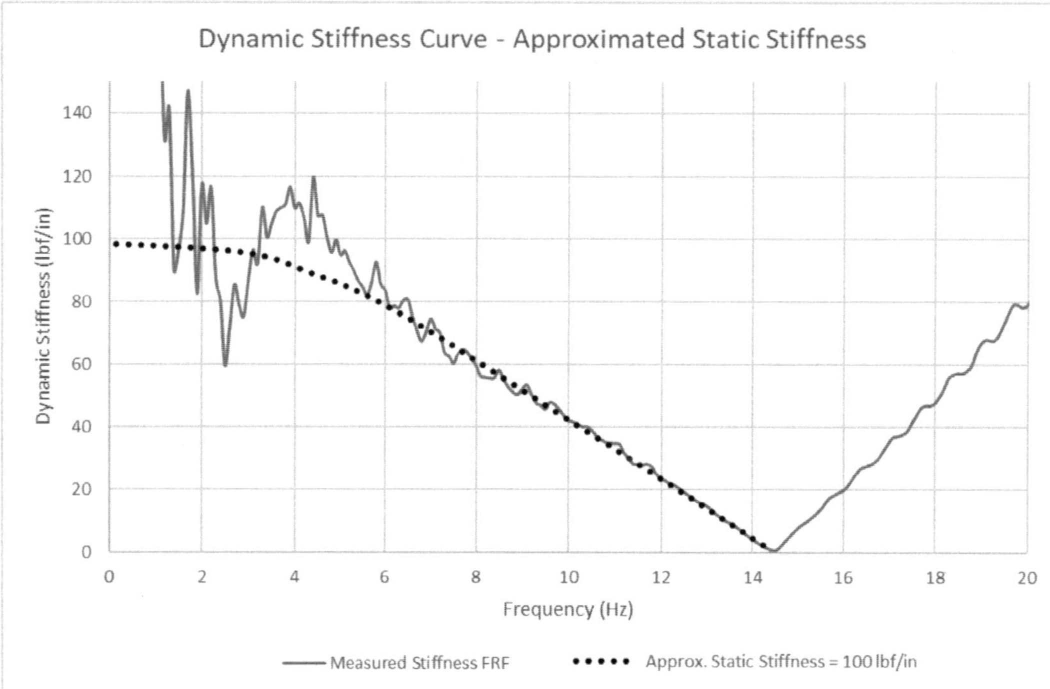

Figure 18. Dynamic stiffness curve from testing used to approximate static stiffness.

$$Dynamic\ Stiffness = \sqrt{(k - m\omega^2)^2 + (\omega c)^2}$$

Measured Values from Dynamic Stiffness Testing		
Natural Frequency	14.5	Hz
	91.11	rad/sec
Dynamic Stiffness at Resonance	0.821	lbf/in
Effective Mass (see Table 1)	0.012305	lbf*sec^2/in
Damping	Negligible Effect	
Calculated Static Stiffness	103.0	lbf/in

Figure 19. Calculated static stiffness using measured values from dynamic stiffness testing.

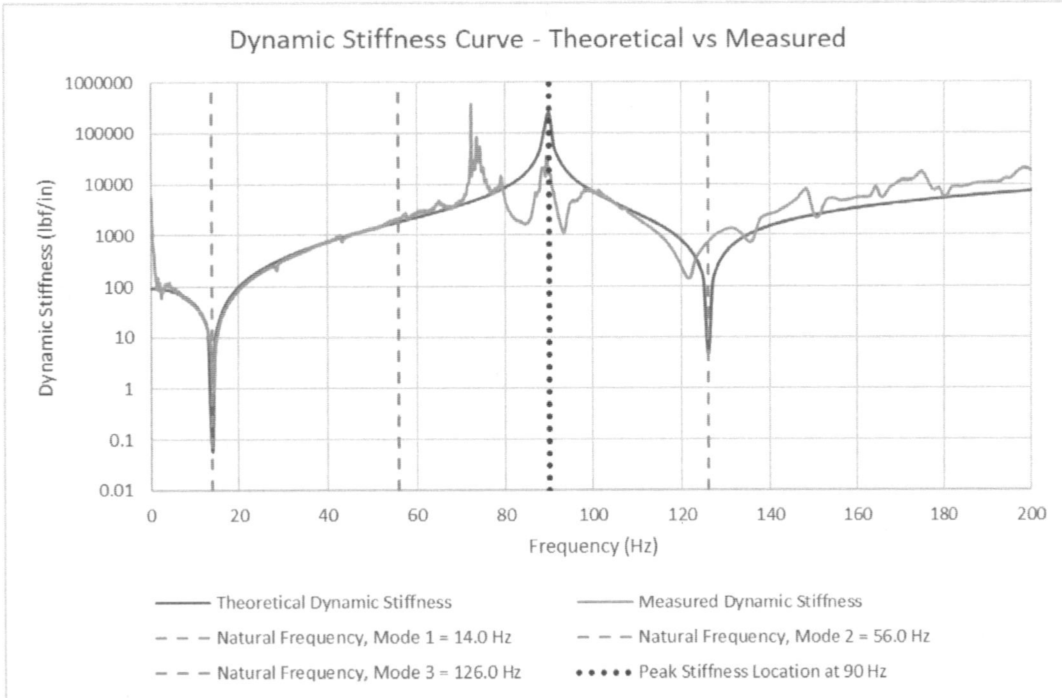

Figure 20. Theoretical vs measured dynamic stiffness curve with resonances and maximum stiffness shown.

Figure 21. Picture of the vertically-oriented "fan" discussed in the case study.

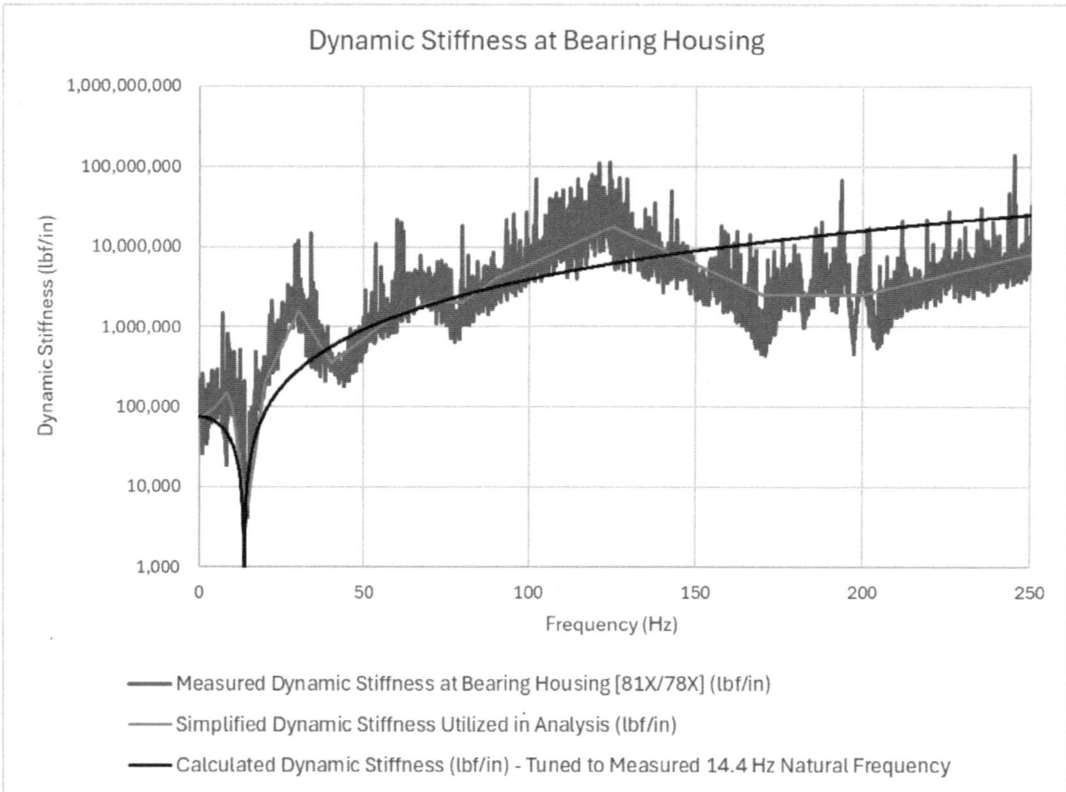

Figure 22. Dynamic stiffness values of the bearing housing determined using measurements and SDOF stiffness calculation.

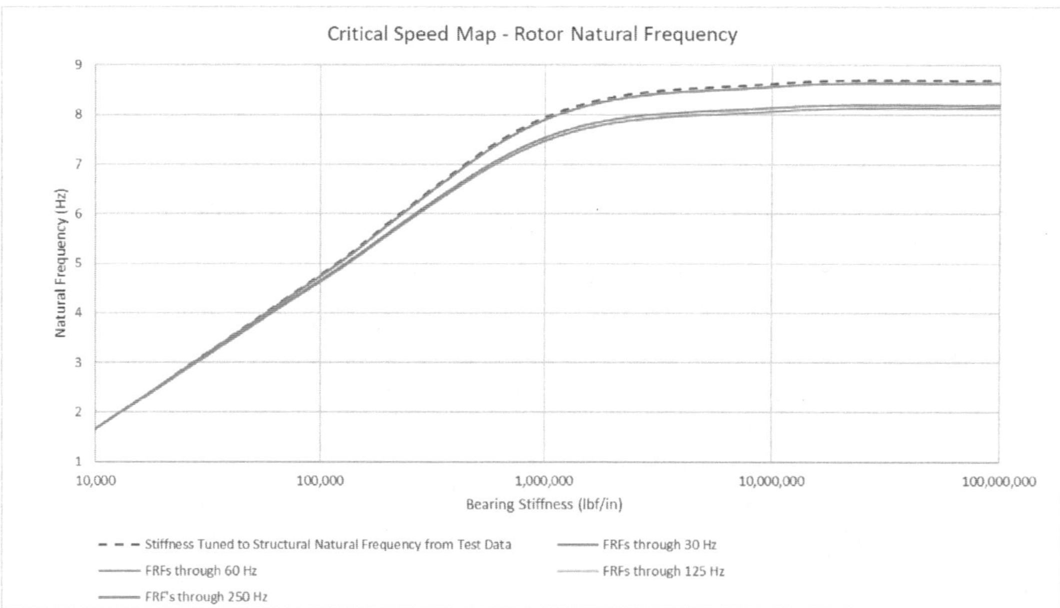

Figure 23. Critical speed map of the fan rotor (1st mode).

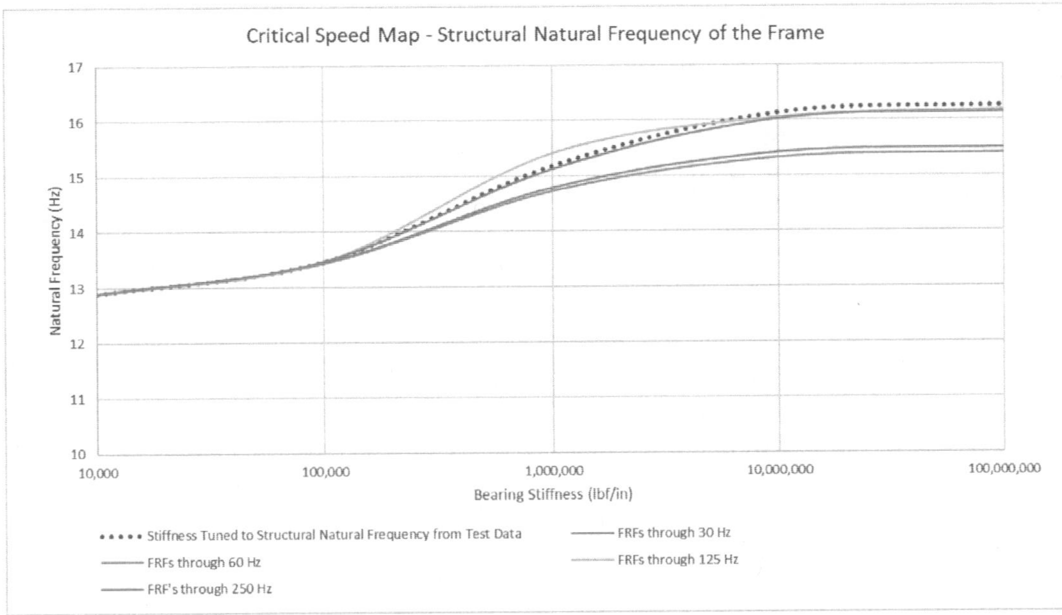

Figure 24. Critical speed map of the natural frequency of the structure (2nd mode).

TABLES

Table 1. T-Section Arrangement Properties

Item	Value
Tee Dimensions	Flange: 1.5 in Web: 1.25 in Thickness: 0.125 in
Modulus of Elasticity	30×10^6 psi
Distance b/t Supports	97.75 in
Moment of Inertia	0.0609 in^4
Calculated Static Stiffness	93.9 lbf/in
Cross Sectional Area	0.344 in^2
Total Weight	9.509 lbf
Effective Mass Factor	0.50
Effective Mass	0.01230 lbf*sec^2/in
Calculated Natural Frequency (SDOF)	13.9 Hz

Table 2. Static stiffness determined by measuring beam deflection with static weight applied.

Measured Static Stiffness		
Weight (lbf)	Deflection (in)	Stiffness (lbf/in)
22	0.21875	100.6
44	0.4375	100.6
Average		100.6

Table 3. Natural Frequency Summary Table

Natural Frequencies (Hz)			
	1st Mode	2nd Mode	3rd Mode
Measured	14.5	N/A	121.8
Calculated	14.0	56.0	126.0
Difference	-4%	N/A	3%
* 2nd mode was not excited since node of the 2nd mode was at the impact and measurement location (see Figure 2).			

Table 4. Stiffness Summary Table

Comparison of Methods to Determine Static Stiffness			
Method	Reference	Stiffness (lbf/in)	Difference
Static stiffness calculation from beam theory	Table 1	93.9	N/A
Direct static displacement measurements	Table 2	100.6	7%
Theoretical dynamic stiffness curve	Figure 8	94.5	1%
Measured dynamic stiffness curve	Figure 17	100.0	7%
Measured dynamic stiffness and SDOF equation	Figure 18	103.0	10%

COVINGTON, KENTUCKY | AUGUST 7-9, 2024

MAKE SERVICE COMPANIES YOUR PARTNER IN RELIABILITY

Forrest Pardue
24/7 Systems – TANGO™
parduef@tf7.com

After earning a BSEE at North Carolina State and then an MBA, Forrest has worked in the field of vibration analysis and predictive maintenance for more than 30 years. As one of the founding members of Computational Systems, Inc (CSI), he was actively involved in the technical and market development of modern condition monitoring technologies. Following Emerson Electric's acquisition of CSI in 1998, Forrest co-founded 24/7 Systems. He recognized that the greatest challenge facing industrial maintenance had shifted to the measurement, management, and improvement of plant asset reliability. 24/7 Systems is focused on the development and delivery of Reliability Information Management software and services.

ABSTRACT:

Collecting asset reliability information is getting easier by the day, prompting reliability service companies and repair shops to up their game with new condition monitoring sensors, tools, processes, and services. Plants rely on these organizations to conduct essential asset reliability tasks that they are unable to fully support internally. Coordinating the existing reliability processes is already a challenge but amassing larger data volumes will simply create more noise and clutter unless the service organization can effectively manage and apply it.

Leaders in this field are turning to off-the-shelf reliability information management systems (RIMS) to help consolidate and gain maximum value from the numerous and growing sources of available information. RIMS solutions improve their ability to eliminate in-service failures, extend the operating life between failures, and increase the productivity and performance of their asset reliability services through more timely and robust analysis, actionable information, and optimal outcomes.

KEY WORDS: Reliability Information, Service Companies, Contractors, Integration, Information Management, Reliability Solutions, Avoided Costs

Introduction:

Increasing asset uptime by performing condition inspections and predictive maintenance is an underutilized reliability best practice. But some manufacturing facilities are too small to have their own dedicated condition-based inspection team. Other facilities have found that they cannot maintain the people and knowledge necessary to provide a dedicated condition-based inspection program. Some facilities choose to maintain in-house expertise for certain condition-based inspections, while contracting out inspection services for other technologies.

In each scenario, third-party reliability services companies can provide the people, equipment, knowledge base, and predictive maintenance experience often needed to assist plants in achieving a successful condition-based maintenance focus, and partner with the facility over the long term.

When a plant decides to use a reliability services provider, several questions must first be asked:

1. What assets are to be inspected?
2. What technologies are to be used?
3. What is the inspection interval?
4. How will results be reported and who is accountable for results?
5. How does the plant know that all machines contracted were inspected?
6. Are results reported in a printed report or online?
7. Who is responsible for creating work orders?
8. Is a weekly status meeting with each operating area about priority conditions part of the contract?
9. If a machine is critical upon inspection, what is the contractor's responsibility?
10. Is the contractor responsible for validating repairs?
11. What program metrics are required?
12. Is it required to calculate and report program savings or avoided cost?

Reliability services companies that are best able to meet expectations are turning to off-the-shelf reliability information management systems (RIMS) to help consolidate their data. In this way, they gain maximum value from the numerous and growing sources of available information.

It improves their ability to eliminate in-service failures and extend the operating life between failures. It also increases the productivity and performance of their asset reliability services through more timely and robust analysis, intuitive communication, and optimal outcomes.

Moreover, asset owners and operators reap the rewards of improved service partner relationships and greater reliability outcomes, without having to purchase or maintain advanced reliability tools or systems in-house.

Read on to learn how several well-established reliability services companies improved how they interact and report inspection results with their partner plants, and why a robust software platform matters to your success.

Section 1: Interaction between the plants and service providers has not always been seamless.

Some plants simply want work orders entered into their computerized maintenance management system (CMMS) when defects are identified, "but we have not found any CMMS that populates and tracks the relevant information needed to optimize your condition monitoring efforts," Dean Stephens, reliability director at Benchmark Reliability Services.

"A simple KPI that gets lost in most CMMS systems is: What was measured? Because it lacks sufficient granularity at the component level and insight into what was *not* measured, a CMMS is not well equipped to reveal hidden risks," Stephens explains.

At the other end of the spectrum, many plants use spreadsheets. "It is inexpensive but would end up a mess if there was more than one person involved and if the equipment count was large," observes the lead reliability engineer at a major North American industrial services provider.

"Some clients also use technology specific systems, but this is not an 'asset approach' and it is also pricey."

Dave Reynolds, owner of Midlands Reliability & Consulting, believes that small clients with fewer than 100 machines to track might get by using Excel spreadsheets and Gantt charts to manage workloads and scheduling, and Word documents for reporting. However, for clients with large numbers of machines, it took longer to track, organize, and report the findings using these methods.

"Challenges arose when numerous files were needed to assign tasks and resources, conduct mobile rounds logging, complete reports, and track asset health metrics. Incorporating information from multiple vendors and condition monitoring technologies into a single file was excessively time consuming," Reynolds explains.

Reliability service providers were also limited by early software options. For instance, a U.S. predictive technologies service company originally used Excel for tracking and PageMaker to build reports that were published in book form or emailed as a PDF. The owner explained, "These manual processes were made much worse when historical trend calculations were needed as there was no archived history, and the lack of information at my fingertips was frustrating."

Vagueness also makes it much more difficult to determine what to do because specifics are what lead to actions to effectively maintain and improve the equipment. Early asset-specific reliability tools could indicate whether an asset was in an alarm state (red) or not (green), but not if the problem was in the asset's pump or, better yet, a certain seal or bearing within the pump.

Section 2: How does a RIMS build a long-term bridge between the service company and plant?

Reliability information management systems exist to solve these very challenges. Whether initiated internally or prompted by a client, leading reliability services companies are transitioning to RIMS-based approaches for the benefits below, and not looking back.

Consolidated, consistent information:

Robust RIMS solutions provide the ability to integrate reliability information from multiple condition monitoring technologies into one, standardized database. Having access to motor circuit evaluation, alignment and balancing, infrared, oil, ultrasound, vibration, photographs, videos, sound files, weblinks, attached office products or pdf files, visual findings and more—both current and historical—all in one place is transformational. It allows users to:

- Replace dozens of disparate information sources with a single source of truth.
- Collect reliability information from every asset, area, and site within an organization.
- Standardize on a common nomenclature for efficient communication.
- Produce reports from a software package with uniformity across technologies.
- Export data for CMMS or alternative solution integration.

Standardized reporting:

RIMS reporting is automated and standardized, providing higher quality deliverables in significantly less time than prior approaches. At the push of a button, integrated condition status reports, assessment task reports, and repair histories with lifecycle metrics and cost tracking can be generated. This provides more accuracy and increased control than simply emailing disparate reliability files.

Some RIMS's web-based reporting tools also improve continuity. Generated reports are finished quality, with cover letters, statistics, details and summaries, along with previous report histories. Custom reports, such as an annual KPI report for major clients highlighting the year's achievements, may also be quickly compiled by snipping key KPI data and trends while preserving standards consistent with the RIMS solution.

"Our condition problem entry process uses preset picklists to enforce standardization, while also offering a comment section for additional in-field findings or explanations. Users benefit from standard fault descriptions and the ability to judge the severity of findings," says Midlands' Reynolds of the Tango RIMS solution used by his company. "Users have snapshots of monitored or missed machines, which component on the machine has the fault, detailed fault trees of condition-based findings, and asset performance comparisons by asset type or location. Furthermore, data is backed up regularly, ensuring secure storage of findings."

Active communication:

Using a RIMS organizes the service provider's schedule and workload, and it automatically notifies providers and clients in advance that work is coming due. All parties have ready access to a historical record of activities, findings, and reports, which is invaluable when service questions arise. Automated delivery and notifications improve communication and ensure nothing is overlooked.

Client actions such as replacing a failing motor must be communicated back to the service provider, and RIMS software promotes this feedback by providing a place to document corrective actions for deteriorating conditions, thus closing the communication loop with real-time repair status updates. Clients also benefit from rapid access to reports, built in KPIs, and remote access to the RIMS for real-time insights and oversight, which further improve communication between the provider and client.

During client meetings, RIMS reports may become central talking points. "We have 'Exit' meetings (some call these 'Reds' meetings) with clients when we are not a full-time site contractor. Going through the most recent defects refreshes everyone's memory and instigates discussion on the craft workflow required for corrective action," says Benchmark's Stephens. "It also provides back-and-forth questions for clarity, accuracy, and verifying the severity of the defect failure mode."

Communicating RIMS information to all parties adds the benefit of providing an expanded view of provider-assessed asset conditions, such as visibility of electric motor testing results to mechanical technicians. The reports and documentation also optimize communication between Maintenance and Operations. "Giving Operations access to RIMS results helps during their rounds logging inspections or with general machine knowledge when they hear, see, or feel things that are out of sorts in the plant," explains Stephens.

Increased accountability:

Tasks can be set up in the service contract and managed in the RIMS to ensure accountability for completion and quality control. Auditable actions provide proof of work, improving accountability by removing the veil.

One major beverage manufacturer allows its reliability service providers to use whatever data gathering tools they want as long as standardized details are entered into Tango. Client-to-provider accountability is increased by automatically generating emails to company management to escalate a correction when an assessed serious problem exceeds 14 days of age.

Some service providers elevate accountability by setting up their RIMS solution to generate an email to a manager the moment a technician hits "save" on a priority 1 or 2 entry. That manager can immediately review the data and trends and interject with expert guidance as needed before the technician's report is even finished.

Periodically, there may be awkward moments where a client did not log in to review the machine history or status dashboard to see if a defect was actively identified before failure occurred. Or they saw the emerging defect but chose not to act. A quick look at the RIMS' archive history will confirm when the defect was first reported and how it progressed over time. In some cases, such incidents lead to a management directive requiring that high-priority defects be addressed within a specified number of days.

Section 3: Best practices enablement:

According to the lead reliability engineer for the major industrial services provider, "The biggest benefit we see from our RIMS solution is that it is asset based instead of technology based. Having a single location for all information is key for root cause failure analysis (RCFA) efficiency. Also, with the drop-down menus for asset status and issues, it allows us to quickly see patterns and pull KPIs for the client."

For example, if there are electric motor problems caused by a bearing issue and pump problems due to cavitation, then processes can be developed for mitigating those issues to avoid future failures. Ready access to such knowledge delivers tremendous value when applied to defect elimination and continuous improvement.

Infrared Technician Damian Roche shares an example: "We take photos and notes on every piece of electrical equipment in a client's complex, and we can look back years and see how the equipment has slowly deteriorated over time and enact improvements." He continues, "Users can see how their equipment is performing on the fly right off of their phone. Having an entire facility's electrical layout in the palm of your hand makes things run a lot smoother."

Providers find particular benefit in the ability to look back in their RIMS solution and see what issues they have been dealing with most. "When maintenance managers are dealing with fire after fire, we urge them to look at the severity levels. Don't be overwhelmed by the forest; focus on the tree and how to prevent that fire," suggests the owner of the predictive technologies service company.

Conclusion: Renewed relationships:

Everybody wants to be relevant. Using a RIMS solution adds newfound relevancy to roles and relationships.

1. For reliability services providers, RIMS boosts program sales when the clients realize that data and statistics missing in their current system are readily available in the provider's RIMS. Its provable value boosts client retention while also opening the door to add-on sales by offering clients the opportunity to buy a user seat, or even a standalone RIMS package to extend the advantages to other plant processes and technologies.

2. For clients, it allows them to really see how their equipment is operating year in and year out to improve planning, because they only have so much time and budget.

3. For technicians, it provides a reason to be detailed, because they can see the value they are providing. It also gives them a sense of pride that they are actually helping to drive change and operational improvements.

Moreover, the entire vendor-plant team becomes more collaborative in decision making and planning when your reliability services partner relationship is supported by robust reliability information management.

# # #

Gap in pagination due to formatting issues.

Pages 130-131

VIATC TRAINING CONFERENCE & EXPO
COVINGTON, KENTUCKY | AUGUST 7-9, 2024

WAR STORIES OF A VIBRATION ANALYST

Monroe Voyles III
Solutions Engineer for ITT Goulds Pumps
monroe.voyles@itt.com

Mechanical engineer with 29 years of experience. Ten years as a maintenance engineer in refineries and chemical plants; nineteen years working for pump manufacturers. Experience with finite element modeling (CREO/Simulate/DyRoBes). Instructor for the Vibration Institute for 8 years. B.S.M.E Louisiana State University 1994. CAT IV Vibration Analyst 2010.

ABSTRACT:

Vibration is the response of a machine or system to a dynamic force. Often we want to know the source of the force to identify and then correct faults in machinery. In many cases the magnitude of the force correlates with the severity of the fault or problem. The frequency is not difficult to obtain, but estimating the magnitude of a force from a vibration response amplitude is far from simple in most cases. Vibration response magnitude is influenced by the force applied, proximity to a natural frequency, and damping. Further complicating the analysis, multiple sources of vibration may create the same frequency of response. This paper discusses four case studies to show how not considering force, resonance, and damping can lead to undesirable results. Vane-pass vibration in centrifugal pumps is discussed in the first two case studies. Disc coupling faults on centrifugal pumps are discussed the last two case studies.

KEY WORDS: Acoustic Resonance, Best Efficiency Point (BEP), Damping, Dynamic Pressure, Frequency Response Function (FRF), Magnitude, Mode Shape, Mechanical lag, Mechanical Resonance, Natural Frequency, Operating Deflection Shape (ODS), Orbit, Vane-pass

VANE-PASS VIBRATION:

Centrifugal pump impellers use discrete vanes to accelerate fluid. This increase in velocity is used to create flow and an increase pressure. The flow and pressure around the impeller circumference is not uniform or steady. When this time varying flow interacts with stationary structures in the volute or diffuser dynamic forces are created at vane-pass frequency and often harmonics of vane-pass frequency. The wake behind the impeller vanes is the primary source of velocity variation relative to the stationary structures [Ref 1]. If the radial clearance (B gap) between the impeller and stationary structures is increased there is more time and distance for diffusion of the velocity gradient; this reduces the dynamic forces, dynamic pressure, and vibration amplitudes at vane-pass frequency.

The flow rate through the pump also affects vane-pass vibration amplitudes. Near the best efficiency point (BEP) flow of the pump the flow exiting the impeller discharge should have a minimum angle difference relative to that of the casing flow path (volute or diffuser vanes), see Figure 1. At lower and higher flow rates the angle between the flow exiting the impeller and the casing flow passage (volute or diffuser) increases and vane-pass vibration amplitudes can increase significantly; this is particularly true at lower flow rates because the pressures are higher than at high flow rates.

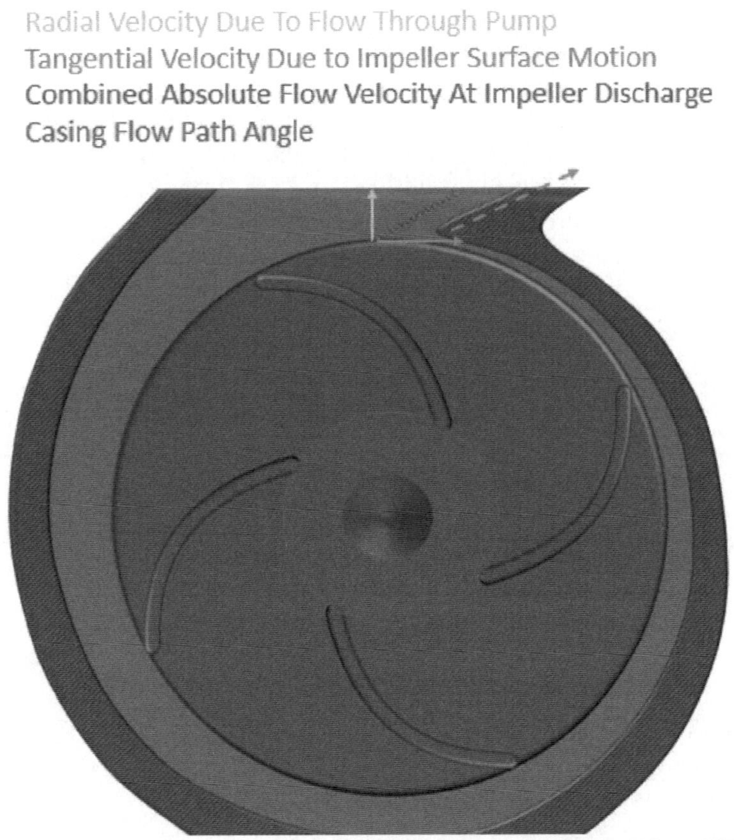

Figure 1: Impeller Flow Vectors At Shockless Flow Rate (Near BEP Flow)

NATURAL FREQUENCY:

A natural frequency is a frequency at which a system or machine readily absorbs energy. This means for a given force magnitude applied the machine will respond with a higher magnitude vibration at or near the natural frequency. Real machines have a continuous mass and stiffness distribution; therefore, there are a very large number of natural frequencies. Typically one or two natural frequencies near forcing frequencies are of concern with respect to large vibration response amplitudes. Natural frequencies also have mode shapes associated with them. It is usually important to know the natural frequencies and their mode shapes when trying to resolve resonance issues.

The mode shape provides the nodes and anti-nodes. Nodes are locations without movement and anti-nodes are locations of maximum motion for a given natural frequency. There are two important reasons to know the mode shapes. First we must know where the motion is maximum to know how to make structural changes that will move the natural frequency higher or lower to avoid resonance at a forced frequency like 1X shaft turning frequency or vane-pass frequency. Second it is helpful when adjusting finite element models (FEM). FEMs have unknown boundary conditions and simplifications that can create errors when comparing the model predicted natural frequencies to the real machine natural frequencies. If the real machine natural frequencies and mode shapes are known the FEM boundary conditions and simplifications can be adjusted so the model predicted natural frequencies and mode shapes have good fidelity to that of the real machine in the frequency range of interest.

The more natural frequencies and mode shapes used in this adjustment process the greater the model fidelity and accuracy. The adjusted FEM is then used to simulate proposed changes. If the FEM is not accurate then proposed changes will not provide the predicted results. Changes usually involve increasing or decreasing either mass or stiffness to move natural frequencies well away from known forcing frequencies or machine generated frequencies. The difference in frequencies is referred to as frequency separation margin. Typically for design a margin of 10-15% above and below all forced frequencies is desired. The actual band of high response amplification may be much less and is on the order of 5-6% above and below a natural frequency. Increased frequency separation margin is often used to account for unknown factors such as FEM errors. In some cases model errors can be much larger than 15% and high vibration amplitudes from resonance amplification can occur even though the FEM predicts adequate frequency separation margin.

WHAT CAUSES LARGE VIBRATION AMPLITUDES:

At frequencies very low relative to the resonance frequency vibration response amplitude is primarily controlled by the force amplitude and stiffness. Hooke's Law, see Eq. (1), is a well known physics principle that represents the force resulting from a deflection of a spring. Hooke's Law provides the response to a force at zero frequency or zero motion. Hooke's Law assumes a linear stiffness.

Hooke's Law $F = -KX$ (1)

F = Force applied

K = Spring constant
X = deflection or deflected position

This relationship holds until forcing frequencies begin to approach a natural frequency. The timing relationship between the force applied and the response or peak vibration motion begin to change. The response peak amplitude begins to lag behind the force applied in time; this is referred to as mechanical lag. Mechanical lag is associated with the resonance mechanism. A simple way to understand resonance is to consider the situation where you are pushing a person on a swing. If you wait until the person on the end of the rope (pendulum) is starting to move away from you before you begin to push your energy will add to the kinetic energy of the person and this is resonance. Each cycle the vibration amplitude will increase until the rope (pendulum) becomes non-linear or the person falls off the rope. If you begin pushing as the person is moving toward you this is analogous to being well above resonance. In this case the persons mass and acceleration are limiting or controlling the vibration response amplitude. If you push when there is no swing motion (starting at rest or very low velocitiy) this is analogous to low frequency well below resonance.

At low forcing frequencies relative to the natural frequency the motion is proportional to the force applied and inversely proportional to stiffness (Hooke's Law). If all systems operated in this low frequency regime vibration analysis would be much less complicated. If resonance becomes relevant then there is a very large change in vibration response amplitude for a small change in the frequency of the force applied. The system is sensitive to forces applied at frequencies near a natural frequency.

Theoretically if you are in the resonance condition energy applied in each cycle would combine and the response amplitude would approach an infinite amplitude over time. In real machines the system becomes non-linear and/or damping limits the response. Damping removes energy from each cycle and moderates the response amplitude. At resonance the only force limiting the vibration amplitude is damping. Damping can be a very effective way to limit vibration amplitudes at resonance. Some systems like journal bearings have significant damping in the oil film, but many structures have little internal damping and have very high resonance amplitudes. The primary issue with damping is it may be difficult to apply to many systems, especially structures. Amplification Factor (AF) describes the ratio of the resonance response amplitude of a force applied to that of the vibration response amplitude well below the natural frequency, see Eq.(2). It describes the gain (Q) of a given natural frequency when the forcing frequency is at the natural frequency and the AF when the forcing frequency is at any other frequency. The AF amplifies or diminishes the response amplitude to a given force by some constant value (AF). For a linear system doubling the force will result in twice the response amplitude, but note it is also multiplied by the AF.

So you can see that force, stiffness, and AF actually control the vibration response of a system or machine. When near a natural frequency the AF will become the dominant factor unless there is a large amount of damping. Near resonance the AF is strongly affected by damping, see [Ref 2] & Figure (2).

$$AF = \frac{1}{\sqrt{\left\{\left[1-\left(\frac{\omega}{\omega_n}\right)^2\right]^2 + \left[2\xi\left(\frac{\omega}{\omega_n}\right)\right]^2\right\}}} \quad (2)$$

ω = frequency of applied force
ω_n = undamped natural frequency
ζ = damping ratio

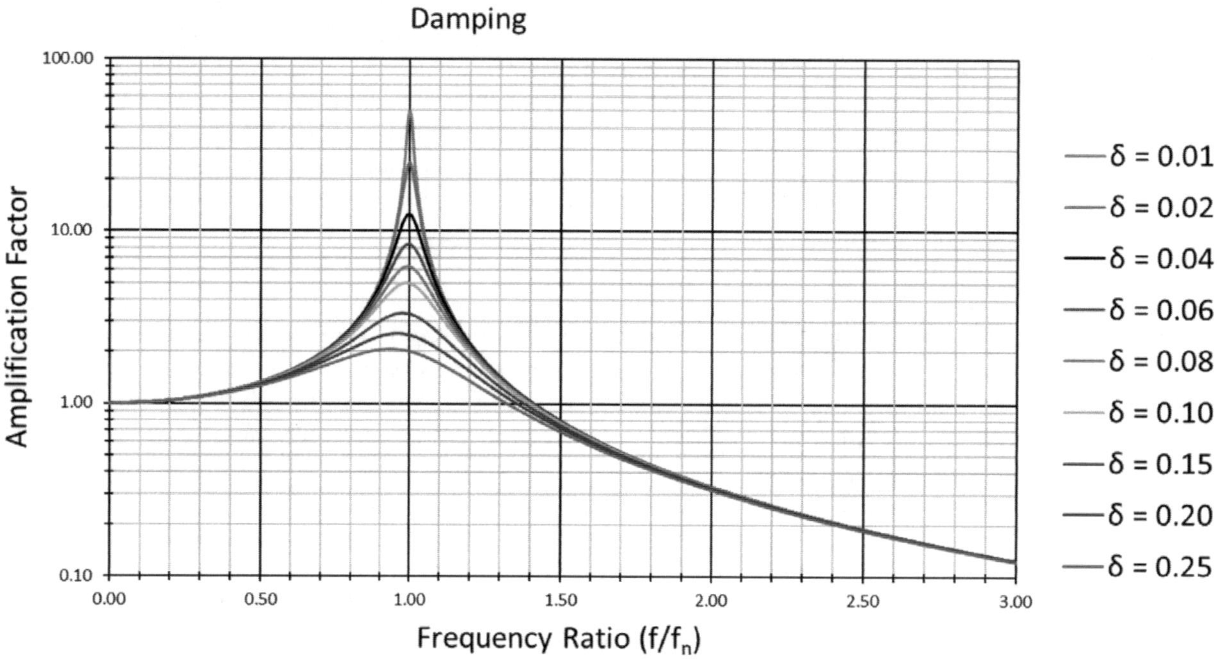

Figure 2: Amplification Factor Plot

If we combine the low frequency Hooke's Law concept with the AF concept we arrive at a comprehensive description of vibration response (X) to forces (F) applied.

$$X = F/K \ * \ \frac{1}{\sqrt{\left\{\left[1-\left(\frac{\omega}{\omega_n}\right)^2\right]^2 + \left[2\xi\left(\frac{\omega}{\omega_n}\right)\right]^2\right\}}} \quad (3)$$

HEARSAY:

It is important to consider reported information, experience, observations, and opinions from people who operate and maintain the machine being analyzed for a vibration issue. However, caution should be exercised as this information can be incorrect or incomplete and it can steer the analysis away from a good conclusion. Keeping a narrow focus on what is anticipated can cause errors and poor outcomes. All four case studies presented in this paper demonstrate some degree of this narrow focus on the part of the author with the resulting less than optimum outcome or uncertainty. It is important to use all available information to make a good analysis. Every fault or dynamic force will create a vibration response; however, it is the author's opinion that many faults do not create meaningful vibration responses. Meaningful is defined here as large enough and clear enough to be measured and noticed in practical vibration measurements. Observing the vibration signature change after the fact is not the same as predicting a fault before failure. The point is some issues are better identified with data other than vibration such as pressure, flow, temperature, sound, visual inspection, pump curves, known characteristics, etc. If a machine fails multiple times it is often useful to install instrumentation during a repair to collect better data. For example accelerometers could be installed on the below grade structure of a vertical pump while being repaired. Finite element modeling has been used to supplement vibration data on vertical pumps; the below grade structure often can't be measured so the model can help interpret the available vibration data above grade.

ANSI PUMP WITH HIGH VANE-PASS VIBRATION AMPLITUDES (CASE STUDY 1):

The owner reported high pump and piping vibrations on several of 12 identical pumps, see Figure 3. Several attempts to add pipe supports were unsuccessful; the support welds broke after a short time of operation. A consultant provided modal animations, see Figures 5 & 6, that indicated horizontal natural frequencies were near vane-pass frequency on some of many identical pumps. A site test was conducted with the process unit operating so modal testing was not possible; the vibration amplitudes and noise were too high for impact testing. Operating data was inconclusive but was consistent with a resonance issue at vane-pass frequency. The pump bearing housing horizontal amplitudes were more than three times higher than the other directions, see Figure 4. Often natural frequencies will vary from horizontal to vertical because of different stiffness in the horizontal and vertical directions. The stiffness major and minor axes could be oriented in any orthogonal orientation but often they are oriented horizontal or vertical.

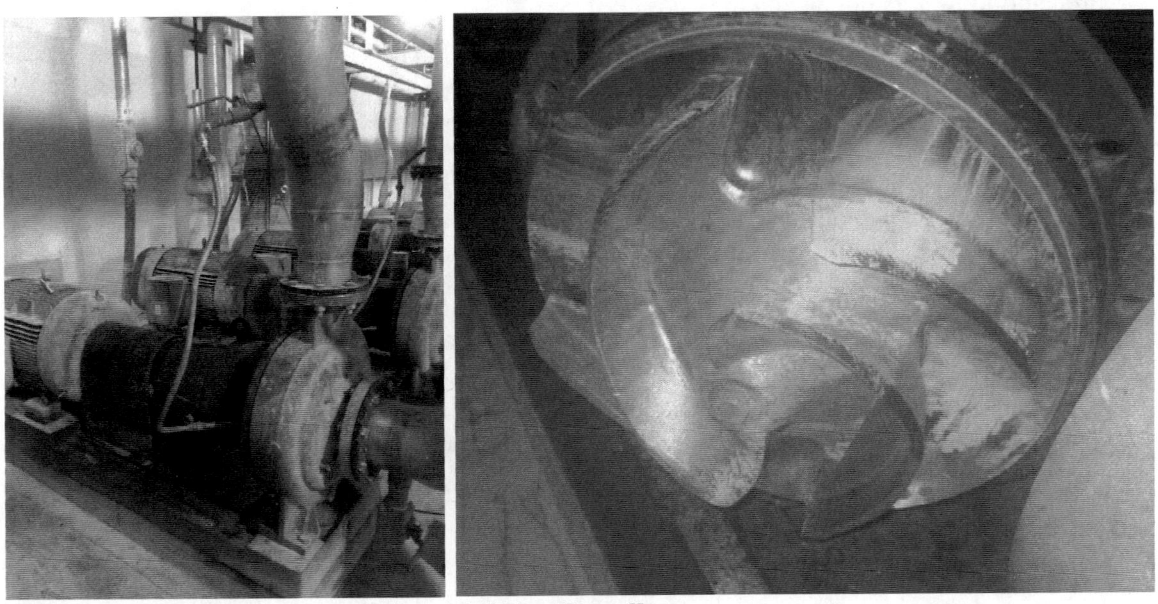

Figure 3: OH1 Pump Installation / 2-4 Van Impeller

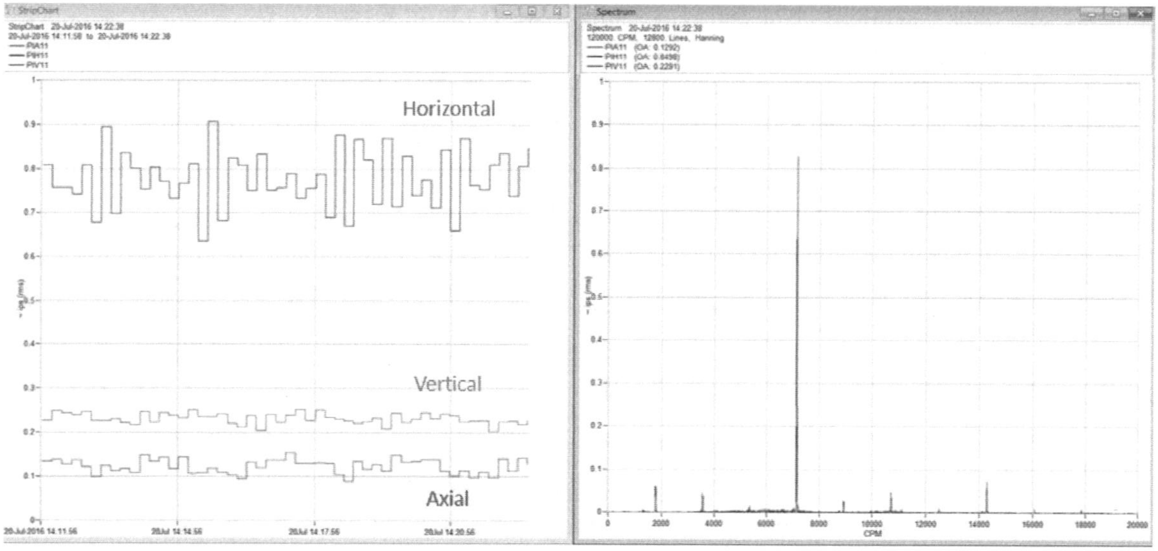

Figure 4: Overall Trends & FFT Spectrums for Axial, Horizontal, Vertical Directions

It appeared to be a simple case of resonance amplification. To reduce time and cost modal animations were used to inform a trial-and-error modification rather than a full finite element model with adjustments to field data. A more rigid bearing housing foot was provided to increase stiffness in a large motion or anti-node location, see Figure 7. A second site test was scheduled at the end of a process unit outage so the pump would be off for modal testing and could then be monitored for a process unit start up. The stiffer foot design provided a moderate improvement; vibration amplitudes were reduced by 25%, Figures 8 & 9. This reduction in vibration amplitude was not adequate.

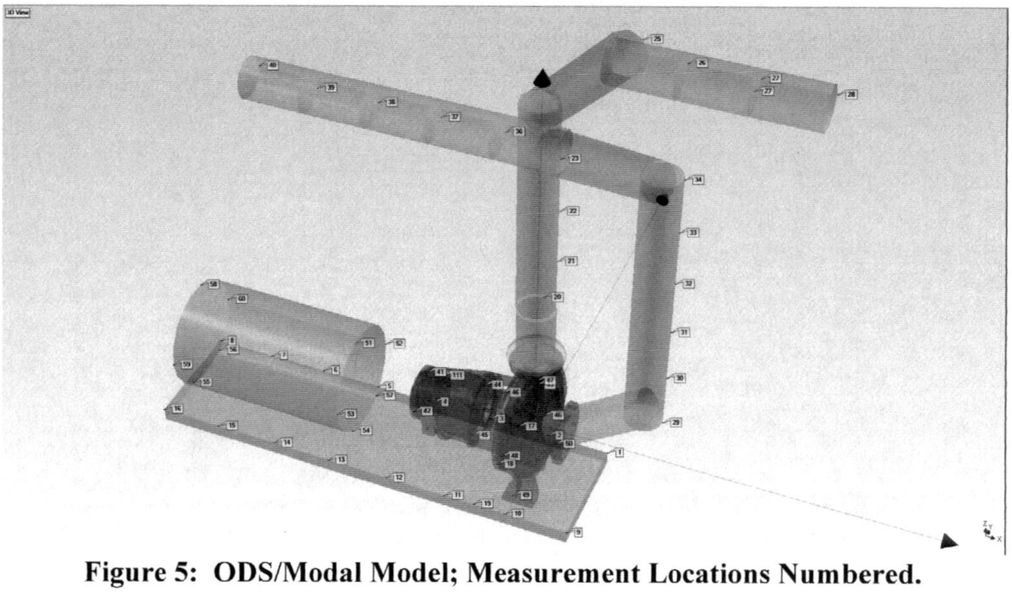

Figure 5: ODS/Modal Model; Measurement Locations Numbered.

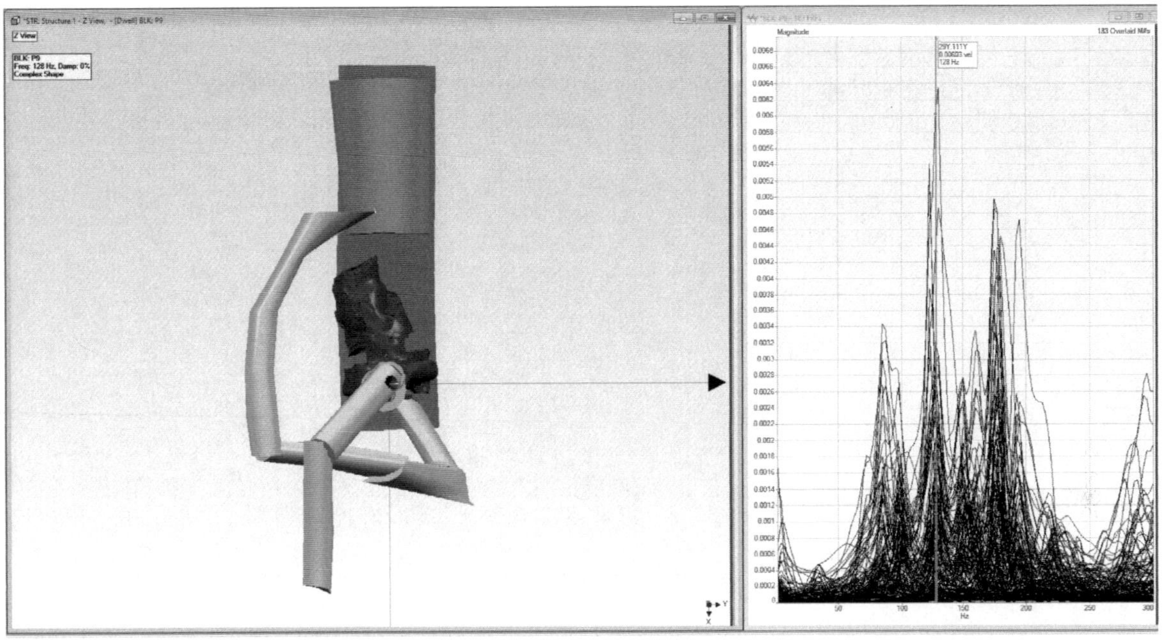

Figure 6: 128Hz Mode Shape

Figure 7: Original Foot Design / Stiffer Foot Design

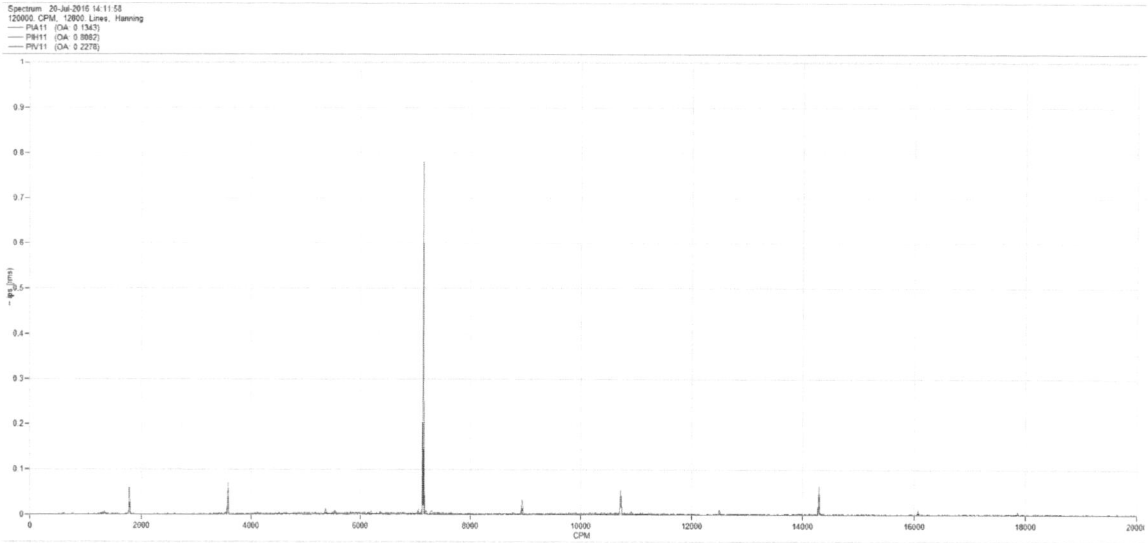

Figure 8: Original Foot FFT Spectrum

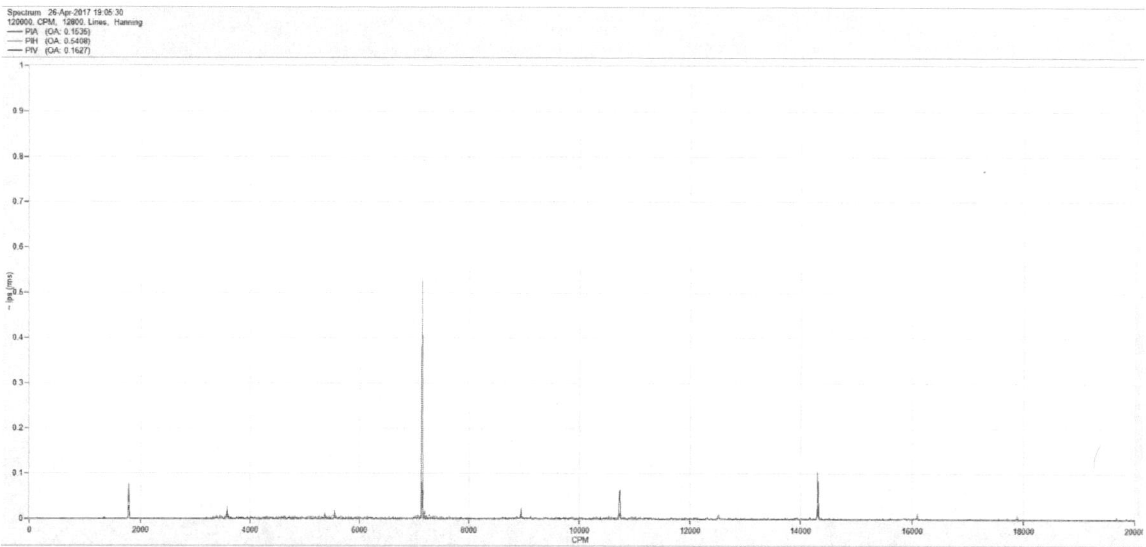

Figure 9: Stiffer Foot FFT Spectrum

During the process unit start-up an unusual overall trend was observed, see Figure 10. Initially the pump vibrations were very low but after approximately 5 minutes of operation at constant speed the vibration amplitude on the pump drive end (DE) horizontal location increased from 0.1ips rms to 0.5ips rms. Discussion with the owner indicated the only change was the water used for start up was replaced with a thick product with higher viscosity (up to 90cP); it is assumed higher resistance caused the pump flow to reduce well below the BEP. Vane-pass data on a similar factory tested pump indicates a similar increase in vane-pass vibration at low flow rates.

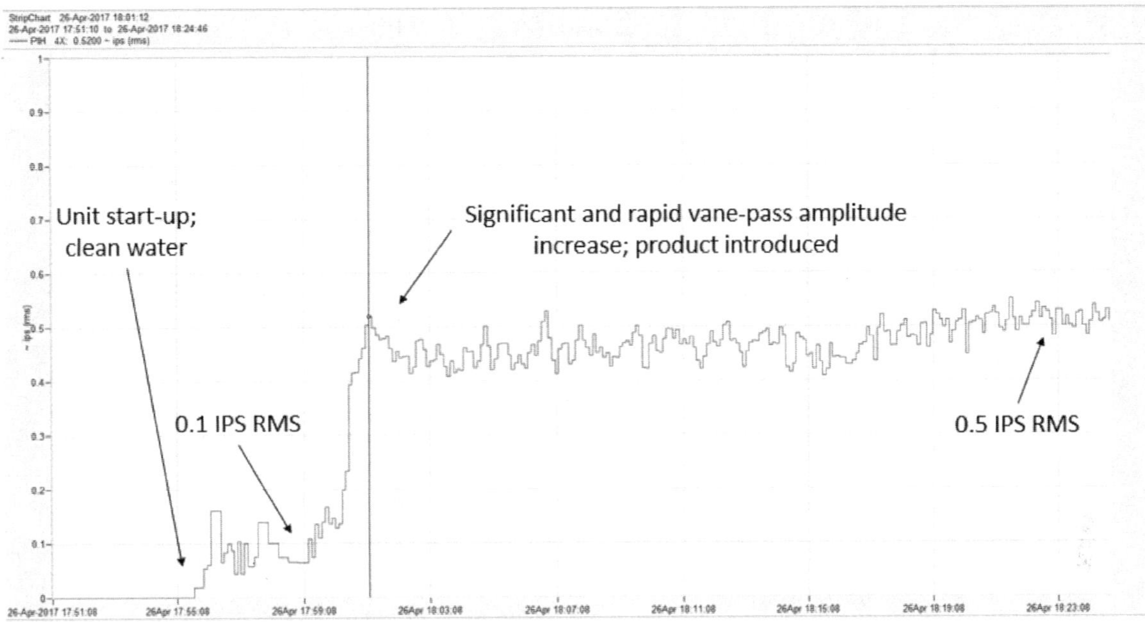

Figure 10: Drive End Horizontal (DEH) Overall RMS Vibration Trend

The stiffer foot increased stiffness (K) and this is expected to have some small effect to reduce motion; however, the natural frequency increase was small, see Figure 11. Without a full adjusted finite element model including the piping it is difficult to know why the natural frequency increase was very small. Modeling may have found a more effective location to increase stiffness; possibly on the piping and the pump. The natural frequencies were dramatically changed when the fluid in the piping was drained, see Figure 12. This demonstrates that modal testing should be done with the system or machine in the same condition as when operating. Fluid mass in pumps and piping can have a significant effect on modal mass and the natural frequencies of each mode. The effect of fluid mass depends on the mode shape and if the fluid mass has high participation for that mode shape.

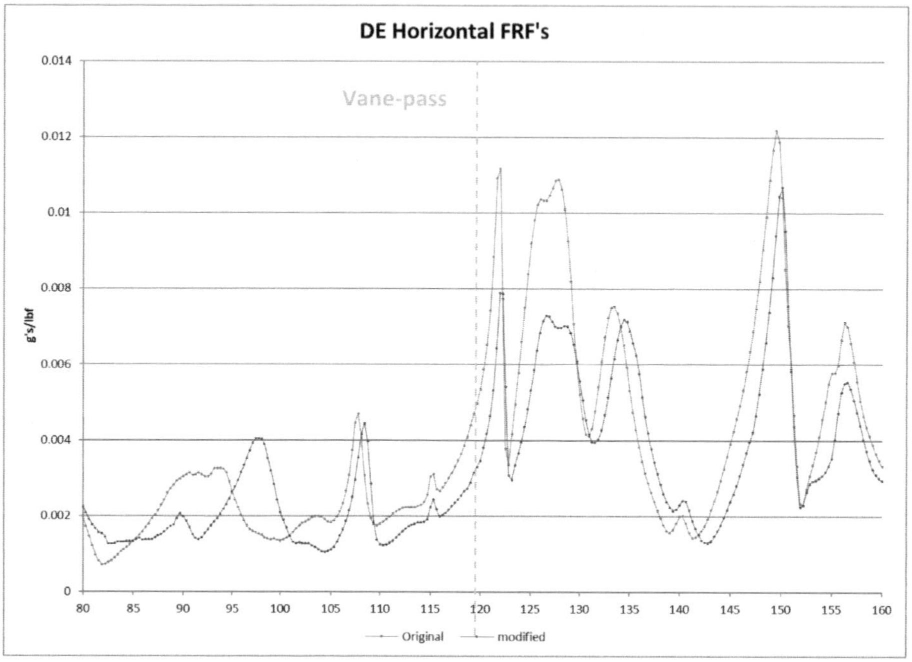

Figure 11: DEH FRFs; Red = Original Foot, Blue = Stiffer Foot

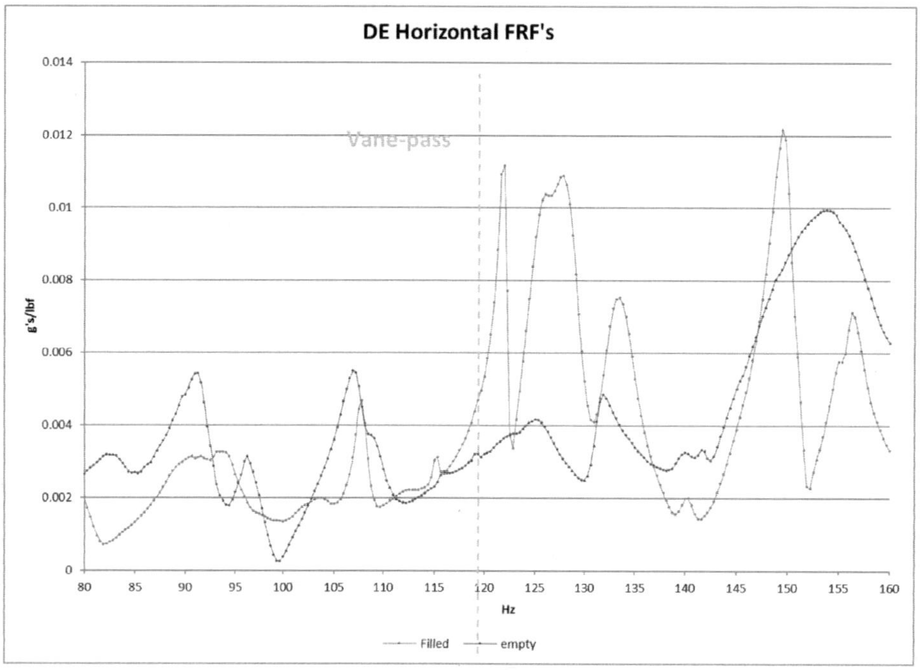

Figure 12: DEH FRFs; Red = Piping Filled With Water, Blue = Empty Piping

Previous factory test data for this general pump type indicated a large increase in vane-pass vibration amplitude at low flow relative to the best efficiency point (BEP) flow rate is a normal characteristic, see Figure 13. This information changed the author's focus from **natural frequency** changes to **force** reduction. Clearly **force** (F), **stiffness** (K), and **amplification factor** (AF) were all involved, but the force seems to be the dominant factor regarding high vane-pass vibration.

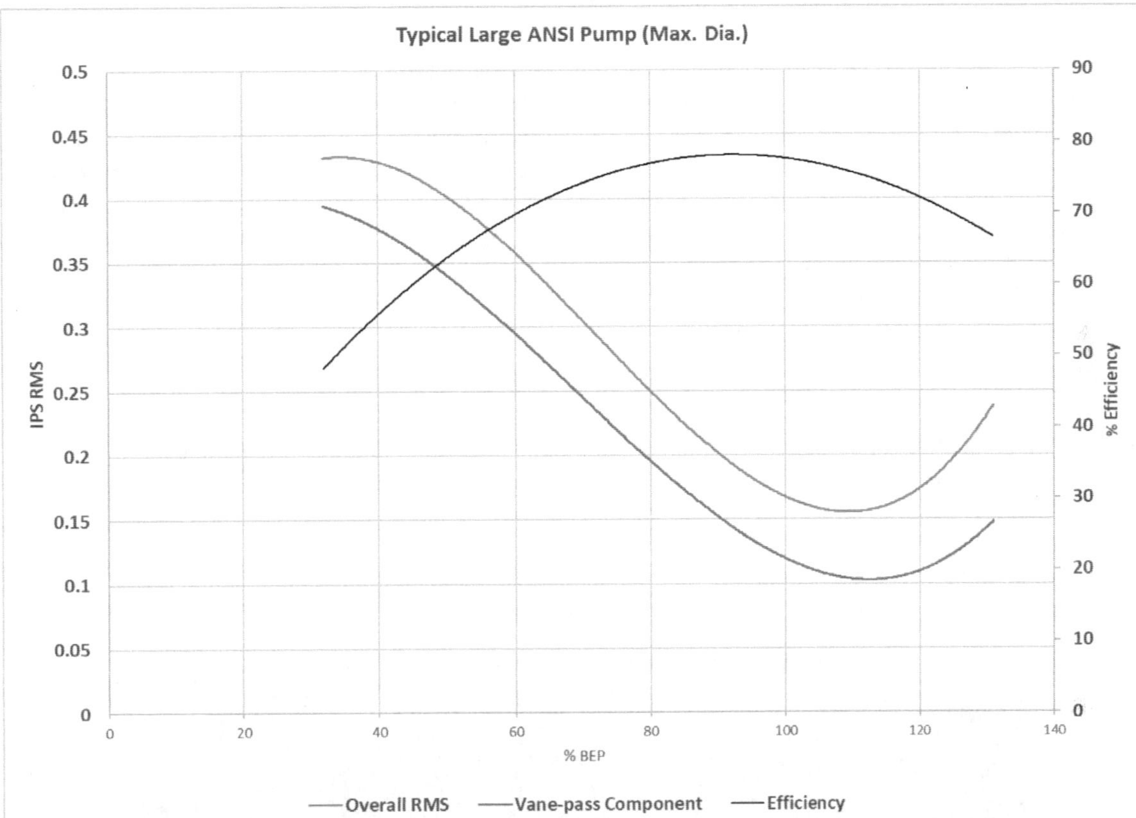

Figure 13: Vibration Amplitude V.S. Flow Rate For A Typical Large OH1 Pump

This example shows how complex the analysis process can be. There is always pressure to reduce time and cost. There are many times when the cost of extensive modal data and an adjusted FEM are not justified. In many cases an informed guess regarding stiffness or mass changes can be successful; the guess is usually informed by modal or ODS data and mode shapes. An accurate and adjusted model of this pump, base, motor, and piping would have been costly. Sufficient bracing on the pump and piping would have also been costly. The start-up vibration data clearly indicates controlling the operating condition would be the most effective solution. Process instrumentation and controls are needed to keep the pump near the BEP; a variable frequency drive (VFD) may also be needed. If the pump can't be operated near BEP a different pump or other costly solution would be needed. Damping would probably not be effective in this case because the natural frequency is only a minor factor and frequency separation margin was increased with the stiffer foot. The author's opinion is the operating parameters need to be controlled for this pump application to reduce vibration amplitudes to a reasonable level. The relatively high vane-pass vibrations on the pump may not have a significant effect on the pump reliability and life, but high piping vibrations could cause piping damage. Vane-pass forces are not related to fault or defect and moderate amplitudes should have little effect on pump reliability. If vane-pass vibration amplitudes are excessive or if other system components are damaged, then corrective action should be taken. The purchaser has to balance cost and performance. Achieving very low vibration at low flow for this hydraulic application may require a more robust design with a large cost increase.

Regarding the original vibration analysis, it would have been beneficial to take extensive static pressure and dynamic pressure data. Understanding the normal characteristics of this machine type may have focused the analysis on the process conditions (flow rate/ % BEP) earlier in the process. A more extensive modal study, ODS study, and finite element modeling could provide more insight, but this would be expensive and may not provide a practical solution. A significant challenge is finding the right balance between testing/modeling cost and useful data to solve the vibration problem. In some cases doing nothing and accepting moderate vibration amplitudes at vane-pass frequency is a justified financial decision. The author has seen many pumps run for decades with moderate vane-pass vibration amplitudes and no significant or measurable increase in maintenance costs or reduction in mean time between repairs (MTBR).

BB3 PUMP ACOUSTIC & MECHANICAL RESONANCE (CASE STUDY 2):
A four stage BB3 pump in a reverse osmosis application experienced high vane-pass vibration amplitudes, see Figure 14. Shaft turning speed is 2980rpm and the discharge pressure is 3000psi. Seven vane impellers created high vane-pass vibration amplitudes on the bearing housings at 348Hz. Vane-pass amplitudes under some circumstances exceeded 20mm/s(0.79ips RMS). The maximum acceptable amplitude for operation is 6.5mm/s(0.25ips RMS).

Figure 14 High Pressure Seawater Pump

The owner's overall vibration trend data indicated some large amplitude variations while at constant speed and flow rate, see Figure 15. For a centrifugal pump the vane-pass amplitudes typically don't vary if the speed, flow rate, and product properties are constant. The initial assumption was that an acoustic resonance may be occurring in the pump cross-over passage. Preliminary simple calculations estimated the cross-over acoustic natural frequency may be near vane-pass frequency. This one-dimensional acoustic model is based on sound waves travelling in a fixed diameter pipe; therefore, some error is expected. The actual cross-over has varying cross-sectional area and curvature. The simple acoustic calculation is often used to determine if further testing is needed or if acoustic resonance can be ruled out as a possible issue. The fluid, temperature, and gas entrainment affect the sound speed and accordingly the acoustic natural frequency. The sound speed is significantly reduced by even small amounts of gas entrainment or vapor.

Dynamic pressure sensors were installed as close to ½ the cross-over length as possible; this is the location of the pressure wave anti-nodes, see Figure 16. The anti-node locations are where the dynamic pressure standing wave is maximum and minimum. There were no pressure taps at exactly ½ the length so the actual peak dynamic pressure may be higher than that measured. Site testing found very high pressure pulsation amplitudes in the cross-over at 25psi pk; this is 11% of the pressure generated by each impeller, see Figures 19-22. A good design should have a dynamic pressure (pk) less than 2% of stage differential pressure. Offline and operating broad band dynamic pressure energy appeared to excite a cross-over natural frequency near 7X running speed frequency or vane-pass frequency. Note the offline test using a fresh water external flush through the cross-over had a different sound speed and pressure than the operating fluid, see Figure 23. The apparent offline natural frequency was slightly above the 7X frequency. The variable amplitudes during steady operation, approximate acoustic frequency calculations, high dynamic pressures, and dynamic pressure noise floor signatures confirmed acoustic resonance was creating large pressure pulsation and vibration amplitudes at vane-pass frequency.

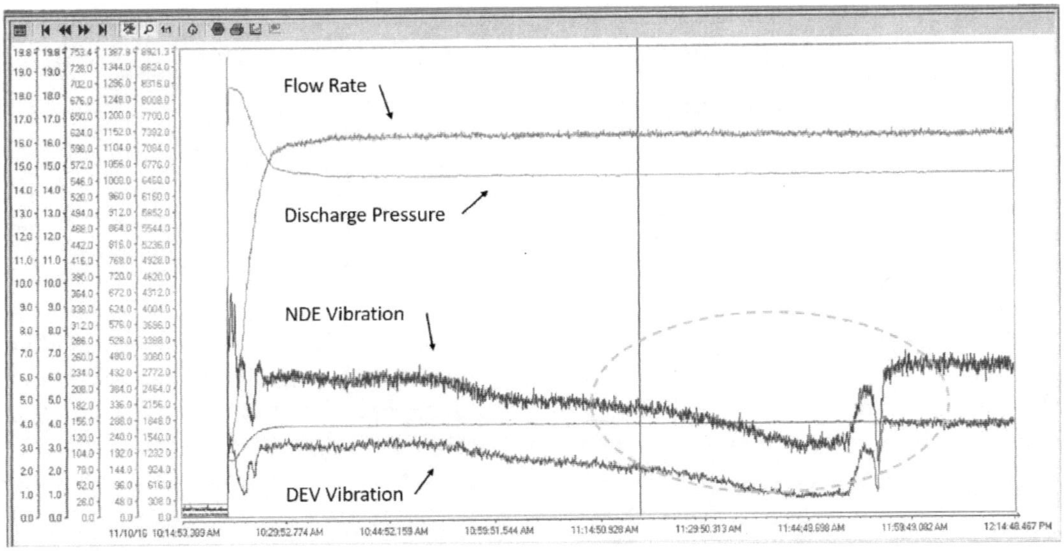

Figure 15 Owner's DCS Trends During Start-Up

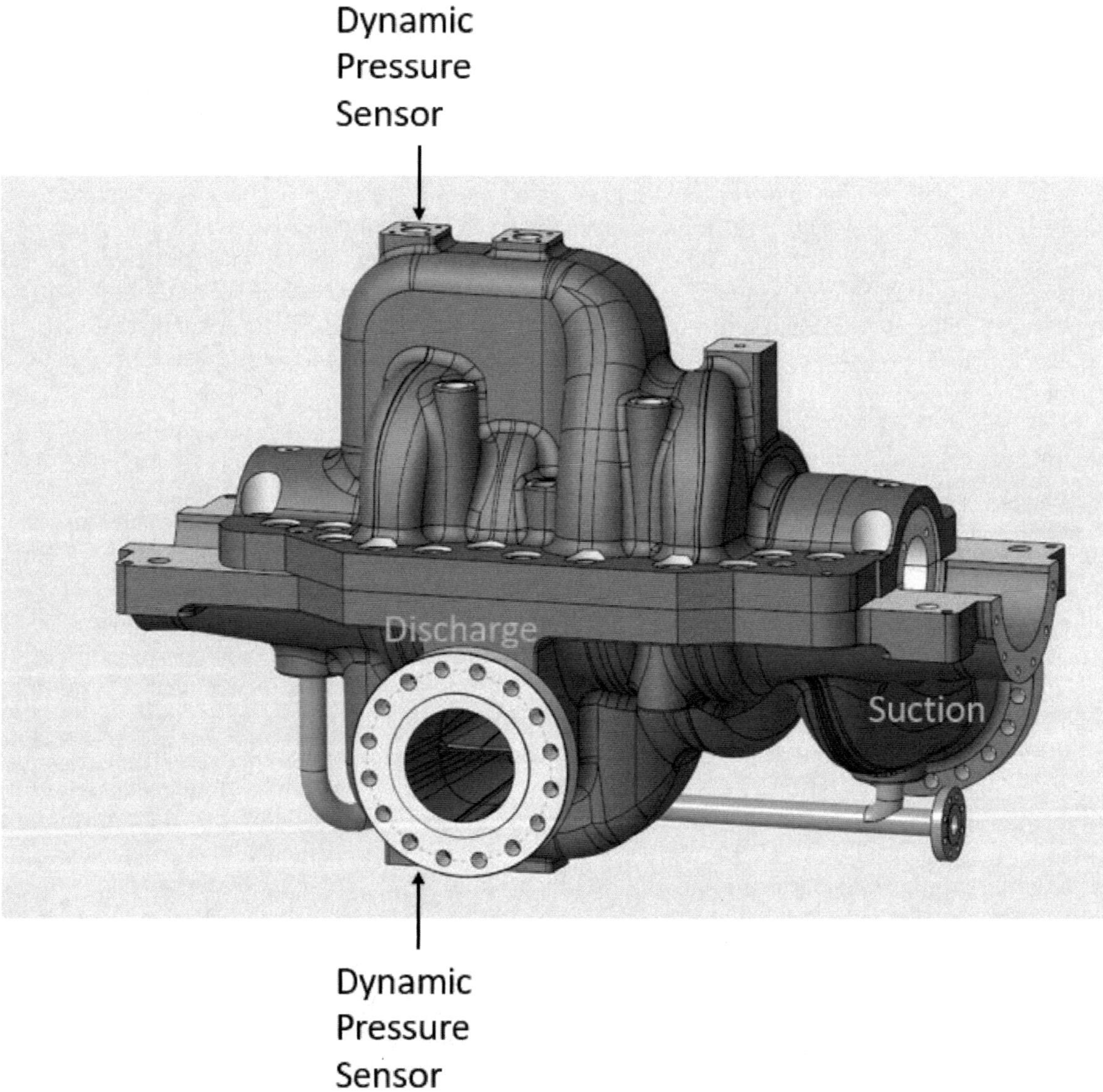

Figure 16: Pump Case Showing Dynamic Pressure Sensor Locations.

The cross-over circuit directs flow from the 2nd stage impeller to the 3rd stage impeller; there is an upper and lower cross-over because flow is collected by two cut-waters (dual volute). Two volutes are used to balance radial hydraulic forces. The cross-over circuit is essentially a circular path with the second stage impeller discharging flow into it and the third stage impeller receiving the flow, see Figure 17. Pressure waves will naturally propagate back and forth from one cross-over to the other. A standing wave will create a positive pressure (compression) in one cross-over while a negative pressure (expansion) is created in the other cross-over. The wave will then oscillate and reverse this process; the positive pressure will move to the opposite cross-over and the negative pressure will move accordingly.

Flow Path

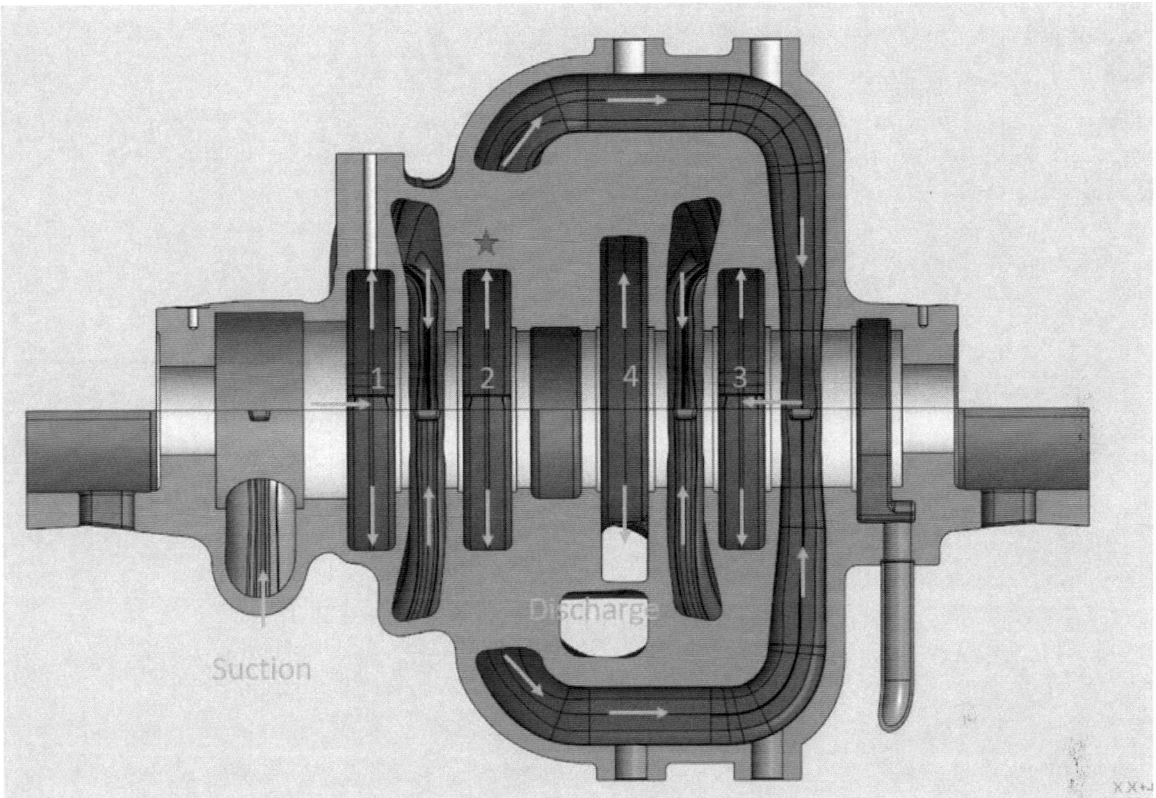

Figure 17: Cross-Over & Impeller Stage Flow Path

A simplified sketch shows this process of alternating pressures in the cross-over created by a standing wave, see Figure 18. This standing wave oscillates at the acoustic natural frequency of the cross-over circuit. This dynamic effect is inherent to the cross-over geometry (primarily length), fluid sound speed, and to a lesser extent the wall stiffness or modulus. When the vane-pass frequency is at or near the acoustic natural frequency the pressure oscillations will be greatly amplified. There is little damping to moderate the resonance response amplitude; very high dynamic pressures may be created. High pressure pulsations can create large vibration amplitudes on the pump casing and bearing housings.

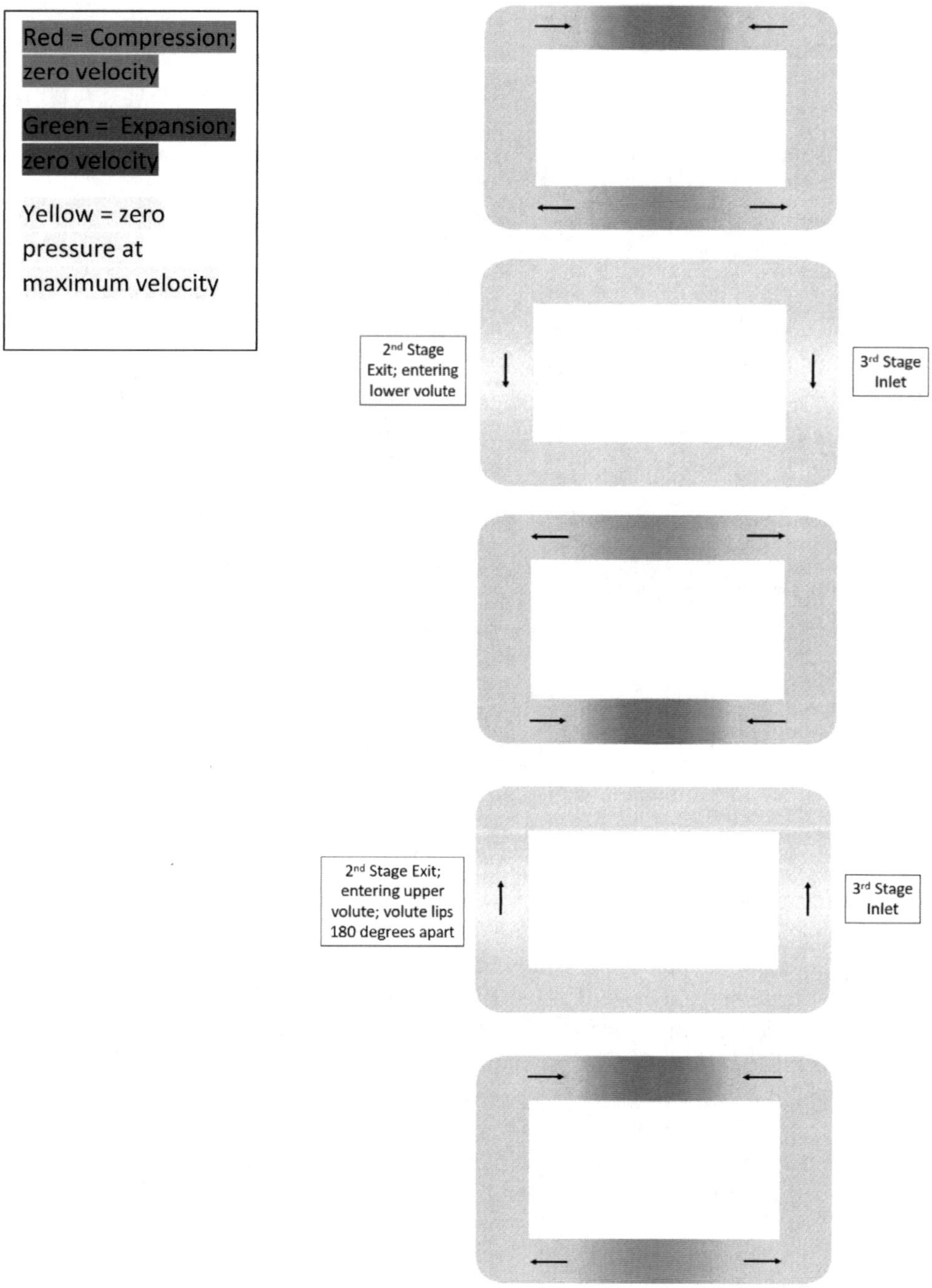

Figure 18: Cross-Over Acoustic Natural Frequency Dynamics

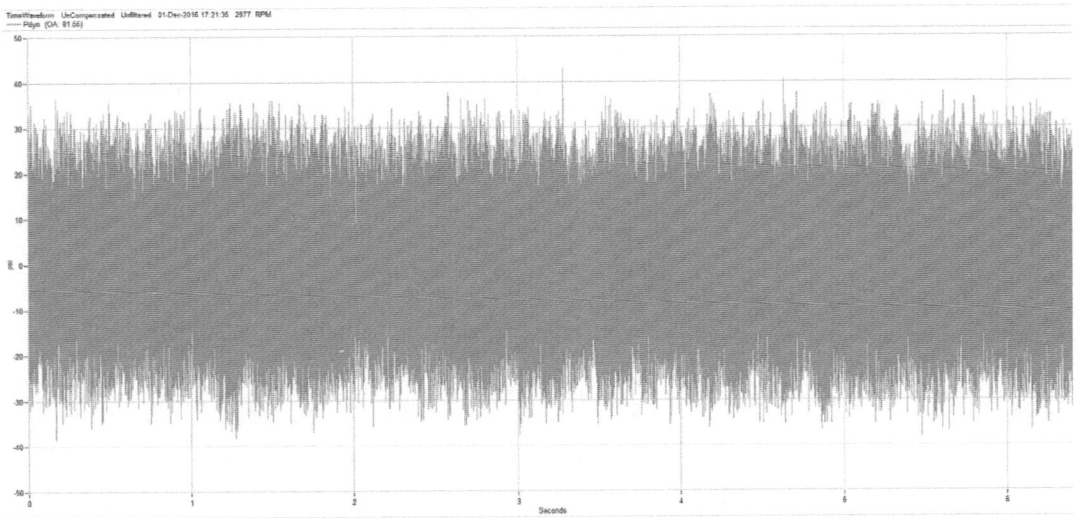

Figure 19: Upper Cross-Over Dynamic Pressure (time waveform); full time sample plotted

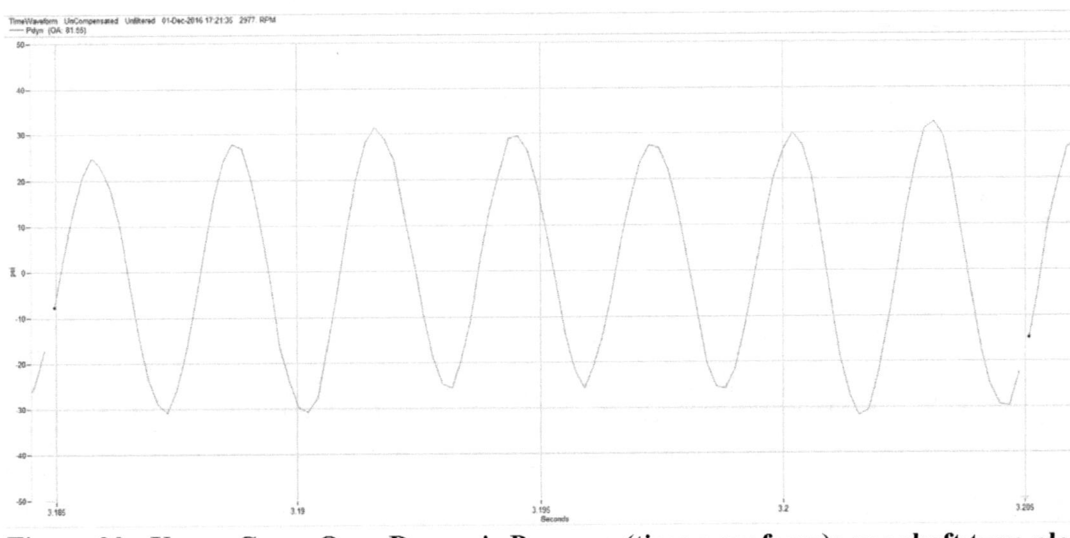

Figure 20: Upper Cross-Over Dynamic Pressure (time waveform); one shaft turn plotted

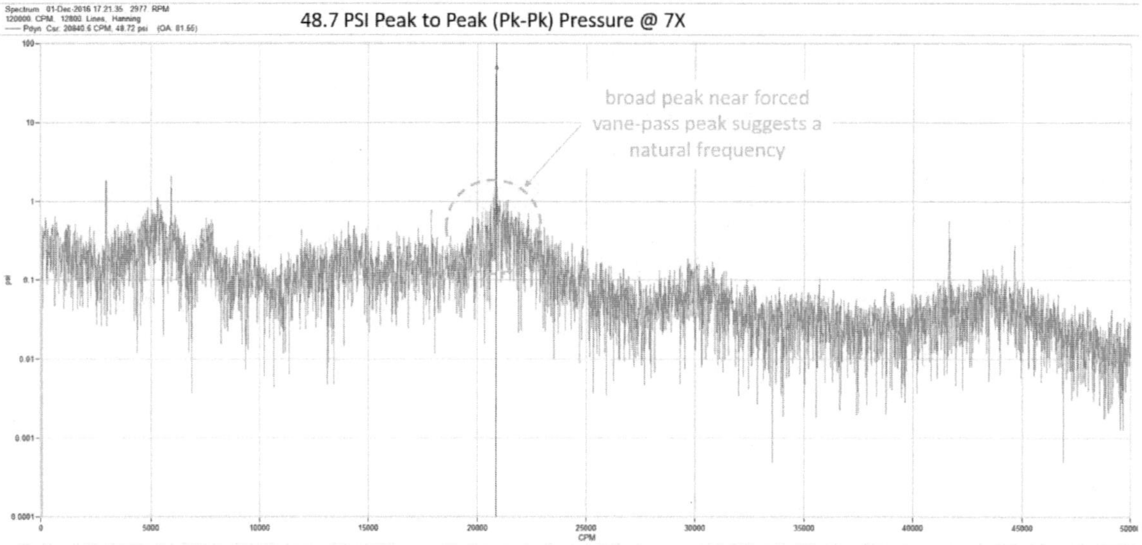

Figure 21: FFT Spectrum Of Upper Cross-Over Dynamic Pressure

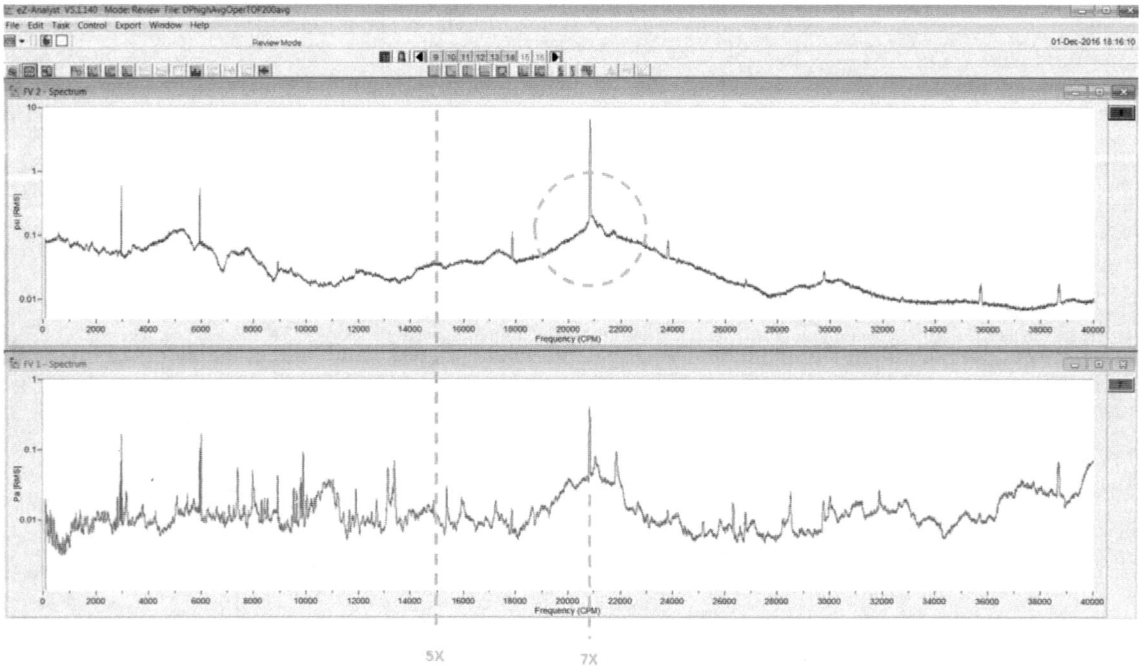

Figure 22: 200 Average Dynamic Pressure FFT And Sound FFT With Pump Operating

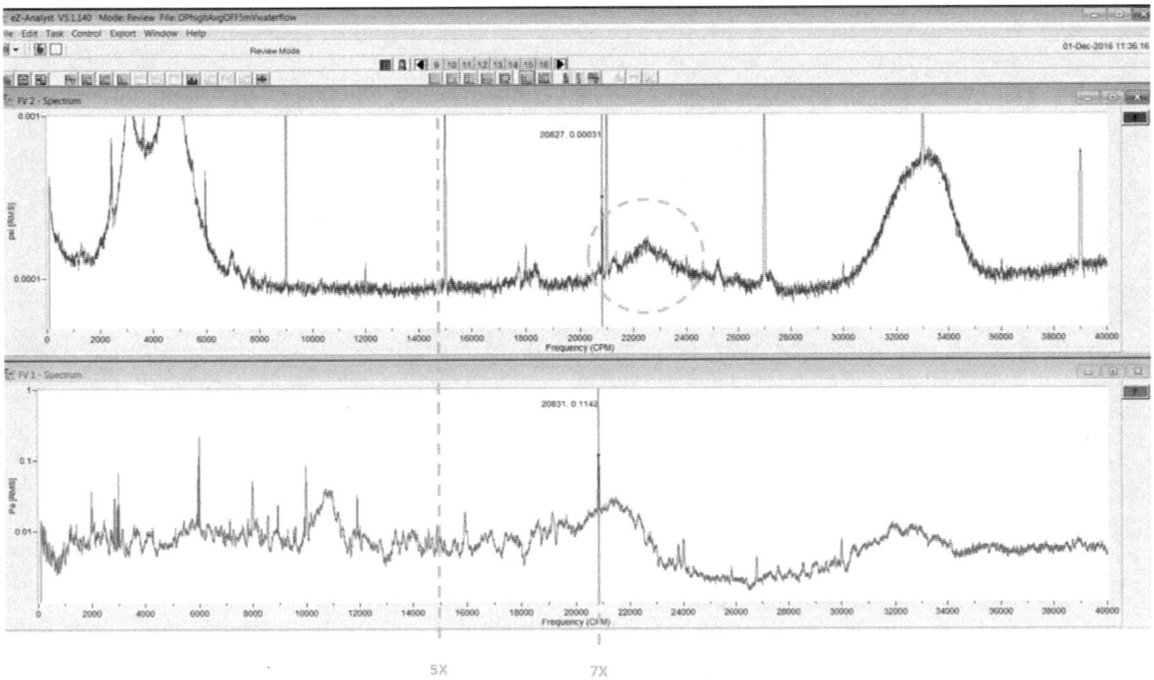

Figure 23: 200 Average Dynamic Pressure FFT And Sound FFT With Pump Not Operating; External Fresh Water Flush Through Cross-Over

Structural natural frequencies were identified near 7X (vane-pass frequency) during the site test, see Figure 24. The initial recommendation was to replace all 4 impellers with a 5 vane design; this would address the acoustic resonance creating large pressure **forces** and possible mechanical resonance amplification due to structural **natural frequencies** near vane-pass frequency.

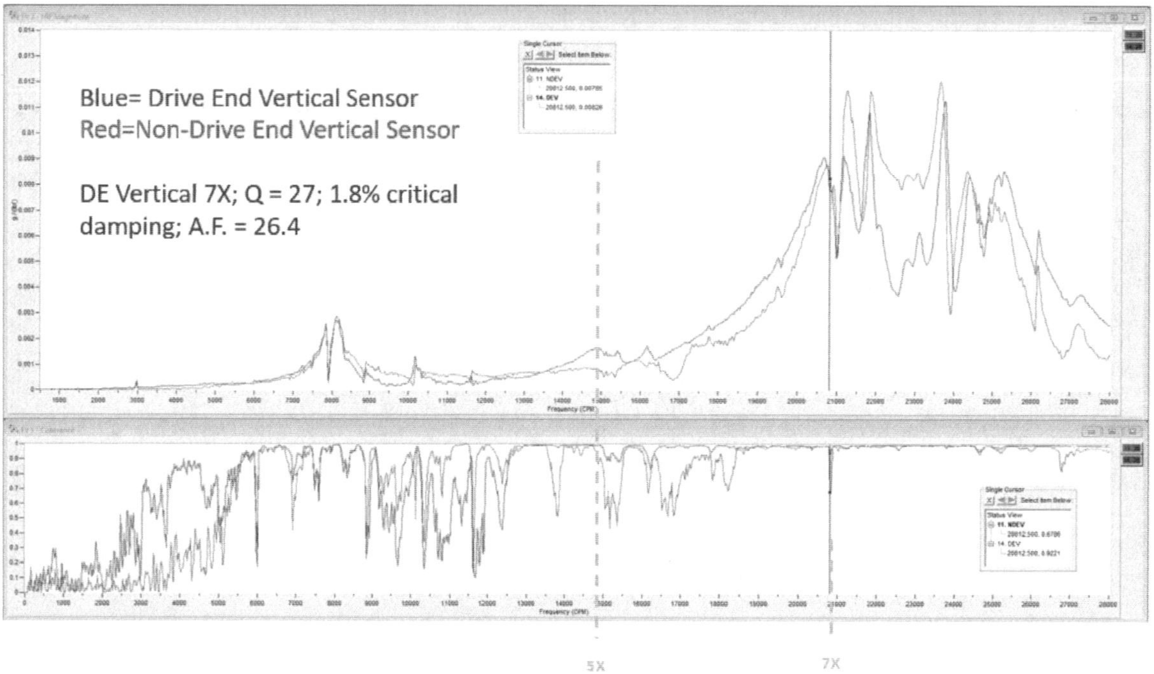

Figure 24: NDE Vertical Impact Test FRFs

The operating deflection shape (ODS) at 7X and mode shapes near 7X indicate the natural frequencies near 7X involve both bearing housings and the pump case in a bending mode, see Figure 25. Trying to shift the structural natural frequencies sufficiently higher or lower than 7X would be impractical since the pump case structure is involved and this is not easily changed.

Figure 25: ODS @ 7X (Vane-pass frequency)

There was a desire to reduce costs and a decision was made to only change the 2nd stage impeller. Theoretically only the stage discharging into the cross-over (2nd stage on this pump) should provide a significant force to excite the cross-over acoustic natural frequency. Only changing the upstream impeller was reported to have significant improvement in other cases of cross-over acoustic resonance. The assumption on this pump was the mechanical resonance amplification would not be excessive if the **force** from the acoustic resonance (high pressure pulsation) was reduced. Only changing the 2nd stage impeller to a 5 vane design did not provide a significant reduction in vibration amplitudes. Neglecting the structural **natural frequency** provided less than optimum results. Moving the structural natural frequencies well above or below the 7X vane-pass frequency was not practical since the mode shape involved the pump case structure; the case structure cannot be easily modified. The motor is fixed speed so the only way to change the vane-pass frequency is to change the number of vanes. Note that moving the natural frequencies would also be complicated by the new 5 vane 2nd stage impeller because that frequency range would not be available; stiffness could not be reduced to change the natural frequencies because this range would have the 5 vane impeller frequency as an excitation. The only practical solution was to replace all four impellers with a 5 vane design, see Figure 26. Vibration amplitudes at all locations and directions were significantly reduced and the vibration issue was completely resolved, see Figures 27 & 28. It is interesting that after the high 7X vane-pass amplitude issue was resolved a band of energy can be seen in the noise floor around the 7X frequency, see Figure 29; random flow energy is being amplified by the structural **natural frequency** (the natural frequency was not changed, the forcing frequency was moved lower to 5X shaft speed frequency). This pump still "likes to vibrate" or readily absorbs energy at the 7X frequency but now there is little energy to provide excitation **force** at 7X frequency. Reducing the number of vanes will typically cause a few percent reduction in total dynamic head TDH. Hydraulic engineers may be able to modify the vane geometry or increase the impeller diameter to compensate for the TDH reduction if required. NPSHR is usually lowered or improved by reducing the vane number.

Figure 26: Rotor with all 5 vane impellers

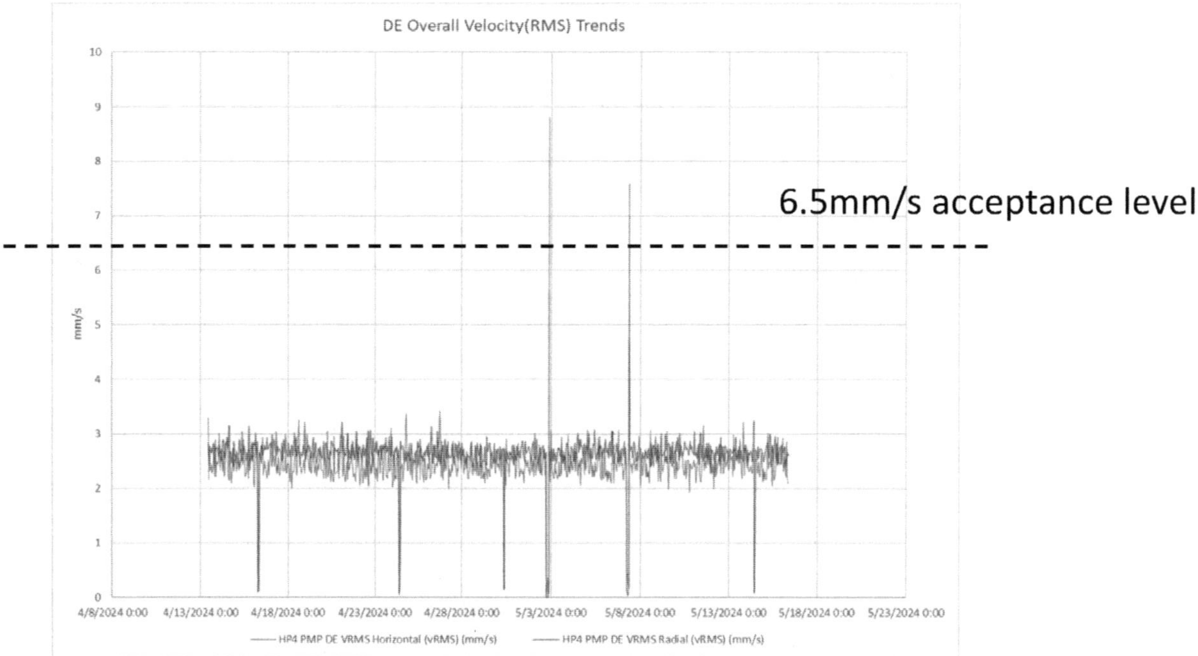

Figure 27: Drive End Wireless Accelerometer Overall RMS Trend Data After All 4 Impellers Were Replaced With A 5 Vane Design.

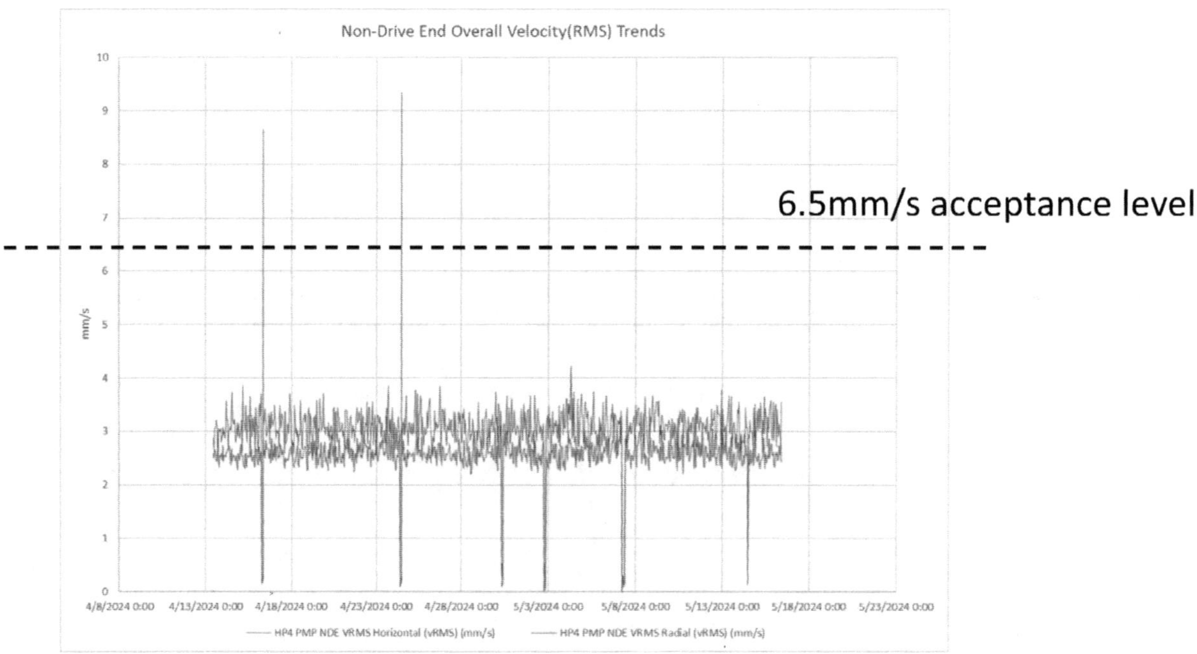

Figure 28: Non-Drive End Wireless Accelerometer Overall RMS Trend Data After All 4 Impellers Were Replaced With A 5 Vane Design.

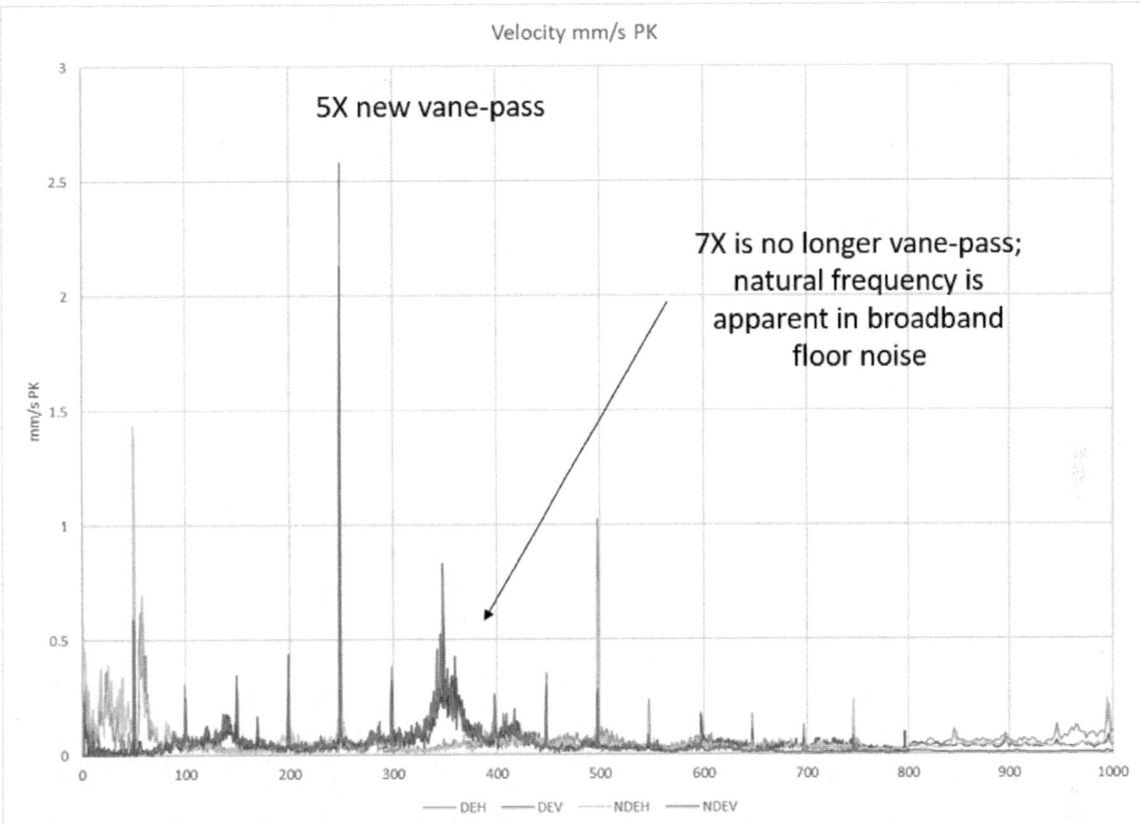

Figure 29: Wireless Accelerometer FFT Spectrum Data After All 4 Impellers Were Replaced With A 5 Vane Design.

In this example all the information was available, but the author did not have enough experience with the factors involved to insist on the best solution. Cost pressures became a focus that suppressed the significance of a potential mechanical resonance problem. The acoustic resonance was also a distraction that contributed to focusing on **force** and neglecting the structural **natural frequency** amplification. It is true that resonance can be less important if the **forces** are very low or if the **damping** is relatively high. In hindsight it is clear the vane-pass forces on a high energy seawater pump should be relatively high and these natural frequencies have low **damping** and a correspondingly high AF.

The drive end vertical accelerance was the largest frequency response function (FRF) at .00879 g/lbf. The drive end vertical AF was 26.4 and the damping was 1.8% of critical. With an amplification factor of 26.4 it should have been reasonable to assume the vibration response of the remaining 7 vane impellers would be significant. An A.F. of 26.4 means the vibration response will be 26.4 times higher due to resonance than it would be without resonance.

OH2 PUMP DRIVE COUPLING FAULT (CASE STUDY 3):

After a failure and overhaul an OH2 pump in hot oil service, see Figure 30, experienced a large 1X vibration amplitude near 0.25ips rms. The owner reported acceptable motor/pump shaft alignment and good rotor mass balance. It was also reported that during a second overhaul the impeller was replaced to rule out hydraulic imbalance. The identical spare pump did not have high vibration amplitudes. It is interesting to contrast the vibration amplitudes on this case study to that of cases 1 & 2. In this case (3) the forcing frequency was at 1X shaft frequency and represented a mechanical fault. The forcing frequency in cases 1 & 2 the was at vane-pass frequency and did not represent a fault. Also vane-pass is at higher frequencies and has less displacement for a given velocity amplitude. Vane-pass vibration transmits primarily through the casing from pressure forces. Running speed faults generate forces on the rotor and bearings because of a mechanical issue; this could affect bearing life. The author has more tolerance of higher amplitudes at vane-pass frequency as compared to 1X fault frequencies. In this case .25ips rms at 1X represents a significant **force** and should be corrected.

Figure 30: 3x4 16 OH2 API 610 Pump; 3600rpm

Figure 31: Triaxial Accelerometer On Bearing Housing

Site testing found a large 1X response consistent with mass unbalance or other rotating force. The 1X phase was relatively constant during coast down as the speed varied, see Figure 37. The rotor was shop balanced within tolerance; therefore, other sources of mass unbalance were considered, see Figure 32. The primary component in the FFT spectrum was 1X, but the 2X component was somewhat elevated, see Figure 33. Impact testing found a structural natural frequency near 2X, see Figures 38 & 39. Therefore, the higher 2X amplitudes are likely due to some degree of resonance amplification of the 2X forces. The amplitudes are low at 2X so no changes were recommended to address this order resonance amplification. The pump 1X amplitudes were much higher than the motor 1X amplitudes; 0.24ips rms v.s. 0.075ips rms, see Figures 33 & 34. This indicated the unbalance force was likely on the pump rotor and not the motor rotor. The pump structure orbit was circular which is also consistent with mass unbalance, see Figure 36. The motor structure orbits were small with moderate ellipticity; this shape is likely due to the foundation/motor stiffness (larger vertical stiffness compared to the horizontal stiffness) and is normal for this machine type.

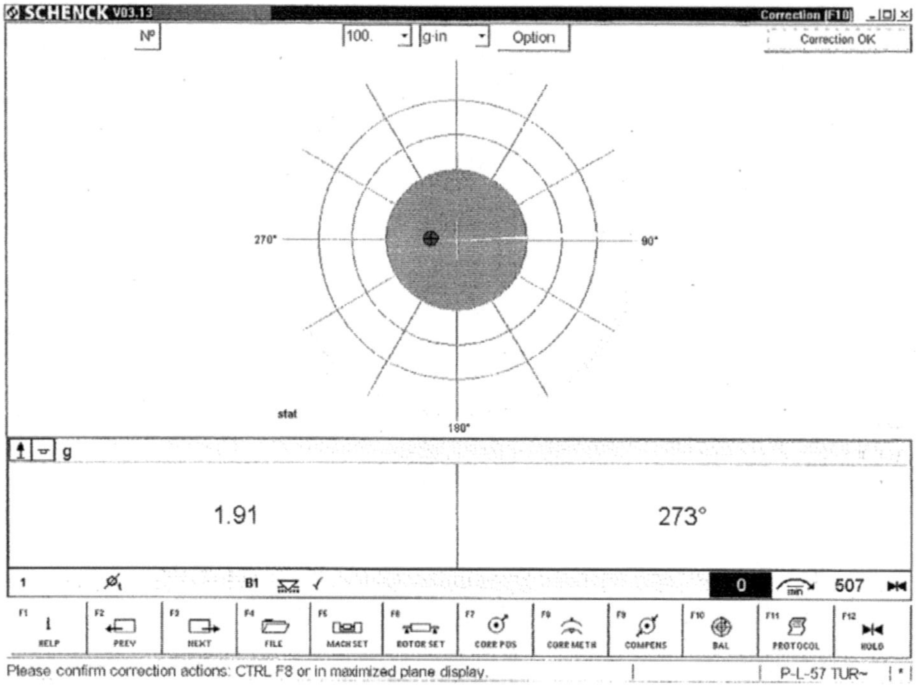

Figure 32: Rotor Final Balance Record

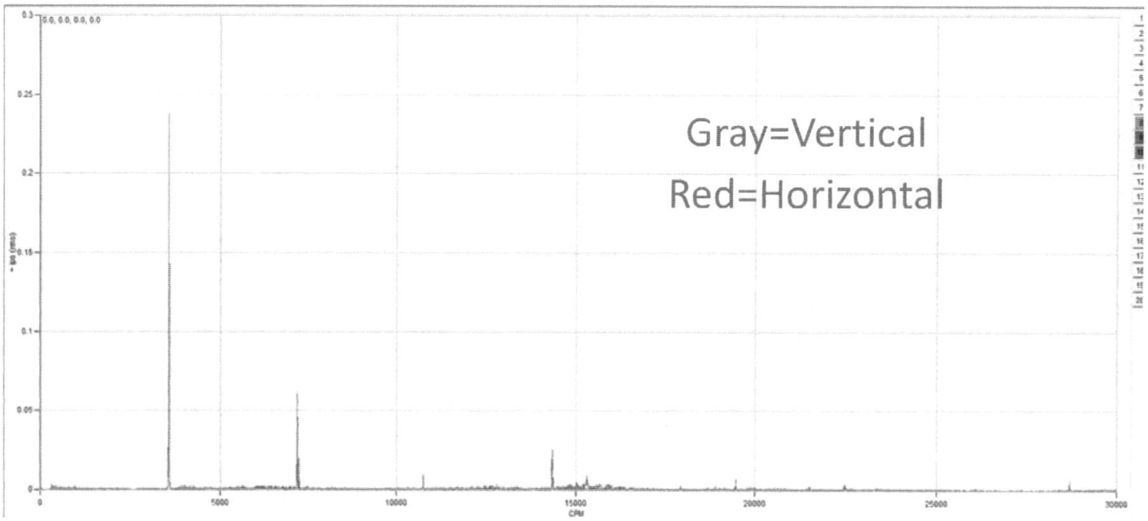

Figure 33: Pump Bearing Housing FFT Spectrums

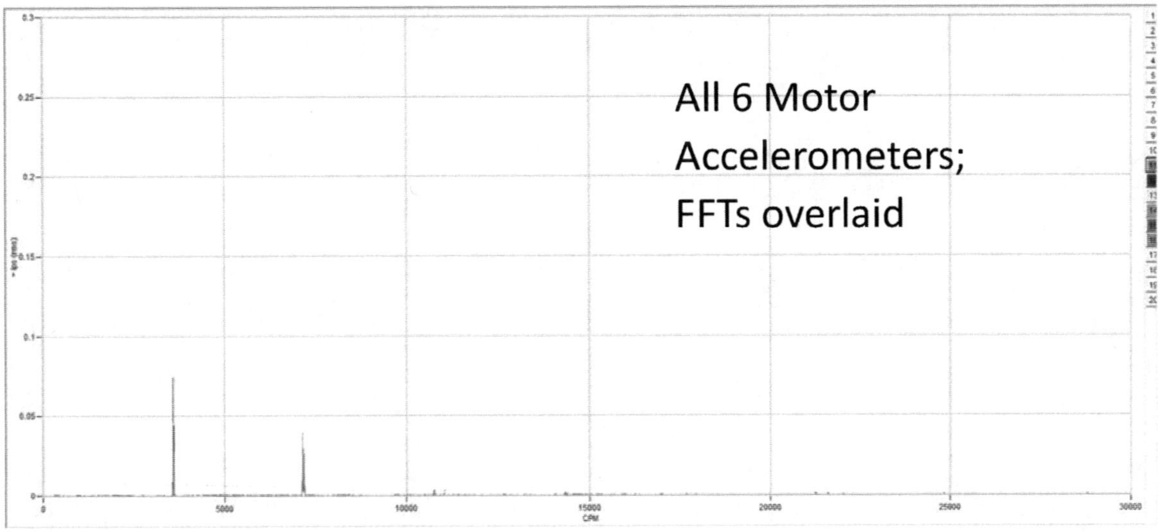

Figure 34: Motor FFT Spectrums

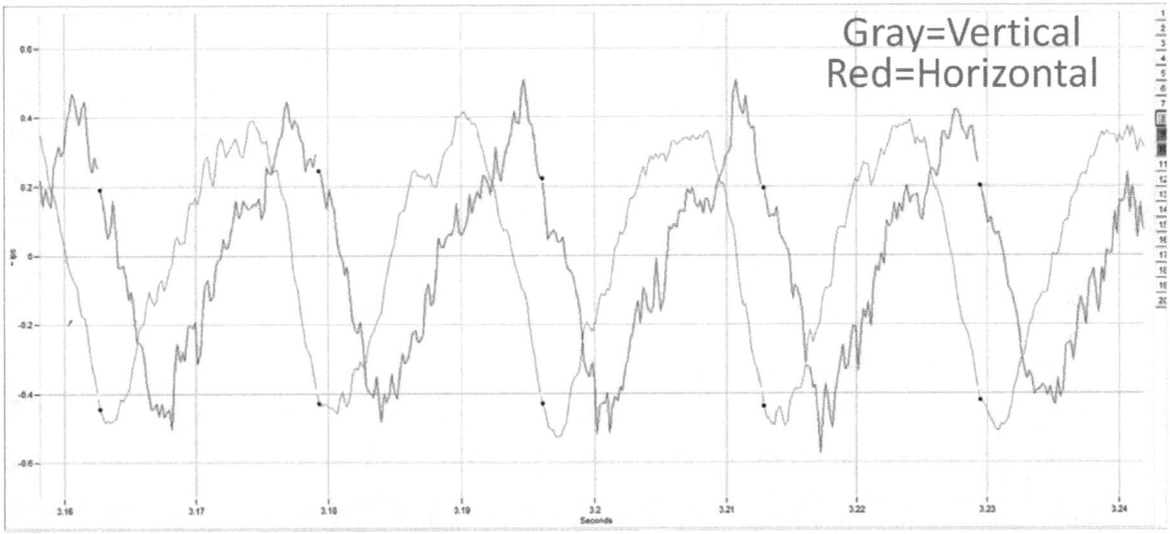

Figure 35: Pump Time Waveforms

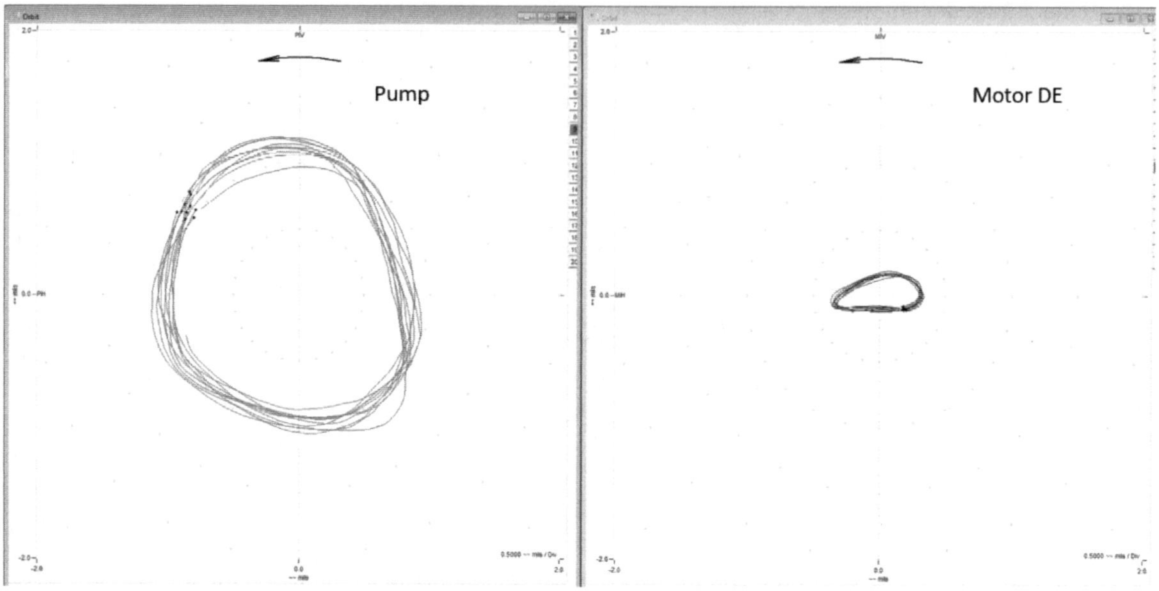

Figure 36: Accelerometer Structure Orbits

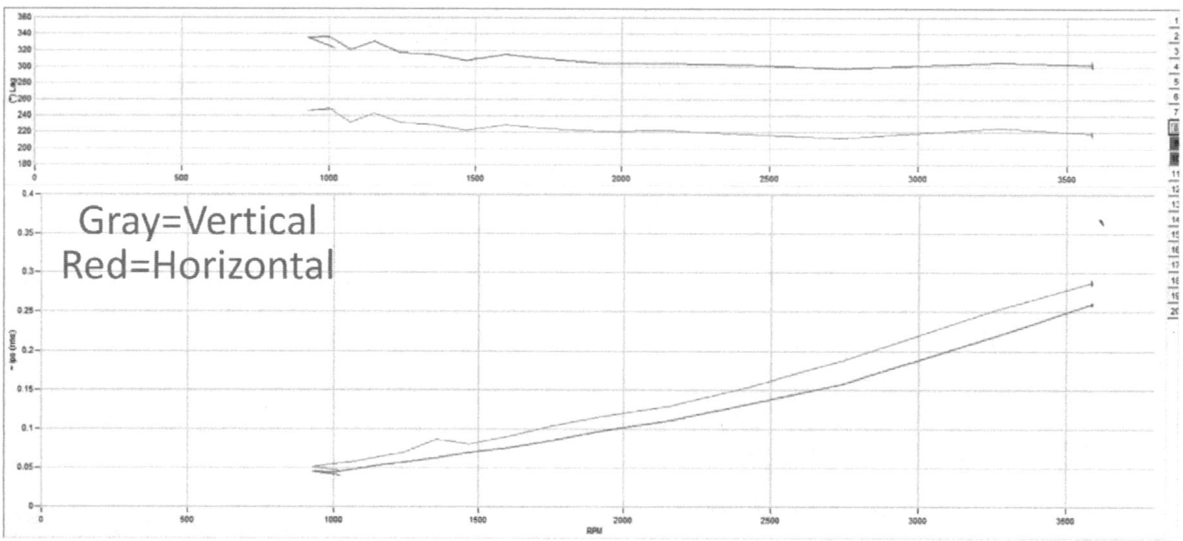

Figure 37: Pump Bode Plots

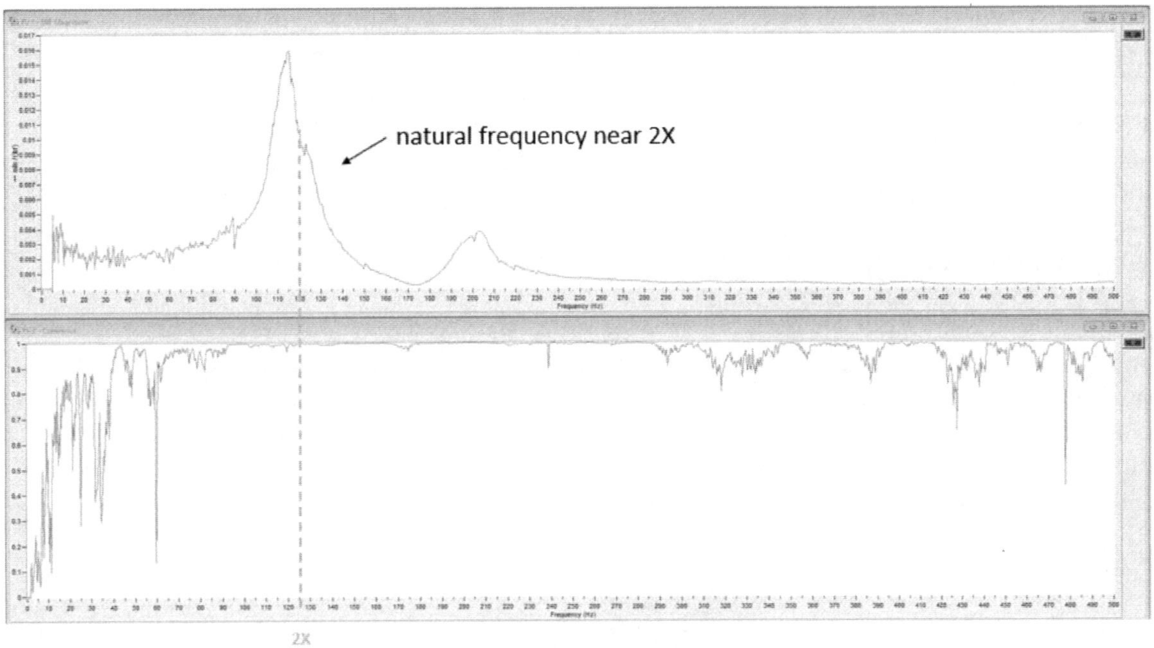

Figure 38: Pump Vertical Impact Test FRF

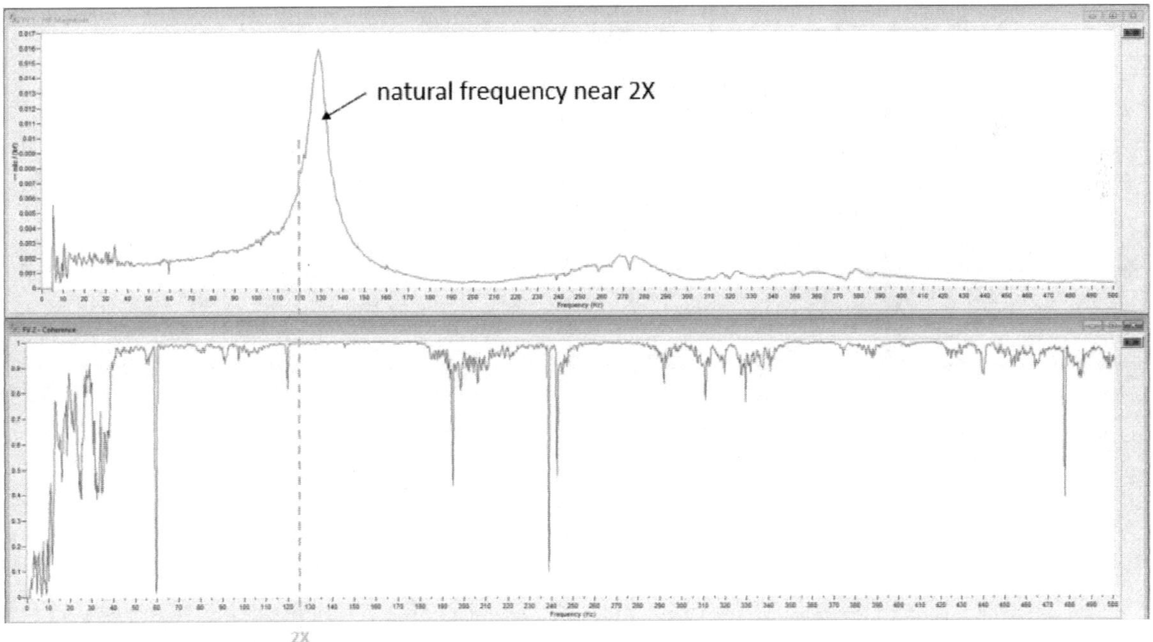

Figure 39: Pump Horizontal Impact Test FRF

Because many of the typical sources of mass unbalance were reported to have been checked a recommendation was made to inspect or replace the coupling and coupling hubs. The spacer coupling design, see Figure 40, could provide a mass unbalance if the hubs are not concentric and parallel to the shaft. It was suspected that run-out or other damage was creating a mass imbalance **force** at the coupling.

The pump coupling hub was found to have some "minor" damage near the bolt holes. This damage was repaired by filing or machining the surfaces. After this corrective action the vibration amplitudes lowered significantly to normal amplitudes. The highest amplitude after the coupling correction was the pump drive end horizontal with a 1X component at .08ips rms, see Figure 41. The 2X component was of similar amplitude; this relatively large 2X component is likely caused by resonance amplification rather than some other force or fault. The coupling is a flexible disk design so it is not expected to generate a large 2X force under reasonable conditions. A **natural frequency** is amplifying the 2nd harmonic response, but the **forces** are low. Low 2nd harmonic **forces** are preventing a large amplitude response at resonance; therefore, no action is needed to change the natural frequency. The horizontal natural frequency has an amplification factor of 16 and 3% critical damping, it would provide a large increase in response if significant 2nd harmonic forces were present.

Figure 40: Image Of Typical Disc Coupling With Spacer; Lovejoy Coupling Solutions

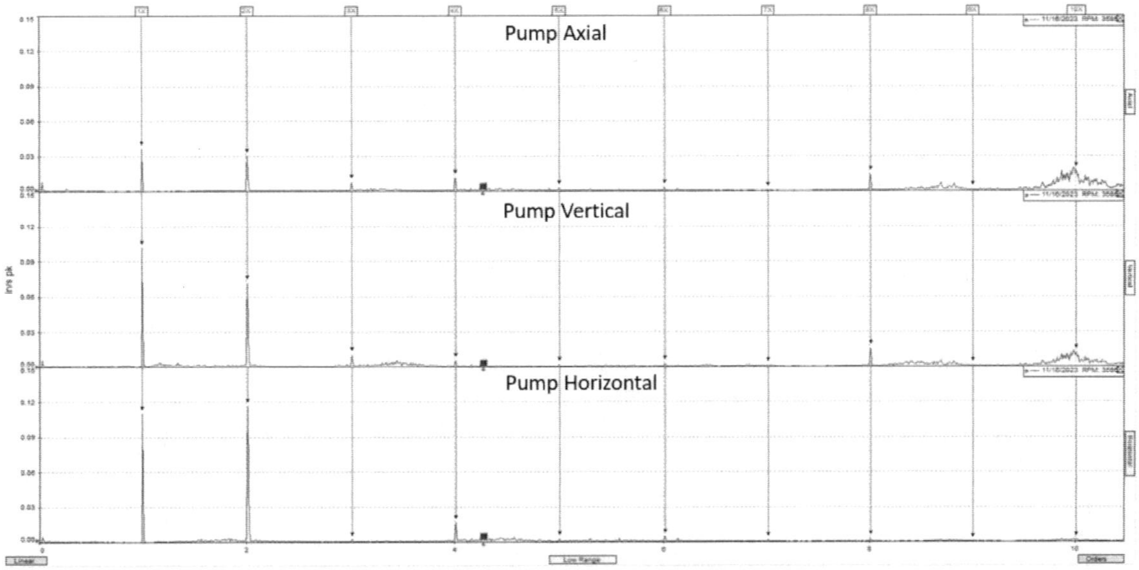

Figure 41: Owner Pump FFT Spectrums After Coupling Repair

The reported history on this machine included all mechanical checks such as shaft alignment, run-out, rotor mass balance, etc. The author assumed a **natural frequency** would probably be involved and impact testing was conducted during the site test. No natural frequencies were found near 1X where vibration amplitudes were large. It was difficult to make recommendations to check mass unbalance again because two previous repairs were reported to have inspected all relevant items. The data was very clear that it was a rotating force without amplification causing the large vibration amplitudes at 1X. It was suspected that a coupling issue may have been overlooked and it could create a mass unbalance if the spacer or hubs were not concentric with the shaft.

The hearsay information was helpful because it ruled out another shop balance of the rotor or replacing the impeller. Assumptions about mechanical checks being sufficient had to be dismissed based on the site vibration test data. It is the author's experience that all information sources should be considered, but the possibility of inaccurate information is a real possibility. Further, some faults require information sources other than vibration data because vibration signatures are not conclusive. Hydraulic issues such as recirculation, cavitation, flow turbulence, etc. may not provide specific frequencies or signatures to determine the exact cause of excessive vibration. Modeling, pump curves, experience, or process data such as pressure and flow rate data may be needed to determine the probable source of broad band vibration. Running speed frequency (1X) vibration can also be created by multiple sources; in this case structure orbits, phase data, and transient coast down data were used to conclude the source of vibration was a rotating force and likely to be mass unbalance. Other hearsay data was used to make an informed guess regarding the area to investigate as the source of mass imbalance.

OH2 PUMP DRIVE COUPLING FAULT (CASE STUDY 4):

An OH2 pump in hot oil service, see Figure 42, experienced very high noise and vibration at 2X shaft speed frequency. After start-up the pump and motor 2X shaft speed component amplitudes increased over 10 minutes from very low to near 0.8ips rms before the author had the pump shut down for safety concerns, see Figures 43-46. Previous operation had similar trends based on noise and foundation vibration; no vibration data was recorded on previous pump runs. Initial inspection used motion amplification video so the time varying aspect of the amplitude was not known. Motion amplification video showed the ODS but not the transient nature, see Figure 47. The operating shape involved motor and pump moving vertically and axially at 2X frequency.

Figure 42: OH2 Pump In Hot Oil Service

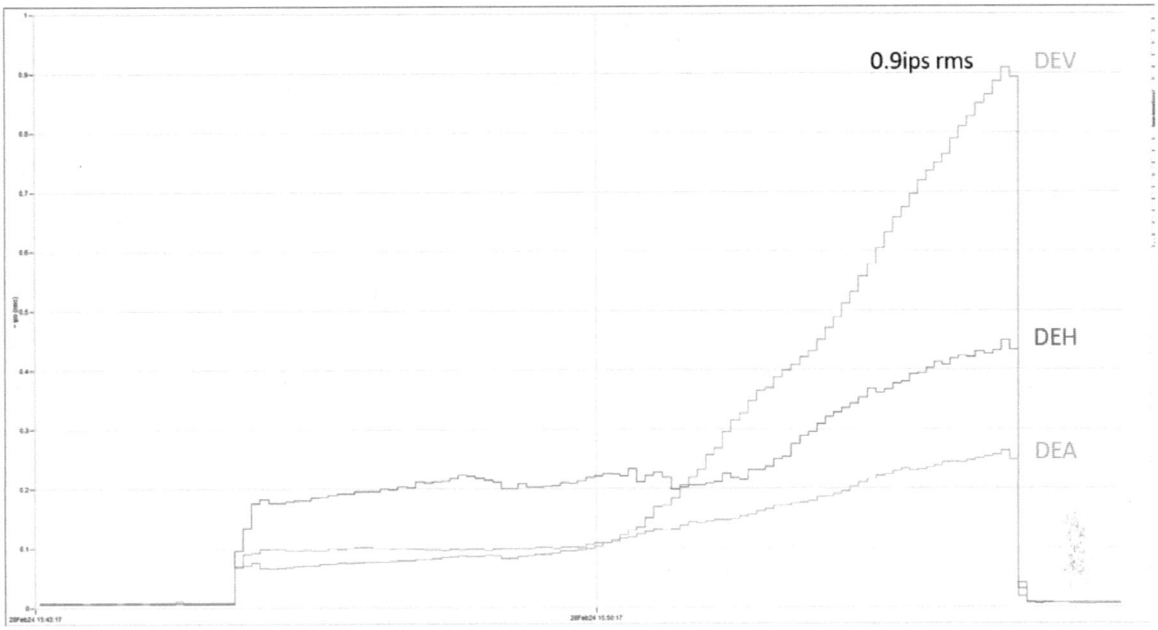

Figure 43: Pump Overall RMS Trends

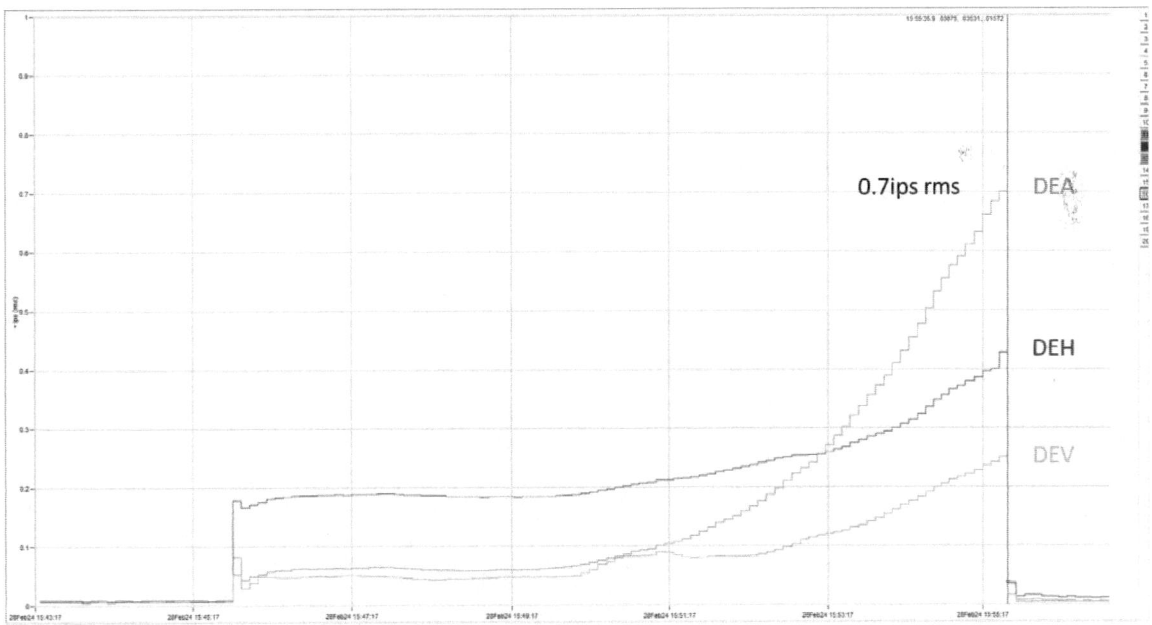

Figure 44: Motor Overall RMS Trends

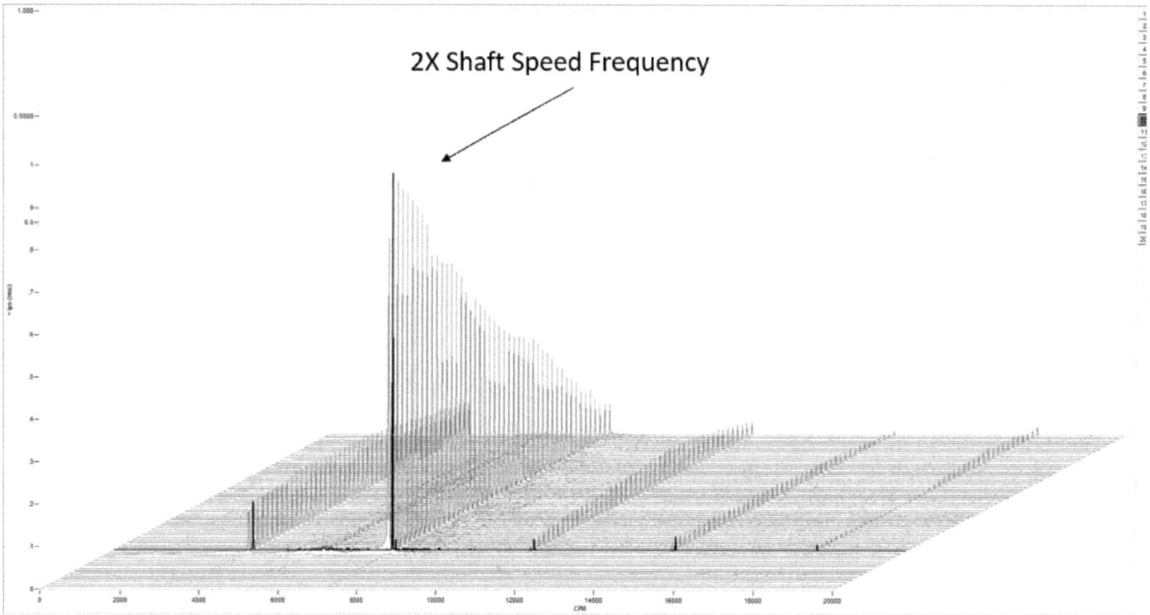

Figure 45: Pump Waterfall Plot (10 span after start-up)

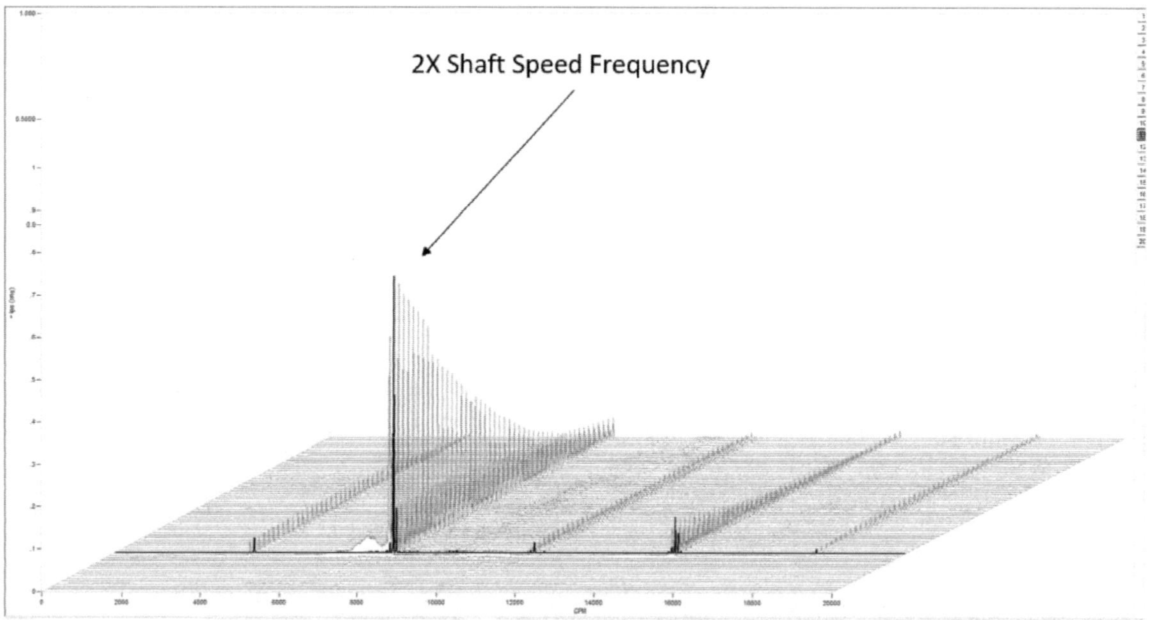

Figure 46: Motor Waterfall Plot (10 span after start-up)

Visual inspection of the coupling indicated distorted shim packs, see Figure 48. It was suspected this may indicate an improper shaft end spacing that could put the coupling spacer in tension or compression. It could also represent shaft misalignment or coupling damage.

Figure 47: Motion Amplification Video

Figure 48: Distorted Coupling Disk Pack

Coast down testing indicated a motor axial resonance may be involved. The time waveform plot shows a large increase in amplitude soon after the motor power was turned off for the coast down test, see Figure 49. The pump vertical vibration did not show a similar increase in amplitude when the motor axial vibration went through resonance. This indicates the motor axial resonance and natural frequency probably does not significantly involve the pump structure. It is unknown why the 2X component amplitude increased steadily after start-up at the motor axial and pump vertical locations. The increasing amplitude with time suggests a thermal effect since flow and speed were constant. Natural frequencies and shaft alignment can change with temperature; therefore, it is unknown why the 2X amplitude increased steadily after start-up.

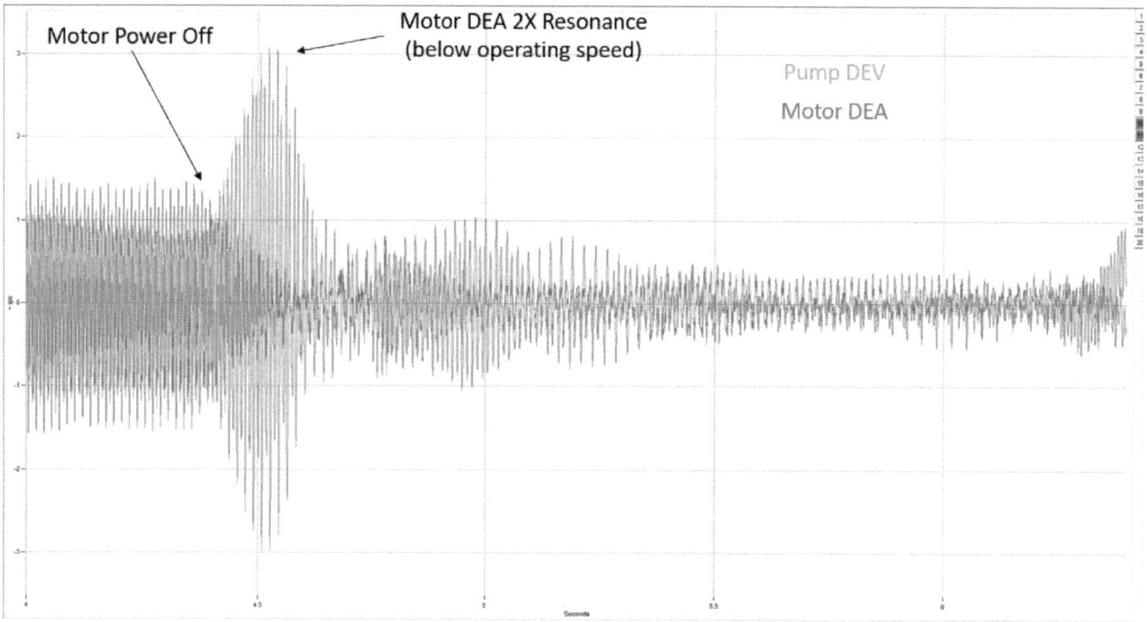

Figure 49: Pump Vertical & Motor Axial Time Waveforms During Coast Down Test

A pump horizontal natural frequency at 120.9Hz was later found with impact testing. The pump horizontal natural frequency will certainly amplify any 2X forces, but the highest pump vibration responses were not in the horizontal direction. Also the disk coupling should not create significant 2X forces if installed properly and is in good condition [Ref 3]. Because the coupling disks had visible distortion it was recommended replace the coupling, ensure proper shaft alignment, and proper shaft end spacing. After installing a new coupling the pump vibration amplitudes at 2X shaft speed frequency were essentially not measurable, see Figures 50 & 51; the 2X line frequency component on the pump remained near 0.014ips rms. The motor drive end vertical 2X line frequency component was 0.09ips rms.

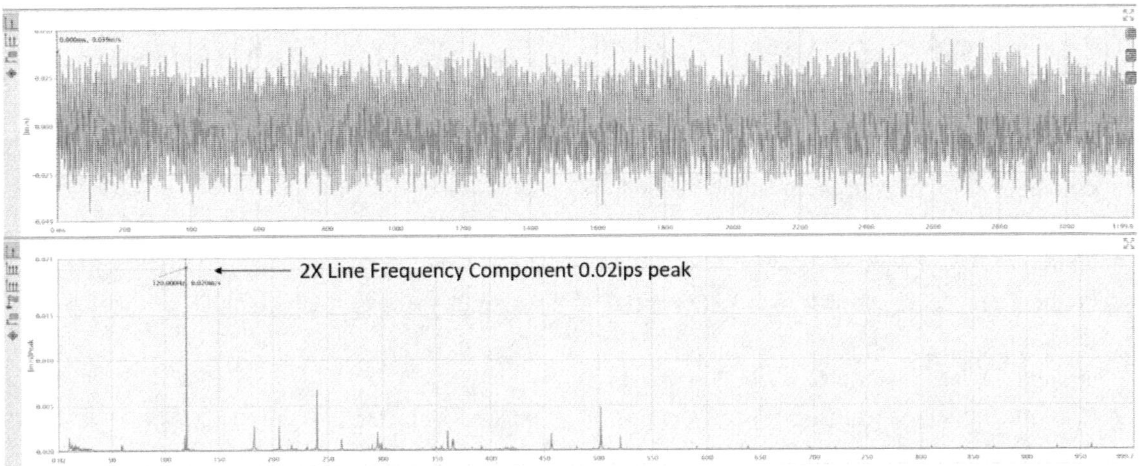

Figure 50: Owner's Pump Drive End Vertical Vibration After Coupling Replacement

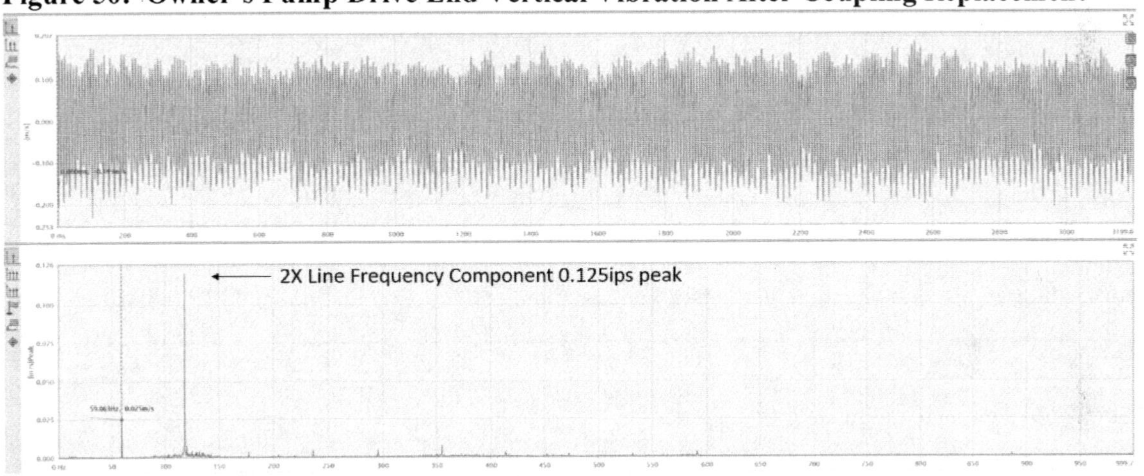

Figure 51: Motor Drive End Vertical Vibration After Coupling Replacement

Impact Test Accelerometer Locations

Channel 8	Pump Drive End Axial	
Channel 9	Pump Drive End Horizontal	Red Trace
Channel 10	Pump Drive End Vertical	Purple Trace
Channel 11	Motor Drive End Axial	Dark Green Trace
Channel 12	Motor Drive End Horizontal	Light Blue Trace
Channel 13	Motor Drive End Vertical	Orange Trace
Channel 14	Motor Non-Drive End Axial	
Channel 15	Motor Non-Drive End 45 Left	Light Green Trace
Channel 16	Motor Non-Drive End 45 Right	Magenta Trace

Note all FRF plots on the next four slides have the same maximum amplitude scale of .06 g's/lbf to compare responses.

Figure 52: Impact Test Accelerometer Locations

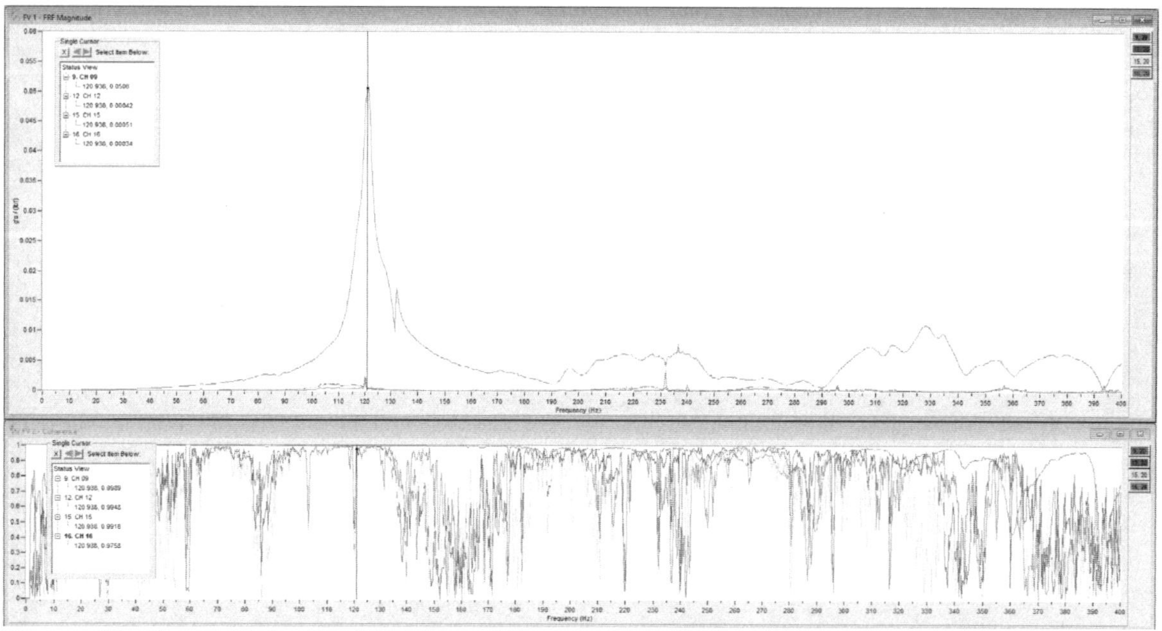

Figure 53: Pump Drive End Horizontal Impact Test FRFs (Accelerance g's/lbf)

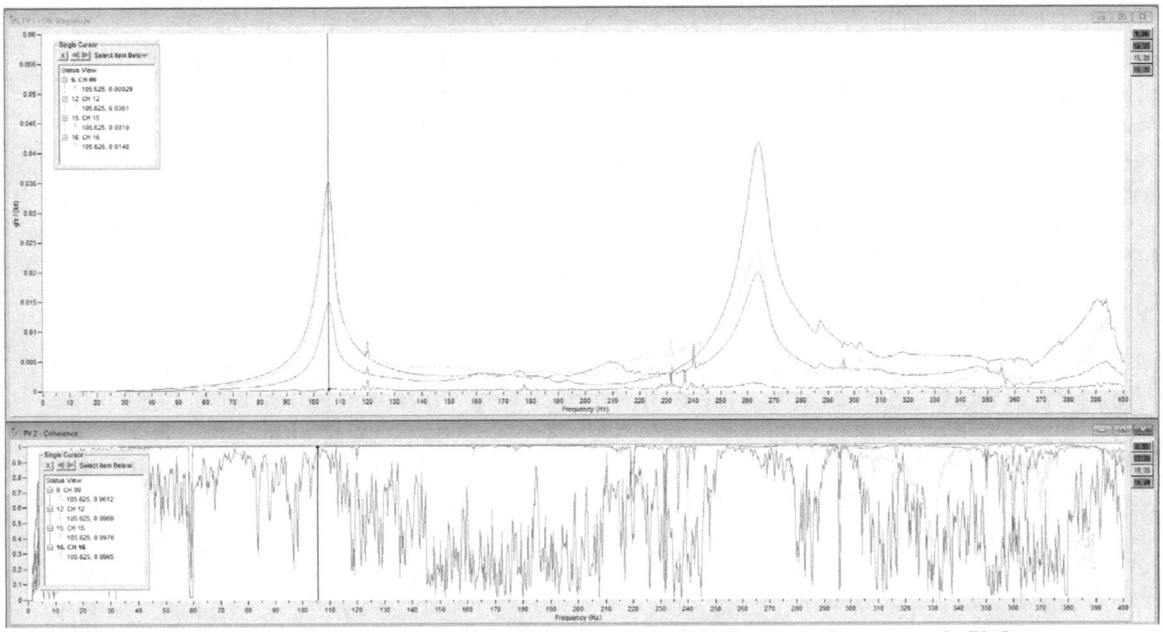

Figure 54: Motor Drive End Horizontal Impact Test FRFs (Accelerance g's/lbf)

Although a pump **natural frequency** was very close to the 2X shaft speed frequency the extremely high 2X component amplitudes on the motor and pump were caused by excessive **forces** from a damaged or improperly installed coupling. Although it is unclear why 2X vibration amplitudes increased steadily after start-up the author suspects it is related to thermal growth and changes in the coupling alignment and/or shaft spacing. Details of the repair were not available but a focus on the coupling installation and coupling replacement completely resolved the vibration issue. Natural frequencies could have been a distraction in this case, but the coast down data suggested the motor natural frequency was far enough below 2X shaft speed frequency (12% below) as to not be a primary issue. A properly functioning disk coupling should not provide significant 2X forces under reasonable conditions of shaft misalignment [Ref 3]; therefore, the author suspected a coupling defect or extreme loading of the coupling that could generate the 2X forces.

CONCLUSION:

This paper discusses four case studies; two involving centrifugal pump vane-pass vibration and two involving drive coupling faults. Vane-pass vibration and coupling related vibration are common issues on centrifugal pumps.

Vane-pass vibration results from the pumping action of a vaned impeller and is not a fault frequency or a defect. If vane-pass vibration amplitudes are excessive then corrective actions need to be taken. The B gap can be increased, the vanes can be staggered on series impellers, the outer diameter can be angle or "V" cut to spread out the pressure pulse period over a few degrees of shaft rotation, or the flow rate can be controlled to reduce pressure pulsation caused by off BEP operation.

Coupling damage or faults can create forces at 1X, 2X, or other harmonics of running speed depending on the coupling type, fault, and shaft misalignment type/severity. The disk couplings presented in this paper are typical for many centrifugal pumps. They do not generate significant 2X forces unless there is a defect or severe loading where non-linear stiffness is created [Ref. 3]. The third case study demonstrated a mass unbalance force created by coupling hub damage and a spacer that was not concentric with the shaft. The fourth case study demonstrated a very unusual and severe 2X force created by coupling damage and/or distortion. The transient increase in 2X amplitude was likely caused by shaft thermal growth and/or shaft misalignment combined with a coupling fault and/or distortion.

All four case studies involved forces generated by some mechanism; all vibration requires a periodic or transient force. This paradigm of forces causing vibration is difficult to use for fault analysis because the forces are not measured. Only the response or vibration is measured and we want to know the force magnitude or fault severity. The vibration frequency often identifies the source of the fault and the amplitude often indirectly indicates the severity. Complicating the issue, some faults do not have unique associated frequencies. Now add resonance and amplification factor; this amplifies or attenuates the vibration response which further complicates and already difficult task of using a machine vibration response to identify a fault and the fault severity. The next question "how long will it run before failure" is a whole other paper to be written by someone else. It is the author's opinion that life prediction is nearly impossible without extensive machine specific experience and even then very difficult.

The four case studies were presented to show how the author had success and failure because of the complex factors involved in vibration analysis. The goal was to stress three factors that should always be considered; **force**, **stiffness**, and **amplification factor**. Most systems are designed with sufficient stiffness to limit deflections from expected forces; therefore, the analyst is usually dealing with the applied **force** and **amplification factor.** Force is not directly measured and amplification factor is not known until a natural frequency or modal test is conducted. Structural natural frequencies can amplify the vibration response to any frequencies that have associated periodic or transient forces; orders of running speed or other harmonics can be amplified. This can cause confusion when orders not associated with a particular fault are amplified. For example, case study 3 had an elevated 2X component due to resonance that could be mistaken a fault other than mass unbalance. Knowing there was a natural frequency near 2X allowed this amplitude to be disregarded in the analysis.

Amplification factor depends on the damping present and the ratio of forcing frequency divided by the natural frequency. A natural frequency near a forcing frequency will cause amplification of the vibration amplitude given a force input, but the resulting amplitude could be insignificant or severe depending how close the forcing frequency is to the **natural frequency** and the amount of **damping** present. In a linear system the vibration response amplitude will be proportional to the **force** amplitude; if the **force** is doubled then the vibration response amplitude is doubled.

To fully understand a system's vibration response the amplification factor should be considered. Without resonance amplification a vibration response amplitude correlates with the force applied; with resonance and mechanical lag the situation is more complicated because the response amplitude is no longer proportional to the force. Now the resonance and/or force could be significant factors. In some cases the natural frequency needs to be changed to increase frequency separation margin, in other cases reducing the **force** may be practical, and in yet other cases both **force** and the **natural frequency** may be addressed. Changing the natural frequency often involves changing the modal mass or modal stiffness. In some cases applying damping is a good solution for resonance, but it may be difficult to effectively apply damping; commercial dampers are available if there is a rigid location to connect to the anti-node location. There are also tuned mass dampers that can be attached to provide damping.

To accurately evaluate a vibration issue the analyst needs to know the machine characteristics, have sufficient vibration signatures (including modal testing), and possibly have other information such as a repair history, pump curves, finite element models, etc. Often an analyst is provided an FFT spectrum and is asked to evaluate an issue. All of the previously discussed factors demonstrate why significant experience and technical training are needed to conduct good vibration analysis. It is important to know what you don't know and when you need to get more information. Vibration data can be a very powerful tool but often more extensive vibration data (examples: phase, modal, transient, operating deflection shape, variable process operation, etc.) and other sources of information are needed (examples: finite element models, pump curves, computational fluid dynamic models, etc.).

REFERENCES:

[1] Bolleter, U., "Blade Passage Tones of Centrifugal Pumps", Vibrations Vol. 4 No. 3, September 1988.

[2] Baxter, Nelson, "Empirical Comparisons of Damping Equations", Chapter 10, Advanced Vibration Analysis Course, Vibration Institute 2024.

[3] Kelm, Ray, "Machinery Vibration Analysis", Category III Training Course Material, Vibration Institute, 2024

VIATC TRAINING CONFERENCE & EXPO
COVINGTON, KENTUCKY | AUGUST 7-9, 2024

Naïve Bayes Classifier Optimization for Bearing Fault Detection

Matthew J. Kenney
QM Analyst/Jr. Developer at Spring Point Solutions, LLC
mattjkenney@protonmail.com

Matthew J. Kenney currently lives in Connecticut with his wife and two kids who are his greatest motivation to succeed in challenging circumstances. Matthew served six years in the US Navy – directly servicing motors and generators as an Electrician's Mate for nuclear powered submarines. After the Navy, he immediately began working in Quality Assurance for nuclear energy critical motor repair at Schulz Electric in New Haven, CT. During his time at Schulz Electric, the era of Big Data was becoming popular, and he quickly realized the potential it had to serve the motor repair industry. He began study at Post University where he earned a B.S. Degree in Data Science. During his study, he developed a method for bearing fault detection with vibration analysis. This paper outlines his work.

ABSTRACT:
Vibration analysis is an attractive option for bearing fault detection since vibration data gathering is non-destructive and can be achieved without equipment disassembly. Although, human interpretation requires much experience and knowledge to accurately interpret. Furthermore, detecting the fault location within the bearing itself is near impossible for slight imperfections in the bearing – complicating manufacturing improvements. Naïve Bayes machine learning classifiers have been proposed to accurately predict bearing health without the need for experienced personnel.

Naïve Bayes classifiers work well with small datasets – a symptom of which the needed training set suffers. However, since vibration and speed depend partly on previous magnitudes of the same parameter, the Naïve Bayes classifiers lack the feature independence assumption. Thereby lowering prediction accuracy.

This paper proposes a solution to improve a Naïve Bayes classifier for bearing fault location detection. Improvements are designed in three ways: 1) limiting predictors to Characteristic Bins, 2) determining the ideal feature-domain set for analysis, and 3) optimizing the number of bins and periods used for aggregating and calculating features. Finally, the optimized method is compared to the same method without predictor limitation and a Multinomial Naïve Bayes method from Scikit-Learn.

KEY WORDS: bearing fault analysis, vibration analysis, machine learning, Naive Bayes

INTRODUCTION:

As of October 2023, wind energy comprises 10.3% of all energy supply in the U.S. [1]. By 2030, the percentage is predicated to nearly double to 20% [1]. Meanwhile, onshore wind energy has the lowest Levelized Cost of Electricity (LCOE) among other sources (including Solar, Geothermal, Gas Peaking, Nuclear, Coal and Gas Combined Cycle) at $24 - 75/MWh [2]. Further, the LCOE of wind energy realized the second largest decline in LCOE between 2009 – 2023 at a 63% drop (the first being Solar PV-UtilityScale with a 83% drop) [2]. With the increasing demand for cheaper alternative energy, it's easy to see quickly rising demand predicted in 2030.

Along with rising demand for supply, comes a rising demand for Wind Power Plant machine integrity. In 2015, the National Renewable Energy Laboratory (NREL – a national laboratory of the U.S. Department of Energy) established a database, now called the Gearbox Reliability Database (GRD), to track turbine gearbox faults from utility-scale Wind Power Plants. The Alliance for Sustainable Energy (the operator of the NREL) describes the purpose of the GRD:

1. Categorize top wind turbine gearbox failure modes,
2. Identify possible root causes, and
3. Direct future wind turbine gearbox reliability research and development (R&D) activities. [3]

From 750 records in the GRD between 2009 – 2015, 76% of the gearbox failures were caused by faulty bearings [3]. Thus, a strong interest has emerged in predictive bearing health analysis in the wind turbine energy industry. Vibration analysis is an attractive option since vibration data gathering is non-destructive and can be achieved without equipment disassembly. Although, such an analysis takes much experience and knowledge to accurately interpret. Therefore, machine learning solutions have been proposed to accurately interpret vibrational data without the need for technical experts.

THE PROBLEMS FACING THE MACHINE LEARNING SOLUTION:

The challenges machine learning engineers face attempting to design bearing fault detection with vibrational data are summed up in three categories: 1) large number of data entries per bearing, 2) low number of bearing samples, and 3) a lack of independent sample data. Each bearing dataset can contain hundreds of thousands of entries. That consumes significant computational resources – therefore, data reduction per bearing sample is much needed. At the same time, every sample dataset is from a single bearing instance in a lot; and every vibration in-part is affected by a previous vibration. That results in low prediction accuracy due to a lack of independent samples and sample data. Expounding on the issue, bearing vibration data is difficult to obtain – it requires specialized test set-ups, trained human resources for operation, and high energy expenses. In consequence, there exists a low number of unique bearing sample data (at least those that are publicly available); and within each sample, the data itself lacks sample independence.

For data reduction efforts, signal features are extracted from datasets. Feature values are statistical calculations that represent a characteristic of a subset of the underlying data. Using features in place of raw vibration signals, new challenges surface: 1) selecting the best feature calculation, and 2) optimizing the subset data size. Additionally, since features utilize the underlying data, they too suffer from sample dependence.

To handle the low bearing sample count, Naïve Bayes (NB) classifiers are an attractive option since they typically perform well with a low number of samples [4]. However, NB classifiers are problematic due to lacking the feature independence assumption.

Despite problem (3), there has been much success with a NB solution in past efforts for bearing fault detection. Zhang et al. [5] overcame the issue utilizing a Decision Tree (DT) to select low correlated features. Yi et al. [6] showed NB can achieve higher accuracy than other models with no feature engineering. Furthermore, Zhang et al. [7] found success with NB and highly correlated features for bearing remaining useful life (RUL) prediction; choosing only features >90% correlation.

This paper proposes another solution to improve a Naïve Bayes classifier for bearing fault location detection. Improvements are designed in three ways: 1) limiting predictors to Characteristic Bins, 2) determining the ideal feature-domain set for analysis, and 3) optimizing the number of bins and periods for aggregating and calculating features. Finally, the optimized method is compared the same method without predictor limitation and a Multinomial Naïve Bayes method from Scikit-Learn.

NAÏVE BAYES INTRODUCTION:

Naïve Bayes equation calculates the *posterior probability*, $P(c|X)$, of the class c according to the predictor x where $x \in X$:

$$P(c|X) = \frac{P(c) \times \prod_{i=1}^{n} P(x_i|c)}{\prod_{i=1}^{n} P(x_i)} \qquad \text{Eq. (1)}$$

The term $P(c)$ is the *prior probability* of the class. It describes the probability the class will occur based on the class size and total population. The term $\prod_{i=1}^{n} P(x_i|c)$, is the *likelihood probability* all the predictors in X will occur given the classification c is true. The *evidence probability* term $\prod_{i=1}^{n} P(x_i)$, is the probability all the predictors in X will occur.

The posterior probability is calculated for each possible class. The object is then classified by the class with highest probability:

$$c_j = c_j \; in \; argmax\big(P(c_1|X), P(c_2|X), \dots, P(c_m|X)\big) \qquad \text{Eq. (2)}$$

where, j = class index and m = the total number of classes.

ALGORITHM DESIGN:

A subsection of the domain containing an equal number of data points defines a *period.* A feature value is calculated with the aggregate of data within the period. Grouping the domain in period increments allows control over the data quantity. Lowering the number of periods will reduce the number of calculations for the Bayesian model, thereby saving computational cost and time. Raising the number of periods should raise the accuracy of the model. Feature – Domain engineering in this way is executed by Pseudocode 1.

Pseudocode 1: Feature – Domain Engineering

Inputs:

1. Fs = Vibration set files, such that each file contains a dataset of vibration velocities at a shaft rotational speed.
2. $nPeriods$ = Number of Periods

Steps:

1. Initialize array $V = [\,]$
2. For each file in Fs:
3. Extract vibration velocity y, speed x, and class label data.
4. If acceleration is preferred, convert x to acceleration.
5. Sort by x.
6. Aggregate equal quantities of data points by calculating a feature across each period for $nPeriods$.
7. Build feature array $A = [class\ label \quad v_1 \quad v_2 \quad \cdots \quad v_{nP}]$ | v_p is the feature value in period p in nP number of Periods.
8. Append A to V.
9. Return V

The range of each period is broken up into vertical *bins*. Bin frequencies for each class are counted for each period. The bins on the extremities of the range for any period include the infinity value, and the bins in the middle are finite. For instance, if a period included 4 bins to cover a range of 20 points, the 3rd highest bin will hold all values between 10 – 15; and the 4th highest bin will hold all values from > 15. Therefore, for each period, all feature values will fall into exactly one bin. Likewise with period quantity, lowering the number of bins will save computational cost and time while raising bin quantity will raise accuracy.

Where n_c is the number of instances in class c and N is the total population including all classes, prior probabilities are calculated:

$$P(c) = \frac{n_c}{N} \qquad \text{Eq. (3)}$$

Where f is the frequency count, the likelihood probability of bin b in period p for class c becomes:

$$P(b_p|c) = \frac{f_{b_p,c}}{n_c} \qquad \text{Eq. (4)}$$

Likewise, the equation for the evidence probability of bin b in period p becomes:

$$P(b_p) = \frac{f_{b_p}}{N} \qquad \text{Eq. (5)}$$

Pseudocode 2 shows the progression for training the model.

Pseudocode 2: Training

Inputs:

1. $V_{training}$ = Pseudocode 1 where Fs = files array for training set
2. $nBins$ = Number of Bins per Period.

Steps:

1. With $V_{training}$:
2. Count N for all instances.
3. Count n_c for each unique class label.
4. Create dictionary P_c | class label = Eq. (3), for each unique class label.
5. Calculate bin edges for each period and $nBins$ per period.
6. Create array $B = [b_{x,p} \quad \cdots \quad b_{nB,nPeriods}]$ | $b_{x,p}$ = bin edges for bin x in period p.
7. With B, count f_{b_p} for all unique class labels.
8. Create dictionary $P_{b_p|c}$ | class label = Eq. (4), for all bins in B.
9. Create dictionary P_{b_p} | $b_{x,p}$ = Eq. (5), for all bins in B.
10. Create dictionary C | class label = array $T_{training}$ containing bin identifiers for all bins occupied by that class.
11. Return $B, P_c, P_{b_p|c}, P_{b_p}, C$

Some bins are not used at all in a particular class, resulting in a 0 probability for Eq. (4). Likewise, some bins are not used at all in any class, resulting in a 0 probability for Eq. (5). Both circumstances show erroneous probabilities. Suppose a vibration set fits in 99 of 100 bins for one known class. Then Eq. (4) becomes 0, where it is actually very likely the bearing belongs to the class since it matches the bin makeup by 99%. Therefore, the model only considers the bins from the intersection of those in the class from the training set and the test samples, called *characteristic bins,* for determining Equations (4) and (5). For instance, let's say bins $\{b_1, b_2, b_3, b_4, b_5\}$ correspond to the bin numbers for periods $\{p_1, p_2, p_3, p_4, p_5\}$. Now, suppose a test sample uses bins $\{1, 2, 1, 3, 2\}$ and class A uses bins $\{1, 2, 1, 4, 2\}$. Then, with *Characteristic Bins*, only the bins in periods $\{p_1, p_2, p_3, p_5\}$ are used for Equations (4) and (5). Without the *Characteristic Bin* utilization, all the bins from the test set are used, regardless of the class characteristic bins. Pseudocode 3 shows the classification algorithm using Characteristic Bins.

Pseudocode 3: Naïve Bayes Algorithm with Characteristic Bin Utilization

Inputs:

1. V_{test} = Pseudocode 1 where Fs = files array for test set
2. $B, P_c, P_{b_p|c}, P_{b_p}, C$ = Pseudocode 2

Steps:

1. With V_{test}:
2. With B, create array T_{test} containing the bin identifiers of all bins occupied by V_{test}.
3. Initialize array $F = [\,]$.
4. For each unique class label in $V_{training}$:
5. Calculate $P(c \mid X)$ = Eq. (1) | $X = T_{training|class\ label=current\ class\ label} \cap T_{test}$.
6. Append $P(c \mid X)$ to F.
7. Class label = Eq. (2) with probability array F.
8. Return class label.

DATA SOURCE:

The data contains vibration amplitude and speed recordings captured by [8]. Five bearing classes were evaluated: healthy, inner fault, outer fault, ball fault, and combination fault. Each class was tested in 4 rotational speed conditions with 3 trials: increasing speed, decreasing speed, increasing then decreasing, decreasing then increasing. The total lot consists of 60 datasets. Vibration readings were sampled at 200,000 Hz for 10 seconds, totaling 2 million datapoints per dataset. Speed was measured by an encoder: EPC model 775, 1024 CPR (Cycles Per Revolution). The encoder scale is not reported but assumed to be x2 based on the reported frequency range per test and the resulting velocity calculations. Vibration amplitudes were measured with an accelerometer: ICP accelerometer, Model 623C01. The experimental bearings were ER16K ball bearings. The test rig is composed of an AC Drive, motor and two bearings for support: one healthy, the other experimental (see Figure 1).

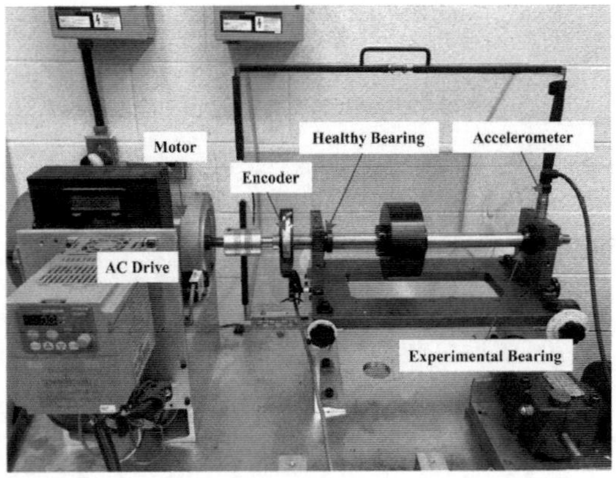

Figure 1: Test Rig [8]

EXPERIMENT 1 – FEATURE SELECTION:

Five features and two domains are compared totaling 10 tests. The domains include shaft rotational Velocity and Acceleration. Where E is an array containing encoder pulse count, CPR is the encoder cycles per revolution, the sampling rate is 2×10^5 and assuming a scale of 2, Velocity is calculated:

$$V = \frac{4E \times 10^5}{CPR} \quad \text{Eq. (6)}$$

Acceleration is calculated by finding the change of Velocity and dividing it by time. Five features are compared: Skewness, Kurtosis, Crest, Shape, Impulse, and Margin (see Table 1).

Table 1: Features

Feature	Description	Equation [9]	
Skewness	A measure of symmetry across the sample mean [10].	$\frac{\frac{1}{N}\sum_{n=1}^{N}(x_n - \bar{x})^3}{\left(\sqrt{\frac{1}{N}\sum_{n=1}^{N}(x_n - \bar{x})^2}\right)^3}$	Eq. (7)
Kurtosis	A measure of tail heaviness of the normal distribution. A heavier tail has more outliers [10].	$\frac{\frac{1}{N}\sum_{n=1}^{N}(x_n - \bar{x})^4}{\left(\sqrt{\frac{1}{N}\sum_{n=1}^{N}(x_n - \bar{x})^2}\right)^4}$	Eq. (8)
Crest	The peak to RMS ratio of a waveform [11].	$\frac{max\lvert x_n\rvert}{\sqrt{\frac{1}{N}\sum_{n=1}^{N} x_n^2}}$	Eq. (9)

Shape	Parameter that affects the general shape of a distribution [12].	$\dfrac{\sqrt{\frac{1}{N}\sum_{n=1}^{N} x_n^2}}{\frac{1}{N}\sum_{n=1}^{N}\lvert x_n\rvert}$	Eq. (10)
Impulse	Height of peak to the mean signal level [13].	$\dfrac{max\lvert x_n\rvert}{\frac{1}{N}\sum_{n=1}^{N}\lvert x_n\rvert}$	Eq. (11)
Margin	Peak amplitude to squared mean of squared roots of absolute amplitudes – also called “clearance factor” [13].	$\dfrac{max\lvert x_n\rvert}{\left(\frac{1}{N}\sum_{n=1}^{N}\sqrt{\lvert x_n\rvert}\right)^2}$	Eq. (12)

Figure 2 shows the mean Feature of all datasets per period for all Feature – Domain sets, in 20 period increments. It’s clear from Figure 2 that a combination of faults among all Feature – Domain sets are distinct. Likewise, inner-race faults are distinct except with the Skewness feature. The goal now is to determine which Feature – Domain set best distinguishes between the remaining conditions: ball fault, outer race fault, and healthy.

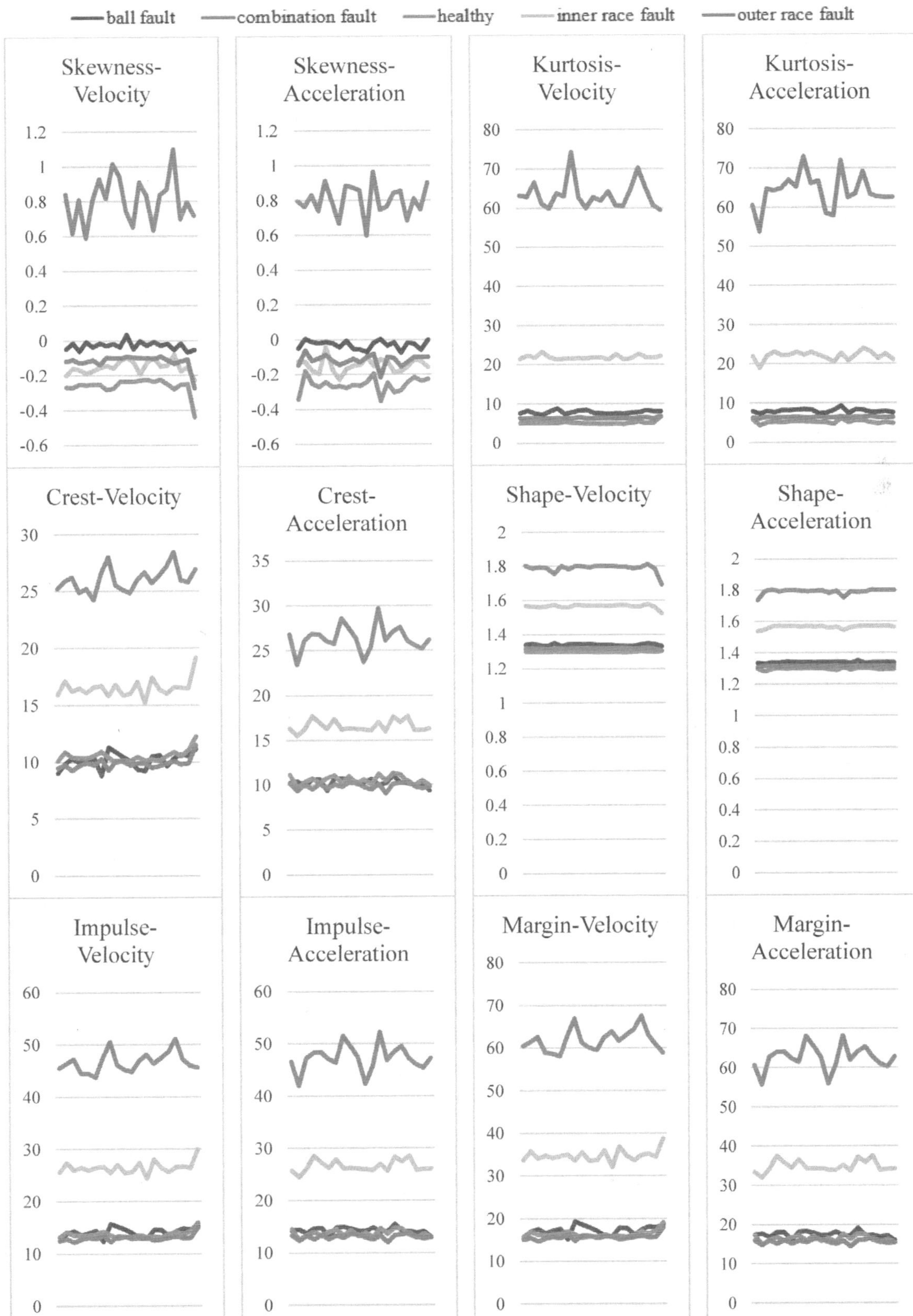

Figure 2: Feature Means in 20 Period Increments, Grouped by Sorted Domains

Each feature was tested for model Accuracy along Velocity, and then Acceleration. Model parameters were set at $nPeriods = 20$, $nBins = 10$, and 50% random sampling. New training and test samples were selected for each test. Table 2 shows statistical data for model Accuracy after running the classifier 50 times. Figure 3 shows the results of Table 2 in a box and whisker plot.

Table 2: Classification Accuracy Statistics by Feature-Domain Sets, n = 50

	Accuracy (%) [a]						
Feature	Mean	Std. Dev.	Min	Max	Median	25%	75%
Skewness							
Velocity	56.933	5.032	50.000	70.000	56.667	53.333	60.000
Acceleration	57.200	5.527	46.667	73.333	56.667	53.333	60.000
Kurtosis							
Velocity	58.933	2.375	53.333	60.000	60.000	60.000	60.000
Acceleration	59.067	2.337	53.333	60.000	60.000	60.000	60.000
Crest							
Velocity	77.867	5.938	63.333	86.667	78.333	73.333	83.333
Acceleration	79.600	4.886	70.000	90.000	80.000	76.667	83.333
Shape							
Velocity	63.987	11.963	50.000	90.000	60.000	53.333	73.333
Acceleration	63.333	12.293	43.333	90.000	56.667	53.333	73.333
Impulse							
Velocity	74.267	5.596	60.000	86.667	73.333	70.000	76.667
Acceleration	73.933	7.429	56.667	86.667	76.667	67.500	80.000
Margin							
Velocity	72.200	8.237	50.000	86.667	73.333	66.667	79.167
Acceleration	73.333	7.284	56.667	83.333	75.000	66.667	80.000

[a] Highlighted fields show where the metrics for mean or standard deviation are the best performing.

Between Velocity and Acceleration, for all features, there appears no significant difference in mean or median. Kurtosis clearly showed the least variation with a near flat-line appearance in a box plot among all others (see Figure 3). Additionally, Table 2 shows Kurtosis-Acceleration held the lowest standard deviation at 2.337% and 60% accuracy across the board spanning the 25 percentiles to the maximum in both domains. In terms of Accuracy, Crest-Acceleration achieved the highest at 79.6%, 95% CI [78.2, 81.0].

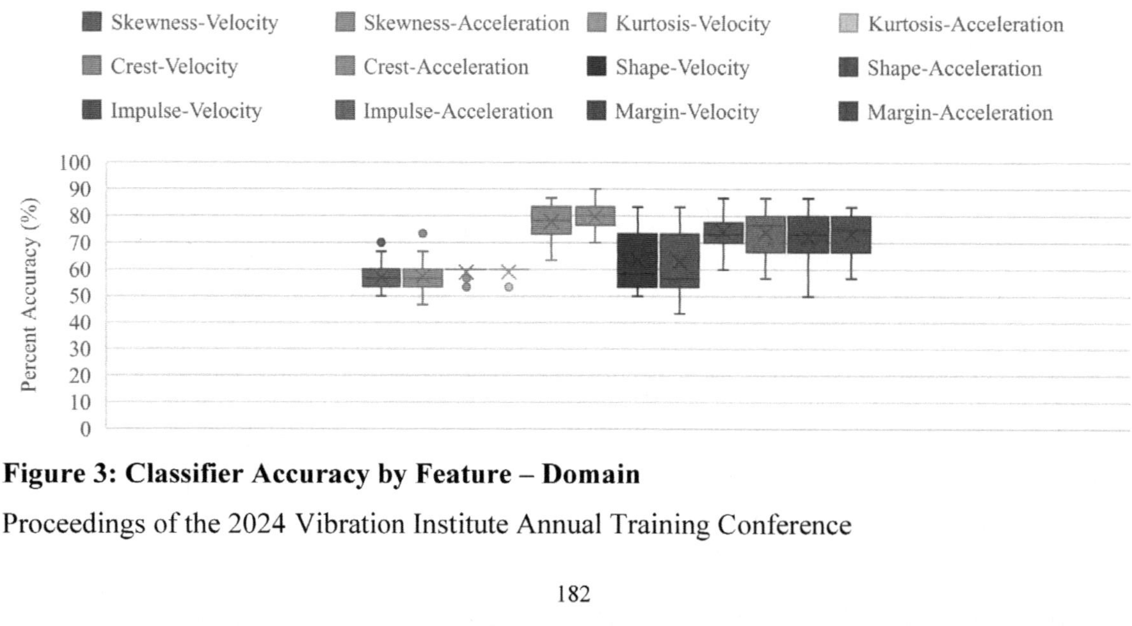

Figure 3: Classifier Accuracy by Feature – Domain

EXPERIMENT 2 – PERIOD AND BIN QUANTITY OPTIMIZATION:
By varying bin and period sizes, we can find the optimal values for each parameter. Since in Experiment 1, Crest-Acceleration had the highest Accuracy and Kurtosis-Acceleration had the least variation, both were selected for Experiment 2. The model was trained and tested, on three occasions, each with the same $nPeriods$ and $nBins$, but with new training and test samples. Where both $nPeriods$ and $nBins \in \{10, 20, \ldots, 100\}$, for each set of 3 tests: method parameters were adjusted such that every combination of $nPeriods$ and $nBins$ was tested. For each set of three tests, mean accuracy was calculated. Figure 4 and Table 3 show the resulting data from Crest-Acceleration. Figure 5 and Table 4 show the results for Kurtosis-Acceleration.

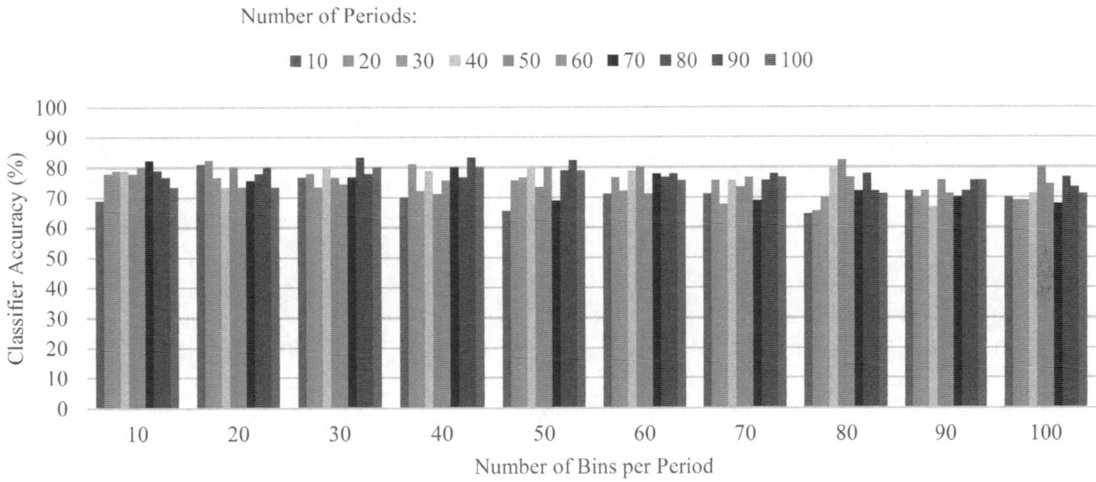

Figure 4: Mean Classifier Accuracy with Crest-Acceleration, Varying Period and Bin Quantity, n=3

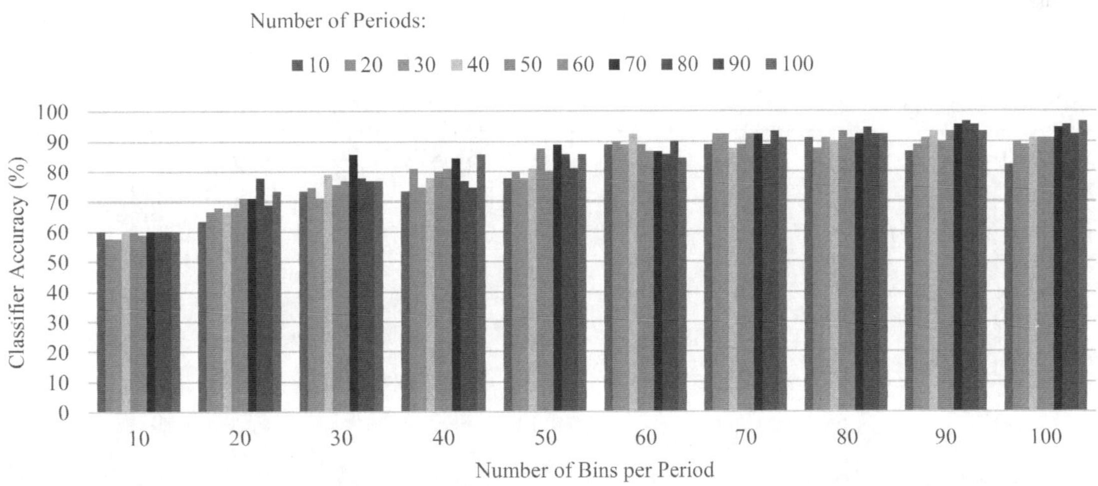

Figure 5: Mean Classifier Accuracy with Kurtosis-Acceleration, Varying Period and Bin Quantity, n=3

Figure 5 shows raising the number of bins per period raises classifier accuracy, plateauing where $nBins > 60$, for the Kurtosis-Acceleration feature set. There is no clear improvement with either feature set while varying period quantity, and the Crest-Acceleration feature set shows no improvement with varying either parameter. The maximum accuracy achieved is 97% with the Kurtosis-Acceleration feature set where $nPeriods = 80$ and $nBins = 90$; and where

$nPeriods = 100$ and $nBins = 100$ (see Table 4). The CA feature set achieved at most 83% accuracy where $nPeriods = 80$ and $nBins = 30$; and where $nPeriods = 90$ and $nBins = 40$ (see Table 3).

Table 3: Mean Classifier Accuracy with Crest-Acceleration, Varying Period and Bin Quantity, n=3 [a, b]

	Number of Bins per Period									
Number of Periods	10	20	30	40	50	60	70	80	90	100
10	69	81	77	70	66	71	71	64	72	70
20	78	82	78	81	76	77	76	66	70	69
30	79	77	73	72	77	72	68	70	72	69
40	79	73	80	79	80	79	76	80	67	71
50	78	80	77	71	73	80	73	82	76	80
60	80	73	74	76	80	71	77	77	71	74
70	82	76	77	80	69	78	69	72	70	68
80	79	78	83	77	79	77	76	78	72	77
90	77	80	78	83	82	78	78	72	76	73
100	73	73	80	80	79	76	77	71	76	71

[a] Highlighted fields show where the maximum accuracy rate is achieved. [b] Data units= %

Table 4: Mean Classifier Accuracy with Kurtosis-Acceleration, Varying Period and Bin Quantity, n=3 [a, b]

	Number of Bins per Period									
Number of Periods	10	20	30	40	50	60	70	80	90	100
10	60	63	73	73	78	89	89	91	87	82
20	58	67	74	81	80	90	92	88	89	90
30	58	68	71	74	78	89	92	91	91	89
40	60	67	79	78	81	92	88	90	93	91
50	60	68	76	80	88	89	89	93	90	91
60	59	71	77	81	80	87	92	91	93	91
70	60	71	86	84	89	87	92	92	96	94
80	60	78	78	77	86	86	89	94	97	96
90	60	69	77	74	81	90	93	92	96	92
100	60	73	77	86	86	84	91	92	93	97

[a] Highlighted fields show where the maximum accuracy rate is achieved. [b] Data units= %

EXPERIMENT 3 – OPTIMIZED ALGORITHM COMPARISIONS:
The optimization test was conducted with only three samples per parameter set to minimize the time required to run the test. Therefore, the test was repeated for the best performing classifier with the lowest parameter constraints and a larger sample size. Table 5 shows summary statistics for accuracy with Kurtosis-Acceleration where $nPeriods = 80$, $nBins = 90$, and $n = 50$. For comparison, the identical training and test samples were classified using 1) the presented algorithm in this paper using only characteristic bins, *Algorithm A*; 2) the same algorithm using all bins where sample weights= 0.01 (prevents division by 0), *Algorithm B*; and 3) the Multinomial Naïve Bayes method from Scikit-Learn [14], also where sample weights= 0.01, *Algorithm C*. Finally, the training time is measured and recorded for *Algorithms A & B* and *Algorithm C* in Table 7.

Table 5: Classifier Accuracy Summary Statistics with KA, 80 Periods, 90 Bins per Period, and n = 50

	Accuracy (%)						
Algorithm	Mean	Std. Dev.	Min	Max	Median	25%	75%
A	93.267	3.531	86.667	100.000	93.333	93.333	96.667
B	59.000	1.684	56.667	63.333	60.000	56.667	60.000
C	46.200	5.871	33.333	63.333	46.667	43.333	50.000

Table 7: Classifier Training Time

	Time (seconds)						
Algorithm	Mean	Std. Dev.	Min	Max	Median	25%	75%
A & B	221.121	21.464	206.639	285.871	219.027	208.281	220.406
C	0.001	0.001	0.000	0.003	0.001	0.001	0.002

DISCUSSION:
It logically follows that raising the bin quantity per period will raise accuracy with nearly all features:

1. More bins per period creates a greater chance for a bearing health class to occupy a bin; and therefore, creating more *characteristic bins* for the class.
2. With more characteristic bins, there is a greater opportunity for a test set to have more matching characteristic bins for any class.
3. Since only the intersection of bins are evaluated, those additional factors can be considered leading to a more accurate result.

One exception, as we saw, was Crest-Acceleration where improvements were insignificant while raising bin quantity, peaking at 40 bins. Kurtosis-Acceleration, on the other hand, outperformed Crest-Acceleration by raising bin quantity. Notice that Kurtosis-Acceleration shows a clearer distinction between flaw types (see Figure 2) which was detected by raising the bin quantity.

Future development can be achieved with these suggestions:

1. Build two Bayesian Models: 1) for all conditions, and 2) for only ball fault, outer race fault, and healthy conditions. First, compare the unknown to the first model. If the label is not an inner race or combination fault, then compare it to the second model. Inner race and combination faults are easily distinguishable among most features. Since the second model removes inner race and combination fault data, bin sizes will be smaller and therefore will more easily capture characteristic bins for the conditions with tighter margins.

2. Determine bin quantities based on the smallest margin between conditions per period. Rather than using a fixed bin quantity per period, dynamically scaling bin quantities controls for variances between periods.
3. Determine methods to lower training time. For example, using best practices in code mechanics.

There is a significant trade-off between training time for accuracy while considering the methods in this paper over a traditional multinomial Naïve Bayes algorithm. Although training time amounts to roughly 3 ½ minutes, the outcome is upward from 90% accuracy (93.3%, 95% CI [92.3, 94.3]) – which is worth the wait for most applications of this nature.

In conclusion, the Naïve Bayes model and Characteristic Bin Utilization algorithm presented in this paper can be a steppingstone for unlocking a powerful tool for bearing fault detection. This tool will not only benefit the Wind Power industry, but all industries that rely on continuous rotational machinery operation.

ACKNOWLEDGEMENTS:

Post University – Here, I was pushed hard to quickly learn the foundational and advanced data science tools utilized in the work presented in this paper.

Schulz Electric (a Timken Company) – The many years spent with Schulz Electric helped me gain appreciation for motor shops and their customers, faced with the challenging feat of troubleshooting and preventing mechanical faults.

REFERENCES:

[1] U.S. Energy Information Administration, "What is U.S. electricity generation by energy source?," 2023. [Online]. Available: https://www.eia.gov/tools/faqs/faq.php?id=427&t=3.

[2] G. Bilicic and S. Scroggins, "LAZARD'S LEVELIZED COST OF ENERGY ANALYSIS—VERSION 16.0," LAZARD, 2023. [Online]. Available: https://www.lazard.com/research-insights/2023-levelized-cost-of-energyplus/.

[3] Alliance for Sustainable Energy, LLC, "Wind Turbine Gearbox Damage Distribution Statistics 2016," 2023. [Online]. Available: https://grd.nrel.gov/.

[4] Z. Zhang, "Naive Bayes Explained," 14 April 2019. [Online]. Available: https://towardsdatascience.com/naive-bayes-explained-9d2b96f4a9c0.

[5] N. Zhang, L. Wu, J. Yang and Y. Guan, "Naive Bayes Bearing Fault Diagnosis Based on Enhanced Independence of Data," *Sensors,* vol. 18, pp. 463-480, 2018, doi: 10.3390/s18020463.

[6] X. J. Yi, Y. F. Chen and P. Hou, "Fault diagnosis of rolling element bearing using Naïve Bayes classifier," *Vibroengineering PROCEDIA,* vol. 14, pp. 64-69, 2017, doi: 10.21595/vp.2017.19153.

[7] N. Zhang, L. Wu, Z. Wang and Y. Guan, "Bearing Remaining Useful Life Prediction Based on Naive Bayes and Weibull Distributions," *Entropy,* vol. 20, no. 12, p. 944, 2018, doi: 10.3390/e20120944.

[8] H. Huang and N. Baddour, "Bearing Vibration Data under Time-varying Rotational Speed Conditions," *Mendeley Data, V2,* 2019, doi: 10.17632/v43hmbwxpm.2.

[9] K. H. Hui, C. S. Ooi, M. H. Lim, M. S. Leong and S. M. Al-Obaidi, "An improved wrapper-based feature selection method for machinery fault diagnosis," *PLoS ONE,* vol. 12, no. 12, pp. 1-10, 2017, doi: 10.1371/journal.pone.0189143.

[10] National Institute of Standards and Technology, "Measures of Skewness and Kurtosis," [Online]. Available: https://www.itl.nist.gov/div898/handbook/eda/section3/eda35b.htm.

[11] Mobius Institue, "Vibration Analysis Dictionary: Terms - Mobius Institute," 2022. [Online]. Available: https://www.mobiusinstitute.com/vibration-analysis-dictionary/.

[12] S. Glen, "Shape Parameter: Definition and Examples," 2023. [Online]. Available: https://www.statisticshowto.com/shape-parameter/.

[13] The MathWorks, Inc., "Signal Features - MATLAB & Simulink," 2023. [Online]. Available: https://www.mathworks.com/help/predmaint/ug/signal-features.html.

[14] F. Pedregosa, G. Varoquaux, A. Gramfort, V. Michel, B. Thirion, O. Grisel, M. Blondel, P. Prettenhofer, R. Weiss, V. Dubourg, J. Vanderplas, A. Passos, D. Cournapeau, M. Brucher, M. Perrot and E. Duchesnay, "Scikit-learn: Machine Learning in Python," *Journal of Machine Learning Research,* vol. 12, pp. 2825-2830, 2011.

VIATC TRAINING CONFERENCE & EXPO
COVINGTON, KENTUCKY | AUGUST 7-9, 2024

CREATING A VIRTUAL ASSISTANT TO VIBRATION ANALYST

Mantosh Bhattacharya
Petrofac International UAE
Mantosh.b@petrofac.com

Mantosh Bhattacharya is designated Subject Matter Expert on Turbomachinery in Petrofac Intl. located in UAE. He is a chartered engineer CEng (MIMechE-UK), Intl. PE (UK), a member of ASME, Vibration Institute and ISCM – (BINDT UK). He is a CAT –IV vibration analyst and member of advisory committee of ATPS [TEES – Texas A&M] He is the member of HI 9.6.4, 9.6.5 and 14.5.1 standard committees. He holds 3 U.S. patents related to structural dynamics. He has published and presented several technical papers in journals and conferences.

ABSTRACT:

Vibration analysis is a multi-layered task and with an enormous number of assets in a large plant, the work becomes quite tedious for specialists to review vibration data for hundreds of machines, sometimes in multiple facilities. With the sheer volume of raw vibration data from data loggers and CMMS (continuous machine monitoring system), it is impractical to properly analyze every routine collection event on every machine. Even simple, four-bearing machines have 12 measurement points for vibration.

This paper proposes to create an envelope based monitoring method with pre alarm set for each frequency band or discrete peaks using expert sources of vibration analysts. Then this data is formatted into csv file and excel then shall be compared and classified as typical fault class. The fault class shall be predetermined by using slicers with band based fault frequencies. The data is then analyzed using a slicer to detect the most probable cause to bring attention to the vibration analyst. The created expert system can pick an anomalous machine using an easy to train machine learning algorithm including process operating parameters. This means a virtual assistant can lower down the quantum of work to vibration analyst.

KEY WORDS: Vibration Spectrum, Envelope Alarm, Slicer, Support Vector Machine

MOTIVATION:

Traditional methods of vibration signal processing were proposed more than 50 years ago and are still actively used. Moreover, vibration-based condition monitoring has traditionally been an expert-oriented field of knowledge, and a human as expert is not physically able to work with a large amount of data. Due to paucity of experienced vibration analyst (one analyst per more than

500 machinery tags) there may be the uncertainty of diagnostics, false alarms, missed opportunities, which can lead to a less than fully successful vibration monitoring program. The paper proposes a method by leveraging Artificial Intelligence with Spectrum based diagnosis which can serve as a "virtual assistant" to vibration analyst. This means the virtual assistant shall filter the machines of concerns based on AI based classification algorithm and then prioritize the cases to be forwarded to specialist for the final diagnosis and remedial action.
Common machinery diagnostic processes use FFT spectrum in day-to-day data acquisition activities and analysis. This paper has considered the FFT spectrum as most of moderately priced data analyzers have the feature to display FFT spectrum with peak amplitude marker.

BACKGROUND KNOWLEDGE AND OBSERVATION

International standards such as ISO 20816, ISO 10816-3 and ISO-13373 are aimed primarily at normal operational monitoring of broad-band vibration. Evaluation criteria are defined in terms of the normal steady-running value of broad-band vibration and any changes that may occur in the magnitude of these steady values' However it is to be noted that conventional vibration-monitoring equipment measures the magnitude of the overall complex signal and does not differentiate between the individual frequency components This criterion has limitations because some changes may only be identified by vector analysis of the individual frequency components. Criteria for assessing changes in broad-band vibration for monitoring purposes is not adequate in operating machines, some changes may not be detected unless the discrete frequency components are monitored. However, these standards are still considered as guidelines to determine to trigger pre-alarm by a reliability centered program as proposed in this paper. Rate of change of vibration amplitude at particular fault frequency also shall be one of major outliers. The pre alarm value is set based on benchmark data vibration of a newly commissioned operating machine or by available data of operating machine of same type and power rating.

VIBRATION SPECTRUM

With an assumption that most of machinery vibration are periodic in nature, spectral analysis offers a convenient way to characterize the vibration profile (the relationship between vibration magnitude and frequency) with both broadband and narrow-band components. One of the important aspects of using the spectral properties of vibration is to fix the Fmax FFT span (Nyquist frequency) and spectral lines (LOR) to capture all possible fault frequencies. Refer figure 1 for sample FFT spectrum set up for a data logger. However, we recommend using more averages for refined data set.

Analysis of the vibration spectrum can be primarily divided into 3 areas:

1. Sub-synchronous is less than 1x turning speed (TS) of the shaft and can identify problems with oil whirl/whip, rubs, loose bearing, cage, or train frequency of antifriction bearing, primary belt frequency, defective tooth-to-tooth repeat frequency, flow recirculation and surge.
2. Synchronous is a 1x turning speed (TS) of the shaft and super-synchronous exact multiple integers of TS and can identify problems with imbalance, misalignment, looseness, bent

shaft, vane/blade pass, gear mesh, belt sheave pitch-line runout, broken gear tooth, sleeve bearing, resonance.

3. Nonsynchronous is not equal to any multiple of TS of the shaft and can identify problems with rolling element defects, electrical, other in the machine, one or more system resonances, belts, noise, cavitation, lubrication faults. Electrical faults occur as a multiple of the frequency of the supply/current.

Machine Type	RPM	BRG Type	Orders x RPM	Fmax - Hz	LOR	Averaging	# Averages
Motor/Pmp	1,200	Roller	50	1000	1600	Normal	6
Motor/Pmp	1,800	Roller	50	1500	1600	Normal	6
Motor/Pmp	3,600	Roller	50	3000	3200	Normal	6

Figure 1 - Set up of FFT spectrum for a typical centrifugal pump.

Narrowband Envelope Alarming -

Basically, the narrowband envelope alarm is a line drawn above the outline of the spectrum collected. A few vibration analysis software packages may automatically generate alarms using an existing spectrum based on built-in configuration or configured based on end user suggested limits by end user. If any peak breaks this line, the next time data is uploaded, the alarm is triggered, regardless of its amplitude. Hence, even a low amplitude changes at any specific frequency will be immediately detected and pointed out by the narrowband envelope alarm. To avoid these from being triggered every time the signature changes, the narrowband envelope alarms can be trend-based. Certain vibration frequencies (depending on machine rotating speed) relate to specific groups of machine faults—for example, rotor mass imbalance will cause 1x rpm vibration in the radial directions. Based on fault matrix of machine of particular type (sample diagram as shown on figure 2), the bands can be created.

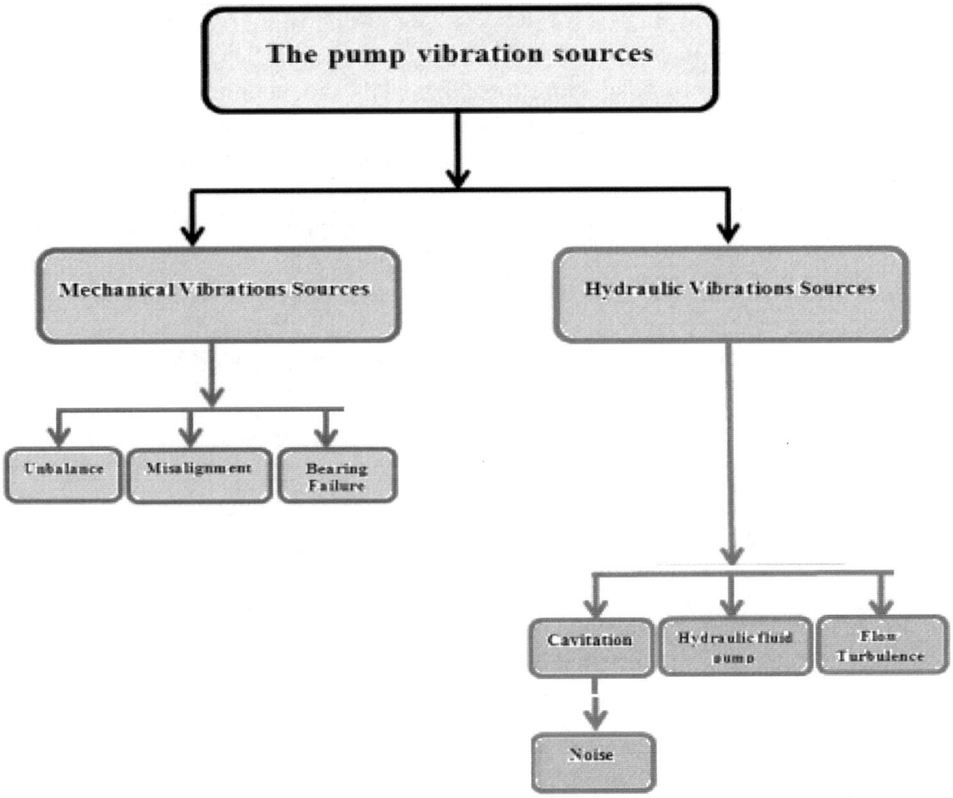

Figure 2 -Generic fault matrix of a centrifugal pump

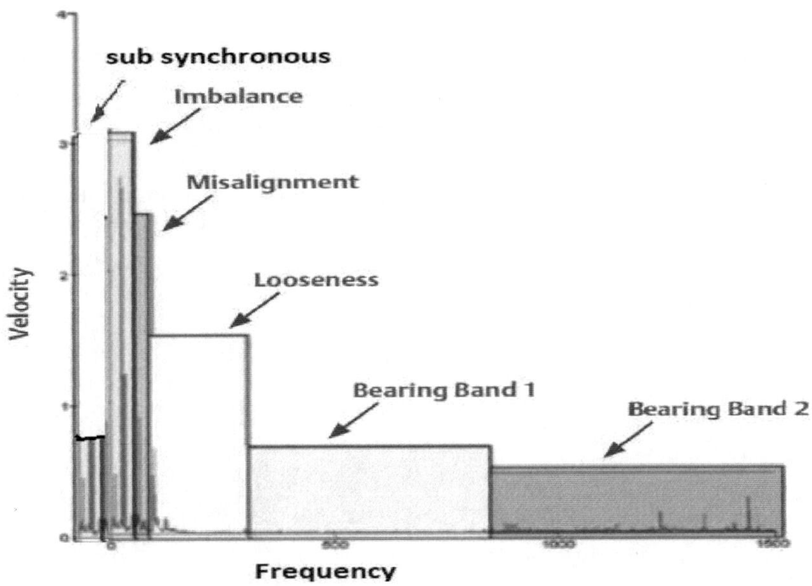

Figure 3- Narrow band spectral alarm set up.

When overall vibration value exceeds the pre- alarm value the software starts trending or based on formatted data acquisition chart, the person who is conducting route based vibration measurement registers this anomaly first. After that the configured system identifies the change in particular band. This avoids huge amount of data handling. Similarly, the technician can take readings and check particular frequency band with a handheld vibration data logger with FFT display feature. A pre alarm can be set for each frequency band, which means low amplitude vibration from an early fault could trigger an alarm, even if the vibration at 1x rpm is much higher. The spectrum data is further converted into a csv file with harmonic and delta annotation. In this way the analyst's focus is not diverted to the highest peak but is focused on the index of incipient damage. In place of bands based on frequencies, bands related to order / harmonics also can be displayed and reviewed. A variant of this type of FFT spectra comparison is called band control. In this method, a maximum level of amplitude can be defined for a given spectrum band as shown in figure 4 as an illustration.

Percentage of Overall Value for each Narrow Band Alarm			
Roller Bearing -			
Band Name	Band No.	Percent	Band Fault
Sub. Synchronous	1	15 – 20%	Cage Defect
1 x rpm	2	80 – 90%	Imbalance/Misalignment
2 – 4 x rpm	3	3	Misalignment/Looseness
5 – 20 x rpm	4	30 – 40%	1ST Bearing Band

Figure – 4 – Range of amplitude heights at various bands

With the use of sensible narrowband envelope alarms, work of vibration data tabulation is reduced by over 80 percent and diagnostics are increased immensely.

Experienced analysts also know the common causes of variation in vibration from certain machine faults and use that knowledge when analyzing vibration data. The below table illustrates common fault of electric motor driven pump with anti-friction bearings.

This knowledge is then reduced to a set of rules for relevant types of machinery, its construction, and operating parameters. Based on the set of rules, a fault-based frequency table for each type of machine is constructed. A sample fault table is shown in figure 5 prepared by an expert vibration analyst.

SN	Fault description	Fault identification in FFT
1	**Imbalance**	1X : Dominating in FFT higher in Horizontal direction
2	**Misalignment - Parallel**	1X, 2X, 3X seen in FFT
		2X : Dominates in FFT (2X is higher than 1X & 3X)
3	**Misalignment - Angular**	1X, 2X, 3X seen in FFT
4	**Bend shaft (OH type)**	1X, 2X in axial direction seen in FFT
5	**Vane/blade passing**	VPF/BPF seen in FFT along with 1X
6	**Rotating looseness**	1X and its harmonics in radial direction seen in FFT
		1X till 10 X : can be seen in FFT.
		0.5X seen along with harmonics in FFT in case of excessive looseness.
7	**Cavitation**	High frequency noise (hump in high frequency area of FFT)
8	**Bearing fault (Roling element bearing)**	Ball pass inner race (BPI) fault, Ball pass outer race (BPO) fault, Cage fault & ball spin fault -Peak at non-synchronous frequency (e.g. 3.9X, 6.4X etc.)
9	**Flow recirculation**	Sub synchrnous vibration

Figure 5 – Sample fault chart Centrifugal pump PROPOSAL

With supervised machine learning algorithms, the first layer of task can be offloaded to an "easy to learn and implement" AI based module. The trained module, then, can be termed as Virtual Assistant to Vibration analyst.

Direct unfiltered numerical values of vibration are proposed be used as marker of pre alarm and alarm as first step of the proposed method. In this case we can use linear regression model to identify good or bad units and another model can advise the severity of unit. This is done with support of expert sources. We select the fault frequencies of particular type of machines in tabular form and allocate allowable vibration amplitude for each fault frequencies. In this way none of the fault frequencies are masked. We create a large data set for such values with multiple columns of fault frequencies based on FFT spectrum generated during the operation of machine. Based on fault matrix of unit, this data set can be considered as training data set of particular class of fault. Hence, based on fault matrix (probable damage index) data sets for many fault classes are created. In a similar method we create a testing data set. We have two choices, either we solve multi class classification problem or break down all fault class into one to one classification problem. Once the data set is ready for testing, we measure the success rate of prediction of algorithm.

In between the process, we can add weightage / modify attributes in decision making tool based on expert opinion. This can be considered with a back-propagation method of refinement. This way the algorithm gets continuously refined and evolved for accuracy. Once the desired success rate is achieved, the system can deploy as Virtual assistant to vibration specialist.

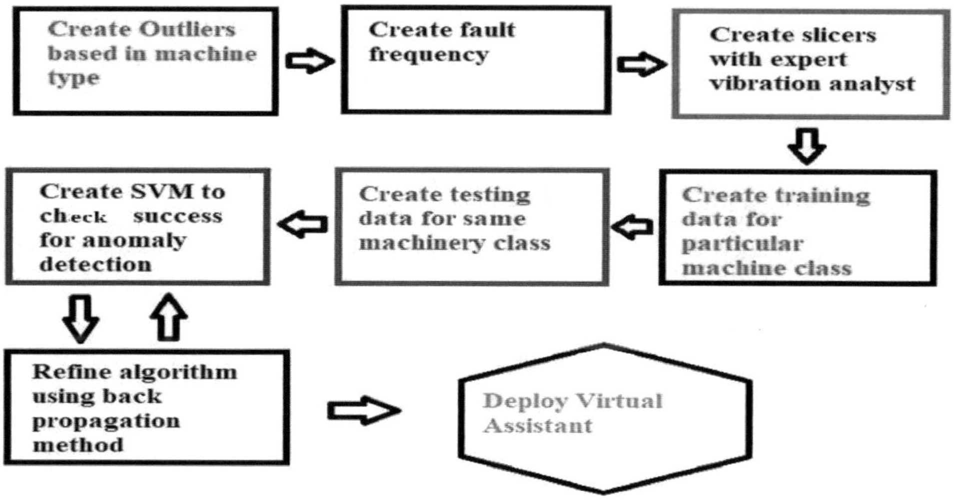

Figure 6 -Proposed workflow

Advance excel method with support of expert vibration analyst is considered to create a tool. This tool utilizes few classifiers - Process parameters, noise, and machinery vibration parameters with spectral peaks. The sample shown below is the fault frequency table from which the most possible fault can be known using Power BI tool using slicers as shown in figure 7. This tool shall serve to create biases and weightage to fine tune the algorithm used as diagnostic tool. However, biases and weightage are highly dependent on skill and depth of the experience of the expert or group of experts. Hence a good measure of the accuracy of output is based on confusion matrix and back propagation algorithm. There are condition monitoring software vendors that provide rules-based (knowledge-based) algorithms that assess the vibration spectra data. Accuracy is highly dependent on depth of the experience of the person writing the rules. A good vibration analyst will always make sure she or he understands the machine configuration, mounting and has good reference of failure cases where vibration signals were precursor of incipient failures. Then only she/he can attempt to analyze spectral data and create the expert library based on various failures modes on similar type of machine.

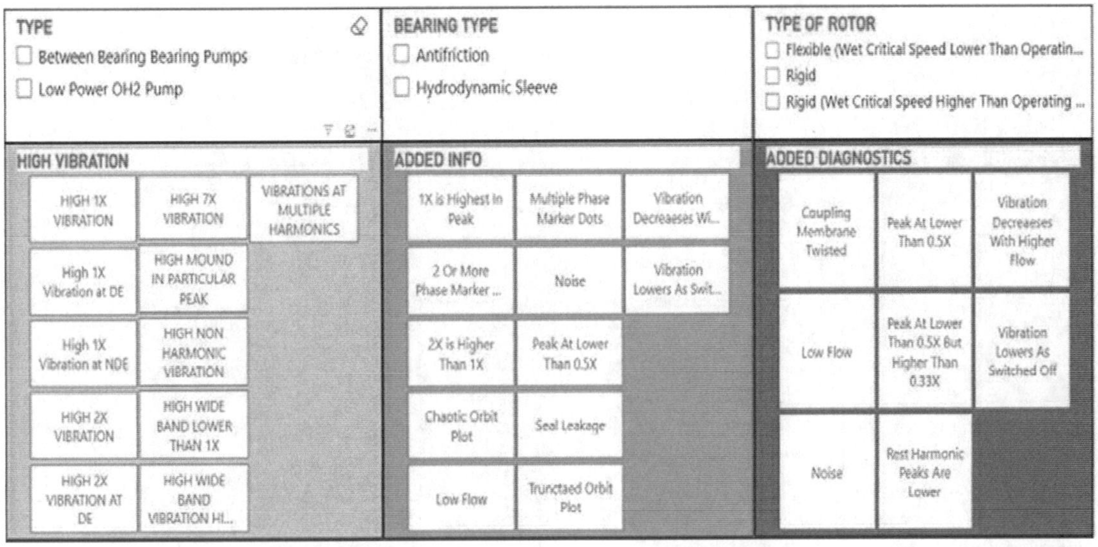

Figure 7 - Slicer used in expert library for creating weightage.

In addition to installed continuous vibration monitoring system, most of handheld vibration meters can list out top five frequencies based on vibration amplitude measured as shown in figure 8. A technician from the reliability department who is assigned to take route-based vibration data can save the screen shot and then populate the data in a table in form of an excel or csv file. Some analyzers have features which can export results to Microsoft® Excel through a USB connection and trend the readings with the built-in templates and plot graphs.

Spectral Band Analysis -

Each spectral band can be considered as rectangular blocks (shown in Figure 3) of various sizes in an FFT spectrum with fixed spectral lines and Fmax FFT span (Nyquist frequency). then the comparison of area of band can be carried out using simple finite integration process. This can be used manually by multiplying difference the start and stop frequency of that band and amplitude at that band. For better accuracy, narrow bands should be created as much as possible avoiding overlapping of signals. Other than manual method, free downloadable software such as Python can perform the task and compare area of spectral band. Increase in area of particular band is an indicator of change in dynamic behavior of unit.

Most of the data acquisition systems provide full spectrum and labels in each Peak for the corresponding frequency as shown in figure 8.

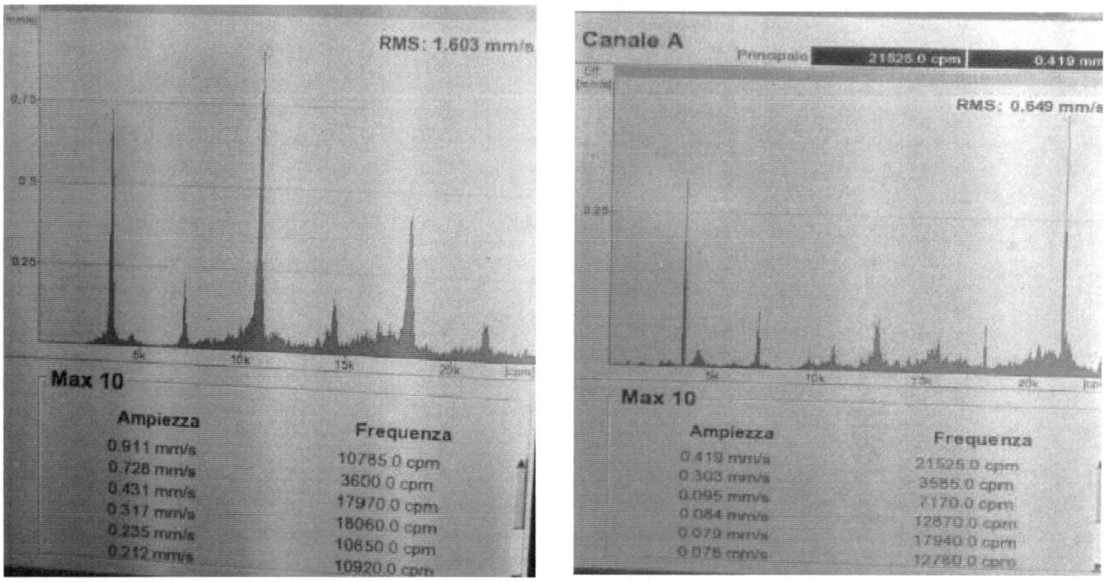

Figure 8 – Fault frequency listing by data logger

Managing vibration data -

The proposed method in this paper is a trigger-based data acquisition system, only useful streaming data shall be generated as a spectrum plot. The spectrum plot shall be configured in terms of identified peaks as numerical values which are segregated in form of distinct bands. Each distinct band shall have its own rectangular area These classifiers are proposed to be used as a data set for training, testing and then employing machine learning algorithms to generate an alert for making the task of vibration analyst easier for corrective action. This means if a high vibration peak at a particular location of a spectrum is detected, this will be compared with existing data set and machinery anomaly can be reported by ML algorithm for further review of experienced vibration analyst. A data set of vibration spectrum using past spectrum plots of similar machine or benchmark data of same machine can be created. If such data sets are not available a machinery simulator can be used. With a dataset created in forms of X-Y coordinates and area of selected spectral bands, the machine learning algorithm can be put into use as supervised mode with coordination of expert system for adding weightage. The proposed expert system consists of data base of similar type assets and a reasoning engine in an excel sheet. The ML algorithm shall use simple language which can be operated with intermediate knowledge of coding. In parallel a decision tree method can be used with data sets provided based on expert system. As the test data provided to module is continuously validated backed by confusion matrix in form of gradual maturity, the module can then later be used as virtual assistant to a vibration analyst.

Deployment of manpower -

One of the foundations of the proposed method is to engage the personnel of various levels of experience and skill sets as listed below:

1. Semi-skilled vibration data collector (technician) is required as part of reliability team. ISO certified CAT -I vibration analyst may not be required, as an induction training to collect data with vibration analyzer is enough.
2. Qualification of person for execution of program -Based on present curriculum of undergraduate syllabus of mechanical engineering, few well practices software and tools are taught in syllabus such as MS excel advanced, Python or MATLAB. Hence a less experienced engineer
3. can be employed to sort out data and then utilize such codes to find the outputs as diagnosis. These codes are available at opensource library and downloadable codes and have good potential to easily implement in proposed cycle of work. However, for simplicity and adaptability we have selected Excel based one to one SVM classification algorithm.
4. Experienced vibration analysts pool for support to create fault frequencies data base and supporting with weightage and biases to back propagation method for tuning the code.

WORKING MODEL

This method is proposed to be utilized first with Balance of plant machinery (criticality rating 2). Since the data set shall be in the form of numbers and adopts binary classification, the processing shall take less machine time.

Supervised machine learning algorithms are the most common algorithms. 3-axis vibration data as well as the measurement specific RPM (revolutions per minute) values are collected for creating a data set. The entire data set of machinery vibration pattern of particular type of machine is divided into various fault class Supervised machine learning is chosen when the acquired labeled data requires skilled resources in order to train it and identify inferred actions to take when the system recognizes a fault pattern. Then the data sets are divided into 80-20 where 80% of data is used in training and 20 percent is used in testing. Training data sets are typically first recognized via human intervention and the system is then taught to recognize patterns associated with each applicable training data set. Then the system compares those training data sets to newly acquired data sets. The system can identify and predict targeted faults after sufficient training data sets are available to compare to. This type of machine learning can further compare its own decided upon action with the corrected intended action and identify errors that improve the applicable model. The paper considered the code using Support Vector Machine (SVM) with supervised Machine Learning platform. An SVM is used for either regression or classification models. When performing classification, an SVM is used to create a Support Vector Classifier (SVC) to determine a hyperplane. This divides a set of data into various classifications. Data on one side of the hyperplane is one class, and the data on the other side of the hyperplane is another class. A hyperplane is a shape that is one dimension less than the number of data features being used to perform the classification.

Refer Figure 9. In this proposal, we implemented the classification of simulated data using one class SVM in Excel. This algorithm identifies outliers by training on a single class of data, making it ideal for spotting anomalies from datasets. SVM supports binary classification and separating data points into two classes. For multiclass classification, the same principle is utilized after breaking down the multiclassification problem into multiple binary classification problems. SVM is fundamentally a binary classifier, it can be extended for multi-class classification . The One- vs-One strategy splits a multi-class classification into one binary classification problem per each pair of classes.

The formula for calculating the number of binary datasets, and in turn, models, is as follows: (NumClasses * (NumClasses – 1)) / 2

Hence if we have 3 fault classes, we need to solve 3 binary classification problems. If we have 2 fault classes, we have to solve 2 binary classification problems. If we have 8 different fault classes, we need to solve 28 binary classification problems.

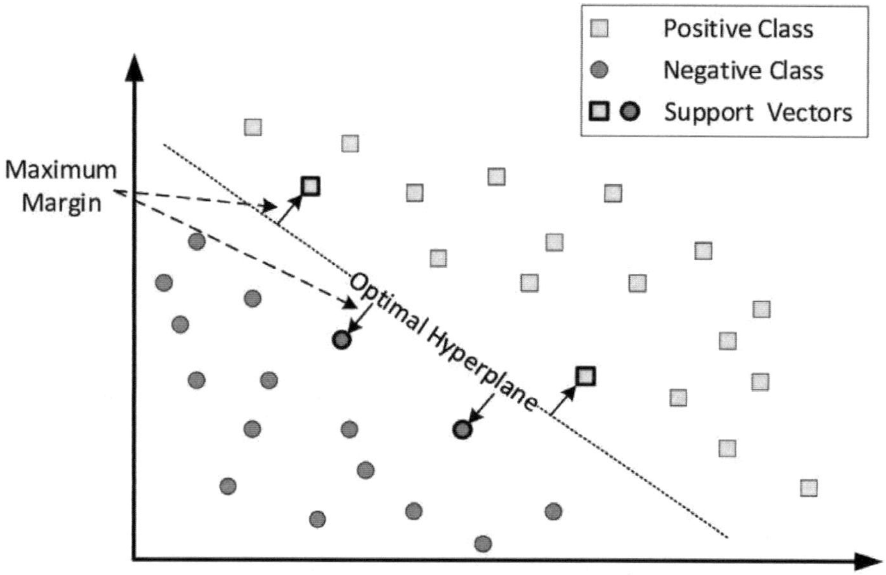

Figure 9 - Typical SVM using binary classification.

The SVM learns to distinguish between the different classes based on the features extracted from the FFT spectrum and can be used to classify fault classes.

We can segregate two different machine conditions using binary classification method in SVM. For example, machine condition identifies two classes based on spectral signature of vibration with training data of various fault frequencies. Using this method, a healthy machine and machine with high 1x vibration is differentiated. The same exercise is carried out with other two machine fault classes utilizing two peaks from two different frequencies using training data. For example, machines with unbalance and with misalignment can be differentiated based on spectral signal. As all fault classes are identified in the training set, then SVM can advise the accuracy level in predicting fault class from training set. It is observed that if the number of iterations is increased, the accuracy level increases, which advises having more training data sets.

The performance of a binary classifier is measured using sensitivity, specificity and accuracy with a term called confusion matrix. Confusion matrix is a term used for defining the accuracy of predictions made on test data. A confusion matrix is a performance evaluation tool in machine learning, representing the accuracy of a classification model. It displays the number of true positives, true negatives, false positives, and false negatives.

TP=Number of True Positive classified cases (SVM correctly classifies) TN=Number of True Negative classified cases (SVM correctly classifies)
FP= Number of False Positive classified cases (SVM labels a case as positive while it is a negative)
FN= Number of False Negative classified cases (SVM labels a case as negative while it is a positive)

A sample is shown in figure 10 where it can be seen that accuracy level increased after further iteration steps.

Accuracy Level is defined as – TP+TN / (TP+TN+FP+FN)

The +1 shows the pump as bad condition whereas – 1 shows condition as acceptable condition. Similarly, +1 and -1 based differentiation can be made for two separate fault classes using two different outliers in columns 2 and 3.

Training Dataset

Iteration	Overall vibration pre alarm	Over all vibration as measured	Pump Status	1/t	B1	B2	Output	B1(t+1)	B2(t+2)	Prediction	Predicted Type	Error	Accuracy
1	2.5	2.7	1	1	0	0	0	4.545454545	4.909090909	0	1	0	
2	2.3	2.2	-1	0.5	4.545454545	4.909090909	-21.2545455	0.181818182	0.454545455	21.25454545	1	1	
3	2.4	2.6	1	0.333333333	0.181818182	0.454545455	1.618181818	0.121212121	0.303030303	1.618181818	1	0	
4	2.5	2.6	1	0.25	0.121212121	0.303030303	1.090909091	0.090909091	0.227272727	1.090909091	1	0	
5	2.5	2.4	-1	0.2	0.090909091	0.227272727	-0.77272727	-0.83636364	-0.69090909	0.772727273	1	1	
6	2.4	1.4	-1	0.166666667	-0.836363636	-0.690909091	2.974545455	-0.6969697	-0.57575758	-2.97454545	-1	0	64
7	2.3	2.5	1	0.142857143	-0.696969697	-0.575757576	-3.04242424	-3.3307E-16	0.155844156	-3.04242424	-1	1	
8	2.6	2.8	1	0.125	-3.33067E-16	0.155844156	0.436363636	0.590909091	0.772727273	0.436363636	1	0	
9	2.7	2.8	1	0.111111111	0.590909091	0.772727273	3.759090909	0.525252525	0.686868687	3.759090909	1	0	
10	2.7	2.5	-1	0.1	0.525252525	0.686868687	-3.13535354	-0.01818182	0.163636364	3.135353535	1	1	
11	2.65	2.7	1	0.090909091	-0.018181818	0.163636364	0.393636364	0.421487603	0.595041322	0.393636364	1	0	
12	2.65	2.5	-1	0.083333333	0.421487603	0.595041322	-2.60454545	-0.01515152	0.166666667	2.604545455	1	1	
13	2.8	3	1	0.076923077	-0.015151515	0.166666667	0.457575758	0.377622378	0.573426573	0.457575758	1	0	
14	2.5	3	1	0.071428571	0.377622378	0.573426573	2.664335664	0.350649351	0.532467532	2.664335664	1	0	
15	2.7	3.1	1	0.066666667	0.350649351	0.532467532	2.597402597	0.327272727	0.496969697	2.597402597	1	0	
16	2.8	2.9	1	0.0625	0.327272727	0.496969697	2.357575758	0.306818182	0.465909091	2.357575758	1	0	
17	2.55	2.6	1	0.058823529	0.306818182	0.465909091	1.99375	0.288770053	0.438502674	1.99375	1	0	90
18	2.7	2.85	1	0.055555556	0.288770053	0.438502674	2.029411765	0.272727273	0.414141414	2.029411765	1	0	
19	3	3.1	1	0.052631579	0.272727273	0.414141414	2.102020202	0.258373206	0.392344498	2.102020202	1	0	
20	2.9	3	1	0.05	0.258373206	0.392344498	1.926315789	0.245454545	0.372727273	1.926315789	1	0	
21	2.9	3.1	1	0.047619048	0.245454545	0.372727273	1.867272727	0.233766234	0.354978355	1.867272727	1	0	

Figure 10 -Training data set from one out of two SVM with accuracy level

Based on above training data for one out of two SVM the test data is verified.

Coefficients	
B1	B2
0.23376623	0.354978355

Prediction on Test Dataset

Overall vibration pre alarm	Over all vibration as measured	Pump Status	Predicted	Actual	Error	Accuracy
2.4	2.6	1.48398268	1	1	0	90
2.7	2.8	1.62510823	1	1	0	
2.65	2.5	1.50692641	1	-1	1	
2.8	2.9	1.68398268	1	1	0	
2.3	3.9	1.92207792	1	1	0	
2.7	2.85	1.64285714	1	1	0	
2.9	3.1	1.77835498	1	1	0	
2.4	2.75	1.53722944	1	1	0	
2.3	2.8	1.53160173	1	1	0	
2.3	3	1.6025974	1	1	0	

Figure 11 -Prediction by testing data set using one vs two SVM

Results, Observations and Future works

In this proposal, it is assumed as a default that all-rotating machines mostly produce cyclo-stationary-periodic signals which are further converted into frequency domain. With a dataset created for training machine learning module the machine learning algorithm can be put into use as supervised mode using test data. The accuracy, confusion matrix is calculated, and back propagation method is employed to further tune the module with coordination of expert system as weightage added. It is to be noted that if we increase the amount of training data and testing data m the accuracy of the result increases. An expert system consists of data base of similar assets and a reasoning engine which can be used for adding weightage / back propagation to refine the algorithm. This approach can provide specialized evaluation for mid-tier plants and/or semi - critical equipment.

CONCLUSIONS

In addition to FFT spectrum analysis, we can use Cepstrum analysis and envelope analysis. Bearing fault index can be found from data base available for each bearing number system. In addition to vibration data, extra numerical values such as temperature, RPM and motor current also can be added for creating a distinct machinery class / group.
For training data, if wireless type vibration sensors are used, they can provide a large amount of data based in csv format to help machine learning algorithm become more accurate.
Bad Vibration Data -Random spikes, peaks, or sudden drops in readings indicate potential sensor malfunctions or measurement errors.
Inconsistent readings, where the same operating conditions produce widely different results, also signal unreliable data.
Ensuring vibration data quality is a multidimensional endeavor that involves identifying and rectifying the root causes of inaccuracies. Proper training on data collection, instrument setup and acquisition practices, and a comprehensive understanding of potential pitfalls contribute collectively. More data beats clever algorithms, but better and cleaner data beats more data. SVM is based on geometrical properties of the data whilst logistic regression is based on statistical approaches.

In Logistic regression exercise, if we increase the iteration then accuracy increases but up to a limit. This can advise the accuracy in test data. Use of Random Forest technique can also give good accuracy.

This is a constantly evolving process as success rate increases based on biases and weightage with back propagation method used. Normally a particular kind of machinery, e.g., horizonal centrifugal pump took two months to create data base, slicers and took two months to see success rate after refinement. Some classification problems may involve more than two classes. We can have two techniques in these kinds of situations: One versus One classification and One versus All classification.

This project task can be offloaded to a software engineer after defining the requirement such as if large and complex data with many attributes are used. A close loop information sharing channel with non-disclosure agreement requires to be established.

This method is utilized first with Balance of plant machinery where antifriction bearings are used. When the algorithm is matured, this can be expanded gradually to critical turbomachinery as test basis.

The implementation of the proposal can greatly lessen the workload of a diagnostic engineer. The paper proposes to employ only those algorithms which are being used in many applications with higher than 90% accuracy rate. Since the data set shall be in the form of numbers and / or in regular geometrical shape in two-dimensional format, the processing shall take less machine time. The AI model can be set to send an alert via text message or email to vibration analyst.

The proposed method is always evolving and requires fine tuning where involvement of a pool of expert vibration analysts is always required. It should not be construed that the presence of a pool of expert vibration analysts shall be phased out.

AKNOWLEDGEMENTS:

Author tenders sincere thanks to Mr. Arnav Bhattacharya and Mr. Ruhul Amin for creating algorithms and slicers as described in this paper and Mr. Narasimham and Mr. Vineet Khanna for providing valuable comments during internal review process.

REFERENCES:

[1] Jim Berry - How to And How Not to Create Meaningful Narrow-Band Spectral Alarm Envelopes. - Mini course Vibration Institute

[2] Eugene Vogel -Techniques for Detailed Screening of Vibration Data Dec 2019 Pump & Systems https://www.pumpsandsystems.com/techniques-detailed-screening-vibration-data

[3] Kenneth R. Piety Mark W. Slemp William F. Bethmann, Jr.- US Patent -Determining narrowband envelope alarm limit based on machine vibration spectra https://patents.google.com/patent/US5922963A/en.

[4] Milo spencer -Harper - How to build a simple neural network in 9 lines of Python code. https://medium.com/technology-invention-and-more/how-to-build-a-simple-neural-network-in- 9-lines-of-python-code-cc8f23647ca1

[5] Angela & Khezan Shi -Linear regression from scratch in Excel

[6] Logistic regression with excel solver.
https://app.myeducator.com/reader/web/1755b/Appendix/fj4cm/

[7] Thin Yin Leong, Leong Yonghui - Simpler Machine Learning Using Spreadsheets: Neural Network Predict
https://dx.doi.org/10.26417/ejef.v4i1.p124-138

[8] Grover V. Zurita and Diego Román Cabrera -A review of vibration machine diagnostics by using artificial intelligence methods - https://www.researchgate.net/publication/314782455
[9] Pump Vibration Analysis by Brian P. Graney, MISTRAS Products & Systems
https://www.pumpsandsystems.com/pump-vibration-analysis

[10] MATLAB for Support Vector Machine https://www.mathworks.com/help/stats/support-vector-machines-for-binary-classification.html

BIBLIOGRAPHY:

[1] ISO 13373- Condition monitoring and diagnostics of machines — Vibration condition monitoring — Part 3:

[2] Vibration Spectrum Analysis: A Practical Approach, Industrial Press, Inc. by Steve Goldman
[3] Maurice Adams - Rotating Machinery vibration from analysis to trouble shooting.
https://doi.org/10.1201/9781439847558

[4] ISO/IEC 22989:2022(en) Information technology — Artificial intelligence — Artificial intelligence concepts and terminology

[5] ISO/IEC 23053:2022 Framework for Artificial Intelligence (AI) Systems Using Machine Learning (ML)

[6] ISO/IEC 24668:2022(en) Information technology — Artificial intelligence — Process management framework for big data analytics

VIATC TRAINING CONFERENCE & EXPO
COVINGTON, KENTUCKY | AUGUST 7-9, 2024

CBM ANALYSIS USING LIMITED VIBRATION DATA: CLASSIFICATION OF PAPER MACHINE BEARING DEFECTS USING ML MODELS & RUL PREDICTION

Lloyd Dewey Lee, Jr. "Buddy"
AI/ML SME, Mesa Associates, Inc.
llee@mesainc.com

After graduating from the University of Tennessee with a BSEE in 1987, Buddy was hired as a General Engineer at the Department of Energy in Oak Ridge. In January of 1989, he joined CSI as an Applications Engineer. Buddy formed MAARS, Inc. in 1996, where he developed online, permanently mounted vibration and process monitoring systems. Buddy is a co-holder on US Patent # 6,276,213, awarded in 2001, for an Analog Devices 5B module redesigned to add vibration capability and mount on an Opto-22 backplane. In 2009, AESSEAL (pump mechanical seals manufacturer) hired Buddy as the National Accounts Manager. Buddy performed consulting work at customer sites from 2013 to 2017 while completing an MBA at the University of Tennessee. In early 2018 Buddy joined Emerson as an Instructor and in 2021 completed a Master of Science degree in Reliability Engineering (with a minor in Statistics). In October 2022, Buddy joined Mesa Associates, Inc. as an Artificial Intelligence/Machine Learning Subject Matter Expert.

ABSTRACT:
Vibration analysis is a complex process that pulls features from waveform data to identify patterns of machine degradation. Heavy-asset industrial equipment has multiple failure modes and opportunities for improvement and optimization that can be addressed using Artificial Intelligence and Machine Learning (AI/ML). Real-time, detrimental machinery events require complex data analysis from multiple sources. For example, AI models can identify machine anomalies by combining lab reports, FMEA templates, and historical records with real-time process parameters and electrical and vibration data. The AI/ML project was developed to show that ML classification could be applied to limited real-world vibration data sets on industrial machinery. The data set was taken from six dryer rolls on a paper machine. Five rolls were used to train and validate the ML models and one was held out for testing. A total of 30 models were evaluated using the MATLAB Classification Learner App, each with five-fold validation. Five of the models (SVM Coarse Gaussian, KNN Weighted, NN Tri-layered, KNN Cubic, and KNN Medium) performed very well (the last four were 90% test accuracy). One model, the SVM Coarse Gaussian, tested at 100% accuracy with 6 features included (9 excluded), classifying the test data set correctly into operational and degraded classes. This paper and presentation will explore the application of AI/ML and models for condition monitoring parameters such as vibration and calculation of equipment remaining useful life (RUL) in heavy asset plants.

KEY WORDS: Artificial Intelligence, Machine Learning, Digitalization, Failure Modes and Effects Analysis, FMEA, Remaining Useful Life, RUL, Downtime Avoidance, ROI, Vibration Analysis

INTRODUCTION:

Current Practices in Condition Monitoring

Predictive Maintenance programs (PdM) in heavy-asset industrial facilities utilize condition monitoring devices to predict upcoming machinery component failures. Several technologies exist for condition monitoring and prediction of remaining life for industrial machines. Condition Monitoring technologies include oil analysis, ultrasonic (high-frequency sound wave) energy emission detection, vibration monitoring, temperature scanning, thermography imaging, and electrical circuit testing (online and off-line) (J. Nicholas, 2016; J. R. Nicholas, Young, & Anzalone, 2008). Vibration monitoring delivers high diagnostic value for the expended effort to collect data (Crawford, A. R., & Crawford, 1992; Moore, 2012). Vibration monitoring programs have two main methods of data collection, online or permanently mounted systems and portable data collection devices.

Permanently mounted systems are applied to high-criticality and high-risk machines due to the expense of acquiring and maintaining such systems. For example, sensors are affixed to critical machines at the shaft bearing locations. Wire cables are attached to the sensors and run via conduit to panel housings, and power (24V-DC) is supplied to the panels for the sensors to operate correctly. Online monitoring support equipment is expensive to install and maintain, limiting applications for permanently mounted systems to critical and expensive plant assets. The vibration and process (temperature, pressure, flow) data delivered from these monitoring systems is temporally dense and rich in measured parameter diversity. Many data samples from these systems may be acquired per day or even per hour, and machine learning classifications and data analytics are run to identify machine status and classify machine condition (such as good; degraded but functional; or functional failure) and operational status (running or not running; slow roll; heat soak; full speed; tripped; and loaded or unloaded). Vibration amplitude a various frequencies are affected by operational speed, status, and machine or component conditions. Rich and dense features from data sets help simplify machine learning classification, and vibration amplitudes and frequencies are examples of features for trending and statistical analysis that are of interest to vibration analysts to determine and predict machine status. Because permanently mounted systems have high data density from diverse sensors, early and recent machine learning research focused on the large and expensive equipment bases where online systems are applied for risk reduction of catastrophic failure. Alternatively, few machine learning and classification model research projects use limited vibration data sets acquired by portable vibration analyzers, so this project was conceived to determine if machine learning classification models can be successfully applied to such data sets.

Periodic data collection devices are used for condition monitoring of machines that are not as expensive or critical to plant operation, i.e., essential balance-of-plant equipment. Although essential equipment may not shut down the entire plant if a failure event occurs, plant functions are limited, or plant sections are shut down, if essential equipment fails. While less expensive than permanently affixed monitoring systems, the data collected for analysis is limited to plant maintenance personnel's ability to physically acquire the data. Data collection events are typically scheduled as monthly walk-down routes at the machines. This project uses vibration data sets collected in the axial direction on paper machine dryer rolls by a vibration analyst using an Emerson^{TM} 2140 Machinery Analyzer.

Data Acquisition and Trending Vibration Amplitudes

Vibration analysis is performed by maintenance personnel in heavy-asset plant production facilities with very limited data sets. The vibration time waveform collected by a vibration analyst is just a few seconds, or even milli-seconds, of data collected once per month. The analyst records waveform data samples relative to zero time by pressing the Enter button on the analyzer. The minimum tools required for a machinery analyst performing route data collection are a data collector capable of acquiring amplitude and phase data, an accelerometer transducer, and a photo-tach or a strobe light to identify and acquire operating speed. Machines are assumed to be operating in a stable manner, with no transient events such as excessive operating speed or load changes.

After data collection, a time waveform can be displayed of amplitude vs. time. For example, Figure 1 shows a typical waveform display of digitized samples consisting of an ordered pair (x, y) of time and amplitude (time, amplitude) for 4,096 points that are connected by a straight line between samples for 10.66 seconds. Sharp reversals in the time trace indicate impacting from an early bearing defect. The Crest Factor of 3.261 indicates moderate 'peaked-ness' in the signal, and the analyst has flagged this point as Priority 3 which is an early problem (Priority 1 flags require immediate action). The waveform in Figure 1 is from the test data set used in this study. Both the RPM (turning speed of the paper machine roll) and the surface speed of the roll in Feet per Minute (FPM) are calculated and listed in the data block under the date and time. FPM is the metric used to measure the surface speed of the paper sheet for production runs. The operational speed in FPM is converted by vibration analysts in the paper industry to RPM for vibration analysis.

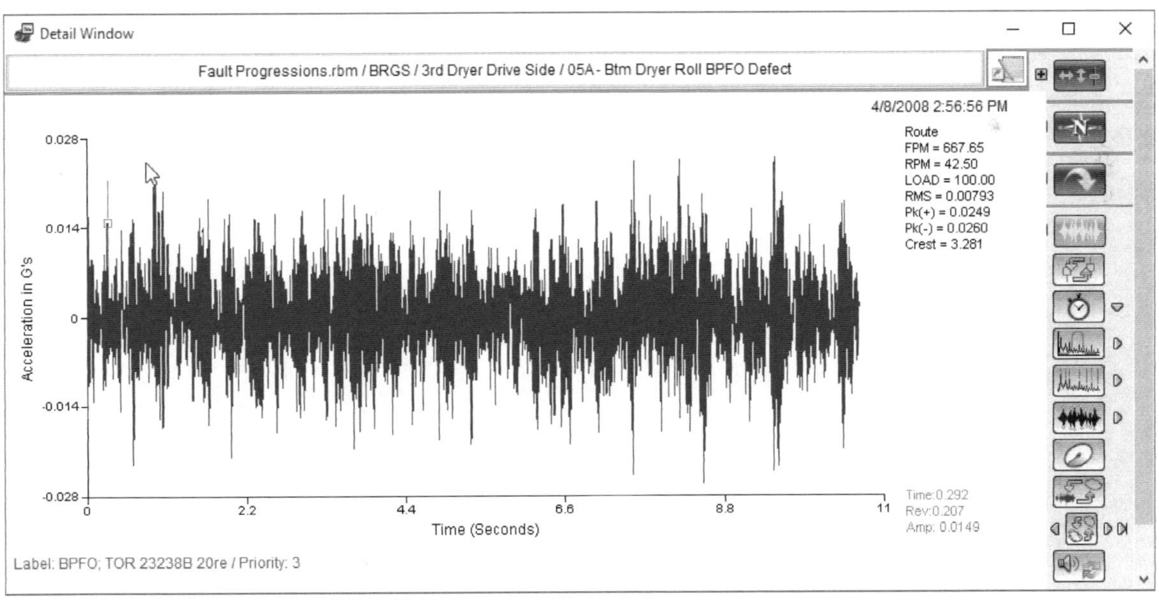

Figure 1: The Time Waveform Display.

From waveforms and Fast Fourier Transforms that are converted to the spectral domain, trends of machinery vibration amplitudes are constructed over several months. Trends identify machine degradation via changing amplitudes under an assumption of non or low-changing operational status. In the real world, such low temporal data density trends are subject to high operational variability. Paper machines are variable speed machines, and the data used for the testing data set varied from a low of 605 RPM to a high of 910 RPM. Variable frequency drives in plant production machinery trains complicates data analysis due to their inherent operational speed variation.

Trend of Paper Machine Rotational Speed Variation.

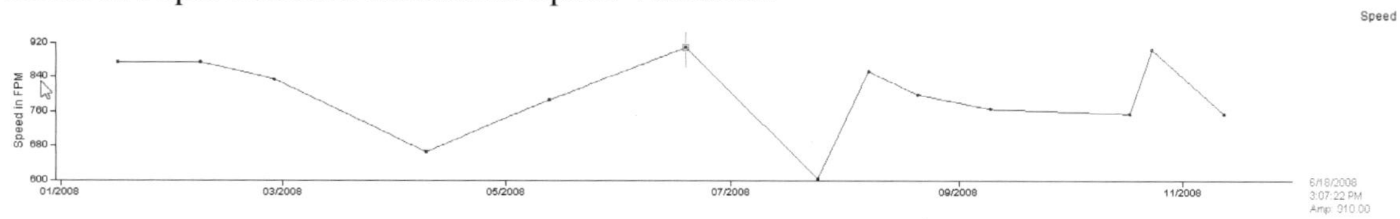

Figure 2: Trend of Paper Machine Rotational Speed Variation.

However, many successful vibration monitoring programs exist in heavy-asset plants where subject-matter experts have *a priori* knowledge of machinery component conditions and past operational history, such as proximity-to-resonance operation. The combination of a subject matter expert's (SME) knowledge with machine condition data drives the success or failure of a PdM program. Unfortunately, sometimes surprise breakdowns result in a significant unplanned machine or process downtime, even with monthly vibration data collection events. When problems arise, the periodic monitoring schedule may increase to daily and even once per shift on critical equipment.

Bearings have four components, the inner and outer bearing raceways, the rolling elements, and the cage that holds the elements in place as the bearing rotates. Figure 2 presents a typical bearing. These four components fail individually but also in tandem with each other at times. When running in a degraded condition, bearing tones (frequencies) are generated at the four characteristic bearing frequencies, called Ball Pass Frequency of the Inner Race (BPFI), Ball Pass Frequency of the Outer Race (BPFO), Ball Spin Frequency (BSF), and Fundamental Train Frequency (FTF) which is generated when a bearing's cage is degraded. Typically, the energy generated by the FTF is sub-synchronous to the shaft turning speed, while the others are above the shaft turning speed and are also not synchronous with the shaft speed. Bearing faults are the most common machinery fault type and account for 30% to 40% of all machinery failures (Zhang, Zhang, Wang, & Habetler, 2020).

Figure 3: Bearing Components.

Vibration Analyst Workload

Consider a typical motor-pump machine train in an industrial facility. Up to 14 data collection locations may be defined on one machine train, typically in the Horizontal, Vertical and Axial directions on each bearing. Every point collected will have waveform and spectral data sets and up to 12 or so trend points. These trend points can be related to the waveform, the spectral data using frequency bands, or statically entered points such as pounds per square inch (PSI) or machine load percentage. Additionally, vibration points may be collected on the foundation, base or pedestal of the machine. Add other technologies in the mix, such as Motor Circuit Analysis

(MCA) or oil analysis and the amount of data that the analyst is expected to evaluate and analyze can easily become overwhelming. Depending upon the application, a vibration analyst may collect an average of 200 to 500 vibration points in a single day.

Astute vibration analysts may use statistics to lighten their workloads. For example, only machines in a Fault or Alert Condition might be flagged for further analysis, if accurate alarm levels are developed and maintained. On some machines, only one point may be in an Alert or Fault condition in any single data collection event. To simplify analysis, bar charts are used to scan an entire machine train quickly. See the Parameter Profile as shown in Figure 4.

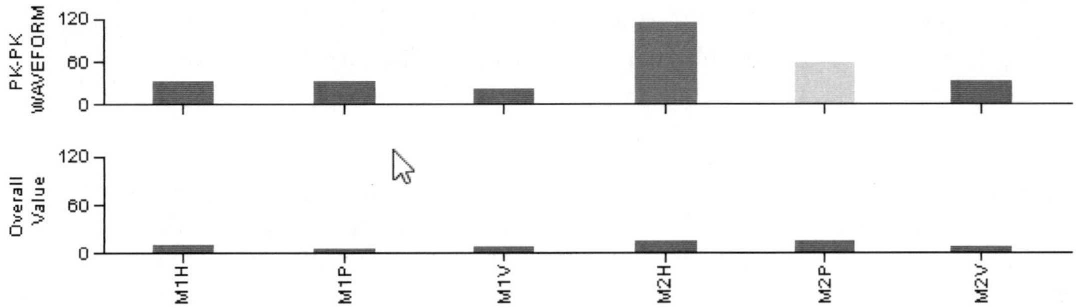

Figure 4: The Parameter Profile.

The Parameter Profile lists the collected measurement points against the various waveform and spectral parameters and color-codes these points according to their alarm status. Red points and parameters are in a 'Fault' condition as specified by the values entered in the database by the analyst or someone else with a working knowledge of vibration. Many sources for appropriate alarm levels exist for like-sized machines, yet each is in a different environment and operated under different paradigms within a given heavy-asset plant. In some cases, descriptive statistics are used by those analysts with training in statistical concepts such as means, medians, modes, standard deviations, and variances. But even these statistical concepts are not perfect. In many cases, a machine is taken off-line and replaced with another machine (this often happens to industrial motors). Another machine is substituted in its place, but the statistical values in the vibration data base are never updated. This causes the statistics used for machinery analysis to be valid only for the previous machine, not the currently installed one, reducing the effectiveness of these statistically generated alarms.

The workload for a machinery analyst is a significant concern as burn-out can and does occur in industrial facilities. Personnel turnover requires expensive re-training. Due to the physical act of data collection and the intellectual exercise of understanding vibration analysis analysts may feel over-committed for their available time. One way to address overload is to employ machinery learning and classification schemes in edge-mounted devices, on equipment such as paper machines to classify if the rolls are in an operational or degraded state. Such 'digital assistants' are being implemented with increasing deployment speed in heavy-asset facilities as adoption of Industry 4.0 characteristics such as 'digital twins' are deployed.

Project Scope and Machine Learning Models

The scope for this project is limited to bearing data collected by a vibration analyst walking a route path for data collection on a typical machine once per month. Because multiple data sets were available for six paper machine rolls with bearing defects, and the Classification Learner App in MATLAB version 2021b provides rapid training, validation and testing for up to 30 different machine learning (ML) models, this project determines which of the MATLAB ML models would correctly classify PM rolls possessing operational or degraded bearings.

Machine learning algorithms may offer good solutions to these issues of overload, alarm settings and statistical updating issues. In heavy-asset plants, the push to digitization and edge-computing devices is fostering acceptance of data streams that flow to data lakes in the cloud. The hype of such systems often makes for empty promises, but certainly the technical capability for the application of machine learning to machinery condition monitoring will be accepted by the next generation of machinery analysts. Those analysts are expecting to use machine learning for reporting not only the current condition of a machine, but also to forecast the machine's condition with a high degree of accuracy. Those forecasts will require temporally dense data sets from sensors and systems mounted permanently on machines. The cost of acquiring and maintaining such monitoring hardware and systems limits their present application to those machines that are considered critical for plant operation. However, these costs are reducing rapidly, with the applications enabled by wireless systems leading the way. The next section describes the application and data used in the ML classification model project.

Paper Machine and Data Descriptions

Paper machines are very massive machines, two to three stories tall, with an average of 22-foot-long rolls that may be up to 5 feet in diameter. Two typical paper machines are shown in Figure 5. Paper Machines have multiple rolls in nip, and a felt cover is present in the portion of the machine which carries the pulp to be compressed and then dried in the dryer section, and then wound up on a take-up winder. Multiple take-up reels allow the machine to be in constant operation. Motors or turbines are connected to gearboxes that are turned from a line shaft down the length of the machine with right angles to turn specific rolls. Vibration analysts monitor the rolls by catwalk access and generally collect data in the axial direction of the rolls. Where possible, radial directions are also collected. For the purposes of this project, only axial directions were used for all six rolls on the paper machine for data consistency. Figure 6 is a graphic of the numbered rolls of the 3rd Dryer section for route data collection. One roll from this section was used for the test data in this project but was renumbered to 05A in the database (as were all the rolls in the database) for privacy concerns. Table 1 has background information for the paper machine rolls in the FaultProgressions.rbm database.

Figure 5: Paper Machines.

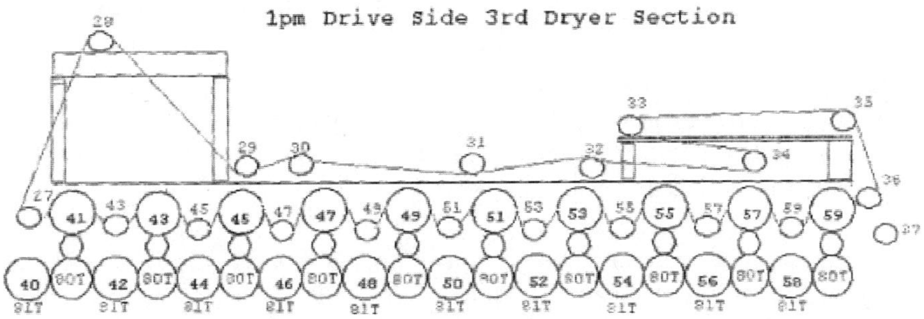

Figure 6: Route Measurement Points for DRYER 3 DS – 3rd Dryer Drive Side.

Table 1: Background Information from Analyst on the FaultProgressions.rbm database.

PRESS ROLL – Press Roll – BPFO Measurement point 01A Training and Validation data	Bottom Press Roll of a Paper machine. The bearing is a 23248 SKF – double row spherical roller bearing with an outer race defect. The progression is steady.
DRYER 1 DS – 1st Dryer Drive Side Measurement point 01A Training and Validation data	60-inch diameter dryer roll operating between 40 and 60 RPM. Torrington 23238KYM Bearing. Common outer race defect progression. Eventually the Inner race becomes defective. Data after the 9/5/12 data shows new bearing. (In all data, the peak at 81 orders is 1 x GMF).
DRYER 2 DS – 2nd Dryer Drive Side Measurement point 02A Training and Validation data	60-inch diameter dryer roll operating between 40 and 60 RPM. Torrington 23238KYM double row spherical roller bearing. Defect begins on the inner race, and eventually the outer race. The last measurement is after replacement. (In all data, the peak at 81 orders is 1 x GMF).
DRYER 2 DS – 2nd Dryer Drive Side Measurement point 03A Training and Validation data	60-inch diameter dryer roll operating between 40 and 60 RPM. Torrington 23238B double row spherical roller bearing. The defect is a typical outer race fault progression up to the point of replacement. The last measurement is taken after repairs. (In all data, the peak at 81 orders is 1 x GMF)
DRYER 3 DS – 3rd Dryer Drive Side Measurement point 04A Training and Validation data	60-inch, diameter dryer roll operating between 40 and 60 RPM. Torrington 23238B bearing. Started out as a common outer race defect progression. Eventually the Inner race defect appears. The bearing defect was present for fifteen months before failure. Rollers fell out of the bearing and down through the drive gears. 3 days downtime. Data on 9/21/2007 shows new bearing. (In all data, the peak at 81 orders is 1 x GMF).
DRYER 3 DS – 3rd Dryer Drive Side Measurement point 05A Test data set	60-inch diameter dryer roll operating between 40 and 60 RPM. The bearing is a Torrington 23238B double row spherical roller bearing. The defect is a typical outer race fault progression up to the point of replacement. The last measurement is taken after repairs. (In all data, the peak at 81 orders is 1 x GMF).

For each of the six PM rolls in Table 1, from 11 to 19 data sets were available and used to generate the 15-member Feature Set described in Table 4.

Classical machine learning algorithms and data mining such as Artificial Neural Networks (ANN) and principal component analysis (PCA), and Support Vector Machines (SVM) have been used for decades (Zhang et al., 2020). These 'shallow' machine learning methods require deep exploration of a data set and perhaps dimension reduction for extracting features. Because the knowledge base of different applications may require specialized expertise within each industry or field, garnering appropriate features and transferability of machine learning models may be challenging for generalizing to other bearing applications in other industries (Zhang et al., 2020). For this study, many machine learning classification models are available through the MATLAB Classification Learner App. Figure 7 shows thirty models run for training, validation, and test on the limited data set for paper machine dryer rolls.

Figure 7: Classification Learner App Models List.

The models in Figure 7 were tested for varying numbers of features, and selections were made to attempt to maximize the accuracy of the test data set. The Classification Learner App allows for rapid testing and re-testing with removed features, as shown in Figure 8.

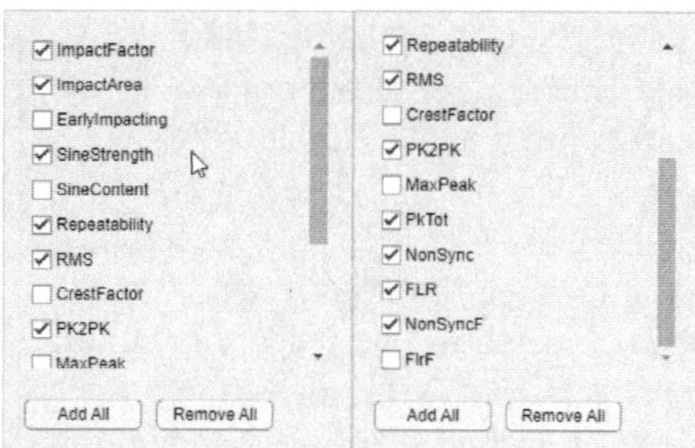

Figure 8: Feature Selection in the Classification Learner App.

All models were run with 5-fold validation, a model assessment technique. By partitioning the known dataset, a subset trains the model algorithm and the remaining data for testing, to prevent over or under training. Cross validation splits the data differently in each fold, to find the model's best algorithm. Two classes, operational (OP) and degraded (DEG), were identified and used in the datasets for training and testing. If these two classes can be identified with a high degree of accuracy for the large number paper machine dryer rolls as shown in Figure 6, the analyst's workload can be significantly reduced.

METHODS:

Methods Overview and Plan

Machine Learning (ML) is known to require very dense data sets, in terms of both temporal and diverse machine condition indicators. This project was conceived to determine if classifications of operational (OP) and degraded (DEG) bearing condition could be properly classified from a limited data set of only one machine condition indicator, vibration data. The plan for the project was to first acquire the data set, then extract the 15 identified features to be used in ML Model development (training, validation, and testing), and finally to determine if a reduced set of features would perform better or worse than the full set of features on the test data set. Once the best five models were identified, they were used to generate MATLAB code for future use, modification, and improvement by reliability engineers in the paper industry.

Data Acquisition

The data required for the machine learning classification study was acquired from an existing database containing paper machine rolls with bearing defects. The database is called "FaultProgression.rbm" and is used in teaching adult vibration analysis students about patterns of machinery failures and how they progress temporally. The database contains six total paper machine rolls, all with bearing defects. Each roll with a defective bearing has several data sets that were acquired monthly, on average, between 2006 and 2013. These data sets were selected to begin with the bearings in good condition and show the failure progression to a failed bearing over a several month period. Each roll had several data acquisitions storing spectral, trend, and waveform data and these rolls and dates are listed in Table 2. The organization of the database is shown in figure 9.

Table 2: Roll Names, Data Sets and Acquisition Dates.

Roll Name	Data Sets	Acquisition Dates
01A Bottom Press Roll	11	November 2008 to September 2009
01A Dryer Side Bottom Dryer Roll	19	July 2011 to September 2012
02A Top Dryer Roll	13	March 2013 to November 2013
03A Top Dryer Roll	16	July 2011 to September 2012
04A Top Dryer Roll	14	April 2006 to September 2007
05A Bottom Dryer Roll	13	January 2008 to November 2008.

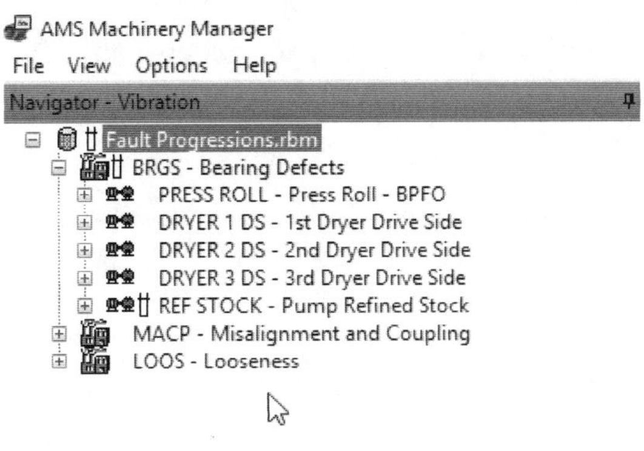

Figure 9: FaultProgressions.rbm Database Hierarchy.

The database contains three Areas, which are: BRGS – Bearing Defects; MACP – Misalignment and Coupling, and LOOS – Looseness. Each Area may have multiple machines under the Area. For this project, only the BRGS – Bearing Defects Area was used, although the original intention was to use all Areas with their associated defects. However, for the final project classifying data into degraded and operational status, the bearing data was used because multiple examples were available for training and validation. The only machine not used in the BRGS – Bearing Defects Area was the REF STOCK – Pump Refined Stock machine. As there is only one data set available for the pump, there were no additional pump data sets for comparison. The pump and other machine defects in the other Areas will be left for future work when multiple data sets become available.

With the data source available and consisting of failure progressions that result in problem abatement through bearing replacement for most of the Paper Machine Rolls, the waveform data was downloaded to spreadsheets. Features from the waveforms and spectral data were extracted for use in training, validation and testing of the required classifications for operational (OP) and degraded (DEG) status on the paper machine (PM) rolls. These waveform and spectral data sets were selected because they are limited to only data acquired with a Machinery Health Manager (MHM) 2140 Machinery Analyzer made by Emerson. Typically, waveform data sets consist of about 1024 to 4096 data points of digitized vibration, collected on average about once a month. When machines are exhibiting failure progressions, analysts will often take additional data with a data collector that is not a part of a normal route data collection event. The additional data was not used for this project as it was not consistently present for the various PM rolls. When an additional vibration measurement was taken with the same route parameters, that data was used for this project.

Next, a series of 15 Features was extracted from the waveform and spectral data for use in the Classification Learner (CL) App in MATLAB version 2021b. The data was preserved in Excel Spreadsheets and imported into the CL App for training, validation, and testing. A total of 86 observations were recorded and the percentages of operational and degraded condition classes was calculated at 53% operational and 46% degraded. Ten data sets (observations) were randomly removed from the list of 86 with the same percentages, so five operational and five degraded observations were held out for testing, and the features were preserved in a separate tab int the Excel spreadsheet. Percentage accuracy data for both the validation and testing events were examined for the 30 analytic models trained, validated, and tested. A list of these 30 models is in Figure 7.

A sensitivity analysis (SA) was carried out by eliminating one of the 15 features at a time, and re-training the 30 models, recording the training validation accuracy and testing accuracy. From the SA event, several features were dropped from the features list and one model (SVM Quadratic) did achieve an 89.5% accuracy level on the training data set. Several models had 90% success rates on the test data set. The top five models based on the test data set accuracy are discussed in Section 4, and do not include the SVM Quadratic model.

One goal of this project was to determine if a reduced vibration data set would have success in classifying machine condition. After analysis, the best five models were identified for use on future data sets. These models may assist analysts to correctly identify whether a PM roll has operational or degraded status based on bearing condition. This would add another tool for use by maintenance analysts in identifying machinery bearing degradation.

Development and Selection of Features

The first step in developing features for training, validation and test is to understand the data set. Fortunately, the MHM Software has many tools available to an analyst to use to better understand vibration data. A few of these tools are: waveform, trend, and spectral displays; difference and harmonic cursors, waveform autocorrelation displays, trend displays of waveform parameters such as Peak-to-Peak values, which are a range of the positive and negative forces on the machine (recorded with an accelerometer); a trend of the Crest Factor ratio, often defined as a "peakedness" waveform parameter; trend displays of frequency bands that provide relative changes in amplitudes for bearing frequencies; spectral displays of frequency and amplitude; pre-calculated fault frequencies for various machinery faults including bearing defects that may be overlaid on spectral displays; and parameter profile displays that provide parameter energy levels and alert and fault percentages. The following figures (10 through 12) are examples of these displays from the 3rd Dryer Drive Side 05A measurement point.

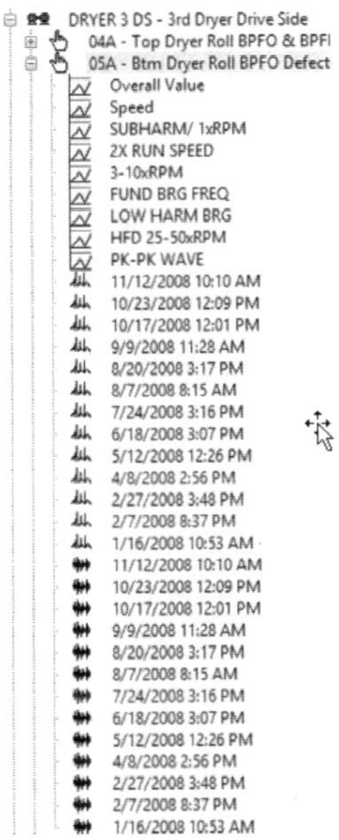

The trends of the vibration data are Overall Value, Speed, SUBHARM/1X RPM, 2X RUN SPEED, 3-10xRPM, FUND BRG FREQ, LOW HARM BRG, HFD 25-50xRPM, PK-PK WAVE. These parameters are a good place to begin looking for features that can be used in the CL App in MATLAB. There also is a listing of the available data of waveforms from 16-Jan-2008 to 12-Nov-2008, and the FFT algorithm was applied to generate spectral data for the same dates.

In Figure 11, Fault frequencies are marked, and peaks of vibration frequencies are presented that line up very well with the pre-calculated fault frequencies (FF). The RPM must be accurately tracked and recorded for these vibration peaks to line up with the pre-calculated dotted lines that indicate where Ball Pass Frequency of the Outer Race (BPFO) peaks should be displayed if a BPFO defect is in the machine. The fundamental BPFO frequency is rarely seen in standard vibration data as shown above in the figure. For the first dotted line, only a very small vibration peak exists.

Figure 10: 05A Bottom Dryer Roll BPFO Defect Vibration Data Set.

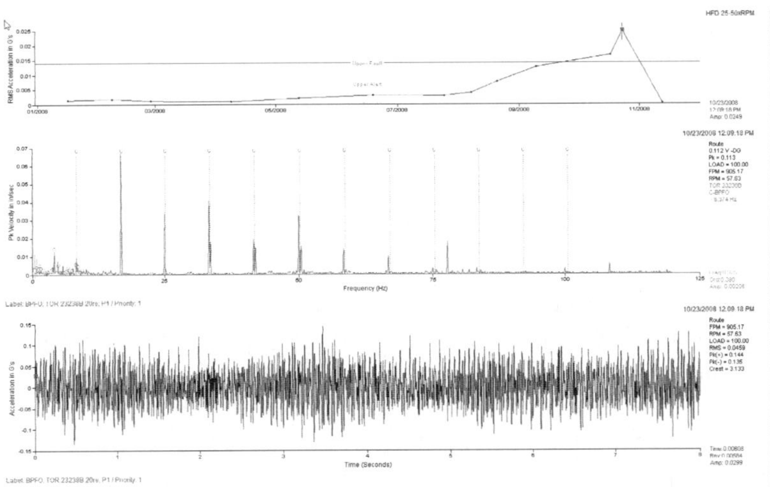

Figure 11: Trend, Spectral and Waveform Data for 23-OCT-2008.

Bearing analysis often begins with tracking the higher harmonics of the fundamental bearing defect frequency, as these higher harmonic peaks are amplified by resonances in the higher frequency ranges. This is the same principle used in ultra-sonic analysis, but that topic is beyond the scope of this project. However, using high-frequency acoustic data with vibration data for classification data analytics would be an expansion on this project for future work.
There is a data block at the top right of the waveform display that contains data about the waveform, as shown in Figure 12.

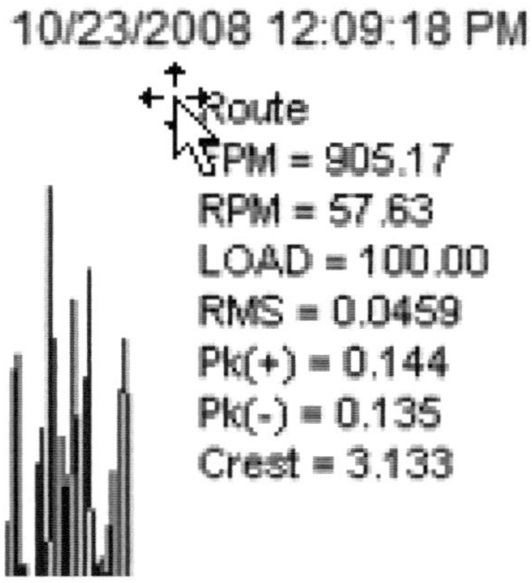

Figure 12: Waveform Data Block.

The waveform has three parameters that are typically used by an analyst to understand the data. First, the Root Mean Square (RMS) value is presented, which is an indication of the equivalent DC power in an AC signal. In the case of vibration analysis, where data was collected with an accelerometer, the RMS value is the average force the machine endured for the recorded measurement. The peak instantaneous forces are the peak positive and peak negative values, and often these are combined to get the total force excursions on the machine, called the "g-swing". By taking the highest excursion from 0, which is the statistical mean in this case, the peak value is determined. By dividing the absolute value of the peak value by the RMS value, the Crest Factor (CF) is obtained. The CF is a unitless ratio that indicates the relative "peaked-ness" of the signal and is considered by most analysts as a secondary indicator of bearing condition. When a bearing starts deteriorating, the bearing clearances begin to open, and when a defect is new sharp impacts cause high instantaneous peaks in the waveform. As the bearing further deteriorates, the clearances open further, and the lowered bearing stiffness causes a reduction in the sharp spikes that drive the CF value high for an early-stage defect. As the bearing gets worse, the CF value may go down slightly. Most analysts use CF ratio values of about 3.5 to 4 to indicate the beginning of a bearing defect, but values as high as 9 to 10 have occurred. The bearing data block is a rich source of potential features, and this information coincides with other sources (Zhiliang Liu, Ming J. Zuo, 2020).

In a 2014 paper, "Remaining Useful Life Prediction of Rolling Element Bearings Based on Health State Assessment", presented at the Annual Conference of the Prognostics and Health Management Society, Zhiliang Liu, et al., specify a statistical feature set for time-domain and frequency domain data of some 33 separate features (see Table 3, Feature Summary in the paper, or Table 4 below for a modified and reduced set of features):

Table 3: Feature Summary of Statistical Features in the Time and Spectral Domains.

Domain	Features
Time	Maximum Absolute Value, Average Absolute Value, Peak to Peak, RMS, standard deviation, skewness, kurtosis, variance, shape factor, crest factor (CF), clearance factor, impulse factor, energy operator and time series entropy.
Frequency	Mean frequency, frequency center, RMS frequency, standard deviation frequency, FFT entropy and Hilbert entropy. The four fault frequencies are Ball Pass Frequency of the Outer Race (BPFO), Ball Pass Frequency of the Inner Race (BPFI), Fundamental Train Frequency (FTF or often called the 'cage frequency', and Ball Spin Frequency (BSF).

Originally, some of the features from Table 3 were calculated from the raw digitized waveform data. Lacking definitions for some of these features such as impulse factor and energy operator made using these features impossible. Further, the BPFO and other bearing frequencies were not a good fit to the data, as the PM roll waveform and spectral data already consisted of a BPFO fault, or in one case, both BPFO and BPFI faults. Features that were available and readily accessible were found in two reports from the MHM software, the Waveform Parameter Report, and the Spectral Analysis Scan Report. Based on these reports, the following features in Table 4 were developed for this project.

Table 4: Feature Set and Definitions Used in the Time and Frequency Domains.

Domain	Features
Time	**Impact Factor (IF)** – Divides the absolute average waveform amplitude into the Maximum Peak-to-Peak value. (Max Pk-PK/ABS Ave.) This parameter is like CF but is more sensitive than CF as the IF retains the highest value found over all revolutions. **Impact Area (IA)** – Quantifies the impact area. Impacts become less distinct and more spread out as the bearing degrades, and this number becomes greater. **Early Impacting (EI)** – Derived by taking the IF and dividing it by the RMS from the waveform. Detects the presence of impacting even at low RMS values. **Sinusoidal Strength** – Examines the waveform on a per revolution basis for the amount of area underneath the waveform pattern. **Sinusoidal Content** – Compares the measured waveform to ideal sine waves using different approximation algorithms. Used to separate out shaft vibrations from bearing vibrations. **Repeatability** – Compares the time slice from each revolution for its amplitude content. If the amplitudes are similar, the waveform is more repetitive. **RMS** – The RMS value of the waveform data. **Crest Factor** – The maximum absolute amplitude peak value divided by the RMS value for the entire waveform. **PK-PK** – The difference between the highest positive waveform amplitude and the lowest negative waveform value. **Max Peak (MaxPk)** – The absolute value of the maximum positive or negative waveform amplitude value.
Frequency	**Peak Total** – The total of the peak energy in the spectrum. **NonSynchronous Energy** – The total of the nonsynchronous energy in the spectrum. Bearings emit nonsynchronous energy, so a high value of this feature should indicate a bearing defect. **Floor Energy** – The amount of floor energy across all of the spectral data. Bearing failure is often seen with a raised noise floor in the data due to randomness in the phase data from nonsynchronous events. **NonSynchronous Percentage Fault** – This is a percentage number of how close to or over the fault level is the nonsynchronous energy. **Floor Percentage Fault** - This is a percentage number of how close to or over the fault level is the Floor energy.

The 15 members of the feature set from Table 4 were used in the training, validation, and testing events for the 30 models used in this project.

Machine Learning Models

The Classification Learner (CL) App in MATLAB supports 30 different models for testing, which was considered an appropriate number for this project. A convenient function of the CL App was the ability to test and evaluate all models sequentially, yet quickly. After training, the Receiver Operating Characteristic (ROC) and Area Under Curve (AUC) graphics could be displayed for the validation and testing events for all models. Also available are Confusion Matrices and scatterplots for the various combinations of classes and features, respectively.

The CL App divides the 30 models into eight groups consisting of Decision Trees, Discriminant Analysis, Logistic Regression Classifier, Naïve Bayes Classifiers, Support Vector Machines, Nearest Neighbor Classifiers, Ensemble Classifiers and Neural Network Classifiers. Individual classifiers are versions within these groups and are found in Table 5.

Table 5: CL App Classification Models

Classification Group	Classifier Models
Decision Trees	Fine Tree, Medium Tree, Coarse Tree
Discriminant Analysis	Linear Discriminant, Quadratic Discriminant
Logistic Regression	Logistic Regression
Naïve Bayes	Gaussian Naïve Bayes, Kernel Naïve Bayes
Support Vector Machines	Linear SVM, Quadratic SVM, Cubic SVM, Fine Gaussian SVM, Medium Gaussian SVM, Coarse Gaussian SVM
Nearest Neighbors	Fine KNN, Medium KNN, Coarse KNN, Cosine KNN, Cubic KNN, Weighted KNN
Ensemble Classifiers	Boosted Trees, Bagged Trees, Subspace Discriminant, Subspace KNN, RUSBoosted Trees
Neural Network	Narrow NN, Medium NN, Wide NN, Bilayered NN, Trilayered NN

Project Task Procedures

After the determination of the feature set and garnering the specific values for the fifteen individual features, the spreadsheet of the values for each feature for all 6 of the available PM rolls was loaded into the CL App, less the 10 observations held out for testing. The last column of the spreadsheet housed the two classification values, Operational (OP) or no significant defect and Degraded (DEG) where a significant bearing defect was present. Figure 13 is the first 23 lines of the 76 lines of data entered for the training and validation set.

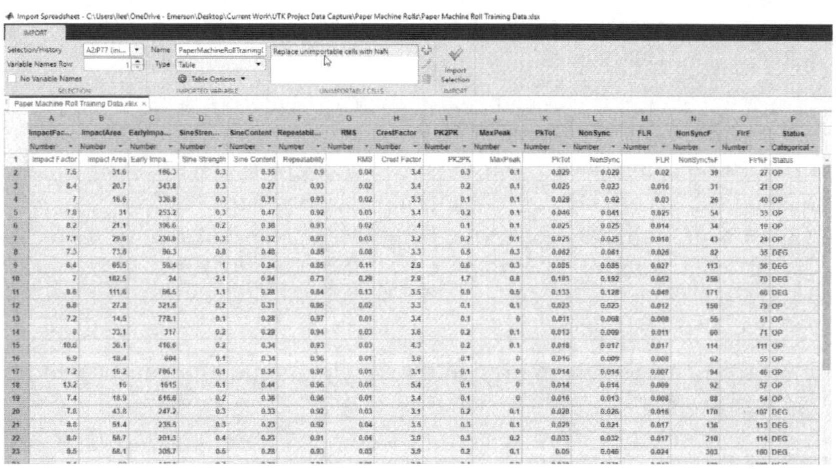

Figure 13: Features and Response Variables for the Training and Validation Set of Data.

When these features were imported, 5-Fold Cross Validation was selected to protect from over-fitting the data, presented in Figure 14.

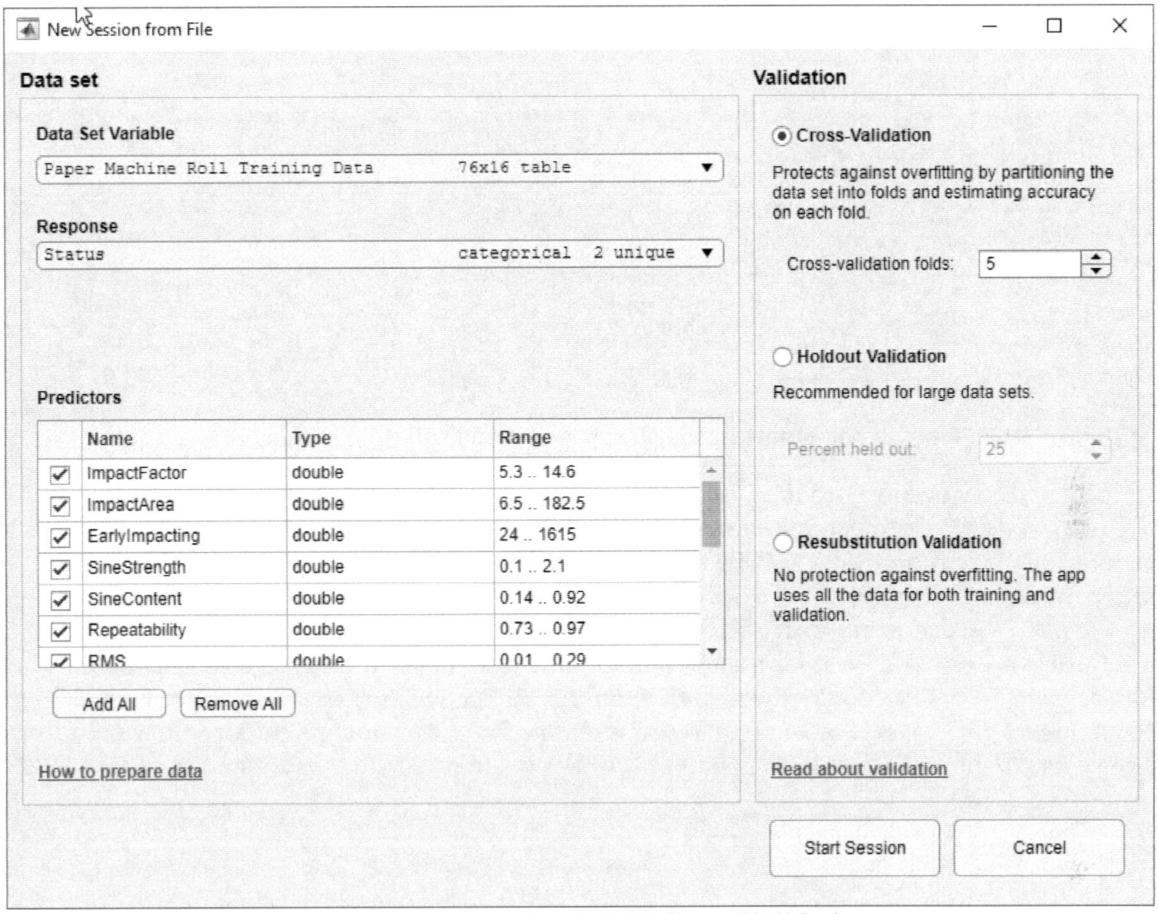

Figure 14: Imported Data Set Features with 5-Fold Cross Validation.

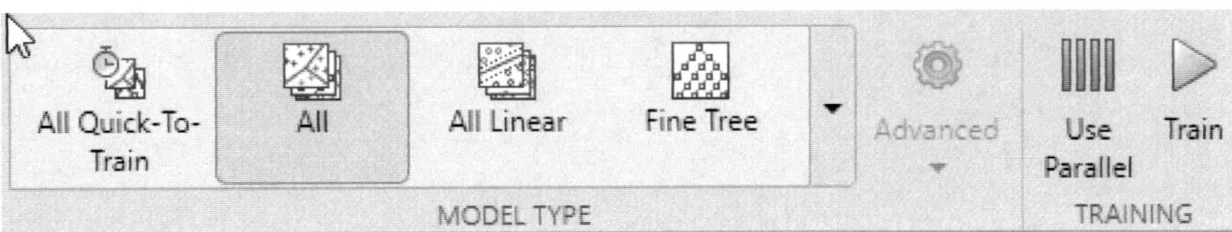

Figure 15: All Models Selection.

A convenient MATLAB method allowed for rapid testing of all ML Classification models sequentially as shown in Figure 15. By selecting the ‘All’ button, all 30 models were trained and validated, and the results were reported as accuracy percentages. Further, Receiver Operating Characteristic curves and AUC plots, along with confusion matrices for both the validation and the testing data are presented for evaluation.

Initially, three classification values were defined for the dataset, which was Operational, Degraded and Failed. However, on running these models, the low incidence of failed data caused unusual ROC curves to be generated, as shown in the third ROC curve in Figure 16.

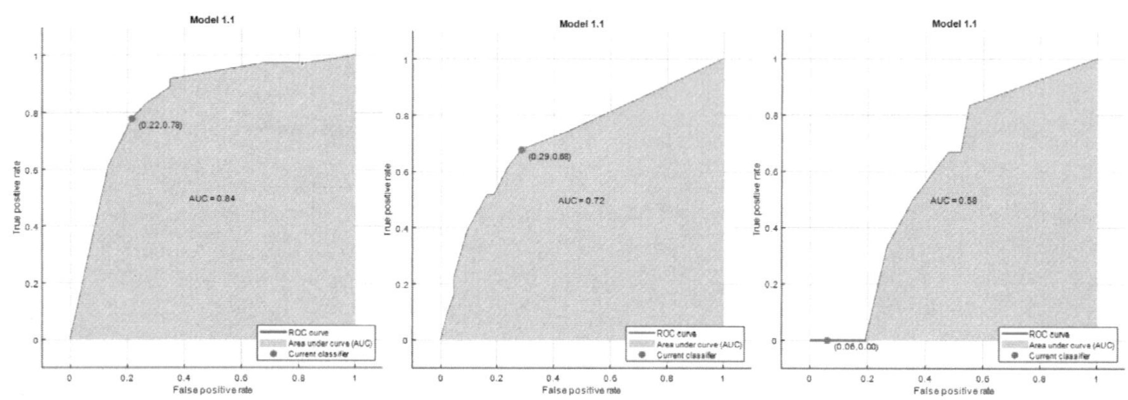

Figure 16: ROC Curves for Operational, Degraded and Failed.

The 'Failed' classification was dropped from the analysis as a result, and only two classifiers were used in the final project procedures. After settling on the final classification classes, a Primary Component Analysis (PCA) was run using all features. After examination of the PCA vs. non-PCA validation accuracy, a decision was made to not use PCA-based analysis for this project, as the validation accuracy suffered with lower numbers across the board for all models. Next, comparisons were made via scatterplots for all the combinations of features. The sample plots below were collected after the non-PCA run for all features for the Naïve Bayes Kernel Model. Figure 17 shows a high correlation between the Impact Factor and the Crest Factor, no surprise due to the way Crest Factor and Impact Factors are calculated, as defined in Table 4.

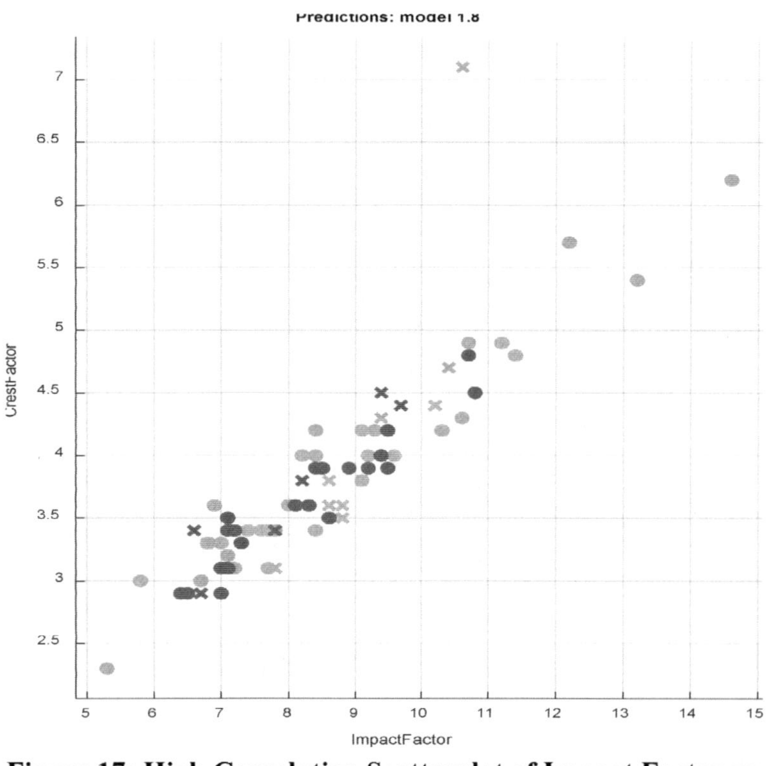

Figure 17: High Correlation Scatterplot of Impact Factor vs. Crest Factor.

Relatively few of the features selected were highly correlated. Figure 18 shows one such relationship between the Impact Factor and PK Total, which is the energy of the total peaks in the spectral data. These two factors should have low correlation, as bearing failures are often low-energy events compared to shaft energies and shaft-associated peaks.

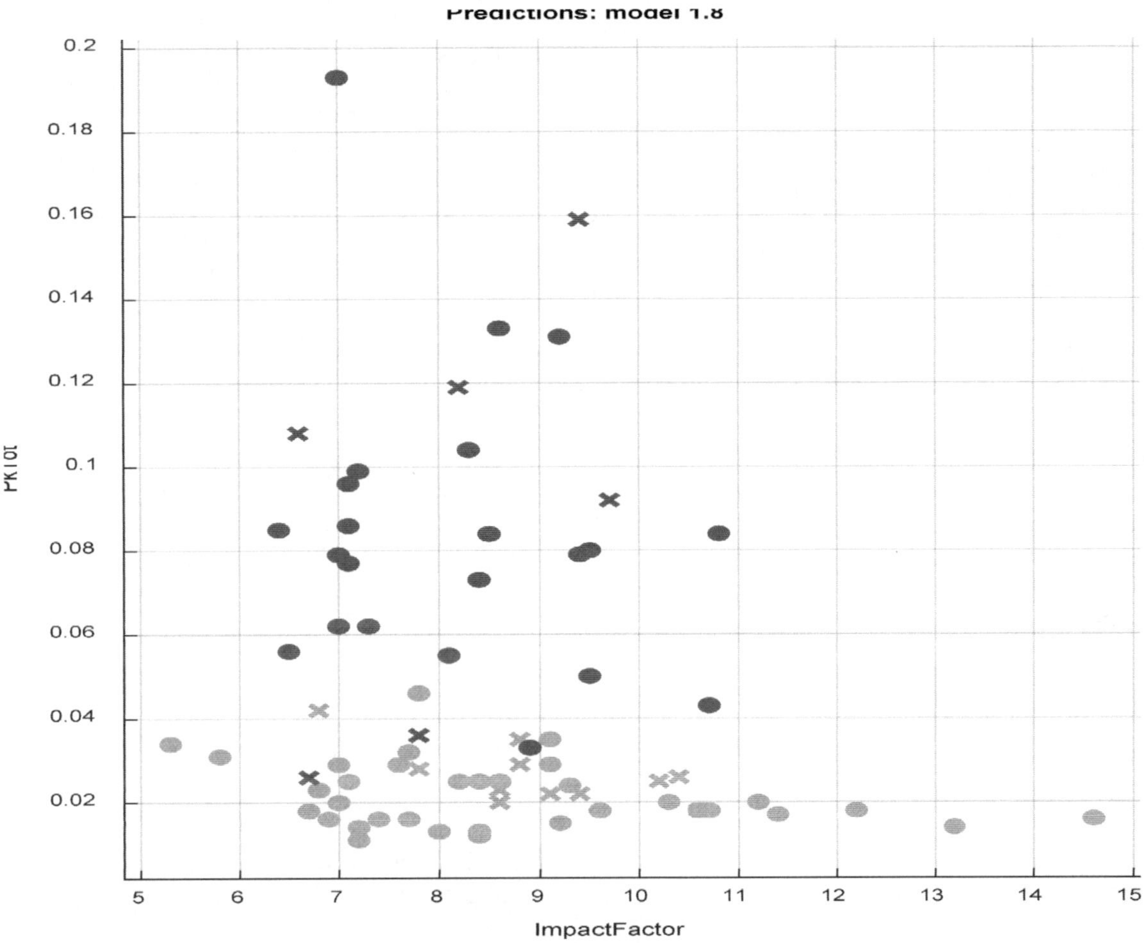

Figure 18: ***Low Correlation Scatterplot of Impact Factor Vs. Peak Total Energy (PkTot).***

After examining the scatterplots, the Sensitivity Analysis portion of the project was begun. Initially, the training, validation and testing was run with all features. Subsequently, another run was made excluding all features except for the Impact Factor, and the results were dismal. Only one model returned 60% accuracy and the rest were in the 40s and 50s. After adding Impact Area, the numbers increased to the high 60s and mid-70s. Adding the third feature, Early Impacting, did not improve these scores. By following this procedure of comparing the accuracy percentage scores for all the features, added one at a time, several features were able to be removed. Individual features were added back when the numbers reduced over the previous set of accuracy numbers for the validation set. The high number of features removed implied that some features were superfluous to the models and were not needed. In the final analysis, six features were removed for the NN Trilayered model, eight were removed for the KNN Weighted, KNN Cubic, and KNN Medium models, and finally, nine features were removed for the SVM Coarse Gaussian model. When run against the reserved test data, these models presented the highest scores with the respective features removed. Table 5 lists the features that were kept and removed for all five models.

Table 5: List of Features Removed and Kept for the Final Model Tests.

Feature List	NN Trilayered Removed	NN Trilayered Feature Kept	KNN Weighted, KNN Cubic, and KNN Medium Removed	KNN Weighted, KNN Cubic, and KNN Medium Kept	SVM Coarse Gaussian Removed	SVM Coarse Gaussian Kept
Impact Factor		X		X		X
Impact Area		X		X		X
Early Impacting		X	X		X	
Sinusoidal Strength		X		X		X
Sinusoidal Content		X	X		X	
Repeatability		X		X	X	
RMS		X		X		X
Crest Factor		X	X		X	
Peak to Peak	X			X		X
Max Peak	X		X		X	
Peak Total	X		X		X	
Nonsynchronous Energy	X		X		X	
Floor Energy	X		X		X	
Nonsynchronous Percentage Fault		X		X		X
Floor Energy Percentage Fault	X		X		X	

The ROC curve and AUC numbers were processed for all the models, along with the confusion matrices and associated numbers and percentages for predicted vs. true classifications.

Results

The results of this project are presented for the trained and validated 30 models, and five top models were selected that worked well at predicting whether the test data set was in an operational or degraded state. The five models selected for the Test Results data shown in Figure 19 are the best overall models based on their test percentage accuracy. The models are: SVM Coarse Gaussian, KNN Weighted, NN Trilayered, KNN Cubic, and KNN Medium. Table 6 compares these top model's training and test classification accuracy scores. The top model selection process is discussed in this section.

Figure 20 presents results for the training and validation data set accuracy percentages. The figure has been color-coded to reveal changes in the model percentages as features were removed. The best-performing eight models based on the validation data accuracy are: Fine Tree, Medium Tree and Coarse Tree, at 88.2% and the SVM Linear, KNN Weighted and NN Medium at 86.8%. SVM Quadratic and SVM Medium Gaussian round out the top eight validation models at 85.5% validation accuracy. All these results were for the full removal of nine features: FlrF, FLR, NonSync, PKTot, MaxPK, CF, Sinusoidal Content, Early Impacting and Repeatability. The highest validation accuracy percentage recorded (89.5%) was for the SVM Quadratic model with five features removed: FlrF, FLR, NonSync, PKTot and MaxPK. Four of the models performed dismally: KNN Coarse, Ensemble Boosted Trees, Ensemble Subspace KNN and Ensemble RUS Boosted Trees. All these models performed near or below the 70% level for validation accuracy and were also poor at predicting the classes for the test data set. KNN Coarse and Ensemble Boosted Trees both had only 50% test accuracy for any combination of removed features. Ensemble Subspace KNN and Ensemble RUS Boosted Trees performed slightly better for some feature combinations, but their scores were very low compared to other models.

	A	E	F	G	H	I	J	K	L	M	N	O	P	Q	R
1	Models - No PCA applied	FlrF, NonSyncF, FLR	FlrF, FLR	FlrF, NonSyncF, FLR	FlrF, FLR, NonSync	FlrF, NonSyncF, FLR & NonSync	FlrF, FLR, NonSync, PKTot	FlrF, FLR, NonSync, PKTot & MaxPk	FlrF, FLR, NonSync & MaxPk	FlrF, FLR, NonSync, PKTot, MaxPk & PK2PK	FlrF, FLR, NonSync, PKTot, MaxPk & CF	FlrF, FLR, NonSync, PKTot, MaxPk, CF, & Sinusoidal Content	FlrF, FLR, NonSync, PKTot, MaxPk, CF, Sinusoidal Content & Early Impacting	FlrF, FLR, NonSync, PKTot, MaxPk, Sinusoidal Content & Early Impacting	FlrF, FLR, NonSync, PKTot, MaxPk, CF, Sinusoidal Content, Early Impacting & Repeatability
2	Fine Tree	82.9	86.8	82.9	86.8	80.3	86.8	86.8	86.8	86.8	86.8	88.2	88.2	88.2	88.2
3	Medium Tree	82.9	86.8	82.9	86.8	80.3	86.8	86.8	86.8	86.8	86.8	88.2	88.2	88.2	88.2
4	Coarse Tree	82.9	85.5	82.9	85.5	81.6	85.5	85.5	85.5	85.5	85.5	88.2	88.2	88.2	88.2
5	Linear Discriminant	82.9	81.6	82.9	81.6	81.6	80.3	78.9	80.3	78.9	78.9	82.9	82.9	82.9	82.9
6	Quadratic Discriminant	78.9	80.3	78.9	78.9	80.3	80.3	80.3	81.6	84.2	78.9	81.6	84.2	84.2	82.9
7	Logistic Regression	82.9	82.9	82.9	84.2	85.5	82.9	82.9	82.9	81.6	81.6	84.2	85.5	85.5	84.2
8	Naïve Bayes (Guassian)	78.9	80.3	78.9	81.6	80.3	81.6	82.9	82.9	82.9	81.6	84.2	82.9	82.9	85.5
9	Naïve Bayes (Kernel)	80.3	78.9	80.3	78.9	80.3	78.9	80.3	80.3	78.9	80.3	76.3	75	73.7	75
10	SVM Linear	81.6	80.3	81.6	81.6	80.3	81.6	81.6	80.3	81.6	81.6	88.2	86.8	86.8	86.8
11	SVM Quadratic	82.9	85.5	82.9	85.5	84.2	78.9	89.5	85.5	88.2	88.2	82.9	82.9	77.6	85.5
12	SVM Cubic	69.7	76.3	69.7	76.3	76.3	76.3	75	76.3	73.7	81.6	80.3	75	73.7	75
13	SVM Fine Guassian	65.8	67.1	65.8	68.4	64.5	68.4	68.4	67.1	68.4	73.7	77.6	81.6	69.7	84.2
14	SVM Medium Guassian	78.9	80.3	77.6	80.3	76.3	80.3	80.3	80.3	78.9	80.3	85.5	85.5	82.9	85.5
15	SVM Coarse Guassian	73.7	73.7	73.7	73.7	71.1	75	75	73.7	75	76.3	76.3	75	75	75
16	KNN Fine	69.7	75	69.7	78.9	73.7	76.3	78.9	77.6	77.6	80.3	76.3	75	73.7	81.6
17	KNN Medium	80.3	80.3	80.3	81.6	80.3	78.9	82.9	82.9	82.9	80.3	84.2	82.9	76.3	77.6
18	KNN Coarse	52.6	52.6	52.6	52.6	52.6	52.6	52.6	52.6	52.6	52.6	52.6	52.6	52.6	52.6
19	KNN Cosine	78.9	78.9	78.9	80.3	77.6	80.3	80.3	80.3	81.6	81.6	78.9	77.6	76.3	78.9
20	KNN Cubic	84.2	84.2	84.2	82.9	81.6	81.6	84.2	85.5	84.2	81.6	82.9	84.2	78.9	81.6
21	KNN Weighted	77.6	77.6	77.6	77.6	76.3	75	78.9	78.9	78.9	80.3	84.2	86.8	81.6	86.8
22	Ensemble Boosted Trees	52.6	52.6	52.6	52.6	52.6	52.6	52.6	52.6	52.6	52.6	52.6	52.6	52.6	52.6
23	Ensemble Bagged Trees	84.2	78.9	82.9	82.9	81.6	82.9	84.2	85.5	86.8	80.3	82.9	84.2	80.3	84.2
24	Ensemble Subspace Discriminant	80.3	80.3	78.9	77.6	80.3	77.6	80.3	77.6	77.6	77.6	84.2	82.9	81.6	82.9
25	Ensemble Subspace KNN	63.2	73.7	65.8	67.1	59.2	71.1	72.4	71.1	65.8	69.7	71.1	68.4	69.7	64.8
26	Ensemble RUS Boosted Trees	63.2	72.4	72.4	75	73.7	82.9	63.2	64.5	63.2	76.3	67.1	64.5	80.3	61.8
27	Neural Network Narrow	81.6	82.9	81.6	81.6	84.2	80.3	85.5	81.6	82.9	85.5	80.3	85.5	81.6	81.6
28	Neural Network Medium	80.3	80.3	82.9	82.9	80.3	82.9	84.2	82.9	85.5	86.8	81.6	84.2	82.9	86.8
29	Neural Network Wide	84.2	81.6	80.3	80.3	77.6	81.6	84.2	82.9	81.6	81.6	85.5	86.8	84.2	84.2
30	Neural Network Bilayered	77.6	81.6	82.9	81.6	80.3	80.3	86.8	81.6	85.5	86.8	80.3	85.5	78.9	80.3
31	Neural Network Trilayered	78.9	81.6	75	77.6	80.3	78.9	82.9	80.3	86.8	80.3	80.3	81.6	82.9	82.9

Figure 19: Results of the Models for Selected Features using Validation Accuracy.

The eight models selected from the validation data set (Fine Tree, Medium Tree, Coarse Tree, SVM Linear, SVM Quadratic, SVM Medium Gaussian, KNN Weighted, and NN Medium shown in Figure 20 are selected for their performance figures of 85.5% accuracy with the fewest features. KNN Cubic did not quite meet the same standard nor did the Logistic Regression model, although they both did when higher numbers of features were included. Although only the NN Medium model was included in the top eight, all the other NN models did perform well with eight features removed except NN Trilayered, which performed best with six features removed.

	A	E	F	G	H	I	J	K	L	M	N	O	P	Q	R
37	Models - No PCA applied - Test	FlrF, NonSyncF, FLR	FlrF, FLR	FlrF, NonSyncF, FLR	FlrF, FLR, NonSync	FlrF, NonSyncF, FLR & NonSync	FlrF, FLR, NonSync, PKTot	FlrF, FLR, NonSync, PKTot & MaxPk	FlrF, FLR, NonSync & MaxPk	FlrF, FLR, NonSync, PKTot, MaxPk & PK2PK	FlrF, FLR, NonSync, PKTot, MaxPk & CF	FlrF, FLR, NonSync, PKTot, MaxPk, CF, & Sinusoidal Content	FlrF, FLR, NonSync, PKTot, MaxPk, CF, Sinusoidal Content & Early	FlrF, FLR, NonSync, PKTot, MaxPk, Sinusoidal Content & Early	FlrF, FLR, NonSync, PKTot, MaxPk, CF, Sinusoidal Content, Early Impacting & Repeatability
38	Fine Tree	80	70	80	70	80	70	70	70	70	70	70	70	70	70
39	Medium Tree	80	70	80	70	80	70	70	70	70	70	70	70	70	70
40	Coarse Tree	80	70	80	70	80	70	70	70	70	70	70	70	70	70
41	Linear Discriminant	80	80	80	70	70	80	80	70	80	80	80	80	80	80
42	Quadratic Discriminant	80	80	80	80	80	80	80	80	80	80	80	80	80	80
43	Logistic Regression	90	90	90	80	90	80	80	80	80	80	80	80	80	80
44	Naïve Bayes (Guassian)	80	80	80	80	80	80	80	80	80	80	80	80	80	80
45	Naïve Bayes (Kernel)	80	80	80	80	80	80	80	80	80	80	80	80	80	80
46	SVM Linear	80	80	80	80	80	80	80	80	80	80	80	80	80	80
47	SVM Quadratic	80	80	80	80	70	90	70	70	70	80	70	50	50	50
48	SVM Cubic	80	70	80	70	80	70	70	70	70	70	70	70	70	80
49	SVM Fine Guassian	70	70	70	70	70	70	70	70	70	80	70	80	80	80
50	SVM Medium Guassian	80	80	80	80	80	80	80	80	80	80	80	80	80	80
51	SVM Coarse Guassian	80	80	80	80	80	80	80	80	80	80	90	90	90	100
52	KNN Fine	70	70	70	70	70	70	70	70	70	70	80	70	80	70
53	KNN Medium	80	90	80	80	70	80	80	90	80	80	80	90	80	70
54	KNN Coarse	50	50	50	50	50	50	50	50	50	50	50	50	50	50
55	KNN Cosine	80	80	80	80	80	80	70	70	70	70	80	80	80	80
56	KNN Cubic	70	80	70	70	70	70	70	70	70	70	80	90	80	70
57	KNN Weighted	90	90	90	90	80	90	90	90	90	90	80	90	80	80
58	Ensemble Boosted Trees	50	50	50	50	50	50	50	50	50	50	50	50	50	50
59	Ensemble Bagged Trees	70	70	70	70	70	70	70	70	70	70	70	70	70	70
60	Ensemble Subspace Discriminant	80	80	80	80	80	80	80	80	80	80	80	80	80	80
61	Ensemble Subspace KNN	80	70	80	70	80	70	70	70	70	70	70	60	60	50
62	Ensemble RUS Boosted Trees	50	50	50	70	50	80	80	70	70	70	50	50	50	70
63	Neural Network Narrow	80	80	90	80	90	80	70	70	80	80	70	70	70	70
64	Neural Network Medium	90	80	80	70	80	70	70	70	70	80	70	70	70	70
65	Neural Network Wide	90	80	70	70	80	80	70	80	70	70	70	70	70	70
66	Neural Network Bilayered	70	80	80	80	80	70	80	90	80	70	70	80	70	70
67	Neural Network Trilayered	80	90	70	80	80	80	70	80	90	80	70	70	70	70

Figure 20: Results of the Models for Selected Features using Test Accuracy.

Five models were selected from the Test Results data: SVM Coarse Gaussian, KNN Medium, KNN Cubic, KNN Weighted, and NN Trilayered. These five models were selected because they have the highest test accuracy of all models using the fewest features. The SVM Coarse model did not score highly on the training and validation test accuracy at 75% yet had the highest overall performance that increased to 100% test accuracy with nine removed features: FlrF, FLR, NonSync, PKTot, MaxPK, CF, Sinusoidal Content, Early Impacting, and Repeatability. No other model achieved a 100% test accuracy rating, but all the Neural Network models achieved a 90% test accuracy rating for various feature removal combinations. NN Trilayered scored 86.8% for six removed features of FlrF, FLR, NonSync, PkTot, MaxPK, and PK2PK, and 90% test accuracy for the same removed features in the test data set. KNN Weighted performed at the same level as NN Trilayered, with 86.8% training set classification accuracy and 90% test data classification accuracy. KNN Medium performed well with 82.9% validation accuracy that resulted in a 90% test data set accuracy for classification. KNN Cubic also performed consistently with 84.2% training set accuracy and 90% test data set accuracy. These five models are compared for consistency in Table 6.

Table 6: Training and Test Performance Comparison.

Model	Features Removed	Training	Test
SVM Coarse Gaussian	Nine: FlrF, FLR, NonSync, PKTot, MaxPK, CF, Sinusoidal Content, Early Impacting, and Repeatability	75	100
KNN Weighted	Eight: FlrF, FLR, NonSync, PKTot, MaxPK, CF, Sinusoidal Content, and Early Impacting	86.8	90
NN Trilayered	Six: FlrF, FLR, NonSync, PkTot, MaxPK, and PK2PK	86.8	90
KNN Cubic	Eight: FlrF, FLR, NonSync, PKTot, MaxPK, CF, Sinusoidal Content, and Early Impacting	84.2	90
KNN Medium	Eight: FlrF, FLR, NonSync, PKTot, MaxPK, CF, Sinusoidal Content, and Early Impacting	82.9	90

The SVM Quadratic model also achieved 90% test accuracy (78.8% training accuracy) with four features removed (FlrF, FLR, NonSync, and PKTot), but its performance degraded significantly with any additional features removed, and with all nine features removed its performance was very low at 50% test accuracy. The SVM Quadratic was not included with the best five models because of the low accuracy scores in the test data set, even though it was a top eight model in the validation test set. The SVM Coarse Gaussian model required inclusion of only six features (Impact Factor, Impact Area, Sine Strength, RMS, PK2PK, and NonSynchF), or excluding the other nine features to achieve 100% test accuracy. The KNN Weighted Model had high performance scores across most of the feature removals and did achieve a 90% test accuracy (training 86.8) with eight features removed (FlrF, FLR, NonSync, PKTot, MaxPk, CF, Sinusoidal Content, and Early Impacting). NN Trilayered also has 90% test accuracy (training 86.8) with eight features removed (FlrF, FLR, NonSync, PKTot, MaxPk, CF, Sinusoidal Content, and Early Impacting). KNN Medium and KNN Cubic also achieved 90% test accuracy (training 82.9 and 84.2, respectively) with the same eight features removed for both models.

Regarding the selected top five models in Table 6, a deeper investigation into their performance is required. An examination of the Receiver Operating Characteristics Curve (ROC) and the Area

Under Curve (AUC) plots are important for visualizing the performance of a multi-class classification. The ROC curve is a probability curve and AUC represented the measure or degree of separability. A poor result for the ROC curve would be a 45-degree line that indicates the results are no better than random probability. AUC reveals the degree to which the model is capable of distinguishing between classes. Additionally, the relationships between Sensitivity, Specificity, False Positive Rate (FPR), True Positive Rate (TPR) and the Threshold must be evaluated.

As an example, starting with the first top model in Table 6, SVM Coarse Gaussian, Figure 21 contains the confusion matrices for the training and validation data sets. The interpretation of the training and validation data set Confusion Matrix for this model is that 19 of the class values were predicted as belonging to the OP class when they were in fact in the DEG class. The right plot in Figure 21 lists the performance as percentages with TPR (OP true to OP predicted =100% and DEG true to DEG predicted = 52.5%) and FNR percentages as marked. The lower plot in Figure 21 lists the Positive Predictive Values (PPV) and False Discovery Rates (FDR). The Positive Predictive Value is the probability that the class with a positive test truly is classified correctly. The False Discovery Rate is the ratio of the number of false positive results to the number of total positive results. In Figure 22, 34.5% of the predicted OP class are in fact in the DEG class. Table 7 is a summary of these values for the training and validation data for these figures.

Likewise, Table 8 is a summary of the confusion matrix values for the test data set.

Table 7: Summary of Confusion Matrices for the Training and Validation Data Sets.

Models	True-Pred.	TPR	True-Pred.	FNR	True-Pred.	PPV	True-Pred.	FDR
SVM C.G.								
	OP-OP	100%	DEG-OP	47.5%	DEG-DEG	100%	DEG-OP	34.5%
	DEG-DEG	52.5%	OP-DEG	0%	OP-OP	65.5%	OP-DEG	0%
KNN Wt.								
	OP-OP	97.2%	DEG-OP	22.5%	DEG-DEG	96.9%	DEG-OP	20.5%
	DEG-DEG	77.5%	OP-DEG	2.8%	OP-OP	79.5%	OP-DEG	3.1%
NN Tri.								
	OP-OP	80.6%	DEG-OP	15%	DEG-DEG	82.9%	DEG-OP	17.1%
	DEG-DEG	85%	OP-DEG	19.4%	OP-OP	82.9%	OP-DEG	17.1%
KNN Cubic								
	OP-OP	91.7%	DEG-OP	27.5%	DEG-DEG	90.6%	DEG-OP	25%
	DEG-DEG	72.5%	OP-DEG	8.3%	OP-OP	75%	OP-DEG	9.4%
KNN Medium								
	OP-OP	86.1%	DEG-OP	32.5%	DEG-DEG	84.4%	DEG-OP	29.5%
	DEG-DEG	67.5%	OP-DEG	13.5%	OP-OP	70.5%	OP-DEG	15.6%

Table 8: Summary of Confusion Matrices for the Test Data Set.

Models	True-Pred.	TPR	True-Pred.	FNR	True-Pred.	PPV	True-Pred.	FDR
SVM C.G.								
	OP-OP	100%	DEG-OP	0%	DEG-DEG	100%	DEG-OP	0%
	DEG-DEG	100%	OP-DEG	0%	OP-OP	100%	OP-DEG	0%
KNN Wt.								
	OP-OP	60%	DEG-OP	0%	DEG-DEG	71.4%	DEG-OP	0%
	DEG-DEG	100%	OP-DEG	40%	OP-OP	100%	OP-DEG	28.6%
NN Tri.								
	OP-OP	40%	DEG-OP	0%	DEG-DEG	62.5%	DEG-OP	17.1%
	DEG-DEG	100%	OP-DEG	60%	OP-OP	100%	OP-DEG	0%
KNN Cubic								
	OP-OP	60%	DEG-OP	0%	DEG-DEG	71.4%	DEG-OP	0%
	DEG-DEG	100%	OP-DEG	40%	OP-OP	100%	OP-DEG	28.6%
KNN Medium								
	OP-OP	60%	DEG-OP	0%	DEG-DEG	71.4%	DEG-OP	0%
	DEG-DEG	100%	OP-DEG	40%	OP-OP	100%	OP-DEG	28.6%

These five models have good agreement between the training and validation numbers and the test set numbers, although the SVM Coarse Gaussian's FNR and FDR rates are higher than the other models. Also, the KNN Cubic numbers seem to diverge between the training and test sets.

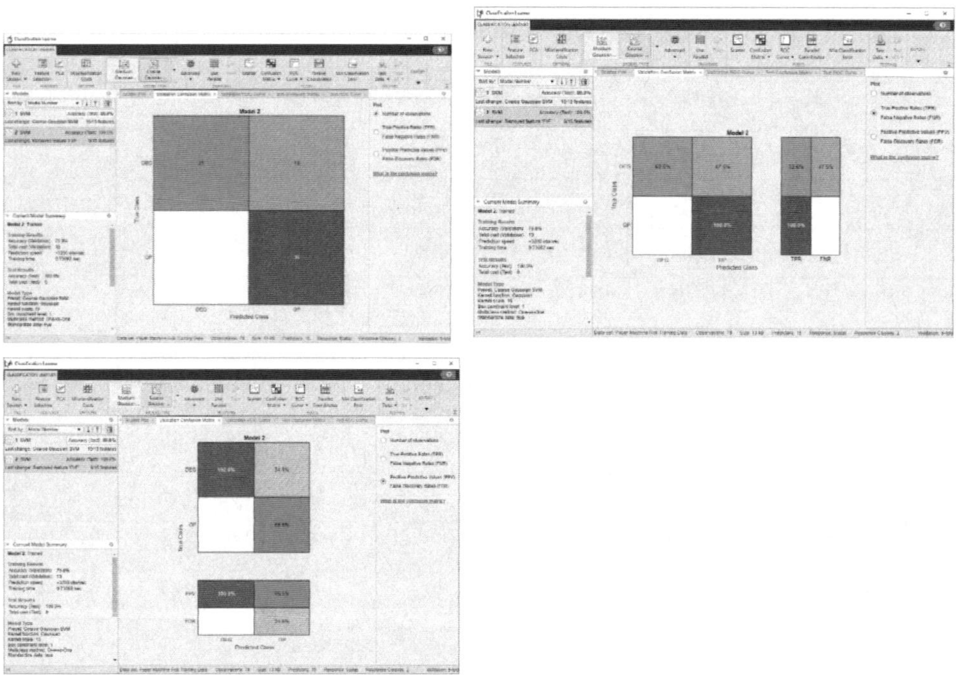

Figure 21: Training and Validation Confusion Matrices for the SVM Coarse Gaussian Model.

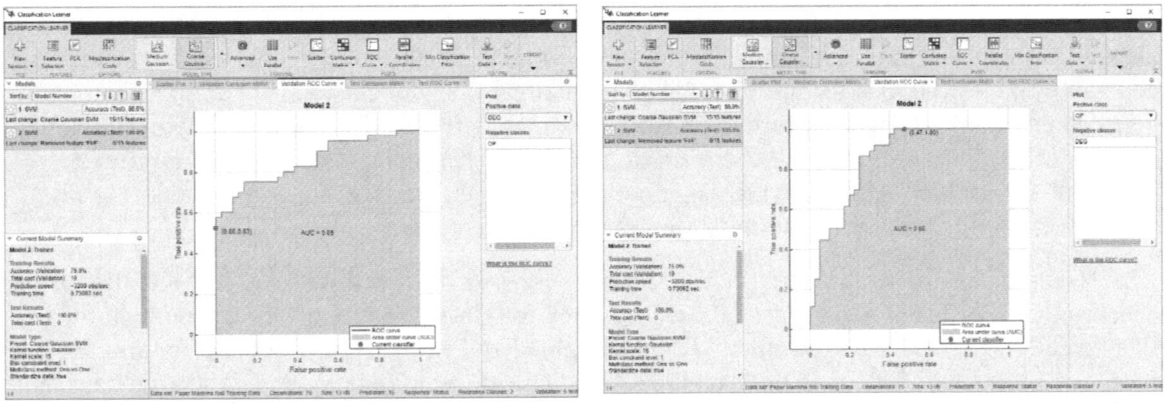

Figure 22: Training and Validation ROC for the OP (L) and the DEG (R) Classes for the SVM Coarse Gaussian Model.

Table 9: Model Performance Summary.

Model	Training Accuracy	Correct /False	Training OP/DEG ROC/AUC	Test Accuracy	Correct /False	Test OP/DEG ROC/AUC
SVM Coarse Gaussian	75%	57 19	OP (.47, 1)/.85 DEG (0, .53)/.85	100%	10 0	OP (0, 1)/1 DEG (0, 1)/1
KNN Weighted	86.8%	66 10	OP (.23, .97)/.89 DEG (.03, .78)/.89	90%	8 2	OP (.0, .6)/1 DEG (.4, 1)/1
NN Trilayered	86.8%	63 13	OP (.15, .81)/.92 DEG (.19, .85)/.92	90%	8 2	OP (.0, .4)/.8 DEG (.6, 1)/.8
KNN Cubic	84.2%	62 14	OP (.28, .92)/.85 DEG (.08, .72)/.85	90%	8 2	OP (.0, .6)/1 DEG (.4, 1)/1
KNN Medium	82.9%	58 18	OP (.33, .86)/.85 DEG (.14, .68)/.85	90%	8 2	OP (.0, .6)/1 DEG (.4, 1)/1

These five selected models are working very well for this limited data set. There is no guarantee of generalizing these models to bearing defects in other applications but experimenting with the correct models will help a vibration analyst determine a good model to use to classify paper machine roll bearings into operational or degraded status.

Conclusions

The original scope of this project was to apply the Classification Learner App in MATLAB software to classify paper machine dryer roll bearing condition into operational and degraded classes. After extensive training and validation of 30 classification models under varying feature sets, five models were selected that performed well on the test data set. Two models were very consistent with eight and six removed feature combinations: KNN Weighted (eight removed) and NN Trilayered (six features removed) each with 86.8 and 90 percentage training and test accuracy rates, respectively.

The highest-scoring test model, SVM Coarse, did not score well on training and validation accuracy at 75%, yet had the highest overall performance that increased to 100% test accuracy with nine removed features: FlrF, FLR, NonSync, PKTot, MaxPK, CF, Sinusoidal Content, Early Impacting, and Repeatability. While this model correctly classified 100% of the Operational and Degraded Classes in the test data set, there was a false positive rate of 47.5% for truly degraded data being misclassified as Operational in the training data set. No other model achieved a 100% test accuracy rating, but all the Neural Network models achieved a 90% test accuracy rating for various feature removal combinations. The conclusion is that the SVM Coarse Gaussian model will perform well for all six included features of Impact Factor, Impact Area, Sinusoidal Strength, RMS, Peak to Peak, and NonSynchronous Percentage Fault Level. The "best" model is SVM Coarse Gaussian because it outperformed all the other models to hit 100 percent test data set accuracy for the fewest retained features.

KNN Medium was the fifth model selected because it performed well with 82.9% validation accuracy that resulted in a 90% test data set accuracy for classification, which ties with all the other selected models for test accuracy. KNN Weighted and KNN Cubic both tied for all the scores for TPR, FNR, PPV, and FDR on the test data set. For the training and validation data set, KNN Weighted scored marginally better than KNN Cubic, i.e., the KNN Weighted model was slightly better for TPR at 97.2% vs. 91.7% for KNN Cubic. Not to be outdone, the NN Trilayered tied with KNN Weighted at 86.8 percent accuracy which slid it into the third best model, just edging out KNN Cubic. The scores were so close for these models that only 1 more correct classification for NN Trilayered (63 vs. 62 correct, and 13 vs. 14 incorrect) separated it from KNN Cubic. In fact, for the bottom four models, no more than 4 correct and 4 wrong classifications separated any adjacent model from another in the training set. In the test set, all the bottom four models (KNN Weighted, NN Trilayered, KNN Cubic and KNN Medium all tied with 8 of 10 correct and 2 incorrect. The only model that performed better was the SVM Coarse Gaussian with 10 out of 10 correct classifications.

Special recognition must also go to the Neural Network (NN) family of models. Although all the NN models (NN Narrow, NN Medium, NN Wide, NN Bilayered, and NN Trilayered) scored very well (in the 80's) on the training and validation percentage scores, most of the NN models performed well on the test data set, with scores for the percentage test accuracy in the 90s for varying feature sets. Two NN models (Bilayered and Trilayered) had scores that tied three of the top four selected models at 90%. However, the Bilayered model was ousted due to the number of features required to achieve that performance. The top three models of SVM Coarse Gaussian, KNN Weighted, and NN Trilayered are concluded to have the best chance of correctly separating dryer rolls into operational and degraded bearing classes. If implemented as a part of an equipment maintenance program, the best five models will outperform the other 25 trained models to classify operational vs. degraded bearing condition for paper machine dryer rolls.

In the final analysis, if one implements any one of the top five models to classify paper machine dryer roll bearings into operational and degraded classes, the chances of successful classification between operation and degraded states are between 80 and 90 percent or better.

Limitations

Several important limitations must be presented for this project. First and foremost, this project did not explore any other degradation condition other than for bearings, as features that pertain to bearings were selectively used. A few features do cross over into other machinery faults, such as Crest Factor which may also be increased by a broken tooth on a gearbox, but these features were in the minority, and not specifically pursued or required for this project. Further, the raw data sets for the paper machine dryer rolls contained bearing defects for only two versions of a double row spherical roller Torrington bearing, the 23238KYM and 23238B, and one SKF 23248 double row spherical roller bearing. More specifically, the only bearing defects considered were Ball Pass Frequency of the Outer Race (BPFO) and Ball Pass Frequency of the Inner Race (BPFI) with BPFI only present in one data set that also had BPFO frequencies present. All the other rolls had only BPFO frequencies in the raw data. Another limitation was the small sample size, which was an intentional design limitation for this study. Only six total rolls from one plant facility were used, five for the training and validation data set and one for the testing data set.

These limitations may severely limit any generalizability of the project results to other paper machine roll defects, other paper machine rolls such as press rolls, breast rolls, calender rolls, suction couch rolls, covered rolls, lump breaker rolls, or any other type of paper machine roll found on sections of a paper machine. Generalizing to other bearing types is another limitation, as the data was collected on rolls with only two specific bearings. However, all rolling element bearings have both BPFO and BPFI frequencies so this may not be as severe a limitation as the paper machine roll type. In a study of this specificity, broader applications may be missed, such as support equipment (pumps, fans, and other equipment with rolling element bearings). Because multiple people collected the data on these paper machine rolls, and a paper machine is a variable speed machine, variations in data collection techniques and the machine's response are unavoidable. The paper machine's mass, stiffness, and damping control the response that creates measurable vibration at different speeds, and therefore base the data for this project on complex and dynamic relationships between machine components and the vibration analysts performing data collection on the machine. These multiple sources of variability limit the application of this very specialized project.

Recommendations

The first recommendation is that further research must be performed before using machine learning classifiers with rolling element bearings in paper machine rolls. Because Predictive Maintenance programs (PdM) in heavy-asset industrial facilities utilize condition monitoring technologies such as oil analysis, ultrasonic (high-frequency sound wave) energy emission detection, vibration monitoring, temperature scanning, thermography imaging, electrical circuit testing (online and off-line) (J. Nicholas, 2016; J. R. Nicholas, Young, & Anzalone, 2008), many of these technologies may be applied to classification of operational and degraded condition of paper machine rolls. Vibration monitoring delivers high diagnostic value and should be a main component of future research on classification of paper machine rolls (Crawford, A. R., & Crawford, 1992; Moore, 2012). Based on this project, the limited data sets did not prevent correct classification on a test data set for five out of the 30 models studied. Certainly, more research is recommended for the best five models of SVM Coarse Gaussian, KNN Weighted, NN Trilayered, KNN Cubic, and KNN Medium on data sets from paper machine dryer rolls. Research that is based on actual field data is preferable to that collected in a lab environment, even with reduced signal to noise ratios, as models developed with noise tolerance will be required in the real world.

Future Work

Future work should test the results obtained in this project, with expansion to many more paper machine rolls. Additionally, other shaft related faults should be modeled and tested, and other features such as acoustic emission (AE) and oil condition may be added for feature enrichment. Defects such as wear, fatigue, corrosion, fractures, deformations, and cracks should be added to the list of classes and variables such as lubrication condition, operating condition and mounting of the bearing on the shaft may be added to the feature list for training, validation and testing in the future. Many deep learning algorithms are available with multiple versions and constructions in the literature. The application of machine learning to the heavy asset machinery condition monitoring field is wide open for additional research and future work.

REFERENCES:

Azergui, M., Abenaou, A., & Bouzahir, H. (2018). Bearing fault classification based on the adaptive orthogonal transform method. *International Journal of Advanced Computer Science and Applications*, *9*(1), 375–380. https://doi.org/10.14569/IJACSA.2018.090151

Chen, S., Meng, Y., Tang, H., Tian, Y., He, N., & Shao, C. (2020). Robust Deep Learning-Based Diagnosis of Mixed Faults in Rotating Machinery. *IEEE/ASME Transactions on Mechatronics*, *25*(5), 2167–2176. https://doi.org/10.1109/TMECH.2020.3007441

Crawford, A. R., & Crawford, S. (1992). *The Simplified Handbook of Vibration Analysis. V. 1, Introduction to Vibration Analysis Fundamentals.* (1st ed.). Knoxville, TN: Computational Systems, Inc.

Gulati, R., & O'Hanlon, T. (2017). *10 Rights of Asset Management* (1st ed.). Ft. Myers, FL: Reliabilityweb.com.

Kitai, M., Kobayashi, T., Fujiwara, H., Tani, R., Numao, M., & Fukui, K. I. (2021). A Framework for Predicting Remaining Useful Life Curve of Rolling Bearings under Defect Progression Based on Neural Network and Bayesian Method. *IEEE Access*, *9*, 62642–62652. https://doi.org/10.1109/ACCESS.2021.3073945

Konar, P., Bandyopadhyay, R., & Chattopadhyay, P. (2009). Bearing fault detection of induction motor using wavelet and neural networks. *Proceedings of the 4th Indian International Conference on Artificial Intelligence, IICAI 2009*, (December), 798–809.

Li, C., Sánchez, R. V., Zurita, G., Cerrada, M., & Cabrera, D. (2016). Fault diagnosis for rotating machinery using vibration measurement deep statistical feature learning. *Sensors (Switzerland)*, *16*(6). https://doi.org/10.3390/s16060895

Lindh, T., Ahola, J., Spatenka, P., & Rautiainen, A. (n.d.). Automatic bearing fault classification combining statistical classification and fuzzy logic. *Technology*.

Moore, R. (2012). *Making common sense common practice: Models for manufacturing excellence* (4th ed.). Terrance O'Hanlon.

Nicholas, J. (2016). *Asset Condition Monitoring Management*. Retrieved from https://books.google.com/books?id=1-wpMQAACAAJ

Nicholas, J. R., Young, R. K., & Anzalone, E. K. (2008). *Predictive Maintenance Management* (3rd.). Ft. Myers, FL: Terrance O'Hanlon.

Noble, W. S. (2006). What is a support vector machine? *Nature Biotechnology*, *24*(12), 1565–1567. https://doi.org/10.1038/nbt1206-1565

Ren, L., Cui, J., Sun, Y., & Cheng, X. (2017). Multi-bearing remaining useful life collaborative prediction: A deep learning approach. *Journal of Manufacturing Systems*, *43*, 248–256. https://doi.org/10.1016/j.jmsy.2017.02.013

Schaeffer, E. (2017). *Industry X.0: Realizing digital value in industrial sectors* (1st ed.). Retrieved from https://books.google.com/books?hl=en&lr=&id=PNy_DgAAQBAJ&oi=fnd&pg=PA1&dq=industry+x.0&ots=p5Qa4ezlGC&sig=CdlscUH0I44BStGHWOWCEEs70C0

Techane, A. W., Wang, Y. F., & Weldegiorgis, B. H. (2018). Rotating machinery prognostics and application of machine learning algorithms: Use of deep learning with similarity index measure for health status prediction. *Proceedings of the Annual Conference of the Prognostics and Health Management Society, PHM*, (September).

Zhang, S., Zhang, S., Wang, B., & Habetler, T. G. (2020). Deep Learning Algorithms for Bearing Fault Diagnosticsx - A Comprehensive Review. *IEEE Access*, *8*, 29857–29881. https://doi.org/10.1109/ACCESS.2020.2972859

Zhang, Y., Zhou, T., Huang, X., Cao, L., & Zhou, Q. (2021). Fault diagnosis of rotating machinery based on recurrent neural networks. *Measurement: Journal of the International Measurement Confederation*, *171*(December), 108774. https://doi.org/10.1016/j.measurement.2020.108774

Zhao, X., & Jia, M. (2019). A new Local-Global Deep Neural Network and its application in rotating machinery fault diagnosis. *Neurocomputing*, *366*(August), 215–233. https://doi.org/10.1016/j.neucom.2019.08.010

Zhao, Z., Li, T., Wu, J., Sun, C., Wang, S., Yan, R., & Chen, X. (2020). Deep learning algorithms for rotating machinery intelligent diagnosis: An open source benchmark study. *ISA Transactions*, *107*(August), 224–255. https://doi.org/10.1016/j.isatra.2020.08.010

Zhiliang Liu, Ming J. Zuo, L. Z. (2020). Remaining useful life prediction of rolling element bearings based on Health State Assessment. *Mechanism and Machine Theory*, *153*. https://doi.org/10.1016/j.mechmachtheory.2020.103967

VIATC TRAINING CONFERENCE & EXPO

COVINGTON, KENTUCKY | AUGUST 7-9, 2024

STRUCTURAL DYNAMIC EVALUATION OF RIGID CONCRETE FOUNDATIONS SUPPORTING DYNAMIC EQUIPMENT

Robert J. Sayer, PE
Sr. Consultant, Mechanical Solutions, Inc.
Past President, The Vibration Institute
rsayerpe@gmail.com

Robert J. Sayer, P.E. is a Sr. Consultant at Mechanical Solutions, Inc. and the Past President of the Vibration Institute. He has over forty-five years of experience as a mechanical and structural engineering consultant to a wide range of industries including pulp & paper, mining & metals, fossil-fuel & nuclear power generation, petrochemical, food processing, glass manufacturing, and aerospace & defense. Mr. Sayer provides analytical and experimental dynamic analysis of mechanical equipment for design projects, structural and mechanical failure studies, and process reliability and efficiency studies. He holds engineering degrees from Ohio University (BSCE), Cleveland State University (MSIE), and Purdue University (MSCE). He had the honor and fortune of collaborating with the late Art Crawford, a pioneer of vibration analysis, on numerous projects. Mr. Sayer is a regular contributor at technical conferences and has published over 50 technical articles and co-authored 2 chapters in "The Simplified Handbook of Vibration Analysis". He is Vice-Chairman of ANSI/HI Standard 10.5.1 "Rotordynamic Pumps for Foundation Design"

ABSTRACT:
The engineering analysis performed to design rigid concrete foundations supported directly upon soil is not a deterministic process. Variability in soil properties from subsurface layer-to-layer and within a single layer will affect the natural frequencies and forced dynamic response of the foundation. However, this is not the case for the evaluation of existing foundations. If the natural frequency and compliance of the foundation can be determined from field test data, the soil-foundation interface stiffness can be back calculated. This paper introduces the basic steps of the design of rigid concrete foundations for dynamic response and presents methods to determine the soil-foundation properties from test data of existing foundations.

KEY WORDS: Block-type foundation, finite element analysis, machine-foundation interaction, rigid concrete foundation, shaft stick, soil dynamics, soil-foundation interface, soil stiffness

INTRODUCTION:

The components of a rigid concrete foundation are shown in Figure 1. One type of foundation is a simple rectangular mat which is placed directly upon soil. The mechanical equipment is then bolted to the top of the mat. The second type of foundation has a concrete pedestal or multiple pedestals that extend from the top of the foundation mat to the base of the equipment. In this type, the mechanical equipment is bolted to the top of the concrete pedestal. This foundation type is referred to in ACI-351 [1] as a block foundation or in the case of multiple pedestals a combined block foundation.

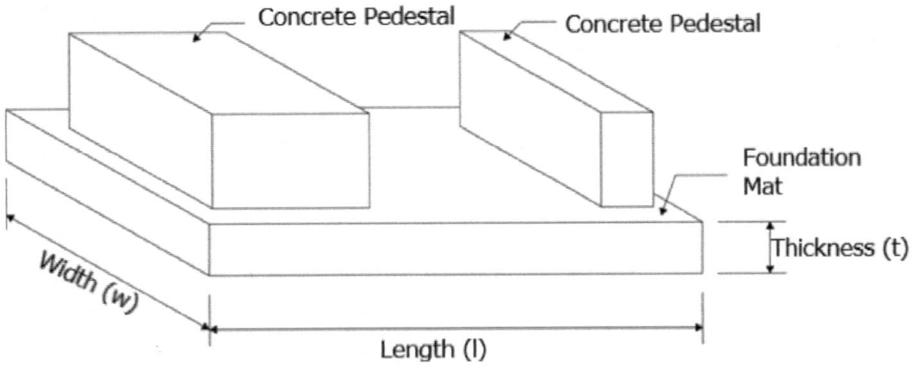

Figure 1: Components of Rigid Concrete Foundation

A foundation is considered rigid if the deformation of the foundation mat and pedestals are insignificant compared to the soil-foundation interface. The main elastic element in the structural-dynamic system is the soil. The concrete foundation participates only as mass; the dynamic response resembles that of a rigid body. The principal natural frequencies of the foundation are associated with rigid body translation or rotation (rocking). Rigid foundations can have elastic modes of natural frequency in addition to the rigid body modes. But these natural frequencies are typically much greater than the operating speed of most large mechanical equipment.

This paper is limited to the evaluation of the dynamic response of rigid foundations that are set directly upon soil. The foundation mat for these types of foundations is sometimes referred to as a spread footing; the load transferred from the concrete to the soil is spread out as pressure between the concrete foundation mat and the subsurface directly below it. The dynamic analysis of foundations supported by piles or caissons (sometimes referred to as deep foundations) is outside the scope of this paper.

DESIGN PROCEDURE:

The basic design procedure for a foundation consists of:

Step 1) Sizing the foundation to accommodate the footprint of the mechanical equipment, providing sufficient edge clearance for proper embedment of all anchor bolts, subject to surrounding clearance restrictions imposed by the site.

Step 2) Sizing the foundation mat such that the maximum soil bearing pressure is below the allowable bearing pressure specified by the geotechnical engineer for all combinations of loading conditions.

Step 3) Perform a preliminary analysis of foundation based upon rules-of-thumb that have been developed through years of experience. Adjust the foundation sizes established by Step #1 and Step #2, accordingly and re-evaluate foundation for allowable bearing pressure.

Step 4) Perform a numerical dynamic analysis to estimate the natural frequencies and forced response (vibration level). Compare estimated natural frequencies to the frequencies of dynamic loads. If an acceptable separation margin is not provided, readjust design. Compare the estimate dynamic response and compare to allowable vibration level. Readjust the design if the calculated dynamic response exceeds the specified allowable vibration level. The dynamic analysis can be based upon simple single degree-of-freedom manual calculations or complex three-dimensional finite element models.

This paper deals with the latter two steps of the design procedure. Reference can be made to texts on soil mechanics and foundation design [2],[3] for information regarding the design of foundations relative to the first two steps of the procedure.

SOIL-FOUNDATION INTERFACE PROPERTIES:

The main elastic spring in a structural-mechanical system that includes a rigid concrete foundation is the underlying soil at the soil-foundation interface. Unlike construction materials like steel and concrete, the physical properties of soils can differ drastically from type-to-type and site-to-site. A densely compacted granular soil will differ from a pre-consolidated clay; both will differ from a silty material. The properties of the same type of material can differ even within a single site. The accuracy of calculations for natural frequencies and forced response of rigid foundations are highly dependent upon the estimations for material stiffness and damping of subsurface conditions. Therefore, a basic understanding of soil mechanics is required.

The following equations for horizontal, vertical, and rotational stiffness provided to the base of a rectangular foundation by the subsurface materials were proposed by Barkan [4] and reaffirmed by Richart [5]:

$$k_h = 32(1-\mu)GR/(7-8\mu) \qquad \textbf{Equation 1}$$
$$k_v = 4GR/(1-\mu) \qquad \textbf{Equation 2}$$
$$k_\varphi = 8GR_\varphi^3/\ 3(1-\mu) \qquad \textbf{Equation 3}$$

where: $R = (wl/\pi)^{1/2}$
$R_\varphi = (wl^3/3\pi)^{1/4}$
G = shear modulus = $E/2(1+\mu)$
E = modulus of elasticity
μ = Poisson's Ratio

These spring constants require an estimate of the Modulus of Elasticity (E) or Shear Modulus (G), and the Poisson Ratio (μ). These properties are not typically provided in a geotechnical engineering report. In the absence of actual material property data, the elastic modulus and Poisson ratio can be estimated based upon the general description, USCS classification, or SPT blow count information. Estimation of the Modulus of Elasticity and Poisson ratio using these methods introduce a level of uncertainty in the analysis. Typically, a range of values is established resulting in a lower bound and upper bound estimate of the natural frequencies and dynamic response (vibration) of a foundation.

The research that established the subsurface spring constants listed above were based upon testing that used a circular plate to distribute loads to soils. Since most foundations are rectangular, an equivalent radius (R), as defined above, is used instead of the actual radius. When the aspect ratio (w/l or l/w) exceeds 2.0, the actual stiffness of the soil will differ from that calculated using the Richart equations.

The spring constants calculated from the Richart equations are for static analysis. Experience has shown that the dynamic stiffness of the soil-foundation interface can differ significantly from the static stiffness. The small strain shear modulus (G_{max}) used in place of the shear modulus (G), will provide a better estimate of dynamic stiffness.

$$k_h(\text{dynamic}) = 32(1 - \mu)G_{max}R/(7 - 8\mu) \quad \textbf{Equation 4}$$
$$k_v(\text{dynamic}) = 4G_{max}R/(1 - \mu) \quad \textbf{Equation 5}$$
$$k_\varphi(\text{dynamic}) = 8G_{max}R_\varphi^3/ 3(1 - \mu) \quad \textbf{Equation 6}$$

Liepins [6] reported a range of values for the elastic modulus based upon USCS classifications. This method is considered a little more accurate than using a general soil description, but still results in a rather wide range of possible values. Values are listed in Table 1 and Table 2. From Table 2, the Modulus of Elasticity for hard clay (CL) is between 4,350 psi - 10,200 psi. From Table 1, the range of values reported for dense sand (SP) is between 7,250 psi - 11,600 psi.

Table 1: Modulus of Elasticity for Granular Soils (ksi)

USCS	Description	Loose	Medium	Dense
GW, SW	Gravels/Sand well-graded	4.35 -11.6	11.6 – 23.2	23.2 – 46.4
SP	Sand, uniform	1.45 - 4.35	4.35 – 7.25	7.25 – 11.6
GM, SM	Sand/Gravel silty	1.00 – 1.74	1.74 – 2.90	2.90 – 4.35

Table 2: Modulus of Elasticity for Cohesive Soils (ksi)

USCS	Description	Very soft-soft	Medium	Stiff-very stiff	Hard
ML	Silts (slight plasticity)	Not Recommended	1.45 – 2.20	2.20 – 5.80	5.8 – 11.6
ML,CL	Silts (low plasticity)	Not Recommended	0.90 – 1.45	1.45 – 4.35	4.35 - 8.7
CL	Clays (low-medium plasticity)	Not Recommended	0.70 – 1.20	1.20 – 4.35	4.35 – 10.2
CH	Clays (high plasticity)	Not Recommended	0.60 – 1.00	1.00 – 2.90	2.9 – 4.6

The Modulus of Elasticity can also be estimated from the blow count (N) information recorded for a Standard Penetration Test (SPT) performed during subsurface soil sampling. Since the blow count data is provided specifically for the materials upon which the foundation will be placed, it is considered a more accurate method of estimating the elastic modulus. However, it is still not a direct measure of the Modulus of Elasticity, but a correlation based upon research. CIRIA 143 [7] proposes the following simple relationship between blow counts and elastic modulus for cohesionless granular materials:

$$E = N\ (\text{Mpa}) \text{ or } E = 145N\ (\text{psi}) \quad \textbf{Equation 7}$$

Callanan and Kulhawry [8] proposed the following relationships for cohesionless materials:

$E = 73.5N$ (psi)	for sands with some fines	**Equation 8**
$E = 147N$ (psi)	for clean normally consolidated material	**Equation 9**
$E = 220N$ (psi)	for clean over consolidated material	**Equation 10**

Wroth, et. al. [9] suggests the following relationship between SPT blow counts and the Modulus of Elasticity for cohesive materials:

$$E = 283N^{0.63} \text{ (psi)} \qquad \textbf{Equation 11}$$

The small strain shear modulus (G_{max}) is a material property that is often used for the analysis of the dynamic response of machine foundations. It recognizes the fact that the soil-foundation interface will be stiffer when subjected to relatively small dynamic loads compared to larger static loads. The value for the small strain shear modulus (G_{max}) can be significantly larger than the standard shear modulus (G). Wroth, et. al. [9] suggests the following relationship between SPT blow counts and the Small Strain Shear Modulus for cohesive materials:

$$G_{max} = 12.14N^{0.77} \text{ (Mpa)} = 1760N^{0.77} \text{ (psi)} \qquad \textbf{Equation 12}$$

Ohsaki & Iwasaki [10] provided the following correlations for various material types:

$G_{max} = 6.374N^{0.94}$ (Mpa) $= 924N^{0.94}$ (psi)	for sandy soil	**Equation 13**
$G_{max} = 13.73N^{0.71}$ (Mpa) $= 1990N^{0.71}$ (psi)	for cohesive soil	**Equation 14**

Atkinson and Sallfors [11] categorized strain levels into three groups; very small strain levels where the stiffness modulus is constant in the elastic range; small strain where the stiffness modulus varies non-linearly with strain, and large strain where the soil is close to failure. Atkinson and Sallfors indicate that G_{max} could be 2-3x greater than the value for G used to design foundations for static loading.

Due to the uncertainty of soil stiffness, estimating or projecting structural dynamic response of a foundation during the design stage can be difficult. It typically involves defining an expected range of soil properties that provides an upper and lower bound of expected natural frequencies.

Example 1 illustrates the wide range of projected soil properties that could be encountered during the design process. This information can be used to establish a worst-case scenario from which expected vibration levels are calculated. Vibration troubleshooting of an existing foundation is different in that the actual natural frequencies can be determined using modal testing methods. Also, the vibration level (forced response) of the foundation can be measured. Sayer [12] presented a method of back calculating the foundation-soil interface stiffness once the natural frequency and damping of a foundation is determined by field testing.

EXAMPLE 1 - SOIL STIFFNESS CALCULATION:

A geotechnical subsurface investigation reports the soil stratum directly underneath a proposed foundation for a large fan to be a medium-compacted, well-graded gravel with a USCS classification of GW. The blow counts obtained from Standard Penetration Tests performed at three separate boring positions located in this soil stratum were recorded as N = 110, 120, and 124. Based upon the USCS GW classification, from Table 1, the modulus of elasticity would be estimated between 11,600 - 23,200 psi and assuming a Poisson ratio of 0.30, the shear modulus

(G) could vary between around 4,500 - 8,900 psi. Using blow count data, from CIRIA 143 the modulus of elasticity and shear modulus would be estimated as varying between:

$$E = 145N = 145(110 \text{ to } 124) = 15{,}950 - 17{,}980 \text{ psi}$$
$$G = E/2(1 + \mu) = 6{,}135 - 6{,}915 \text{ psi}$$

Applying an Atkinson and Sallfors multiplier of 3 for the G_{max}/G ratio provides an estimated range of:

$$G_{max} = 3\ (6{,}135 - 6{,}915 \text{ psi}) = 18{,}405 - 20{,}745 \text{ psi}$$

Using the Ohsaki & Iwasaki relationship for G_{max}:

$$G_{max} = 924N^{0.94} = 924(110 \text{ to } 124)^{0.94} = 51{,}770 - 85{,}800 \text{ psi}$$

The above example illustrates the level of uncertainty involved in evaluating dynamic response of rigid foundations in the design stage. The lower bound of the estimated small strain shear modulus varies by 51,770/18,405 = 2.8x. The upper bound of the estimated small strain shear modulus varies by 85,800/20,745 = 4.1x. Since the stiffness of the soil-foundation interface is directly proportional to this modulus per Equation 4 – Equation 6, and the natural frequency is a function of the square root of stiffness per Equation 15, the estimated natural frequency using one method could be double that using the other method.

Fortunately, the stiffness of an existing foundation can be estimated with a high level of certainty, based upon measured structural dynamic response. Case Study 1 includes a method of determining soil-foundation interface stiffness, after the natural frequencies and forced response has been determined from experimental testing of an existing foundation. Either a manual calculation using the single spring-mass model, or a computer-aided solution can be employed on more complex foundation systems to back calculate foundation stiffness.

SINGLE SPRING-MASS MODEL FOR DYNAMIC ANALYSIS:

An estimate of the natural frequency of a foundation and its forced dynamic response can be done manually if the concrete foundation mass is rigid and the equipment supported thereon is assumed to be rigid. A finite element analysis (FEA) may be required to estimate the natural frequencies and forced response of rigid foundations supporting flexible machines.

In the single degree-of-freedom approach, the only restoring spring considered in the analysis is the elasticity of the soil-foundation interface. The concrete foundation and mechanical equipment are estimated as lumped masses. However, since the center-of- gravities for these lumped masses are offset from the location of the spring, a simple linear mass-spring model cannot be used. The mass moment of inertia must be considered.

The end view of rigid body rotational response of a foundation (dotted line) due to a horizontal dynamic force ($F_{x(dynamic)}$) applied at a distance above the soil foundation interface is shown in Figure 2. The rocking response depicted in this figure would be for the case wherein the sliding of the foundation is neglected. The undamped natural frequency for the rocking mode ($f_{n\varphi}$) can be estimated using Equation 15 as:

$$f_{n\varphi} = (1/2\pi)[k_\varphi(\text{dynamic})/I_\varphi]^{1/2} \qquad \textbf{Equation 15}$$

where k_φ(dynamic) is defined in Equation 6, I is the moment of inertia of the soil-foundation interface, and I_φ is the mass moment of inertia of the foundation and equipment about the soil-foundation interface.

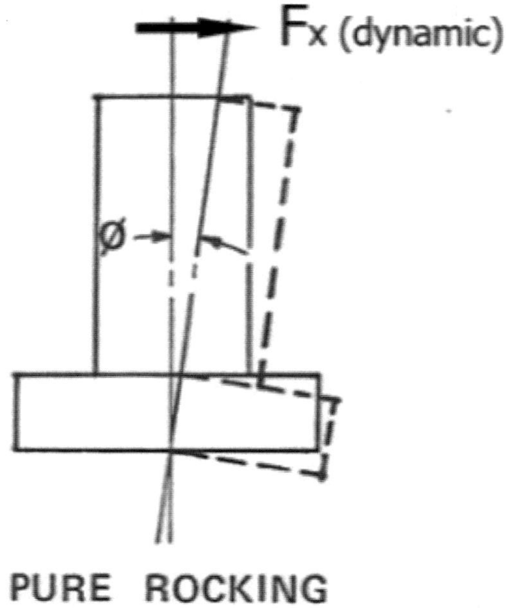

Figure 2: Pure Rigid Body Rocking Mode of a Foundation

The overturning moment from the dynamic force is:

$$M_{dynamic} = F_{x\ (dynamic)}\ y$$

Where y is the vertical distance from the force to the soil-foundation interface. The angle that the foundation would rock due to the overturning moment from the horizontal force would be:

$$\Phi = AF\ (M_{dynamic}/k_{\varphi(dynamic)}) \qquad \textbf{Equation 16}$$

$$AF = 1/[\{1-(f_{dyn}/f_n)^2\}^2 + \{2\zeta(f_{dyn}/f_n)\}^2]^{1/2} \qquad \textbf{Equation 17}$$

The horizontal peak displacement at the elevation of the dynamic force is simply:

$$x = y\ (\tan \varphi) \qquad \textbf{Equation 18}$$

Another case of the rigid body response contains rocking in combination with sliding of the foundation. The end view of rigid body response of a foundation (dotted line) for this case is shown in Figure 3. This motion has a rotation defined by Equation 16 combined with horizontal sliding at the soil foundation interface (x).

The undamped natural frequency for the pure sliding mode (f_{nh}) can be estimated using Equation 19 as:

$$f_{nh} = (1/2\pi)[k_h(dynamic)/m]^{1/2} \qquad \textbf{Equation 19}$$

where k_h(dynamic) is defined in Equation 4, m is the modal mass of the foundation and equipment about the soil-foundation interface.

Equation 15 is used to calculate the natural frequency for pure rotation. Equation 19 is then used to calculate the natural frequency for pure sliding. The combined natural frequency can then be estimated as:

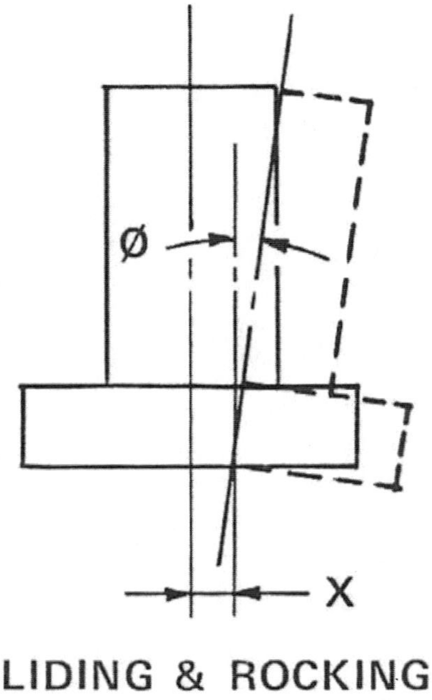

Figure 3: Combined Sliding and Rocking Mode of a Foundation

Equation 20 was proposed by Barkan [4] to obtain the natural frequency of the combined sliding and rocking mode from the individual calculations.

$$f_n^2 = (1/2\Upsilon)[\, f_{nh}^2 + f_{n\varphi}^2 - [(f_{nh}^2 + f_{n\varphi}^2)^2 - 4\Upsilon f_{nh}^2\, f_{n\varphi}^2]]^{1/2} \qquad \textbf{Equation 20}$$

where $\Upsilon = I_\varphi/I_{cg}$ = mass moment about the soil-foundation interface/mass moment about the center-of-gravity of the foundation.

Once the natural frequency of the combined rocking and sliding mode is determined, the amplification factor can be calculated using Equation 17. The horizontal component of the vibration response due to rocking is calculated as before using Equation 16 and Equation 18. The horizontal vibration response due to sliding is calculated using Equation 21.

$$x = F_{x(dynamic)}AF/k_h \qquad \textbf{Equation 21}$$

CASE STUDY 1 -EVALUATION OF EXISTING FOUNDATIONS (RIGID ROTORS): An actual case study will be used to introduce and illustrate the use of the simple spring-mass model to determine the soil-foundation interface stiffness of an existing foundation. A series of large industrial fans used to recirculate air in a traveling grate iron ore drying furnace had experienced high vibration levels since plant start-up. The vibrations resulted in premature bearing failures and required numerous unscheduled maintenance outages for balancing and cleaning. The fans were double-wide, double-inlet (DWDI) type driven by 4,000 hp motors operating at 890 rpm. Each fan and motor were supported upon a concrete foundation (see Figure 4) with approximate exterior dimensions of 34' x 22'.

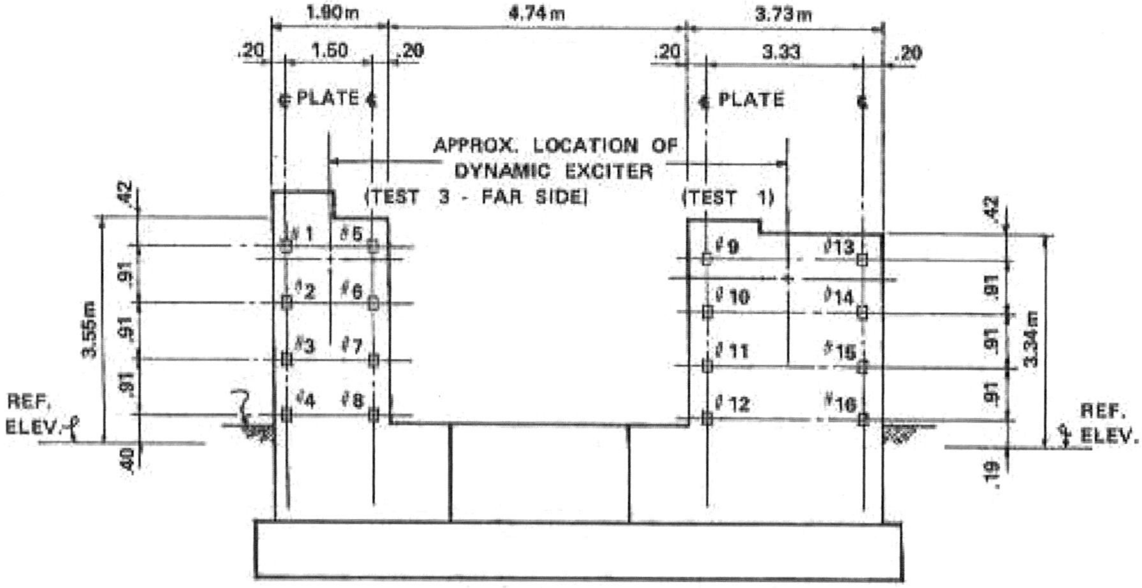

Figure 4: Foundation for Case Study 1 Showing Data Pt Locations

The testing for this case study was done during a period preceding personal computers and modal/ODS software using single-channel (tunable filter), pre-FFT vibration analysis equipment with limited resolution. It is presented to illustrate that modal testing still could be done. Even though test equipment and software has developed, making these types of studies much easier to perform and to document, the basic theory behind the forced response of a vibrating system has not changed.

The first step of the evaluation technique required determining the actual natural frequency, mode shape and damping of the foundation. This was accomplished by affixing a large variable-speed exciter to the foundation and measuring the foundation response at different speeds. Figure 5 is the result of the excitation test, performed when the fan was not operating. This data clearly identified a natural frequency at around 880 cpm. It was very close to the operating speed of the fan. This test confirmed resonance as being the root-cause of excessive vibrations.

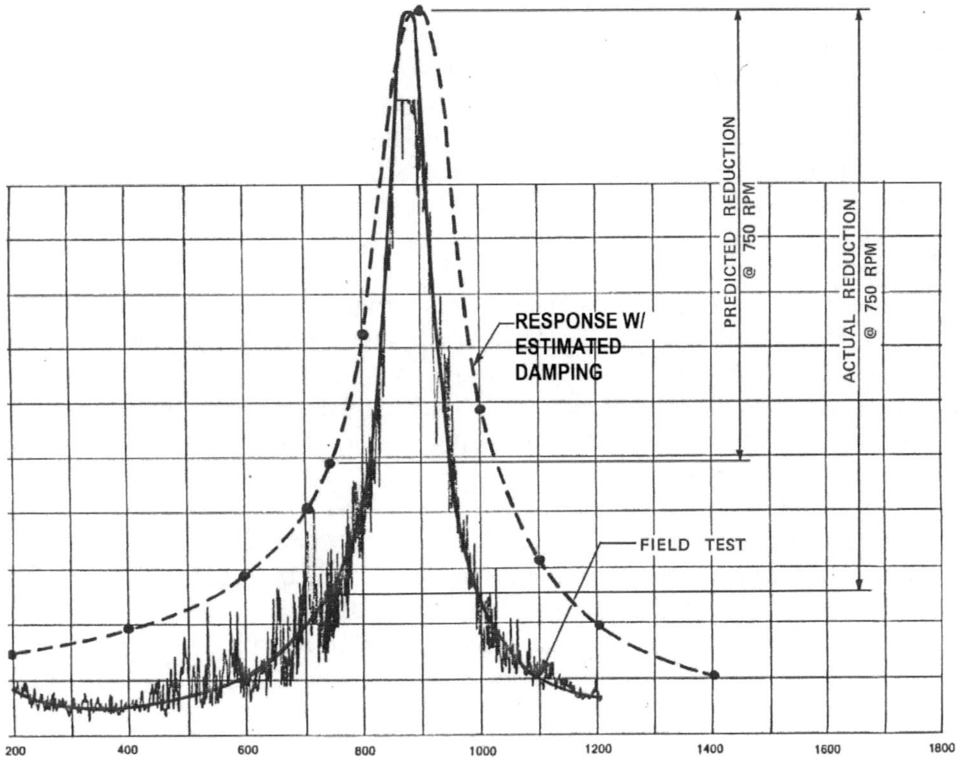

Figure 5: Measured Response of Foundation with Variable Frequency Exciter

Damping was estimated using two different methods. The first method used equations from texts on foundation design. These equations are not listed in this paper because experience has demonstrated them to be too conservative. The calculated damping ratio using these conservative equations was $\zeta = 0.08$ and the predicted response curve using this damping was plotted as a dotted line in Figure 5. Clearly the actual damping ratio was much less than the calculated damping. The second method used the half-power calculation for damping given in Equation 22:

$$2\zeta = f_{P2} - f_{P1} / f_n \qquad \textbf{Equation 22}$$

For the data in Figure 5, the damping ratio is calculated as follows:

f_{P1} = Frequency @ Lower End Half-Power FRF: 850 cpm
f_{P2} = Frequency @ Upper End Half- Power FRF: 920 cpm
$2\zeta = (920 - 850)/880 = 0.0795$
$\zeta \sim 0.04 \sim 4\%$ of critical

Although this damping ratio is considered very low for a typical soil-foundation interface, the subsurface condition for the fan in this case was a densely compacted leveling course of gravel on top of rock. This damping ratio is consistent with this soil type that is very rigid with small deformation and a low amount of hysteretic damping.

The next task was not only to determine the mode shape for the natural frequency, but also the dynamic compliance of the foundation from which the soil-foundation interface stiffness could be approximated. If this stiffness could be established, a numerical model of the system could be developed to evaluate possible corrective actions. The dynamic compliance of the foundation was obtained using a procedure similar to balancing. This test was done while the fan was operating. An initial set of data was acquired consisting of an amplitude of foundation vibration and phase relative to a point on the rotating shaft. This set of data defined the original "as found" response vectors.

Data was measured with a velocity transducer roving amongst the fan and motor bearings. Data was also collected in the horizontal and vertical directions from 16 different steel pads that were cemented onto the concrete foundation at positions shown in Figure 4. An additional 8 steel pads were attached at the end of the foundation to obtain axial data. This data represented the response of the structural system to whatever forcing function was present in the fan, the magnitude of this force being unknown.

Next, a known amount of additional imbalance was introduced by attaching a weight to the fan impeller. A second set of amplitude and phase readings were acquired at each of the data point locations. This set represented the response of the structural system to a combination of the known additional imbalance and the original unknown dynamic force.

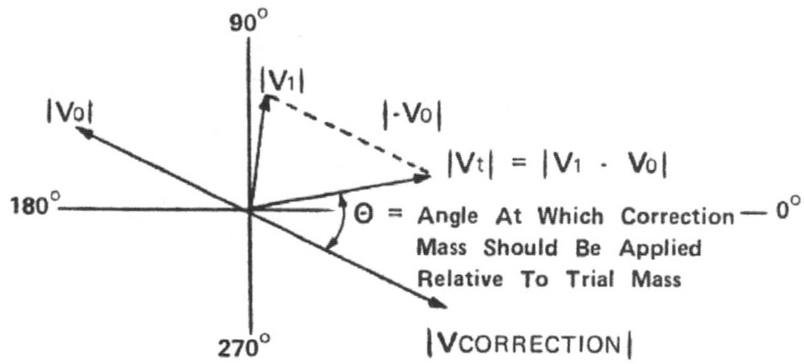

Figure 6: Vector Diagram for Balancing

The response to the known additional force could then be determined, at each data point location, by the following vector subtraction as illustrated in Figure 6:

$$|V_t| = |V_1| - |V_0| \qquad \text{Equation 23}$$

where $|V_t|$ = vector representing effect of trial weight, $|V_0|$ = vector representing original vibration reading, and $|V_1|$ = vector representing second vibration reading.

The known dynamic force developed by the trial weight was:

$$F_{dynamic} = (W/g)r\omega^2 \qquad \textbf{Equation 24}$$

where W = trial weight (lbs), g = acceleration of gravity (in/sec^2), r = radial distance of trial weight (in), and ω = speed of the fan (rad/sec).

For this test, a trial weight of 3.0 lbs was added at the outside diameter of the 900-rpm fan wheel at radius of 50". The dynamic force produced by the trial weight was:

$$F_{dynamic} = (3.0\ lb/386.1\ in/sec^2)(50\ in)(2\pi\ 900/60\ rad/sec)^2 \sim 3{,}450\ lbs$$

The vertical distance (y) from the center of rotation to the soil-foundation interface was 284 in, resulting in a dynamic overturning moment of:

$$M_{dynamic} = 3{,}450\ lbs\ (284\ in) = 9.80E05\ in\text{-}lb$$

The vector of response and phase representing the response due to the trial weight ($|V_t|$) was plotted for each data point location. Since the magnitude of the dynamic force was known, this plot represented a scaled mode shape. An end view of the mode shape is shown in Figure 7 that confirms the response as being very close to pure rigid body rocking.

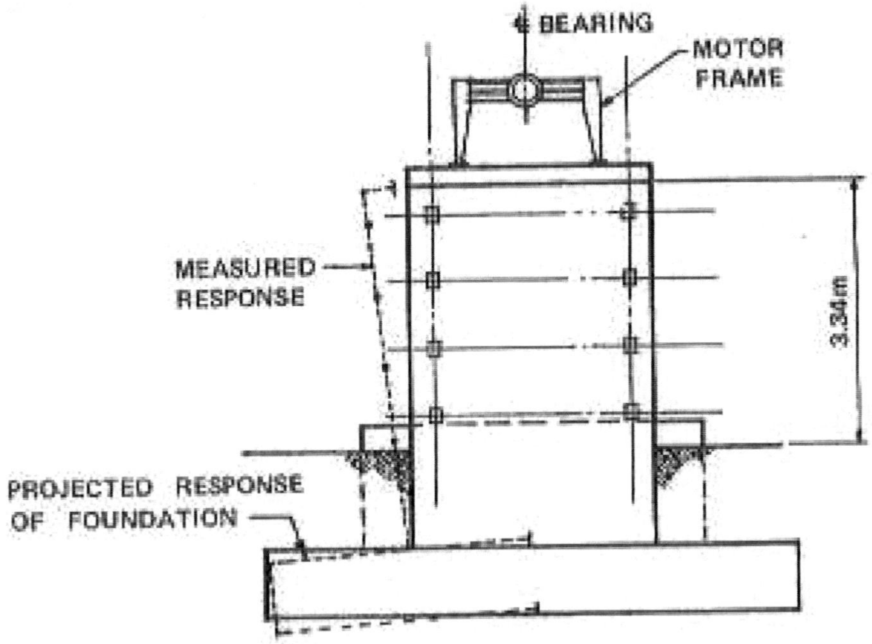

Figure 7: Measured Response of Foundation shown to Linear Exaggerated Scale

Since the mode shape confirmed rigid body motion, Equation 16 was appropriate for calculating the rotational motion of the foundation. Rearranging the terms in this equation, gives the following expression for stiffness:

$$k_\varphi = AF\ (M_{dynamic})/\varphi$$

From test data, the frequency ratio for the foundation was 890 rpm/880 cpm = 1.011, the damping ratio was 0.04, and the peak displacement caused by the trial weight was 0.001 in. The angle of rotation would be:

$$\tan \varphi = 0.001/216 = 0.0000046$$
$$\varphi = 0.0000046\ rad$$

The amplification factor and actual stiffness was calculated as:

$$AF = 1/[\{1-(1.011)^2\}^2 + \{2(0.04)(1.011)\}^2]^{1/2} = 12.4$$

$$k_\varphi(\text{dynamic}) = (12.4)9.80E5 \text{ in-lb}/(4.6E\text{-}6 \text{ rad}) = 2.64E12 \text{ in-lb/rad}$$

Equation 6 is used to estimate the dynamic stiffness of the soil, assuming $\mu = 0.3$, as:

$$k_\varphi(\text{dynamic}) = 8G_{max}R_\varphi^3/\ 3(1 - \mu)$$

$$R_\varphi = (lw^3/3\pi)^{1/4} = [(408 \text{ in})(264 \text{ in})^3/3\pi]^{1/4} = 168 \text{ in}$$

$$G_{max} = 3(1 - 0.3)(2.64E12 \text{ in-lb/rad})/8(168 \text{ in})^3 = 146 \text{ ksi}$$

Using an Atkinson and Sallfors multiplier of 3 for the G_{max}/G ratio yields:

$$G = 146 \text{ ksi}/3 \sim 49 \text{ ksi, from which } E = 2(1 + \mu)G = 127 \text{ ksi}$$

This estimated modulus is significantly higher than the upper bound of 46.4 ksi for dense gravel given in Table 1. However, for this case the concrete foundation was poured on a relatively thin, well compacted layer of gravel that was placed upon a rock surface. The calculated modulus from test data of 127 ksi appeared reasonable for the stated conditions. Another check of the accuracy for the estimate of spring stiffness was made using Equation 15. The natural frequency of the foundation (880 cpm/60 = 14.67 Hz) was known, and the mass moment of the foundation and equipment was estimated as 3.65E08 in-lb-sec^2.

$$f_{n\varphi} = (1/2\pi)[k_\varphi(\text{dynamic})/I_\varphi]^{1/2} = 14.67 \text{ Hz}$$

$$k_\varphi(\text{dynamic}) = (14.67 \text{ x } 2\pi \text{ rad/sec})^2\,(3.65E08 \text{ in-lb-sec}^2) = 3.09E12 \text{ in-lb/rad}$$

The rotational stiffness (2.64E12 in-lb/rad) calculated using forced response data was close to that calculated based upon the natural frequency (3.09E12 in-lb/rad).

Once the simple spring-mass model was correlated to actual field data and the compliance of the foundation was established with a high degree of confidence, the model was used to study various corrective action plans. Natural frequency tuning options were to either increase stiffness to increase the natural frequency or increase mass to decrease the natural frequency.

Since the main spring was the soil-foundation interface and the existing subsurface conditions were considered very rigid, it was concluded that increasing the natural frequency would be very difficult. The corrective action plan consisted of encasing the pedestals with a concrete jacket as shown in Figure 8. The concrete jacket was anchored to the sides of the existing concrete pedestals and rested on top of and anchored to the foundation mat. The new concrete did not contact the soil-foundation interface and thus, did not add any stiffness to the soil-foundation interface. It purely added to the mass and mass moment of inertia of the foundation. This modification successfully reduced the natural frequency of the foundation. Also, after the natural frequency was separated from the operating speed of the fan, it was possible to successfully balance the fan. Vibration levels were reduced by lowering the resonant amplification factor by tuning the natural frequency and reducing the dynamic force by balancing.

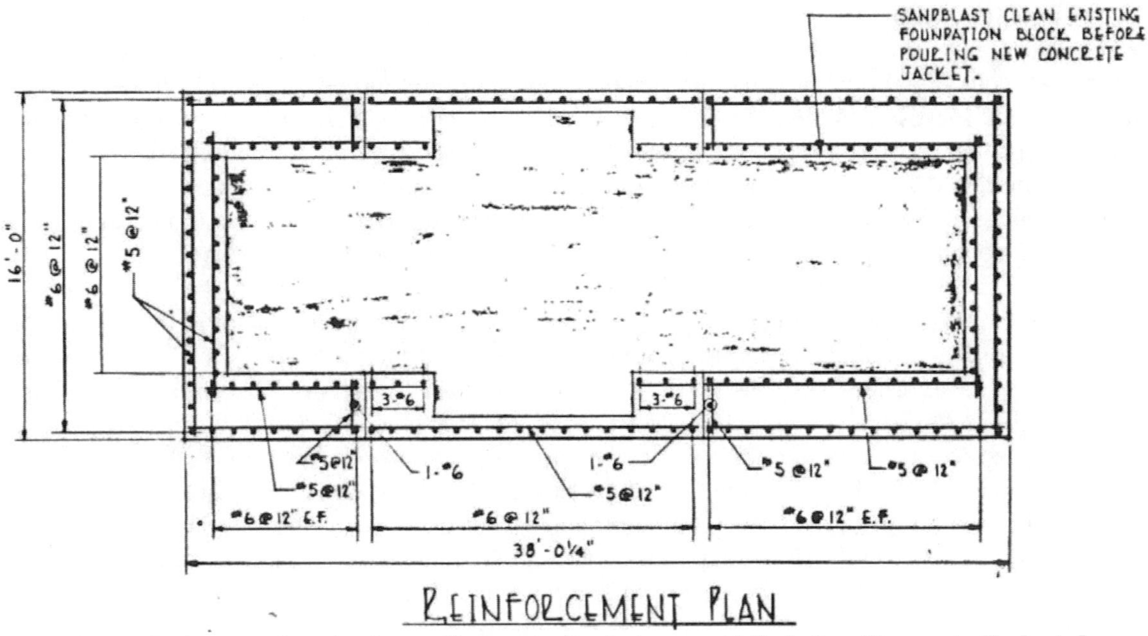

Figure 8: Corrective Action – Concrete Jacket around Existing Concrete Pedestals

This case study, done with pre-FFT vibration analysis hardware and pre-commercially available modal software, illustrates that even complex vibration problems can be solved if the fundamentals of vibration response are understood. The test procedures for this case study used simple balancing theory and vector algebra to produce a fully scaled modal of the rigid concrete fan foundation.

COMPLEX SYSTEM ANALYSIS:

The single degree-of-freedom method assumes that the only spring in the system is that of the soil-foundation interface. It assumes that the concrete mat and pedestals are rigid, as well as the components of equipment supported by the foundation. However, the equipment can be flexible, and this flexibility can affect the accuracy of the single degree-of-freedom calculation. The rotors of many types of machines can be very flexible compared to the rigidity of a concrete foundation. The flexibility of rotating elements is even more significant when supported upon oil-film journal type bearings.

Finite element analysis (FEA) [13] is a numerical analysis technique that can be used to approximate the structural dynamic characteristics of complex mechanical systems. FEA can be used to project the response of rotating mechanical components, such as fan rotors, based upon data acquired from stationary components, such as bearings and support structures. Numerical models can also be used to evaluate the potential benefits of design modifications prior to implementation, thus, eliminating costly trial-and-error procedures. The FEA model can serve as a digital twin for the actual physical system [14]. Once correlated to field data, the effectiveness of proposed modifications can be evaluated digitally. The following case study illustrates using FEA to assist in a diagnostic study of a complex system.

CASE STUDY #2:
A centrifugal fan that provided air to a rotating kiln experienced high vibration levels. The fan was direct driven by a variable frequency drive (VFD) and a 1,000 hp motor. The fan was a double inlet, centrifugal type with the fan wheel located between the bearings. The desired operating speed for much of the process was 1000 rpm (16.67 Hz). Figure 9 is a plot of bearing vibration versus operating speed. The vibration levels approached 0.60 ips as the operating speed was increased toward 950 rpm. The equipment manufacturer indicated that warranty would be voided if the fan was operated above 0.50 ips, and the VFD was programmed to limit the operating speed to 900 rpm. This adversely affected the productivity of the rotary kiln.

The vibration spectra were dominated by response at 1X rotational speed, meaning the source of the dynamic force was unbalance in the rotating components of the motor and fan. The vibration level at 700 rpm was around 0.12 ips. It increased to 0.59 ips at 950 rpm. The ratio of vibration level was 0.59 ips/0.12 ips = 4.92. The ratio of speed increase was 950 rpm/700 rpm = 1.36. Since the dynamic force was an unbalance in the rotor, it would have increased by the square of the speed ratio ($1.36^2 = 1.84$). This indicated a structural sensitivity, most probably resonance since the vibration response increased at a rate much higher than the dynamic force. The nature of Figure 9 suggested that the operating speed was approaching resonance but was not quite there. The actual natural frequency was slightly greater than 950 cpm.

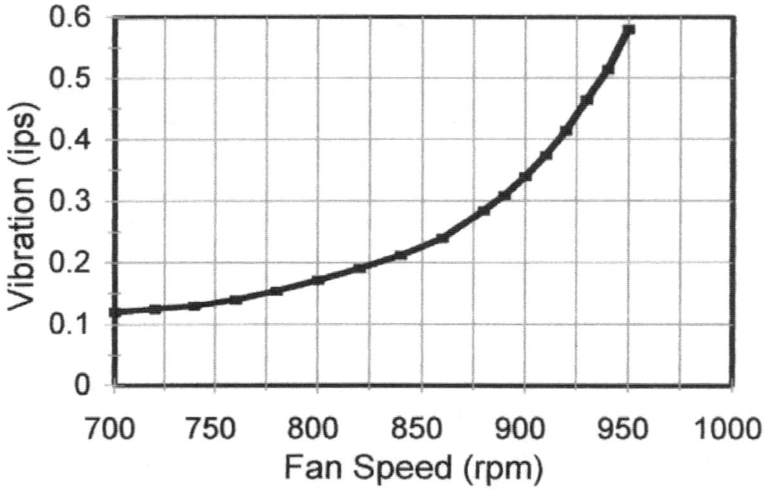

Figure 9: Bearing Vibration versus Operating Speed

The first numerical effort was to study the structural dynamic properties of the rotor. A finite element analysis of the rotor was performed. For this initial preliminary calculation, the support that the concrete foundation provided to the oil-film bearings and rotor was considered infinitely rigid. The calculated natural frequency for the rotor in the rigid bearing support condition was 22.3 Hz (1,338 cpm), the mode shape for which is shown in Figure 10.

As noted, the flexibility of the foundation (soil interface) was not considered in this calculation. The elastic properties associated with the bearing and flexure of the shaft dominated the mode shape. If the flexibility of the soil-foundation interface were to be considered, the calculated natural frequency would be significantly lower. Since the desired operating speed of the fan was 1,000 rpm (16.67 Hz), the frequency ratio for the calculated natural frequency of the rotor not including the foundation would be 16.67/22.3 = 0.75. Based upon this calculation, the separation margin between operating speed and natural frequency would be twenty-five percent (25%), which would have been considered acceptable. Rotor resonance would not have been expected.

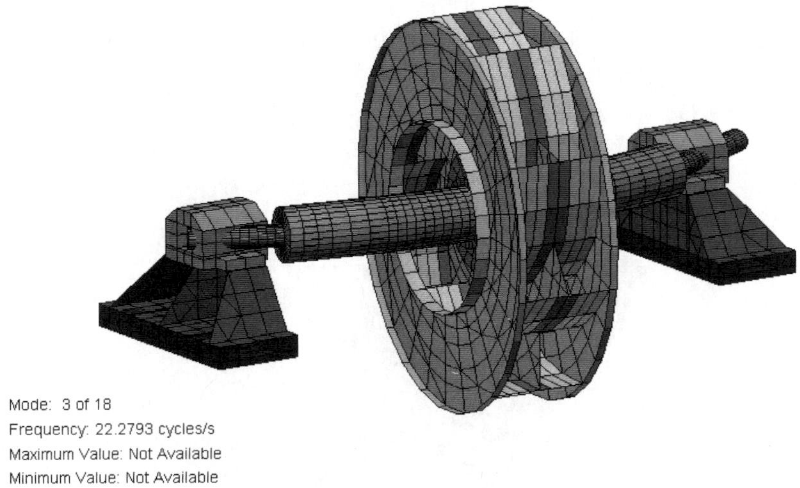

Figure 10: Mode Shape for Natural Frequency of Rotor Calculated as 22.3 Hz

Now consider including the flexibility of the soil-foundation interface and the mass of the concrete foundation in the numerical model. If this flexibility of the soil-foundation interface were to be considered, the calculated natural frequency should be significantly lower than the 22.3 Hz when neglecting the flexibility of the concrete foundation. A finite element model of the complete system was developed to investigate the combined effect of the flexibility of the rotor with the flexibility of the soil-foundation interface. Figure 11 shows the FEA model of the complete system which includes all components of the structural-mechanical system, including the massive fan housing. Elastic springs were defined at all node points on the bottom of the concrete mat to simulate the elastic response of the soil-foundation interface.

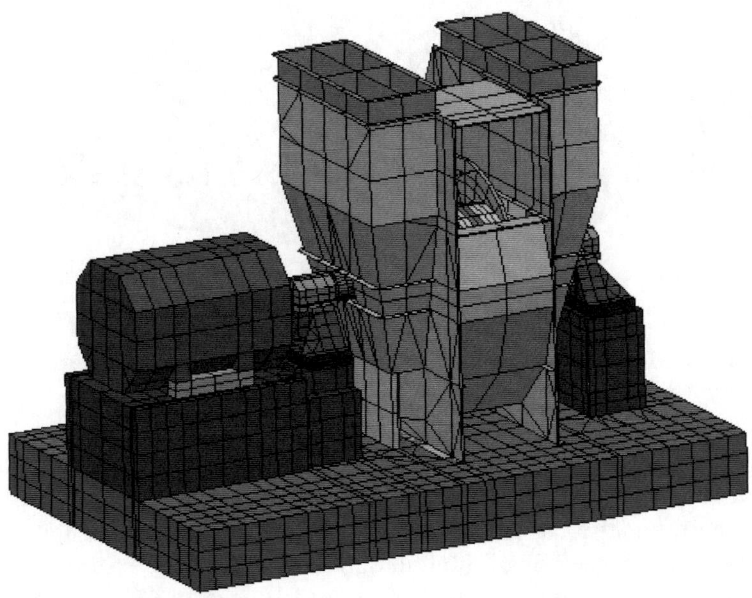

Figure 11: FEA Model of Complete Fan-Foundation System.

Figure 12 (End View) and Figure 13 (3D View) contain the result of the finite element analysis. Although the fan housing was included in the model to account for the mass thereof, it has been hidden from the result to more clearly show that the fan shaft flexure combines with the rocking of the concrete foundation. The calculated natural frequency is 16.656 Hz which is almost coincident with the operating speed of 16.67 Hz.

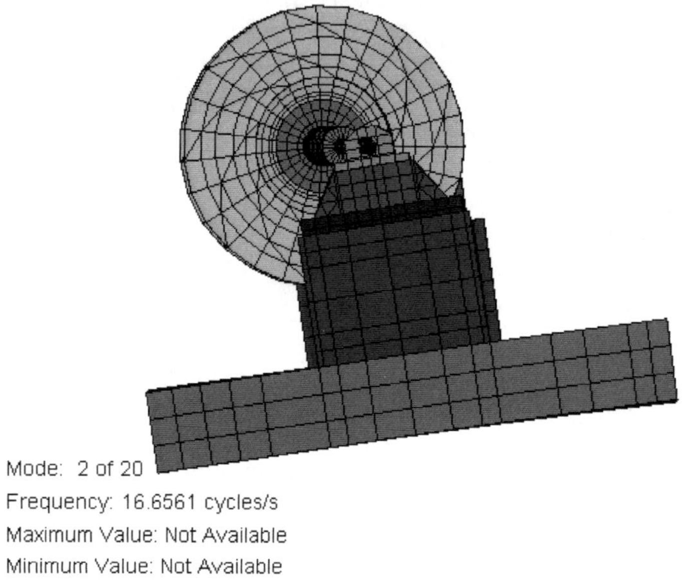

Figure 12: End View of Mode Showing Rotor Flexure Combined with Foundation Rocking

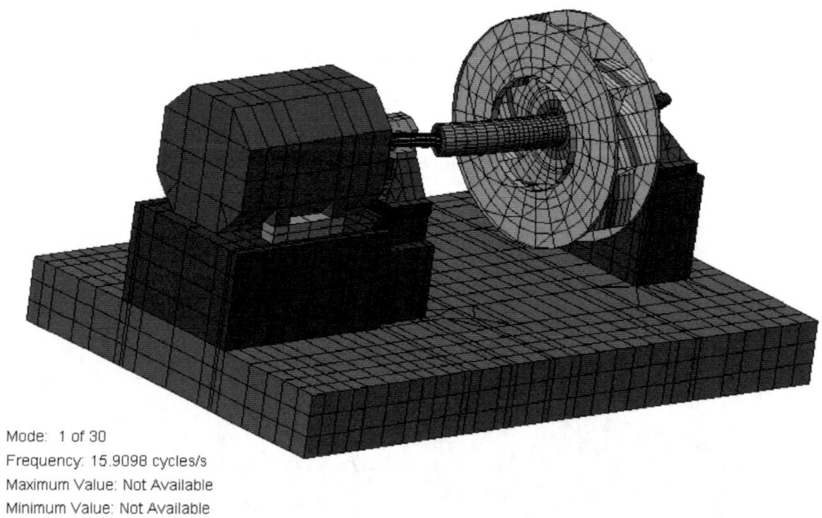

Figure 13: 3D View of Mode Showing Rotor Flexure Combined with Foundation Rocking

The problem with this scenario is that since the flexibility of the rotor and foundation resulted in a combined mode of natural frequency, the dynamic force produced by the fan that was transmitted to the foundation was subjected to an amplification factor. The amplification factor for the response of the foundation and fan rotor was much higher than calculated using the model that neglected the dynamic properties of the foundation. For this Case Study, the frequency ratio for the rotor and foundation, using the finite element result, was 16.667/16.656 = 1.0007, which is nearly 1.0. The only factor that would limit vibration is damping. Assuming $\zeta = 0.02$ for the rotor, the amplification factor for the force would be $AF \sim 1/(2\zeta f_{dyn}/f_n) = 49.97$. Assuming $\zeta =$

0.10 for the soil-foundation interface associated with the rocking mode of the foundation, the amplification factor for the response of the rocking mode would be AF ~ $1/(2\zeta f_{dyn}/f_n) = 4.997$.

An operating deflection shape test was performed to confirm the diagnosis made from the FEA study that the resonant mode shape was a combination of rigid body foundation rocking together with shaft flexure (critical speed). Operating deflection shape [15] is a vibration testing and analysis procedure that can produce animations of the deformation characteristics of a vibrating mechanical system at any discrete frequency. The ODS analysis provides information regarding movement during operation that can be made into an animation. In cases of resonant excitation, the operating deflection shape animation will provide an estimate of the mode shape for the natural frequency that is being excited [16].

ODS data was acquired at multiple locations of the concrete foundation (mat and pedestals) using an internally integrated accelerometer (velocity sensor). ODS data was also acquired from the fan shaft (horizontal and vertical) using a shaft stick with the same type of velocity sensor. Figure 14 is a snapshot of the animation. The shaded gray areas are the foundation, and the red line is the fan shaft. Data from the fan shaft was acquired on either side of the drive-side bearing, one close to the coupling and one just outside of the housing. (Note: safety guards had to be removed to acquire the shaft data, this should only be done by an experienced analyst under supervision of the safety department). Shaft vibration just outside of the non-drive side bearing was also measured.

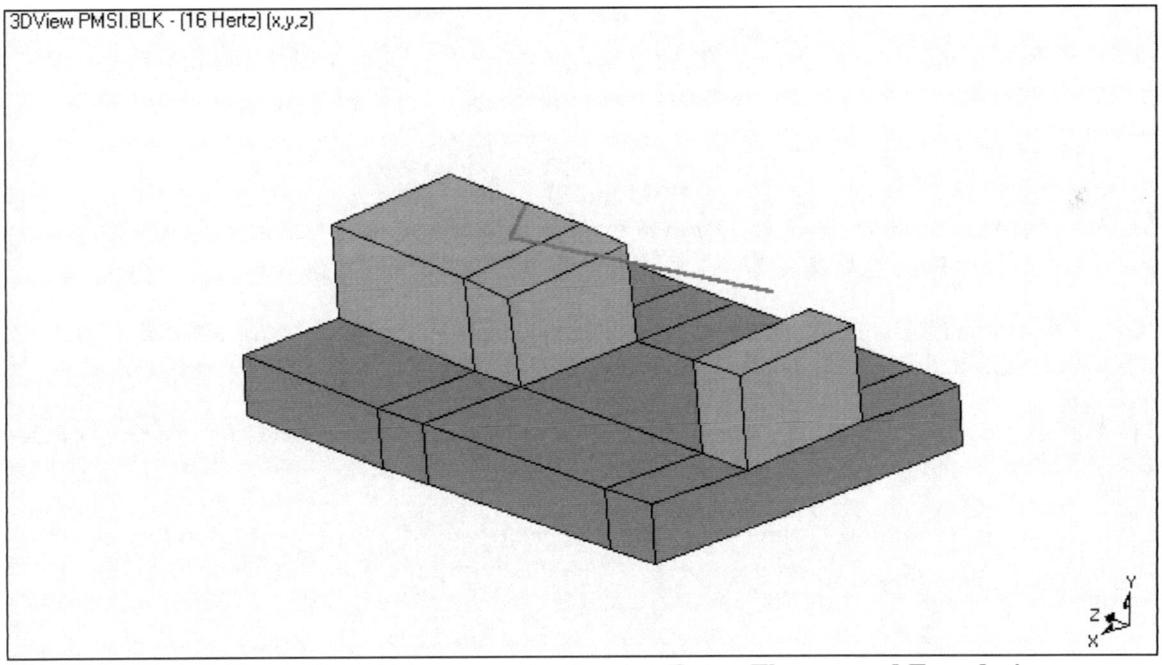

Figure 14: Snapshot of ODS Animation Confirming Rotor Flexure and Foundation Rocking

Since only a limited number of data points could be obtained from the rotating shaft, the animation of the rotor resembled a rotating 'v' instead of the expected curvature from flexure. Regardless, the ODS animation confirmed that the mode shape for the excited natural frequency contained both flexure of the fan rotor and rocking of the rigid concrete foundation. The natural frequencies of the rotor and foundation were not separate from each other.

REFERENCES:

[1] ACI 351.3R-18, *"Report on Foundations for Dynamic Equipment"*, American Concrete Institute, Farmington Hills, MI, 2018.

[2] George B. Sowers and George F. Sowers, "*Introductory Soil Mechanics and Foundations*", 3rd Edition, The Macmillan Company, New York, NY, 1970.

[3] Joseph E. Bowles, "*Foundation Analysis and Design*", McGraw-Hill Book Company, New York, NY, 1968.

[4] D. D. Barkan, "*Dynamics of Bases and Foundations*", McGraw-Hill Book Company, New York, NY, 1962.

[5] F.E. Richart, Jr., "*Foundation Vibrations*", Transactions of the American Society of Civil Engineers, Volume 127, Pages 863-925.

[6] A.A. Liepens, , "*Settlement Analysis of a Seventeen Story Reinforced Concrete Building"*, Thesis, Massachusetts Institute of Technology, Cambridge Massachusetts, 1957.

[7] CIRIA, *"The Standard Penetration Test, Methods and Use"*, Construction Industry Research Information Association, Report 143, London, England, 1995.

[8] J.F. Callanan and F. H. Kulhawy, *"Evaluation of Procedures for Predicting Foundation Uplift Movements"*, Electric Power Research Institute Report EL-4107, Palo Alto, California, August, 1985.

[9] C.P. Wroth, M.F. Randolph, G.T. Holsby, and M. Fahey, *"A Review of the Engineering Properties of Soils with Particular Reference to Shear Modulus"*, CUED/D – Soils TR 75, University of Cambridge, 1979.

[10] Y. Ohsaki and R. Iwasaki, *"On Dynamic Shear Moduli and Poisson's Ratio for Soil Deposits",* Soils and Foundations, Volume 13, Issue 4, December 1973.

[11] Atkinson and Sallfors, *"Experimental Determination of Soil Properties",* Proceedings of the 10th ECSMFE, Volume 3, Florence, Italy, 1991.

[12] R. J. Sayer, "*Structural Dynamic Design and Performance Evaluation of Foundations for Large Centrifugal Fans*", Proceedings of the 39th Annual Training Conference of the Vibration Institute, Indianapolis, IN, July, 2015.

[13] R.J. Sayer, *"Finite Element Analysis: A Numerical Tool for Vibration Analysis*", Proceedings of the National Technical Training Symposium and 27th Annual Meeting of the Vibration Institute, New Orleans, Louisiana, July 9-11, 2003.

[14] R. J. Sayer, "*Using Finite Element Models as a Digital Twin for the Development of Vibration Corrective Action Plans*", Proceedings of the Vibration Institute Annual Training Conference and Society for Machinery Failure Prevention Technology Conference, VIATC/MFPT2021, Arlington, Texas, July, 2021.

[15] R. J. Sayer, *"Operating Deflection Shapes: Principles, Test Procedures & Case Studies"*, Proceedings of the National Technical Training Symposium and 27th Annual Meeting of the Canadian Machine Vibration Association, Banff, Alberta, October 22-24, 2013.

[16] R. J. Sayer, "*Operating Deflection Shape and Experimental Modal Case Histories*", Proceedings of the National Technical Training Symposium and 30th Annual Meeting of the Vibration Institute, Louisville, Kentucky, June 20-22, 2006.

VIATC TRAINING CONFERENCE & EXPO
COVINGTON, KENTUCKY | AUGUST 7-9, 2024

FOUNDATION INTERFACE PROBLEMS AS IT RELATES TO VIBRATION INVESTIGATIONS

Thomas B. Lantrip
Consulting Engineer – Southern Company Services
tblantri@southernco.com
tblantri@icloud.com

Tom Lantrip is a consulting engineer with Southern Company Generation and works in the Performance and Reliability Department - Balancing and Vibration Analysis group. Tom has worked at Southern Company for 41 years, primarily in the field of balancing and vibration analysis, with an emphasis on turbines, fans, and pump components. He has a background in structural dynamics and in structural and seismic design, vibration analysis, and balancing of rotating machinery. Tom graduated from the University of Alabama with a B.S. in Civil Engineering, attended the University of California at Berkeley, and has an M.S. in Civil Engineering from the University of Alabama at Birmingham. Tom is a member of ASCE and is a professional engineer in the state of Alabama and Florida and holds a Category IV certification from the Institute.

ABSTRACT:
As a part of vibration analysis and problem solving, it becomes more apparent that many problems stem from improper foundation designs and/or miscommunicated foundation and/or support assumptions. The interaction of foundation to machine has a large opportunity on many machines to alter the successful operation and the maintenance life of a machine design. Large machines require massive foundations that may characteristically be very difficult to design, construct and/or detail properly. Foundations themselves may require maintenance, just like machines require maintenance.

This would include the following:

- Generator Frame Improper Support Conditions
- Gas Turbine and Generator Foot Soft
- Hydro Unit Resonance at Blade-Pass Frequency
- ID Fan Foundation Looseness
- Generator Base Considerations
- Vertical Pump Base Rotation Contribution

KEY WORDS: Base, Concrete, Flexibility, Interface, Looseness, Relative Movement, Resonance, Structural Interaction

INTRODUCTION: Machines require a support to the earth. The structure or support of the machine is what we commonly denote as a foundation. The purpose of the foundation is to control and limit the machine's movement and provide stability, which is normally our mission as well as vibration analysts. The foundation should provide for a stable connection to the ground, to limit the differential motion of the machine to the supporting connections, and to reduce and attenuate the displacement (vibration) of the machine. Sometimes out of omission we create the vibration higher which is sometimes occurring in resonant structures.

DESIGN PHILOSOPHY: As a part of a basic dynamic design, the physical equation of motion provides the guidance for the fundamental aspects of the approach used to control vibration and motion. The equation (1) below provides; that with the key terms as Mass (M), Stiffness (K), and Damping (C).

$$M\ (a) + K\ (x) + C\ (v) = F\ (t) \qquad (1)$$

M is the mass of the system
K is the stiffness of the system
C is the damping
F(t) is the forcing vibratory force

a is the acceleration of the system
x is the displacement of the system
v is the velocity of the system

One of the simplest approaches to a foundation design is to increase the mass of the foundation base to an extreme proportion of support mass to rotating machine mass. This approach is generally a safe method but could be an expensive alternative. The other approach is to increase the stiffness of the supporting foundation to an extreme order of stiffness, utilizing significant anchorage to supporting soil, base, or roof structural interfaces. In addition to providing significant mass and stiffness, it is imperative that no combination of mass and stiffness create a resonant condition of the foundation system. This analysis of the mass and stiffness needs to consider carefully incorporating this interaction with the forcing frequencies to be assured that no amplification of the vibration is created by the foundation. When it comes to structures that support rotating or dynamic machinery, this question of structural integrity becomes even more of a question mark, as the cyclic nature of the forces wear and tear on the structural and load bearing components. With respect to the use of equation (1) above, and structural integrity, the effectiveness of these elements is only as good as the structural connections and details that comprise an efficient, robust foundation.

OPERATING SURVEY: When a machine is exhibiting excessive vibration, it is typically vital to understand the amplitudes and character of the vibration at various locations, typically at the bearings, supports, etc. Also, by the nature of some conditions, it may also be noted that the foundation or base has some irregularly high vibration response. By collecting the vibration data at various locations and particular frequencies the vibratory motion can be displayed to evaluate the motion of the structure while the machine or structure is vibrating. The raw data used for this study is collected by measuring vibration amplitudes and phase angles at many discrete locations along the structure and machine. This amplitude and phase angles are unique to each frequency therefore a very extensive collection of data can be compiled for various points and at different frequencies. This data and collection of data is typically denoted as an operating survey. The data is typically referenced to a single data point, with the understanding that the reference with the understanding that the vibration will repeat as steady amplitudes cycle to cycle.

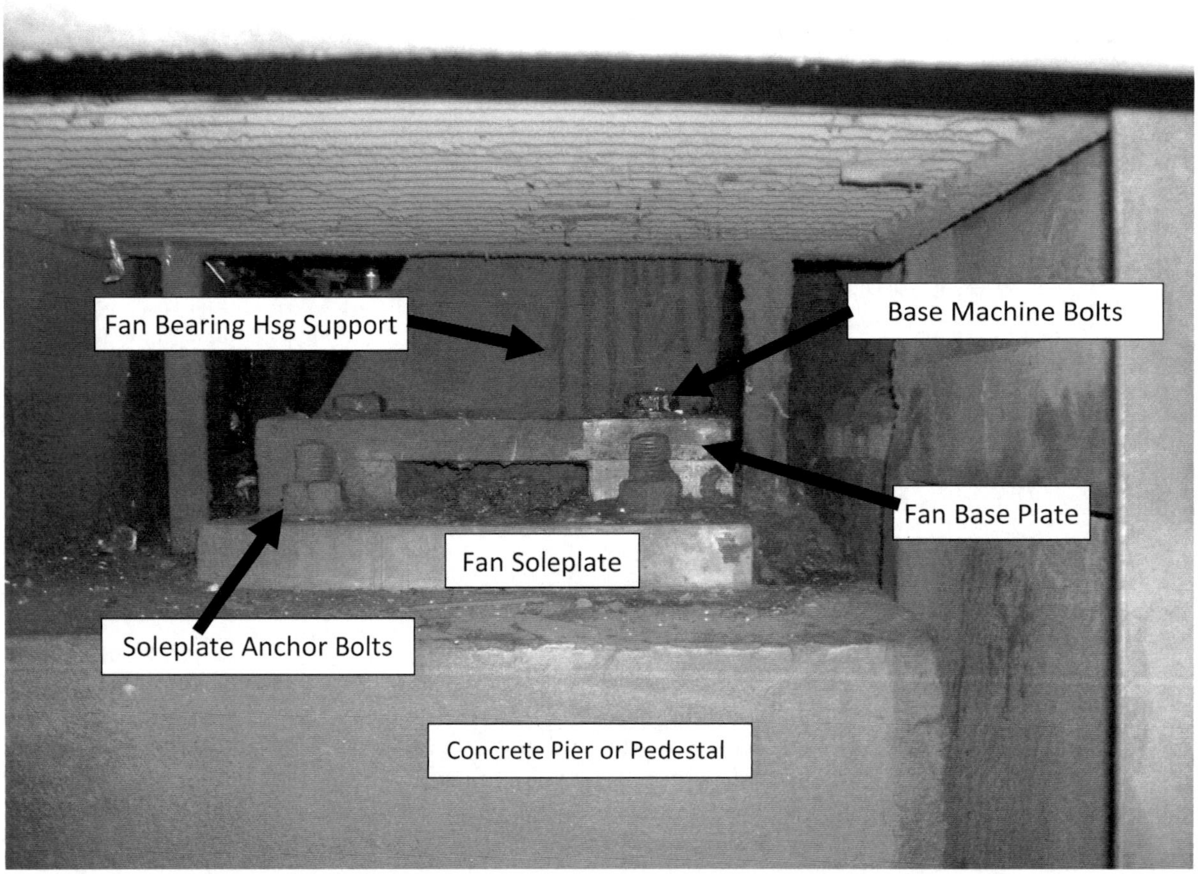

Photo 1 : Forced Draft Fan (Power Generation) Foundation Photograph

The photo provides key features of a typical foundation to include the concrete pedestal, the soleplate (bed plate or foundation plate), the equipment support baseplate, and the machine pedestal or bearing support structure. Generally, the vibration is highest on the bearing, with progressively lower amplitudes as the location of the measurement is closer to the foundation. The concrete pedestal typically has embedded bolts (anchor bolts) buried deep in the concrete foundation that support the soleplate. The soleplate is leveled and grouted into place however the top surface of the soleplate will mate with the machine base plate, therefore this surface interface will be machined to high precision. The soleplate will have through holes typically with exposed projected anchor-bolts/nuts on top to hold the soleplate down. The fan base plate rests on top of the soleplate and is connected with machine bolts that are torqued into the soleplate pre-drilled and tapped blind holes. The base plate supports the upper bearing housing pedestal frame upon which the machine bearing housing rests. Some manufacturers suggest that the soleplate on top of the concrete pedestal is leveled and supported by thick shim plates surrounding and capturing the embedded anchor bolts and later backfilled with grout after initial operational and testing checkouts as shown in Figure 2. In this detail, the grout is there strictly for protection and cosmetic purposes.

Other manufacturers recommend the temporary shimming of the soleplate or skid structure and the use of grout to be dry packed into place to provide the permanent supporting interface for the equipment as shown in Figure 3. In either case, the use of expansive or epoxy-based grouts has some important details to consider before using for equipment base detailing. Some of the concerns with using these type grouts are provided in the following Table 1 below.

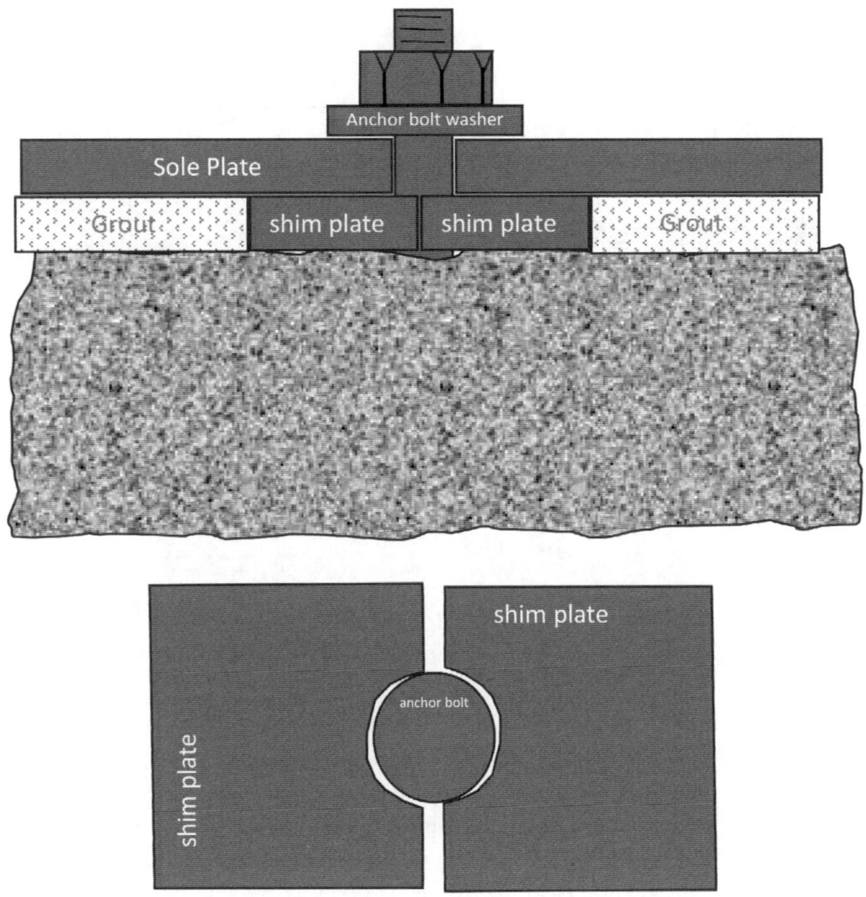

Figure 2: Sole-Plate at Concrete Interface with Shim Plates Surrounding the Anchor Bolts

The Figure 2 approach to the soleplate shimming method around the perimeter of the bolt has the utility of a permanent support adjacent to the bolt to make sure the machine soleplate can become stable no matter what the condition of the grout backfill may be. The grout serves the purpose primarily to keep materials and liquids away from the soleplate to concrete foundation including weather, dirt, debris, insects, etc. This approach also limits the primary concern with the breakdown of grout materials over time, particularly if the grout material is unstable or degraded for some reason. This approach doesn't require the grout to act as a definitive load path between the grout, the soleplate and the foundation.

Figure 3 provides another approach that allows the soleplate interface to be formed up and shimmed in a way that the flowable grout can be pushed into place and allowed to set and cure prior to final alignments. This detail also requires vent holes along the soleplate and/or skid frame to allow air to escape during grout placement. This approach and the stability of the soleplate interface is dependent on the grout integrity over time.

Table 1: Some Considerations for Grouting of Machine Bases

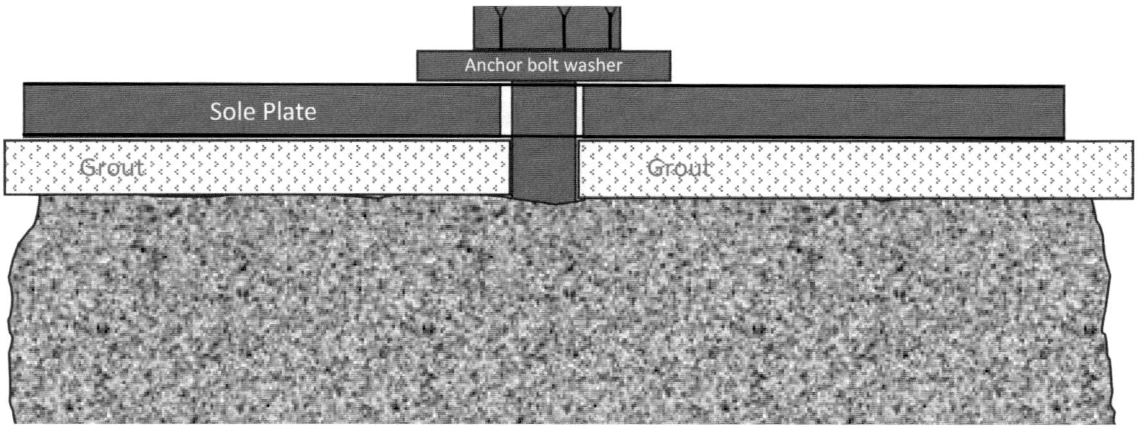

Figure 3: Sole-Plate at Concrete Interface with Grout Placement w/o Shim Plates

	Grout Specification Considerations
1	Setting and Curing Time
2	Expansion/Shrinkage
3	Stability (Short Term vs Long Term)
4	Strengths
5	Workng Time
6	Cost
7	Impact Resistance

ANALYSIS BLUEPRINT: No matter which detail for base plate and soleplate details are utilized, when determining the foundation interface and condition, the measurement of the base components is vital. With that, the understanding of the base details can sometimes be the difference in making good analysis and assessment determinations. The evaluation of the foundation or base assembly requires vibration data and measurements to be collected in such a way to organize the data in a logical way. The measurements typically begin with the bearing housing vibration data. The bearing housing data on the machine, motor, turbine, etc. is collected to determine the amplitude and frequency. This begins the process of evaluating the conditions primarily as to the severity of the amplitude level. Secondly, the frequency of the severe vibration level and the fault signature analysis. Once these are determined, the next thought may be to determine the extent of the location of the vibration and the direction of severe vibration. Figure 4 below provides an example of a generator frame from a turbine-generator installation where the vibration levels were excessive on the generator bearings. Vibration measurements were collected on the elevated pedestal floor that supports the generator feet and base of the generator. This was primarily vibration measurements in the vertical direction of the floor and supporting foundation.

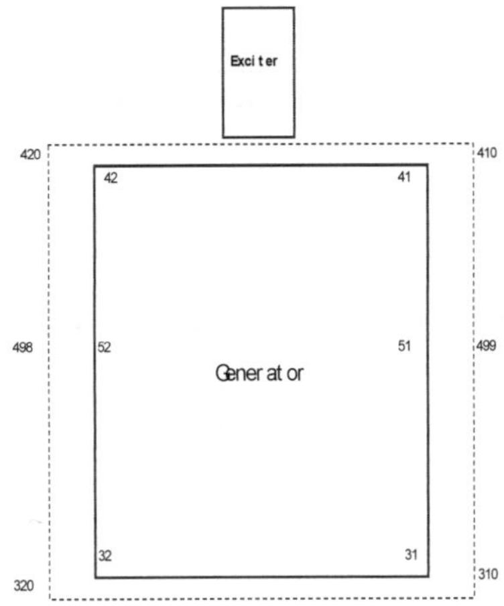

Figure 4: Generator Frame Data Point Layout and Numbering

Figure 5 provides a view of the deflected shape of the generator feet and generator pedestal with the hold-down bolts loose at the operating speed of 60 Hz. (3600 rpm). Even though the generator hold-down bolts are loose (to be explained in detail later), the operating deflected shape shows that there is no relative motion between the generator frame and the generator pedestal interface. This suggests that the generator frame is resting and riding on the generator pedestal and is very compliant with the concrete surface.

Figure 5 provides the deflected shape of the generator feet and pedestal with 12 of 16 bolts tight. The tightening of the bolts reduced the vibration of the bearing housing on Bearing 4 as well as reducing the vibration of the generator feet and pedestal. The shape suggests a change to the operating shape and indicates some twisting of the generator frame. This may indicate that the generator frame when tight may alter the deflected shape of the generator and pedestal and could suggest that the generator is functionally stiffening the generator pedestal foundation, or there is another mechanism involved, possibly resonance or alignment changes.

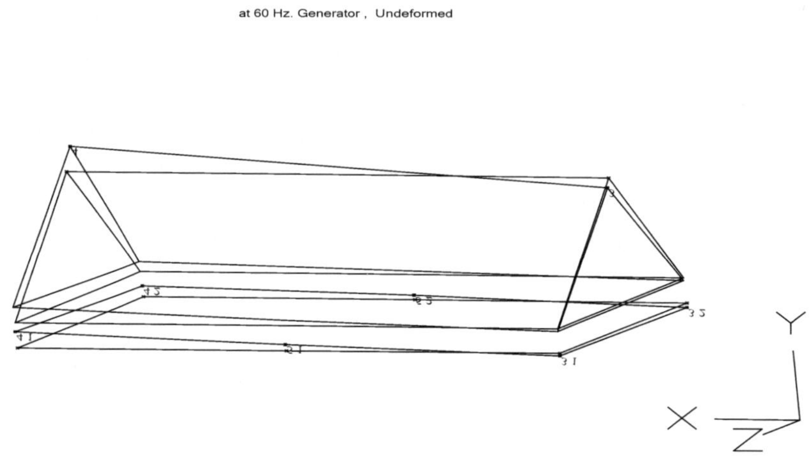

Figure 5: Generator Operating Deflected Shape with Generator Bolts Loose

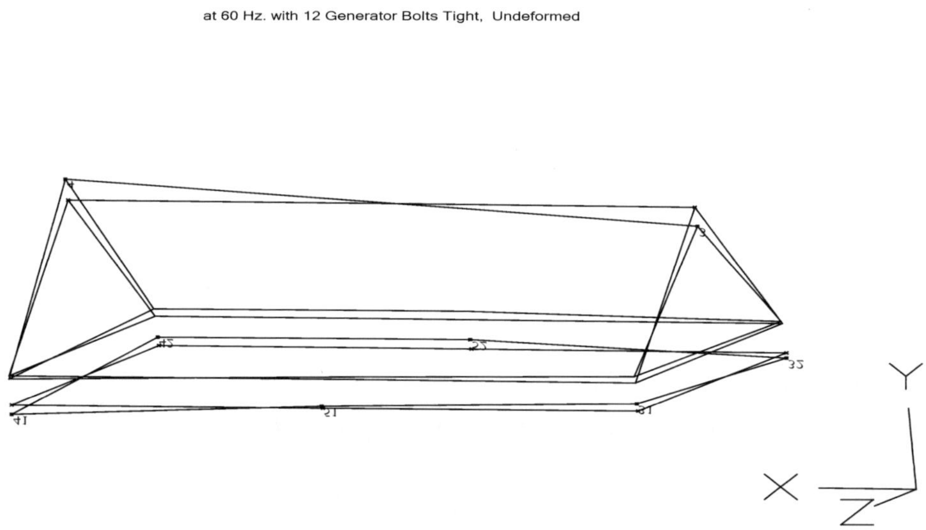

Figure 6: Generator Operating Deflected Shape with 12 Bolts Tight

Figure 7 provides the layout of the hold down bolts for the generator. The generator frame has 12 bolt locations. A test was conducted to compare the differences in the generator and support vibration levels when the hold down bolts were loose (as designed) or when the hold down bolts were tightened. The results indicated that the bolts tight configuration created a lower amplitude in the generator bearings, however this was not a permanent solution as the hold down bolts are required to have some looseness to allow for expansion.

Note: the typical installation detail on large generator support/foundation has the hold down bolts outfitted with washers and nuts, however the nuts will be loose by 10 mils with features to lock the nuts. The loose nuts allow the generator stator to expand as needed and the nuts are still retaining the generator from overturning during emergency conditions. As to the permanent alignment of the generator to the turbine drive component, axial and lateral keys are designed into the generator frame system to keep the alignment stable and still allow for thermal expansion. [1]

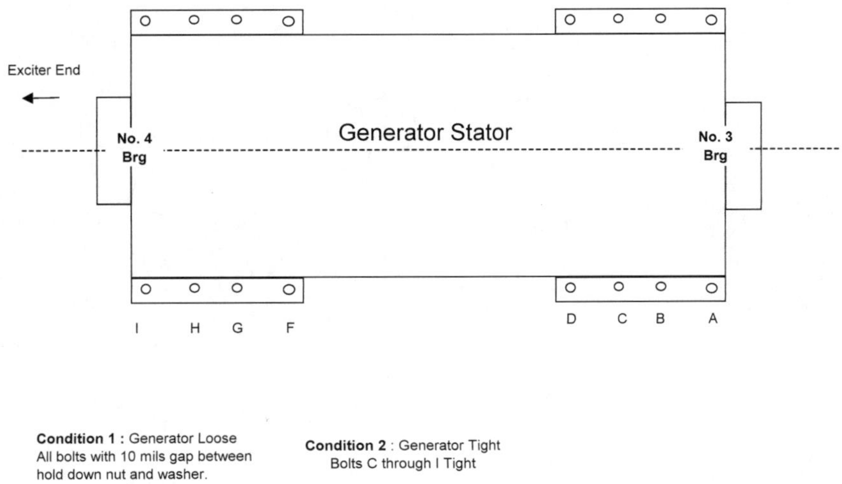

Figure 7: Generator Hold Down Bolting and Layout Configuration

In this case the bolts tight condition (Condition 2) was not workable for the permanent solution because of the loose bolt design requirement. Ultimately shim adjustments and balancing of the generator was conducted to reduce the levels.

GAS TURBINE GENERATOR SUPPORT FOOT LOOSE: A gas turbine and generator was tested for vibration after a report of high levels of vibration on the gas turbine bearing closest to the generator in a 4-bearing arrangement (Figure 8 Layout). The vibration levels on the gas turbine were low during the initial startup however over several hours the vibration levels increased to alarm levels. Vibration analysis was conducted and determined that the primary vibration was occurring at the operating speed frequency and the highest levels were obtained on the gas turbine bearing closest to the generator and the next closest vibration was on the far generator bearing closest to the exciter component (collector end – CE). Figure 9 provides the layout for the operating survey that was conducted to measure the concrete foundation, soleplate, and generator feet. The collector end generator feet and the soleplate are exhibiting a disproportionate level of movement as shown at 0.145 in/s on the left side foot and the 0.112 in/s movement on the right-side soleplate and base foot. This is also provided in the tabular form in Figure 10 with the displacements (mils) also listed. Operating speed (1X) phase wasn't available during the evaluation of the generator feet and soleplate, however typically this would be collected to help determine the complexity of the motion. In this case, the rocking of the generator was evident as shown in the deflected condition depicted in Figure 11, as well as the uplift that was occurring on this end of the generator. High-speed video is a powerful tool in capturing the motion, including some details not easily collected with accelerometer due to space and safety concerns.

Figure 8: Gas Turbine and Generator Layout

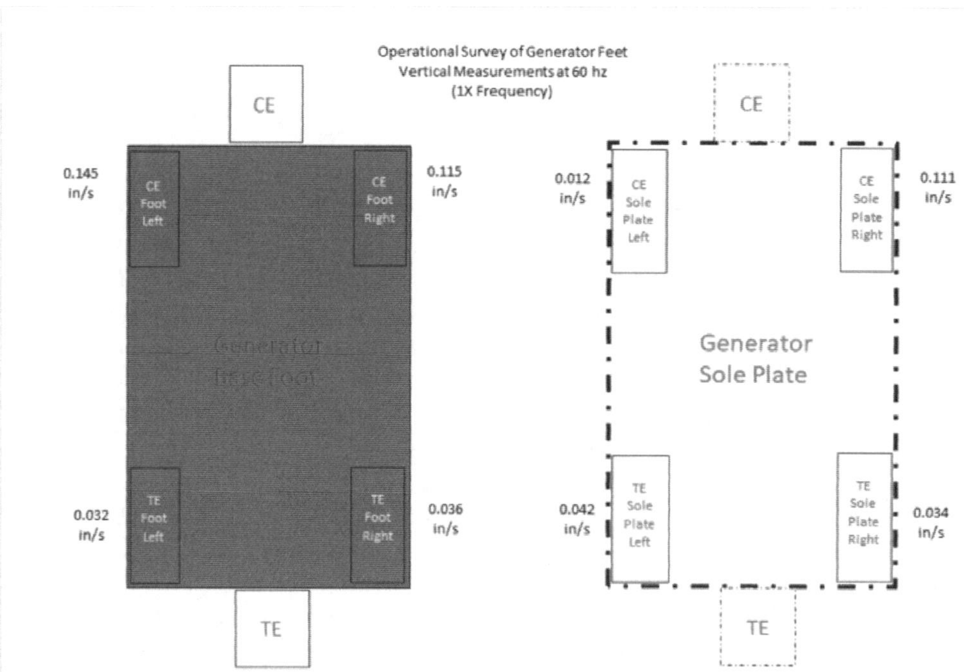

Figure 9: Operating Survey Results collected on Soleplate and Generator Feet (Vertical)

Vertical Displacement Data on Generator Feet and Sole Plate

				1X (60 hz) In/s peak	1X (60 hz) mils p-p
Generator	Foot	Collector End	Left	0.145	0.77
	Foot	Collector End	Right	0.115	0.61
	Foot	Turbine End	Left	0.032	0.17
	Foot	Turbine End	Right	0.036	0.19
Generator	Sole Plate	Collector End	Left	0.012	0.06
	Sole Plate	Collector End	Right	0.111	0.59
	Sole Plate	Turbine End	Left	0.042	0.22
	Sole Plate	Turbine End	Right	0.034	0.18

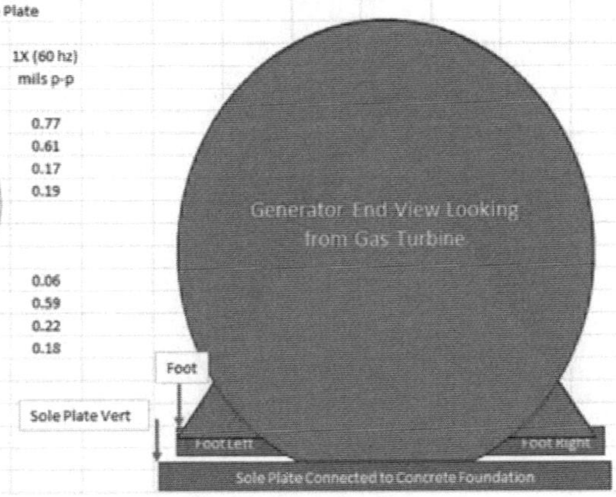

Figure 10: Operating Survey Results collected on Soleplate and Generator Feet (Vertical)

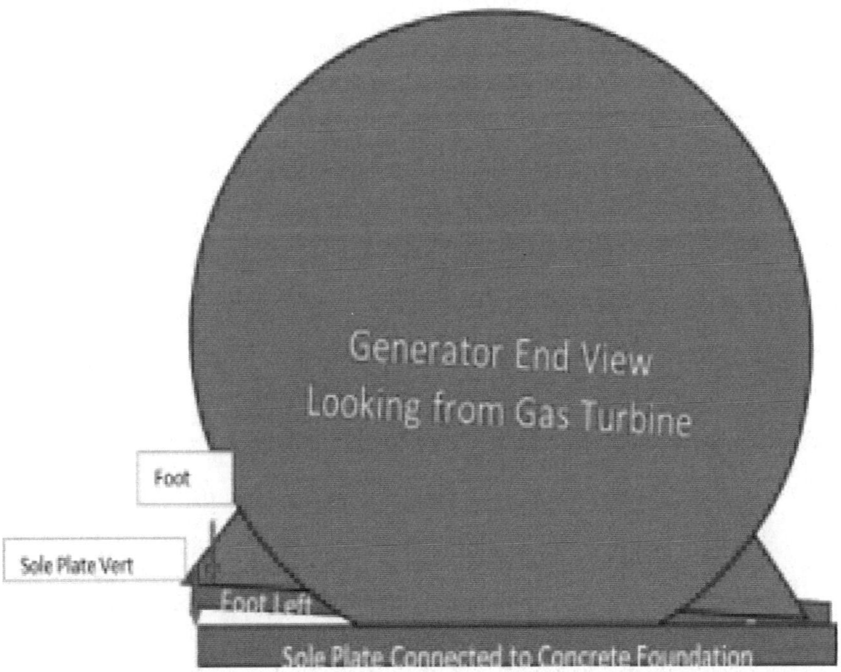

Figure 11: Deflected shape on Soleplate and Generator Feet (Vertical)

Recommendation was made to the owner to check the foundation integrity of the sole plate locations and re-shim and re-align, as necessary.

HYRDO UNIT NATURAL FREQUENCY: Complaints at a hydro-generating facility about a high vibration on the operating floor of a vertical-shaft hydro unit. The vibration levels observed were on the order of 0.25 in/s peak on the floor. This vibration was determined by the operations staff to be excessive and different. Human perception and vibration severity classified this level of vibration unsatisfactory for continuous personnel vibration tolerance.

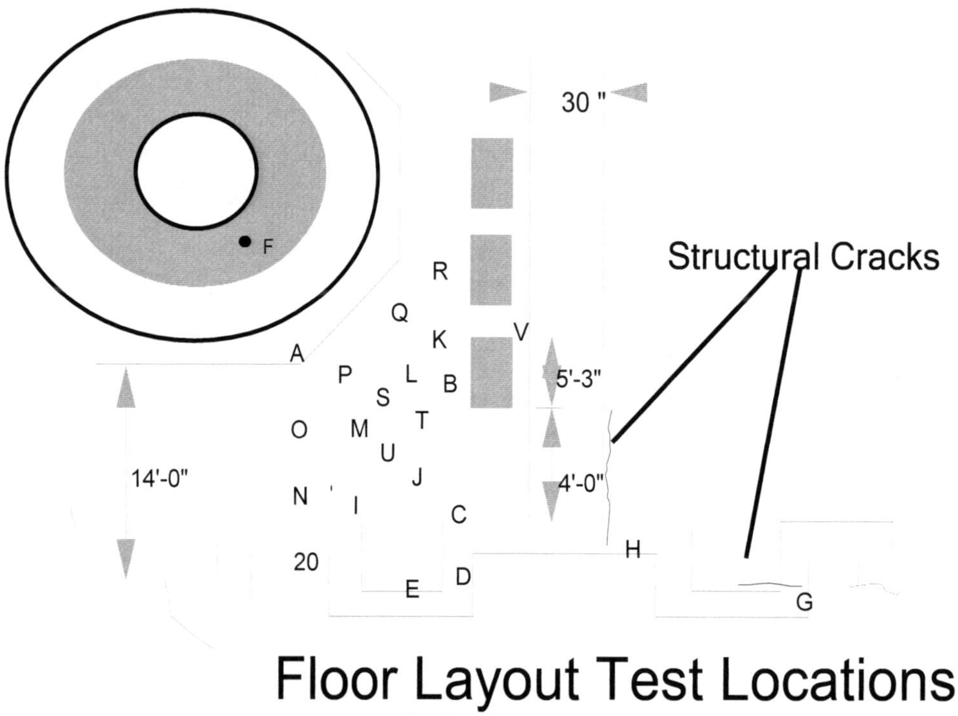

Figure 12: Hydro Operating Floor with Observed Vibration at 19X Operating Speed

Measurements were taken at the critical locations and grid locations on the operating floor where the nuisance vibration was observed. The condition existed only during loaded generation condition. The unit has 19 vanes with a 120 rpm (2 hz) operating speed, therefore the 38 hz vibration response.

Table 2: Operating Floor Vibration Measurements During Operation at 38 Hz

Location	(38 HZ)	Phase	Location	(38 HZ)	Phase
A	0.034	84	L	0.208	-69
B	0.192	-73	M	0.173	-70
C	0.014	-180	N	0.033	-73
D	0.127	107	O	0.073	-64
E	0.056	114	P	0.135	-65
F	0.008	123	Q	0.107	-62
G	0.0003	33.6	R	0.04	-53
H	0.001	3.5	S	0.206	-80
I	0.07	-72	T	0.246	-81
J	0.19	-73	U	0.245	-82
K	0.134	-66	V	0.011	-80
100	0.067	-86	200	.035	-82
101	0.251	-83			

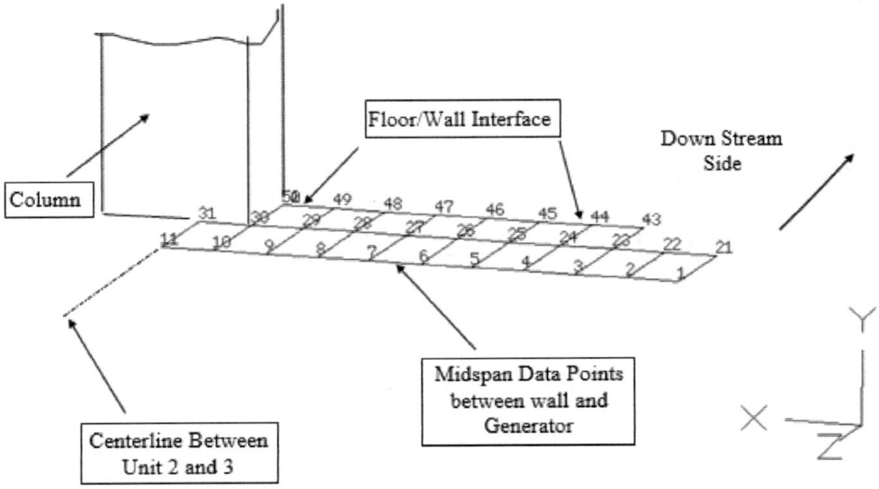

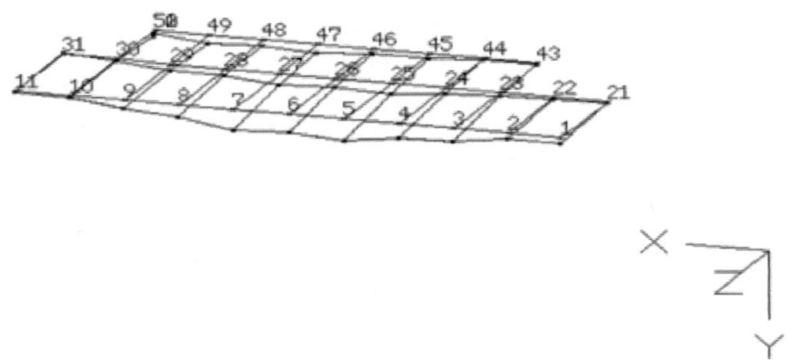

Figure 13: Operating Deflected Shape at 38 hz (blade-pass) Frequency

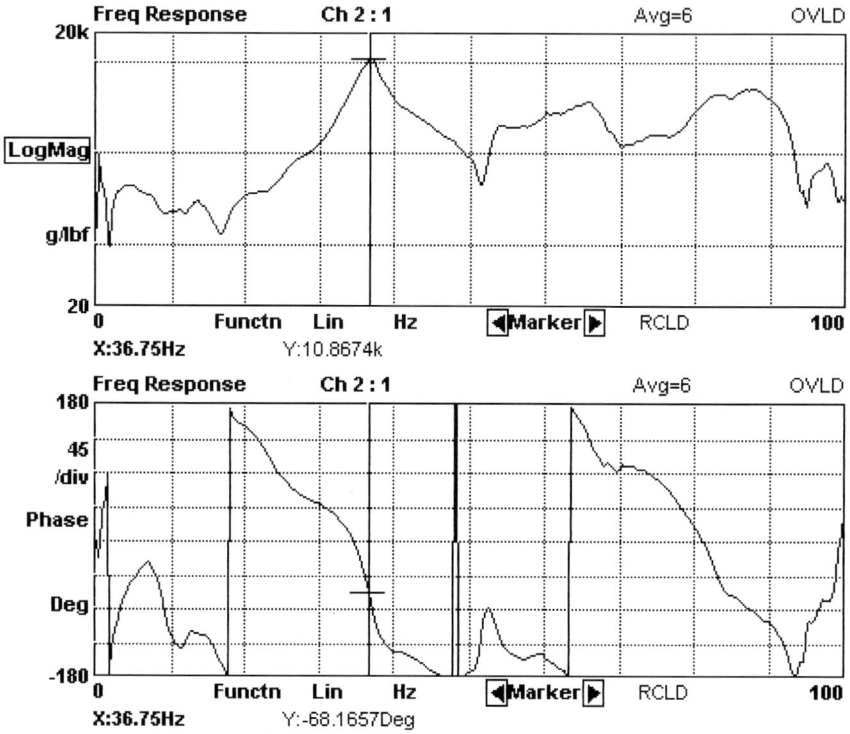

Figure 14: Natural Frequency Determination on Operating Floor at 36.75 hz.

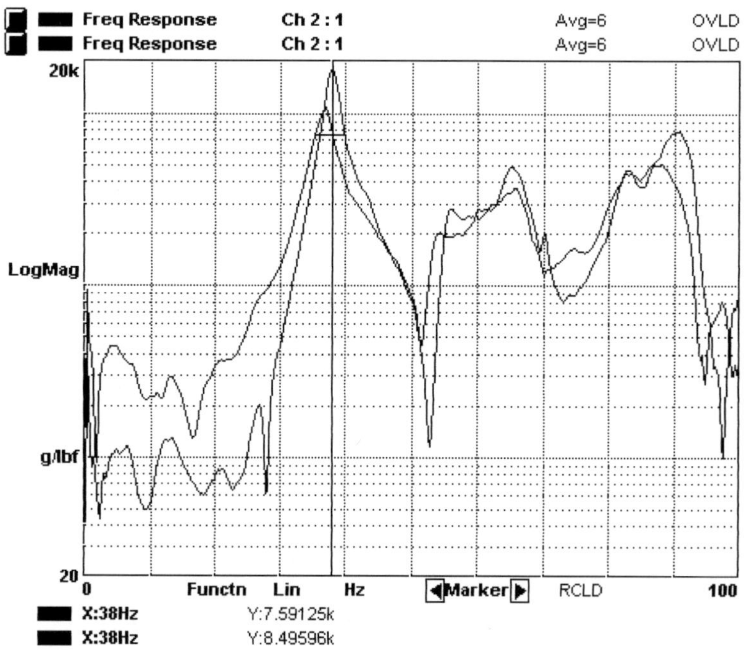

Figure 15: Natural Frequency Determination on Operating Floor at 36.75 hz Existing and Modified Temporary with Floor Jack at 38 hz.

Table 3: Comparison of Natural Frequency Original vs Temporary Jack

Configuration of Floor Slab System	Natural Frequency (Hz)	Percentage Away from Vane-Passing Frequency
Existing (In-Situ) Condition	36.75	**3.3**
With Temporary Jack	38.0	**(0.0)**

Conclusion: Natural frequency (36.75 hz) was amplifying the vibration at 38 hz or (19X Blade-Pass) frequency of the floor system. Additionally, temporary stiffening with floor jack was able to alter and increase the natural frequency of the floor. However, the temporary jack created a less desirable natural frequency at 38 hz, which placed the temporary modified condition exactly on resonance. Future repair and/or stiffening was needed to eliminate the condition. The vibration severity for human perception is not conducive to normal habitation, whereas the vibration severity was considered acceptable for longer-term conditions.

FAN LOOSENESS CASE HISTORY: A boiler fan was repaired during a maintenance outage. The fan bearing was found to have excessive clearance but was reworked with a new bearing. Even after replacement of the new bearing, much higher vibration was still observed on the outboard end fan bearing. The fan is equipped with permanent vibration sensors on the inboard and outboard bearings, however the outboard bearing still had excessive vibration. Upon examination of the vibration and collection of data on the foundation, soleplate, and equipment base, it was determined that the anchor bolts were tight but still the vibration indicated that the vibration was still lifting the soleplate vertically and horizontally as seen in Table 4 below.

Table 4: Vibration Collected on Inboard and Outboard Fan Bearing

	Fluidizing Fan Boiler 2 - 101 (operating speed 3590 rpm)				
	4/13/2018				
			1X	Phase	Overall
			(in/s)	(deg lag)	(in/s)
Fan	Inboard	Horizontal	0.16	199	0.27
		Vertical	0.01	-	0.05
		Axial	0.054	137	0.15
Fan	Outboard	Horizontal	0.33	235	0.48
		Vertical	0.11	138	0.25
		Axial	0.16	323	0.56

Figure 16: Fan Outboard Bearing Pedestal and Foundation with Anchor Bolt Indicated

The anchor-bolt nut holding the soleplate down was torqued to requirements, but the bolt was still moving vertically indicating loose bolt within the concrete pedestal. This vertical looseness coupled in rotation on one side of the base to create a vertical and horizontal rocking motion at the fan outboard bearing. In addition, harmonics were present on the outboard vertical and horizontal vibration spectra.

Figure 17: Fan Outboard Bearing Pedestal and Foundation with Tight Nut on Anchor Bolt but apparent Looseness on Bolt

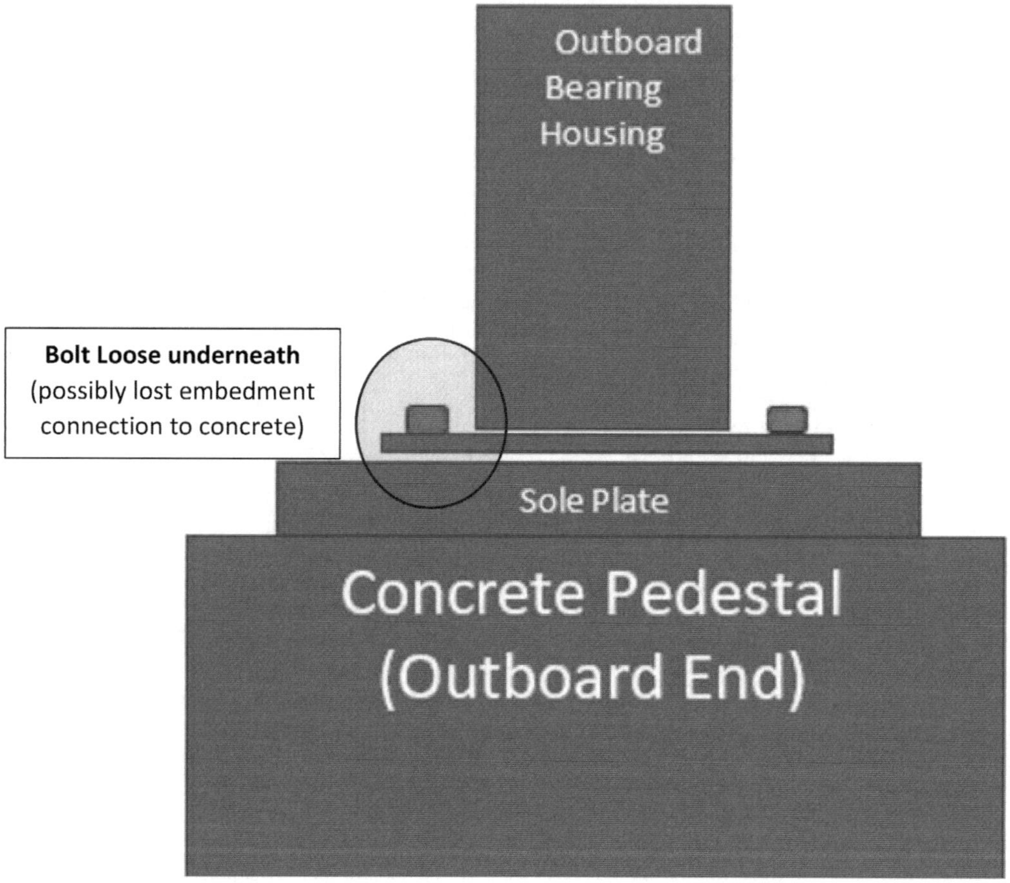

Figure 18: Elevation View of Outboard Bearing and Pedestal

CIRCULATING WATER PUMP VIBRATION ANALYSIS: Circulating water pumps in the common vertical configuration are prone to base stiffness concerns either through degradation, improper design margin against resonance, or environmental changes. As part of this experience, it is common to question the integrity of the supporting base. To evaluate that concern, a standard approach is used again in an operating survey approach. To determine if interaction of the foundation is causing the elevated vibration at the top of the motor, a geometrical analysis of the data was performed. If looseness in the base was the cause of the elevated vibration readings, then the vibration amplitude at the base will be proportional to the vibration at the top of pump. This is determined in the following manner using a simple fixed base test by measuring the vibration (and phase) at the various interface locations from the base of the concrete structure to the top of the motor. The approach follows the assumption of some rigid body kinematics. If we assume the motion of the top of pump (in this case) is due to rotation at the base, then we can use the dimensions at the base in the horizontal plane x the vibration amplitude in the vertical direction to calculate the rigid body rotation. Understanding that the actual rotation could by a higher value if we have additional flexure of the machine or pump base as seen in Figure 19 below, this should provide an upper bound of the rotation at the base. This rotation could be due to the soleplate or base plate, but at this check point condition a general test of the contribution of the base rotation is available. This rotation can be computed and multiplied by the height of the pump to determine the amount of lateral or horizontal vibration due to the rotation of the base due to rigid body motion. If this computation yields 20% (+) of the horizontal vibration level at the top of pump as provided in Figure 20, then there could be reason to interrogate additional measurements at the base of the machine or pump.

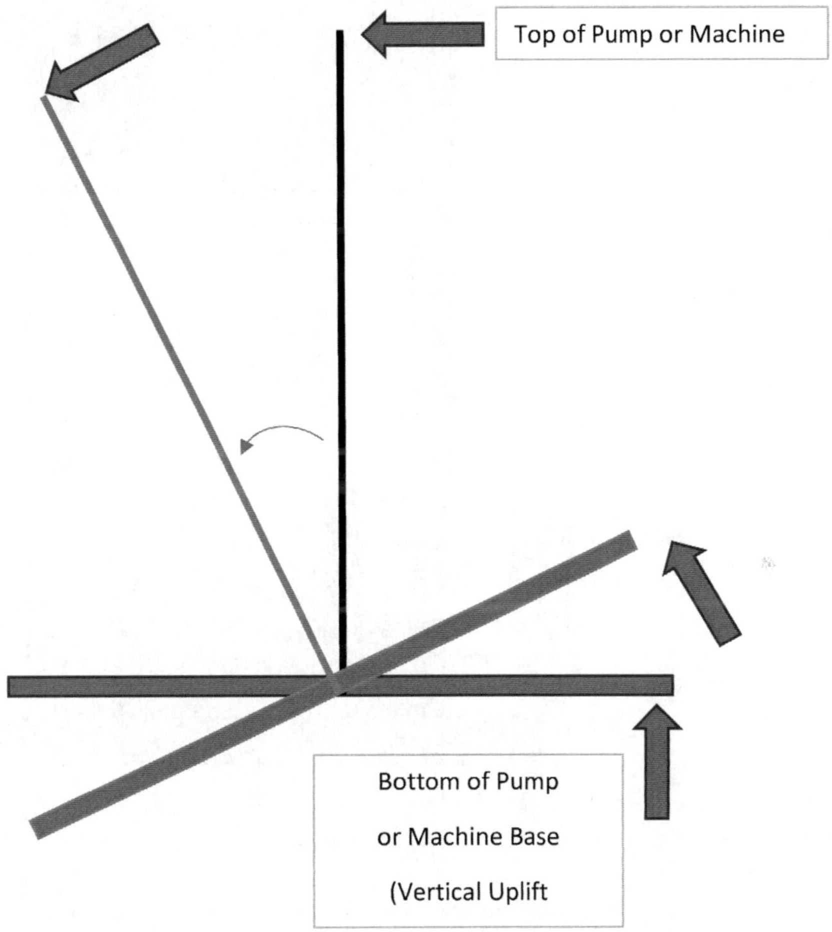

Figure 19: Kinematic Model of Fixed Base Rotation with Resulting Motion at Top of Machine

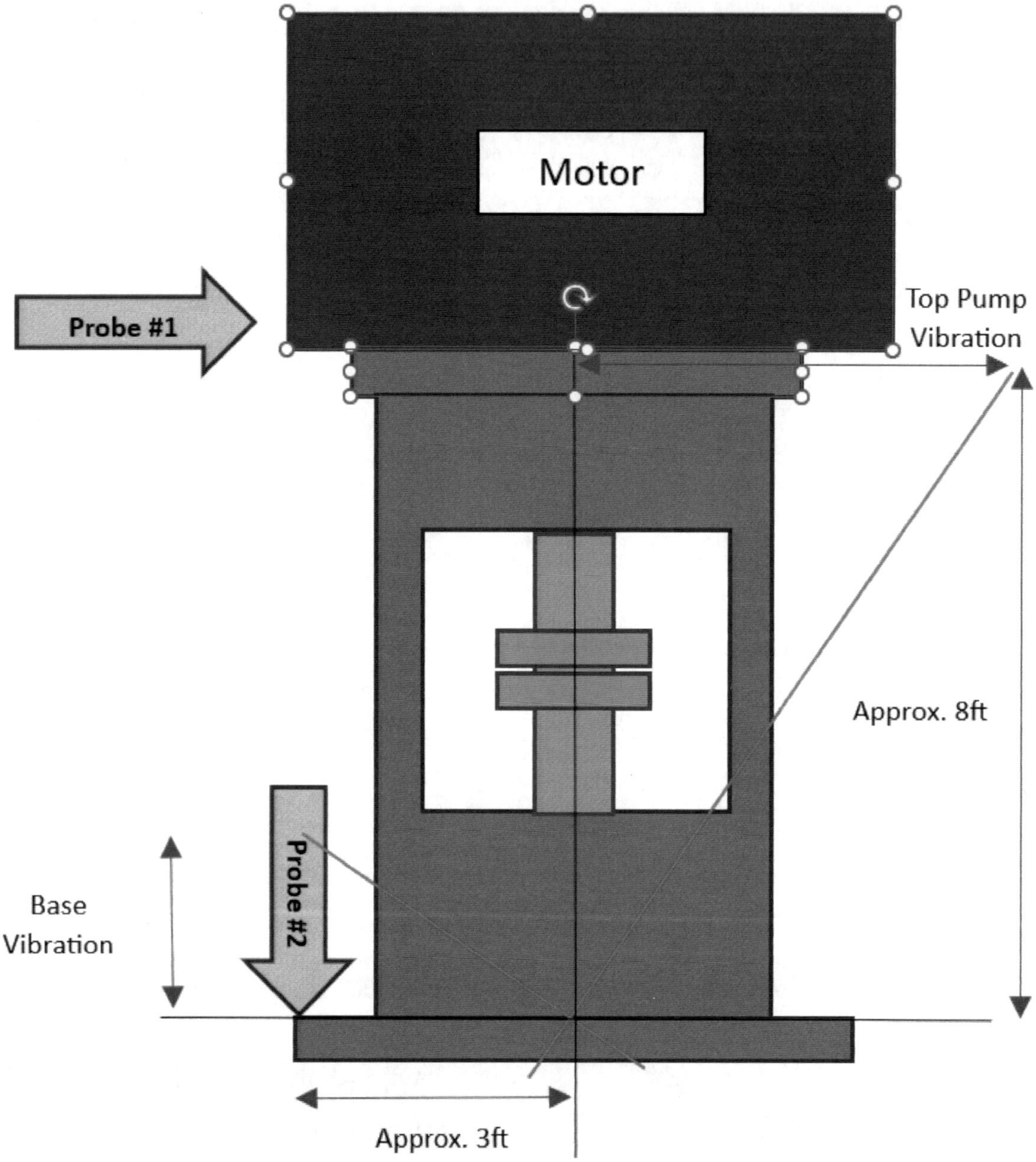

Figure 20: Layout and Measurement Locations on Circulating Water Pump and Base Support Structure

FAN FOUNDATION VIBRATION ANALYSIS: Fan foundations for large centrifugal fans are candidates for loose or resonant foundations. The large rotating inertias, coupled with external locations create opportunities for common problems with fan foundations. Normally, the larger the fan, the taller the foundation and fan centerline are located with respect to the base elevation. The figures below and the operating survey results demonstrate some analysis of a process fan site with some concerning foundations conditions. As part of the analysis methods, comparison contributions from the rocking and translational displacement was utilized to try to understand the character of the condition. Operating data was collected as shown in Figure 21 and the results compared to determine what the normal characteristics were present. After some review, it became apparent that the rocking and translation component contribution was much different on the A fan as shown below. Structural modifications and repairs were recommended and provided below in Figures 23 and 24.

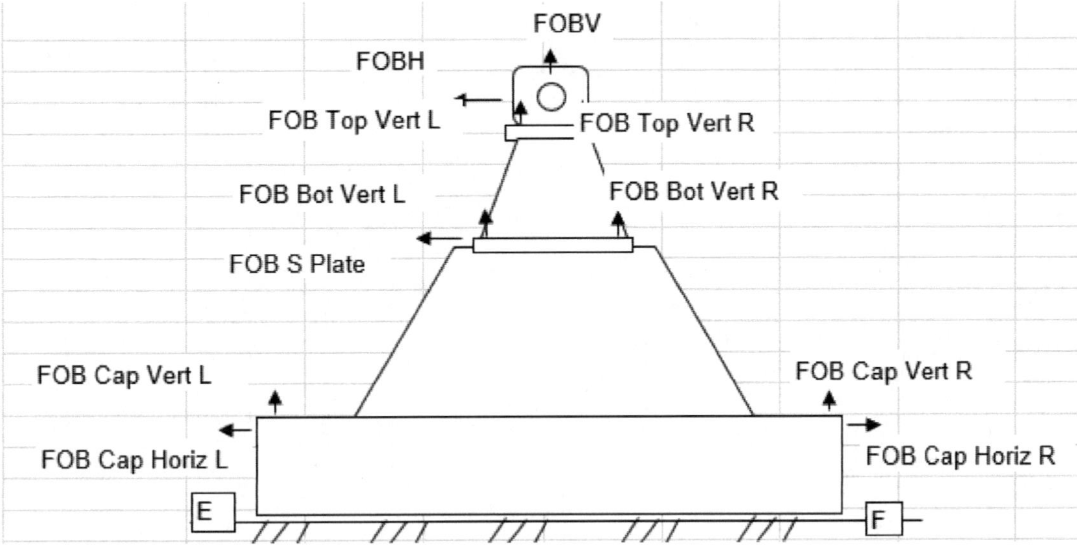

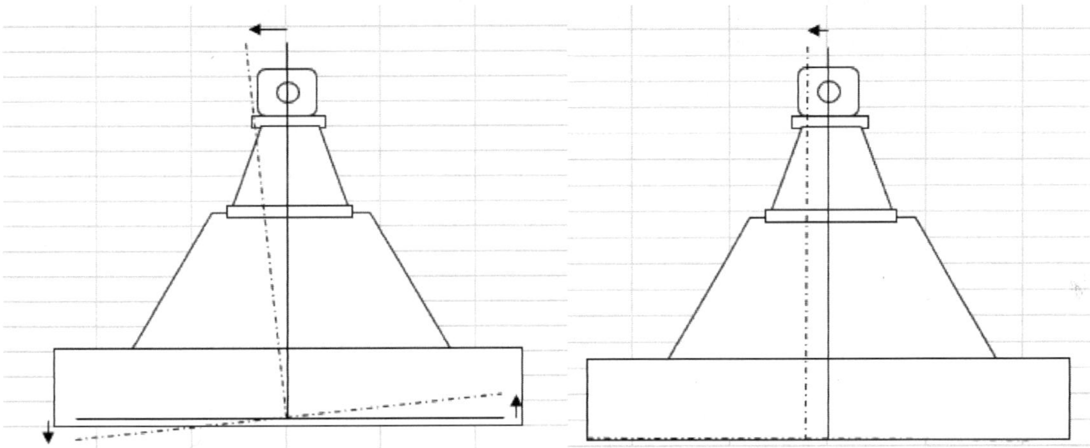

Fan Location							
A	Vertical Left	0.030		C	Vertical Left	0.024	
	Vertical Right	0.051	**% Ratio**		Vertical Right	0.024	**% Ratio**
	A Rotation Left	0.072	**29.5**		**C Rotation Left**	0.043	**42.4**
	A Translate Left	0.171	**70.5**		**C Translate Left**	0.058	**57.6**
	Total Brg Left	0.215			**Total Brg Left**	0.103	
B	Vertical Left	0.037		D	Vertical Left	0.026	
	Vertical Right	0.060	**% Ratio**		Vertical Right	0.018	**% Ratio**
	B Rotation Left	0.086	**48.7**		**D Rotation Left**	0.039	**54.8**
	B Translate Left	0.091	**51.3**		**D Translate Left**	0.032	**45.2**
	Total Brg Left	0.177			**Total Brg Left**	0.083	

Figure 21: Fan Foundation Operating Survey Results for Rotation and Translation

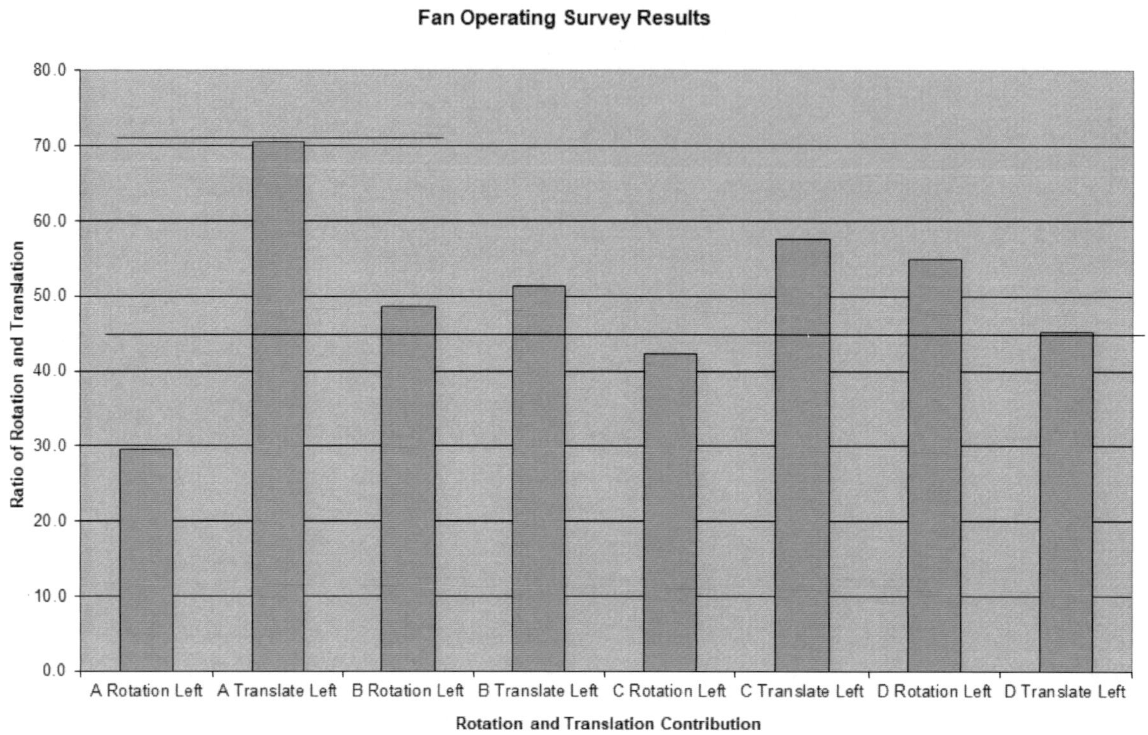

Figure 22: Fan Foundation Operating Survey Results for Rotation and Translation

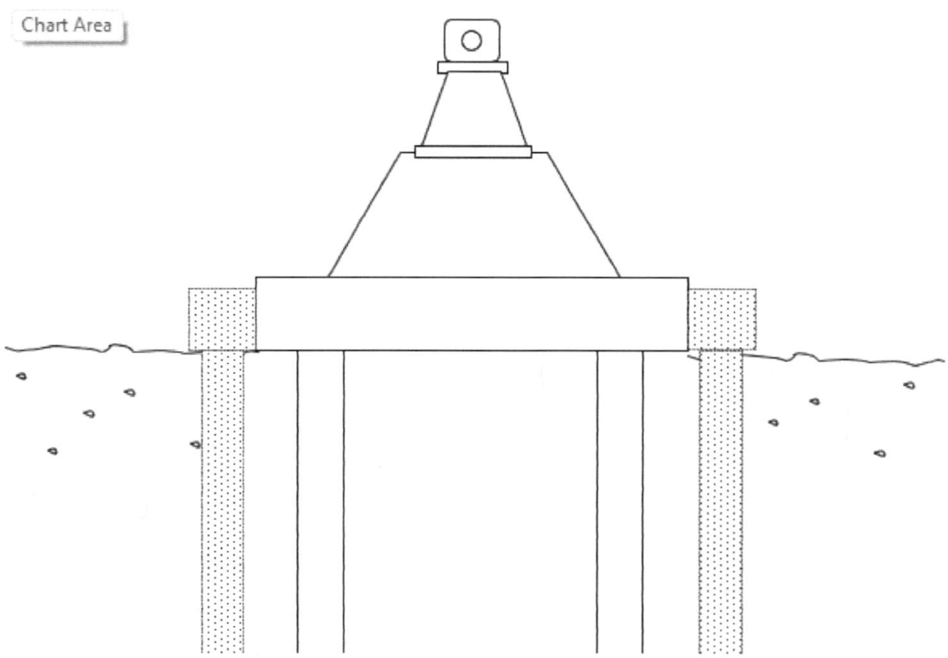

Figure 23: Fan Foundation Modification Elevation View

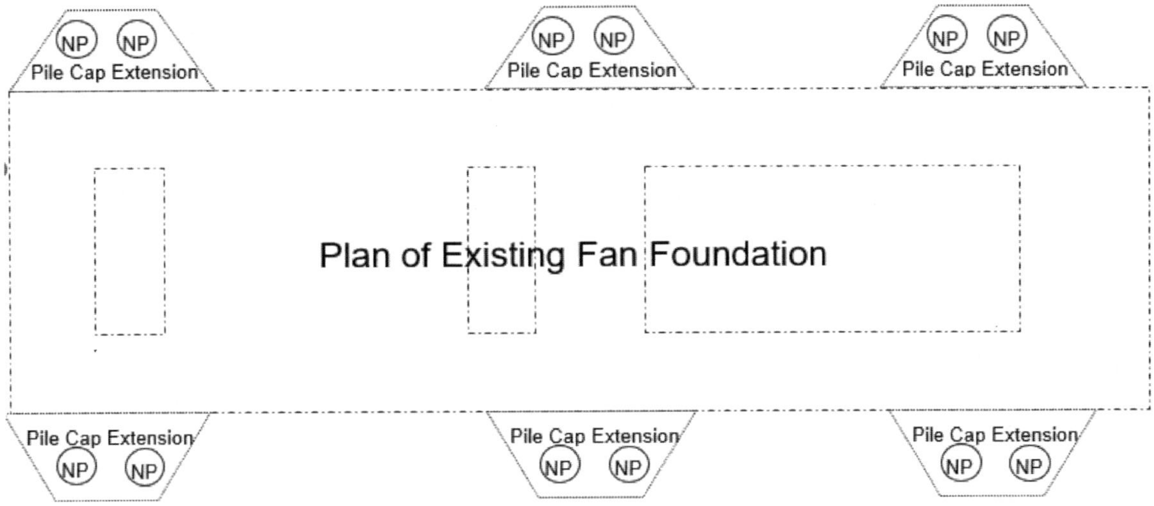

Figure 24: Fan Foundation Modification Plan View

Conclusions: Only the careful design, maintenance and construction of a foundation can ensure adequate support and success for a machine vibration attenuation. With that said, the structural integrity, the effectiveness of these elements is only as good as the structural connections and details that comprise an efficient, robust foundation.

Knowledge of the details of design and construction are as important as the vibration measurements themselves. In most cases, the vibration amplitudes at the base should exhibit much lower amplitudes than the vibration levels at the bearing locations. With machines/foundation systems that are extremely tall for their base widths it becomes quite apparent that the base details are critical to the level of vibration at the bearings.

ACKNOWLEDGEMENTS:
I would like to thank the Southern Company for the opportunity to experience great challenges and work with great people and partners throughout the world.

BIBLIOGRAPHY:

[1] Clough, Ray W. and Penzien, Joseph: Dynamics of Structures – 1975

[2] Jackson, Charles: RESONANCE: Theory and Practice - Proceedings of the National Technical Training Symposium and 27th Annual Meeting of the Vibration Institute, New Orleans, Louisiana, July 9-11, 2003.

[3] Baxter, Nelson L. and Henson, Greg: Steam Turbine Balancing – Case 6 -Mini-Course Notes – 16th Annual Meeting Vibration Institute – Williamsburg, VA July 9-11,1992

.

VIATC TRAINING CONFERENCE & EXPO
COVINGTON, KENTUCKY | AUGUST 7-9, 2024

ANALYSIS AND MITIGATION OF ACOUSTIC-MECHANICAL FCCU FEED PIPING AND NOZZLE VIBRATION

Michael Ackman	Noel Hart	Scot McNeill, Ph.D., P.E.*
Senior Advanced Analysis Engineer	Noise & Vibration Principal	Noise & Vibration Engineer
ExxonMobil	ExxonMobil	ExxonMobil
		(* Registered in Texas)

Michael Ackman is an Advanced Analysis Engineer in ExxonMobil with a total of 26 years of experience including 21 years as a Machinery Engineer supporting ExxonMobil's Downstream and Chemicals assets in Baton Rouge, LA. In his current role, he performs detailed analysis (stress, modal, vibration, acoustic, etc.) of fixed and rotating equipment using finite element analysis techniques. He received his B.S. and M.S. degrees from Louisiana State University.

Noel Hart is the Noise and Vibration Principal at ExxonMobil. In addition to providing technical support to the various downstream, midstream and upstream sites, his duties include managing the noise and vibration technology discipline for the corporation, including resourcing, training and development, standards, and new technology deployment. He is currently a voting member of international standards committees, ANSI/ASA S12 and ISO TC 43/SC 1 for noise. Member of INCE.

Dr. McNeill has over 20 years of experience in vibration measurement, modeling, and analysis of fixed equipment. He specializes in flow induced vibration and analysis of random processes. Dr. McNeill contributed to industry standards including API 579/ASME FFS and Energy Institute Guidelines. He is a member of the API 579/ASME FFS committee and SEM (Society of Experimental Mechanics) and is also a certified category III vibration analyst and a registered professional engineer in Texas.

ABSTRACT:
The Fluid Catalytic Cracking Unit (FCCU) is a critical piece of refining equipment used to convert the heavy portion of crude oil feedstock into lighter petroleum products, including liquified petroleum gas and gasoline. Vibration limits are often imposed on process equipment, including FCCUs, to ensure structural integrity and safe operation.

Changes in the configuration of an existing FCCU feed nozzle resulted in coincidence of acoustic and mechanical piping resonant frequencies, resulting in high vibration levels. Vibration and pressure pulsation measurements were taken over varying feedstock temperature. It was noted that both the pulsation/vibration frequency and vibration amplitude varied with temperature. As temperature increased within the range tested, frequency decreased, and vibration levels increased.

The nozzle acoustic and feed piping mechanical systems were modeled using a combination of Finite Element (FE) software and closed form calculations. Modeling revealed that the observed temperature dependence was due the inverse relation of the sonic velocity (and acoustic natural frequencies) in the feedstock to temperature. As temperature increases, the acoustic natural frequencies in the FCCU nozzle decrease, approaching an existing mechanical piping natural frequency. Modifications to the nozzle design were proposed to be implemented in the next shutdown and turnaround. The modifications were designed to keep the acoustic frequencies above the mechanical natural frequency over the range of operating temperatures. The model was also used in a predictive manner to determine an interim temperature limit on the feedstock, until the permanent nozzle modifications could be implemented.

Vibration/pulsation measurements repeated post-mitigation revealed that the modifications performed as intended, leading to sufficient separation of acoustic and mechanical natural frequencies. As a result, vibration levels were reduced significantly and are well below allowable limits. The modifications enabled the lifting of production-limiting operational restrictions.

KEY WORDS: FCCU, piping, vibration, acoustic modes, mechanical modes, flow-induced

BRIEF EQUIPMENT OVERVIEW:
The Fluid Catalytic Cracking Unit (FCCU) is a critical piece of refining equipment used to convert the heavy portion of crude oil feedstock into lighter petroleum products, including liquified petroleum gas and gasoline. Figure 1 depicts a simplified FCCU schematic. The feed nozzles are located at the bottom of the riser and mix feed, catalyst, and steam. Steam and oil are mixed at high velocity at the nozzle tip to atomize the oil. Hot catalyst from the slide valve is conveyed upward around the outside of the feed nozzles using steam. The catalyst and atomized oil mix at the exit of the nozzles and flow upward through the riser to the reactor vessel. Cracking the heavy feedstock into lighter molecules takes place in the riser. Feed and steam velocities in the nozzle are very high to promote thorough atomization and contact with the catalyst. Due to the high temperatures, a significant amount of thermal expansion must be allowed for, resulting in relatively soft supports. Reference [1] discusses three case studies involving FCCU piping vibration.

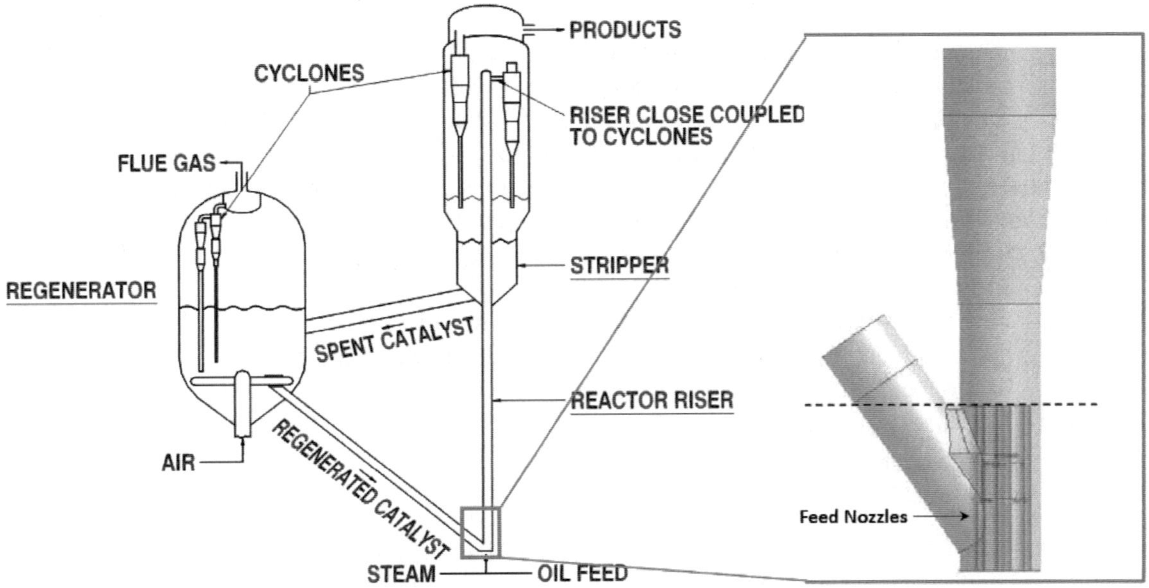

Figure 1: Simplified FCCU Schematic

INITIAL FIELD OBSERVATIONS:

The feed nozzle assembly and inlet piping in an FCCU in an ExxonMobil facility began experiencing intermittent high vibration after modifications were made to the nozzles during a unit turnaround. The elevated vibration was most prominent on the nozzle piping and feed inlet piping on the right side of the wye shown in Figure 2. Note that the upper portion of the nozzle piping above the large flange is contained within a conical casing and is not accessible.

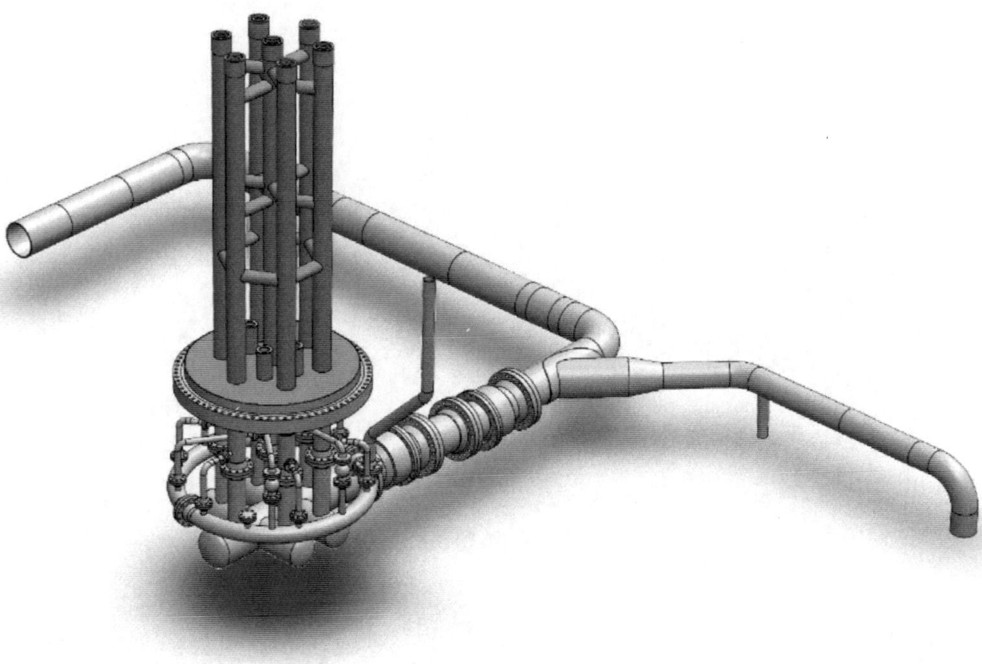

Figure 2: FCCU Feed Piping and Nozzles

Vibration measurements were recorded after elevated vibration levels were observed in 1Q2019 and 1Q2020. The dominant vibration frequency fell in the range between 70 – 75 Hz. Note that the 2nd, 3rd, and 4th harmonics also appeared in the FFT spectrum, with the 3rd harmonic dominating and sometimes higher than the 1st harmonic. In each instance of higher vibration, vibration levels were observed to be reduced with a decrease in feed temperature. The frequency of the dominant spectral peak was also noted to increase with a decrease in feed temperature. These two observations indicated an important temperature dependency.

Continued intermittent high vibration caused the site to limit the unit's operating envelope (primarily maximum feed temperature), prompting further investigation to:

- Determine the vibration root cause
- Recommend mitigation options
- Re-expand operating envelope

HYPOTHESIS RESULTING FROM BASIC MODELING AND CALCULATIONS:
A finite element model of the feed piping was built. All piping and nozzle assembly components were modeled explicitly. Valves were modeled as concentrated masses. Boundary conditions were applied to mimic pipe support behavior. A modal analysis indicated a mode with natural frequency of 69.4 Hz, which was close to the dominant frequencies measured in the range of 70-75 Hz. It can be seen in Figure 3 that the feed piping to the right of the wye exhibits the largest motion for this mode.

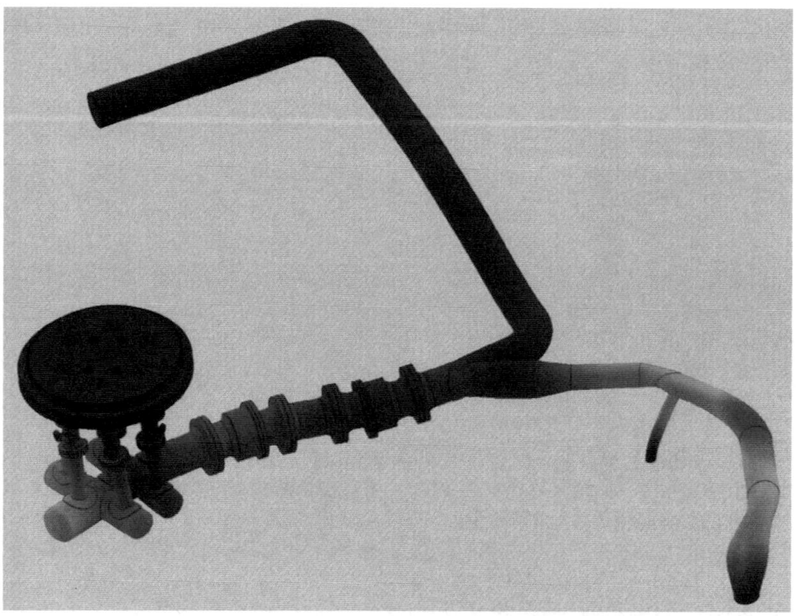

Figure 3: Feed Piping Mode at 69.4 Hz

Screening calculations performed for the feed and nozzle piping indicated that the length of the nozzle pipe leads to an acoustic natural frequency around 70 Hz for the range of sonic velocities of the feedstock fluid. The presence of the mechanical and acoustic natural frequencies near 70 Hz led to the following hypothesis on the cause of excessive vibration.

- The turbulent mixing of process fluid and steam at the nozzle exit is a source of broadband pressure pulsation.
- This broadband energy is always present and excites some of the mechanical natural frequencies of the system but only creates large amplitudes when an acoustic resonance is also present in the nozzle and/or feed piping at the same frequency.
- When the acoustic resonant frequency is near the mechanical natural frequency of the feed piping, the result is high vibration.

DETAILED SONIC VELOCITY AND ACOUSTIC FREUENCY CALCULATIONS:
The acoustic natural frequencies for 1D longitudinal standing waves in a fluid contained in a pipe depend on the sonic velocity (speed of sound), length of the pipe and boundary conditions on the pipe ends. The sonic velocity of the fluid depends on bulk modulus and density and these properties change with stream composition, temperature, and pressure. The relationship is shown in Eq. (1).

$$c = \sqrt{\frac{K(T,P)*g}{\rho(T,P)}}, \text{ where,}$$
$$c = sonic\ velocity, K = bulk\ modulus,$$
$$g = gravitational\ constant, \rho = density, \qquad (1)$$
$$T = temperature, P = pressure$$

Boiling point curves for the feedstock were obtained from the refinery data historian and used to compute the stream properties of the fluid in the nozzles. When the sonic velocity is known, the nth order acoustic natural frequency, f_n, can then be calculated using the well-known formulae in Eq. (2).

$$f_n = \frac{nc}{2L} \quad \text{for half-wave multiples (open-open or closed-closed ends)}$$
$$f_n = \frac{(2n-1)c}{4L} \quad \text{for quarter-wave multiples (open-closed or closed-open ends)} \qquad (2)$$

Here, L is the length of the pipe plus a correction factor proportional to the pipe diameter for the open or closed ends. Since the nozzles have open ends, a factor of 0.4D, where D is the diameter, was added for each end. In addition to the correction for end effects, the effect of the pipe wall flexibility was applied according to reference [2].

Because the mechanical piping natural frequency at 69.4 Hz was known from modeling and vibration measurements, the sonic velocity in the nozzle that leads to alignment between the mechanical natural frequency and first half-wave acoustic natural frequency, $c_{critical}$, was solved for. Then the margin to resonance (alignment of acoustic and mechanical natural frequencies) can be computed using Eq. (3).

$$M_{res} = \frac{c_{actual} - c_{critical}}{c_{actual}} \qquad (3)$$

The margin to resonance was plotted for seven one-day time periods, during which vibration levels had been measured. The result is shown in Figure 4. Time periods with known high vibration are outlined in red; time periods with low vibration are outlined in green. It can be seen that when the margin is above 5% low vibrations occurred and when the margin was below

3%, high vibrations occurred, supporting the hypothesis that high vibrations are caused by alignment of the acoustic and mechanical natural frequencies.

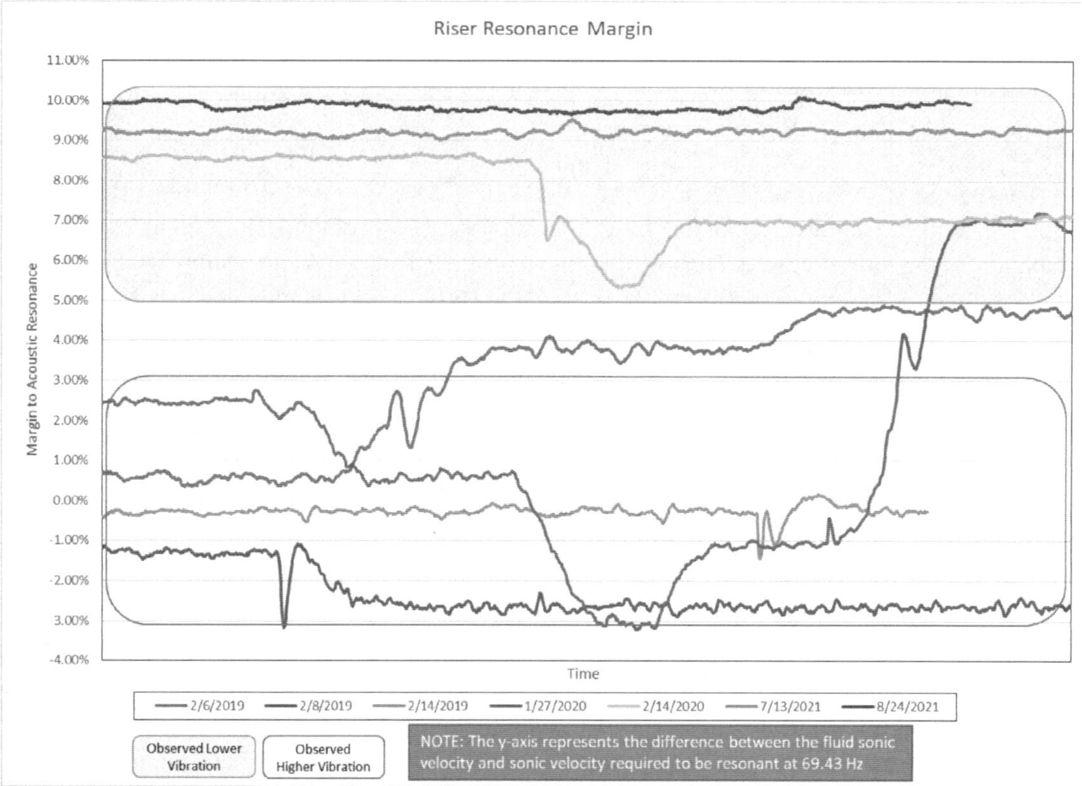

Figure 4: Margin to Resonance

ADDITIONAL DATA COLLECTION IN 2021:

Additional vibration and pressure pulsation measurements were taken in 2021. Pulsations were measured at one location (point 1 in Figure 5) using a dynamic pressure transducer. Vibration was measured in three orthogonal directions at 7 locations (points 2 through 8).

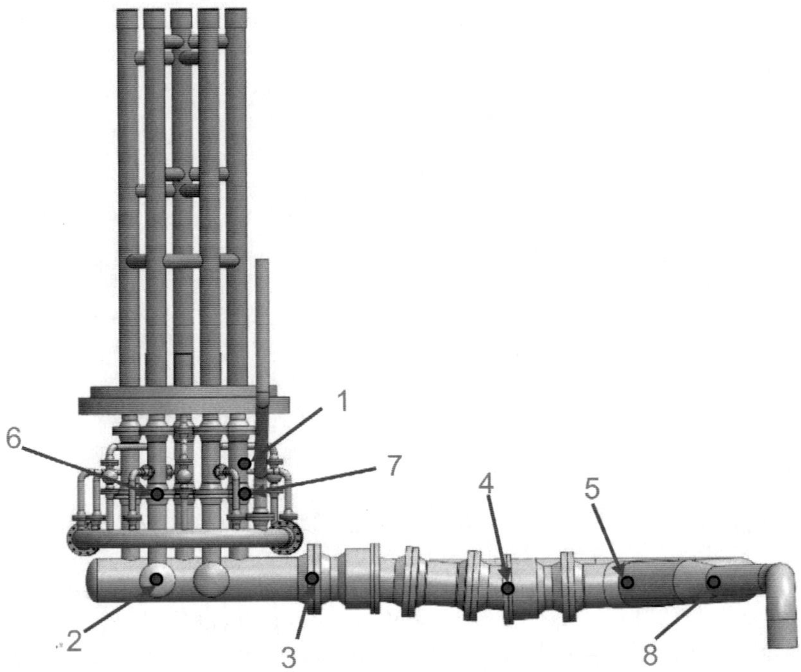

Figure 5: Vibration and Pulsation Measurement Locations in 2021

The pulsation waveform is shown for a 60 second duration in Figure 6 and for a 0.1 second snapshot in Figure 7. It can be seen that the pulsation amplitude envelope grows and shrinks intermittently. With pulsation levels of 110 psi pk-pk, clearly significant pressure pulsation energy is present in the nozzle, exciting an acoustic mode (standing wave). The source of this energy is likely turbulent mixing of the feed, steam, and catalyst at the exit of the nozzle.

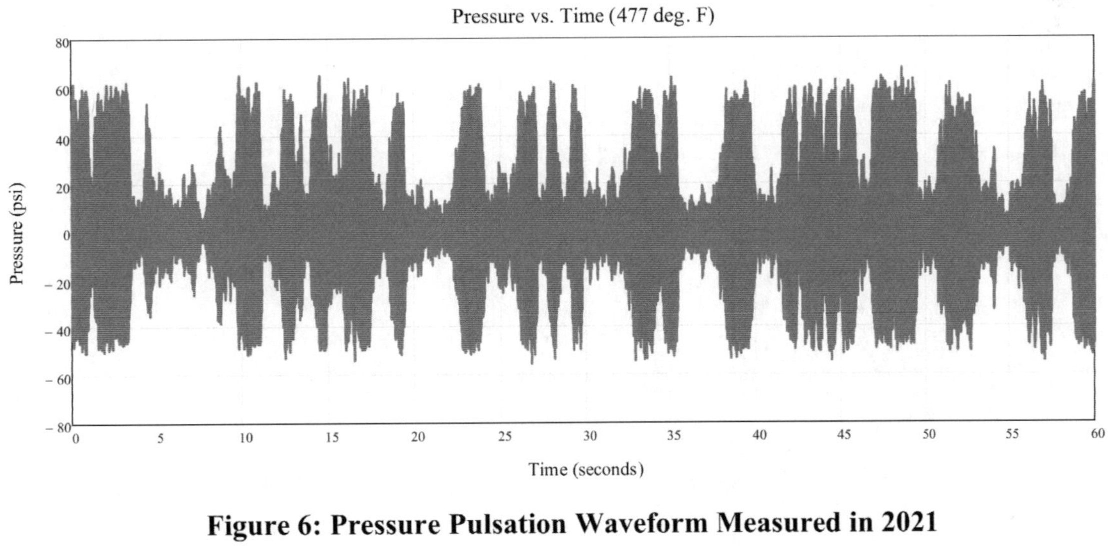

Figure 6: Pressure Pulsation Waveform Measured in 2021

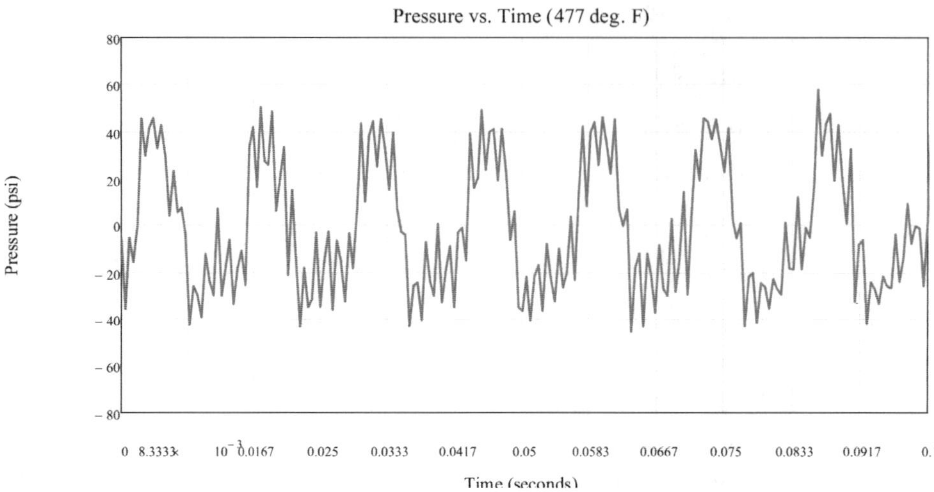

Figure 7: Pressure Pulsation Waveform Measured in 2021 (Zoomed)

The pulsation spectrum is shown in Figure 8 for data collected at two different operating temperatures, 477° F and 527° F. The primary peak frequency is at 74 Hz at 477° and 69 Hz at 527°. The decrease in acoustic natural frequency with increasing feed temperature is expected as the fluid sonic velocity decreases with rising temperature therefore reducing the acoustic natural frequency.

The corresponding vibration spectra at the highest vibration location (location 5), shown in Figure 9, shows a similar spectral shape with a primary peak and two harmonics, as well as the same shift with varying temperature. It can also be seen the peak vibration level is higher at the higher temperature (lower frequency).

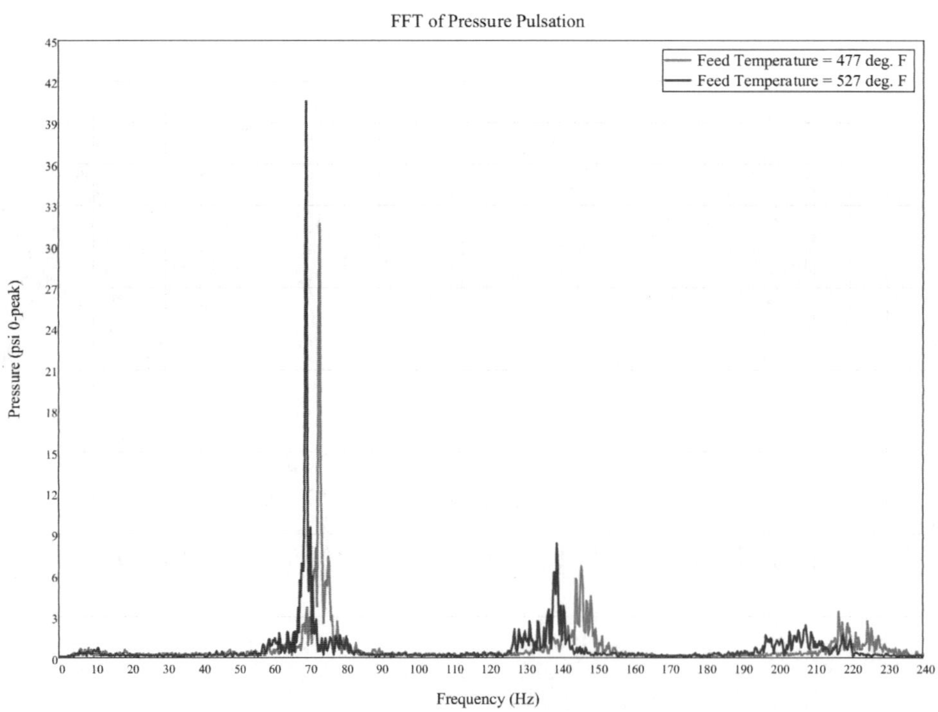

Figure 8: Pressure Pulsation Spectra Measured in 2021

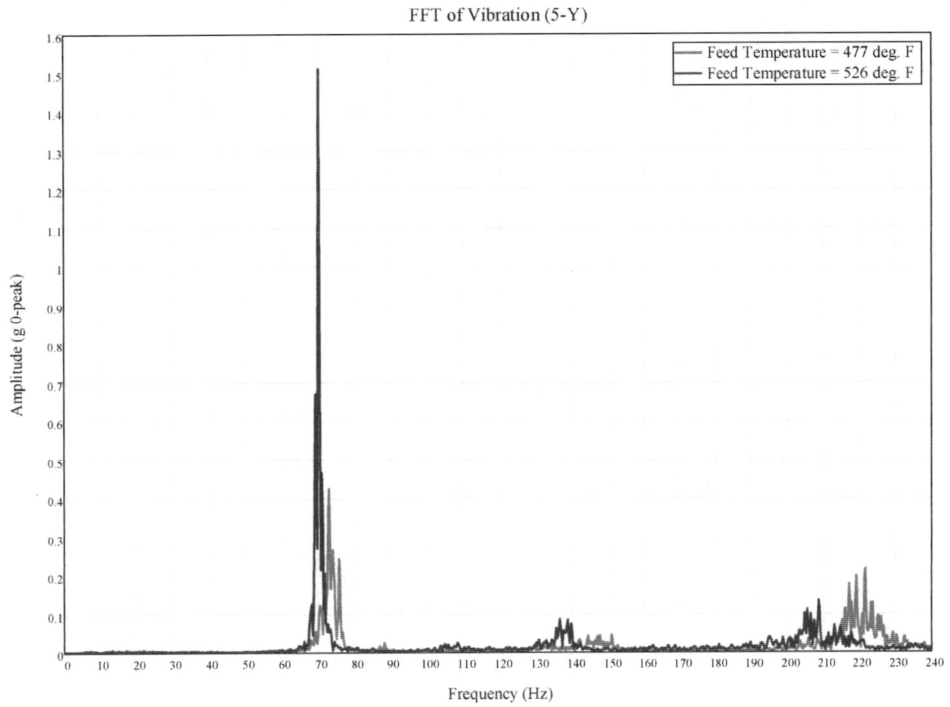

Figure 9: Vibration Spectra Measured in 2021

Figure 10 compares the peak pulsation/vibration frequency trend with temperature from the measured 2021 data to modeling predictions (calculated acoustic natural frequency). Measurements are in line with predicted values. Measured values are shown as points and modeling predictions as lines.

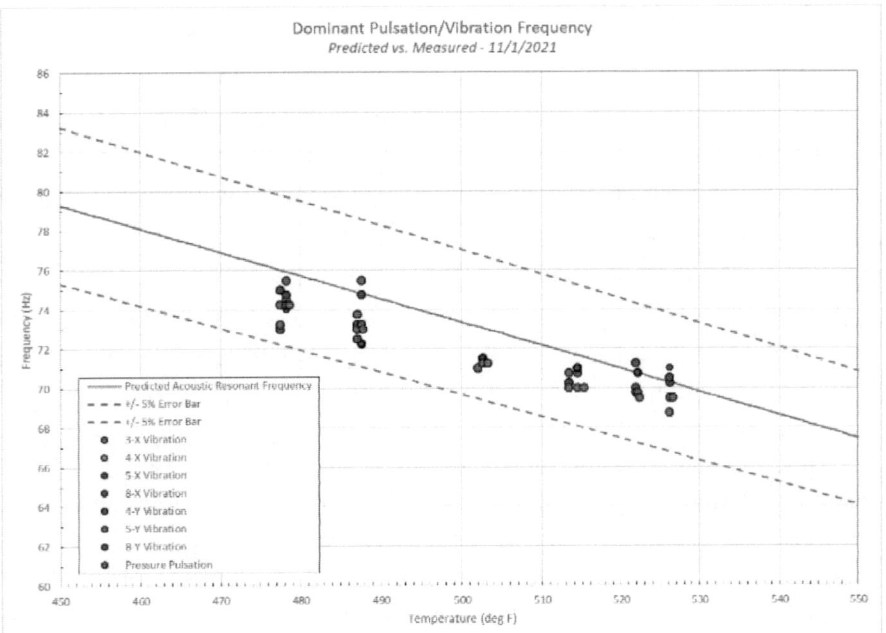

Figure 10: Pulsation/Vibration Peak Frequency vs. Temperature in 2021

The overall vibration level trend with pulsation/vibration frequency at several measurement locations/directions can be seen in Figure 11. Vibration levels clearly decrease with increasing frequency.

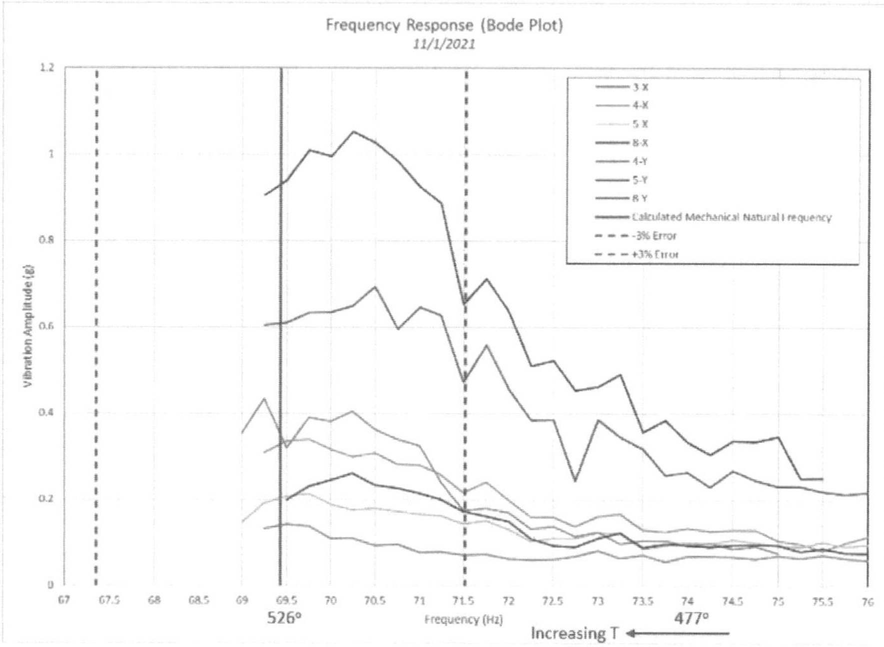

Figure 11: RMS Vibration Trend with Feedstock Temperature in 2021

From the previous figures, it can be seen that the vibrations are highest when the temperature increases to 526° and the acoustic natural frequency is near the 69.4 Hz piping natural frequency. As the temperature decreases to 477°, and the acoustic natural frequency increases to 74 Hz, the vibration levels decrease significantly.

The additional 2021 data set confirms the hypothesis that excessive vibrations result from alignment of acoustic and mechanical natural frequencies, resulting in resonance. This occurs when the feed temperature increases to 526° and the acoustic natural frequency drops to 69.4 Hz.

RECOMMENDED MITIGATION STRATEGY:

The strategy to reduce the vibration levels was to detune the nozzle acoustic natural frequency away from the mechanical piping natural frequency. Addressing the acoustic nozzle frequency was preferred over bracing the piping due to the presence of other piping modes with nearby frequency that may end up aligning with the acoustic natural frequency. It was proposed to shorten the nozzle length inside of the riser to increase the nozzle acoustic natural frequency above the mechanical piping natural frequency for all likely feed compositions and temperatures. Modeling indicated that shortening the nozzles by 24" will provide a significant shift in the acoustic natural frequency and margin to resonance. Figure 12 and Figure 13 show the margin to resonance before and after the proposed nozzle shortening. The proposed 24" shortening was predicted to completely avoid high vibration in the range of operating parameters. However, due to other operational considerations, such as slide valve fouling, the decision was made to shorten the nozzle by 18" in the 2023 turnaround (shutdown). Until the 2023 turnaround, operational temperature limits were kept in place to avoid excessive vibration.

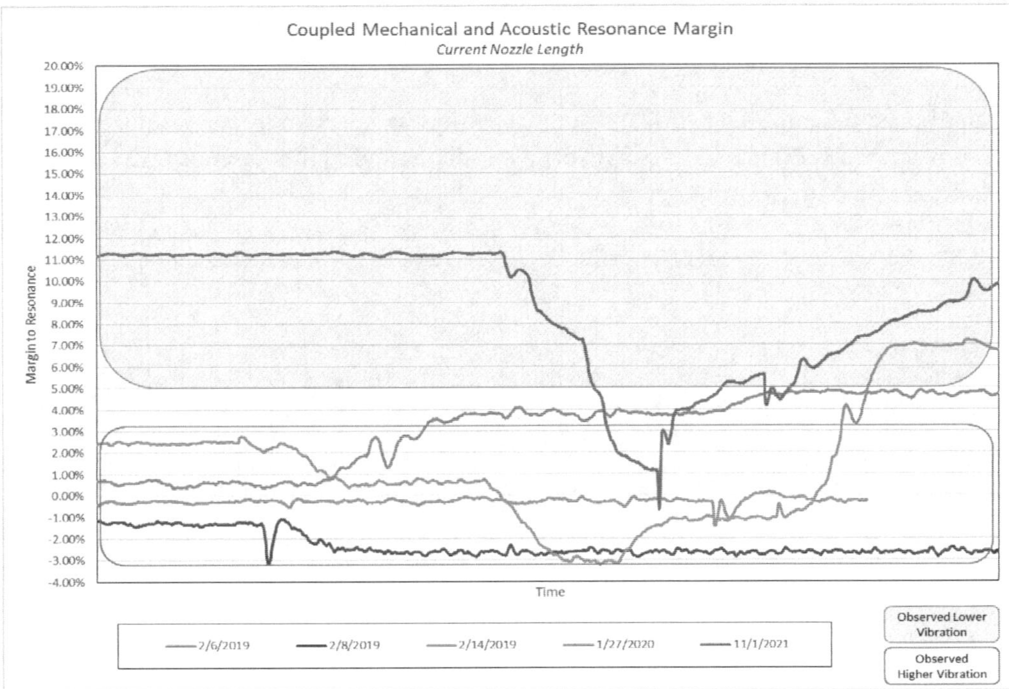

Figure 12: Margin to Resonance – Current Nozzle Length

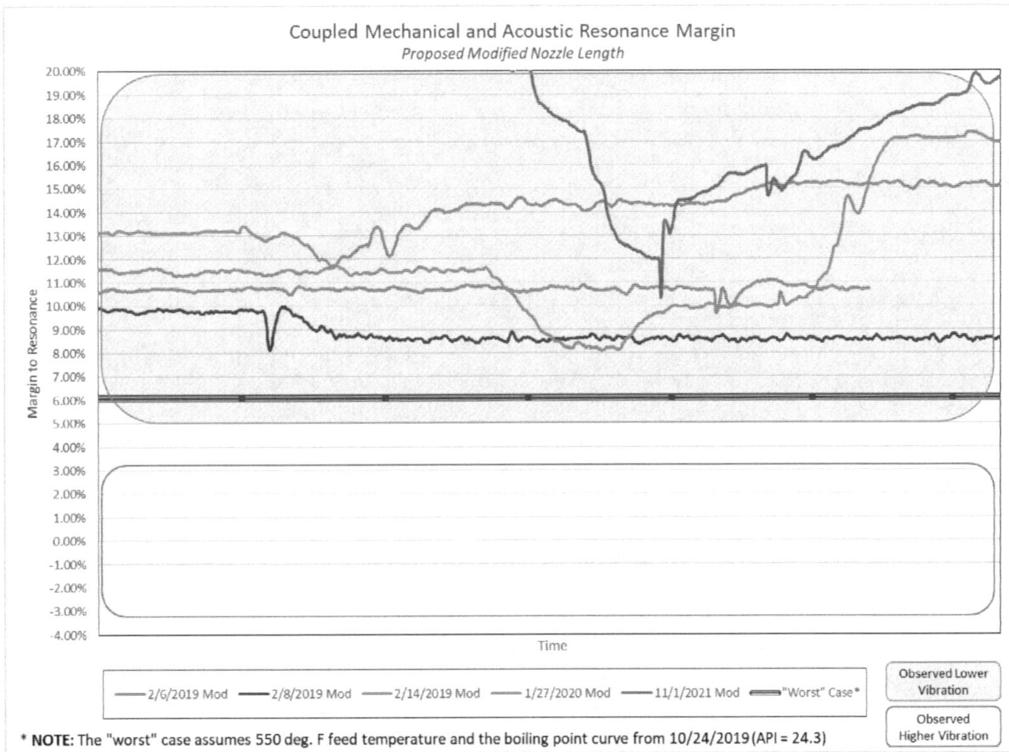

Figure 13: Margin to Resonance – Proposed Shorter Nozzle

POST MITIGATION VIBRATION MEASUREMENTS IN 2023:

Vibration and dynamic pressure data were collected in 2023, following the nozzle shortening activity during the turnaround. Feedstock temperatures were varied over the expected operating range. During field measurements, it was immediately apparent that:

- Vibration levels were significantly lower compared to data taken in 2019-2021.
- Peak frequencies increased by about 10 Hz, from 69-74 Hz to 79-84 Hz.

The pressure pulsation amplitude shown in Figure 14 is steadier than in 2021, though the RMS amplitude is higher (see Figure 15). However, it should be noted that pressure was measured in a different nozzle pipe in 2023 than in 2021.

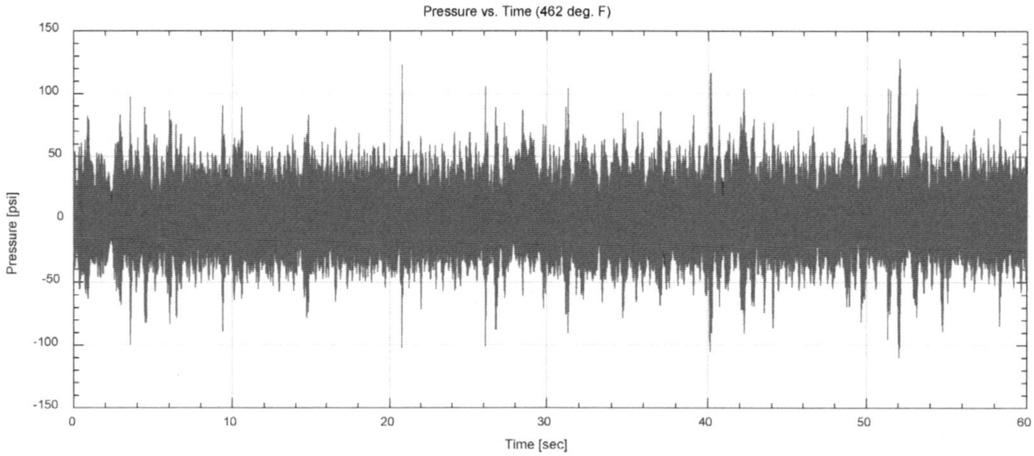

Figure 14: Pressure Pulsation Waveform Measured in 2023

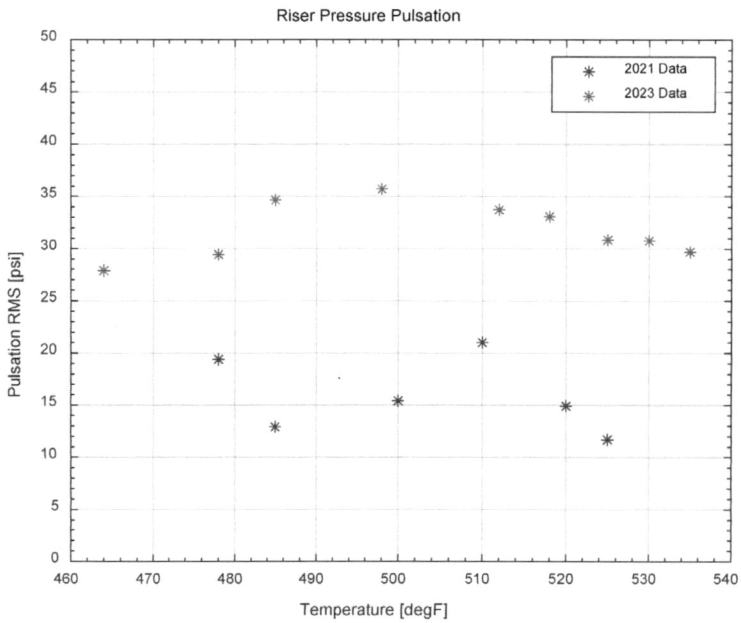

Figure 15: RMS Pulsation Level Comparison – 2021 and 2023 Data

Due to nozzle shortening, the acoustic natural frequency has shifted above the mechanical natural frequency. This is clearly shown by comparing the 2021 and 2023 pulsation spectra in Figure 16 and Figure 17. In the figures, the range of pulsation frequencies to avoid is indicated in gold as the piping natural frequency of 69.4 Hz with a +/- 5% band. It can also be noted that the maximum feed temperature in the 2023 data set was higher (535° compared to 527°).

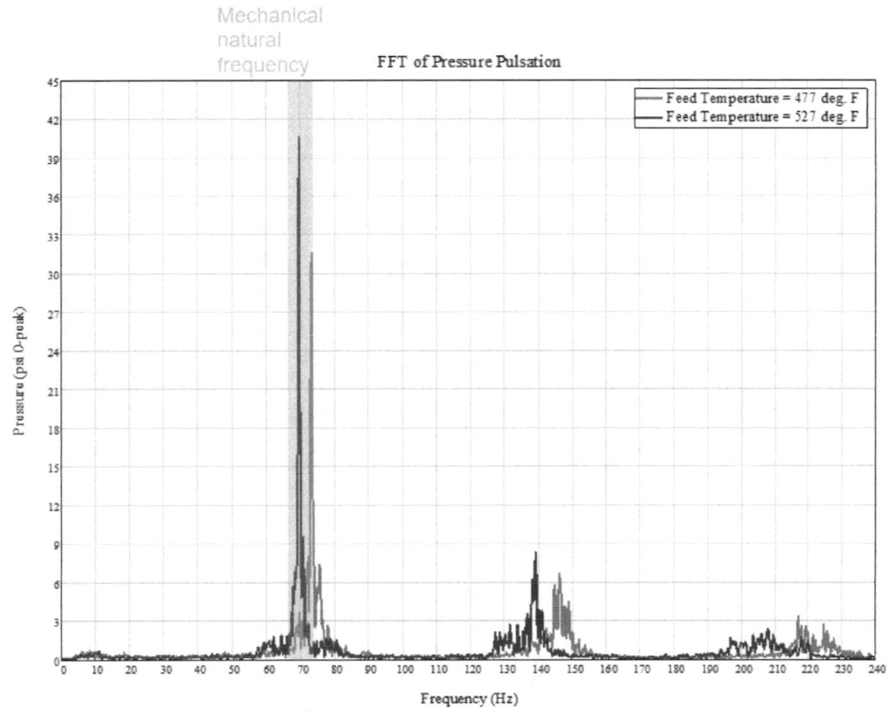

Figure 16: Pressure Pulsation Spectra Measured in 2021 – Shown with Range to Avoid

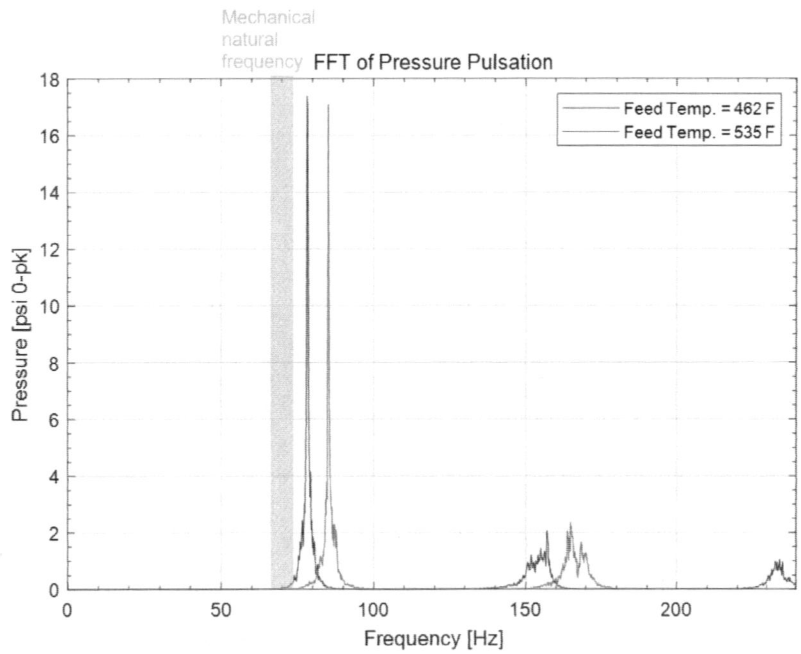

Figure 17: Pressure Pulsation Spectra Measured in 2023

Figure 19 indicates that the peak pulsation/vibration frequency has shifted sufficiently above the mechanical piping natural frequency for the range of feedstock temperatures during the measurement period. The data from Figure 10 is repeated along with the region of frequencies to avoid in Figure 18, for comparison. It can be concluded from the measurements that the acoustic natural frequency increased by around 10 Hz.

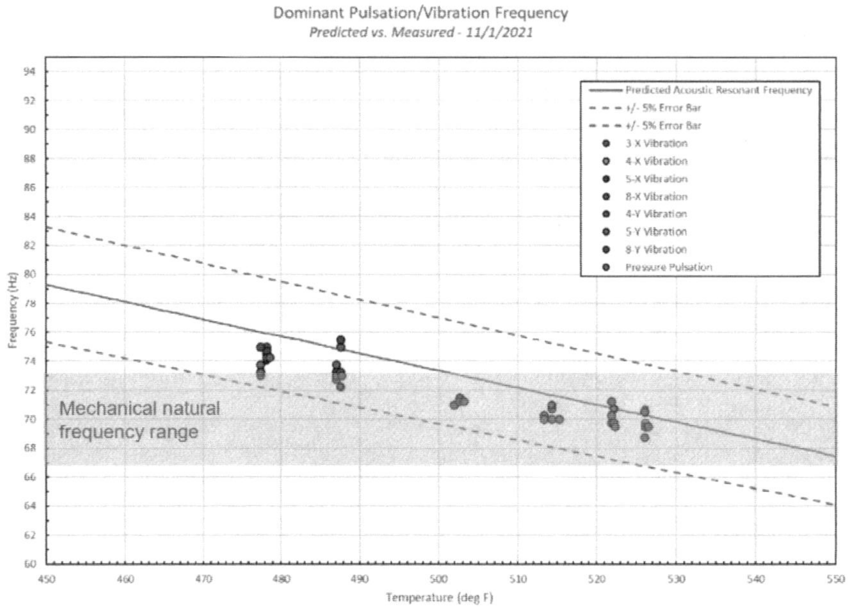

Figure 18: Pulsation/Vibration Peak Frequency vs. Feedstock Temperature in 2021 – Shown with Range to Avoid

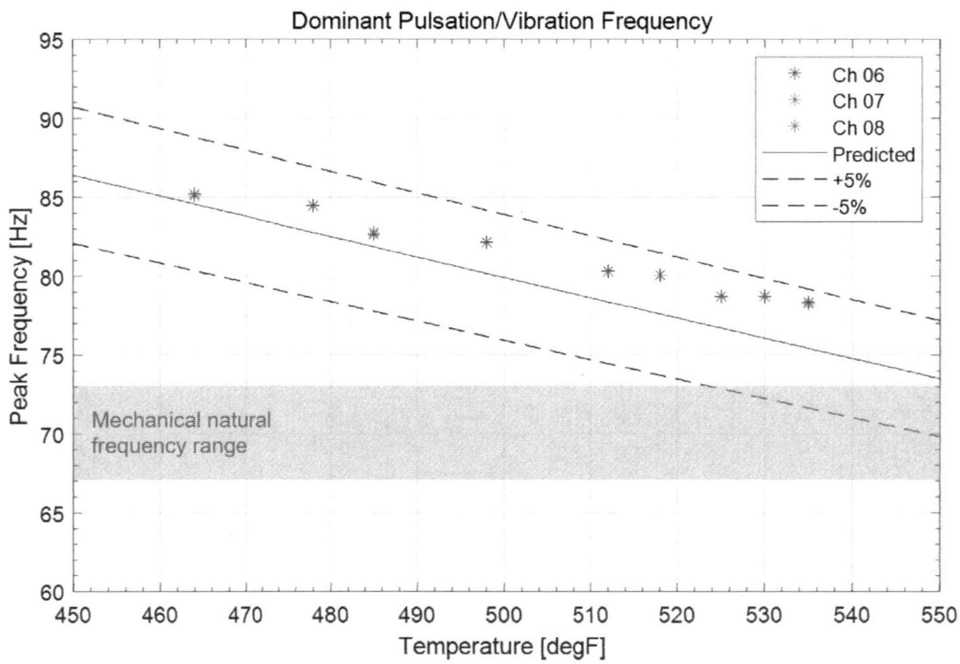

Figure 19: Pulsation/Vibration Peak Frequency vs. Feedstock Temperature in 2023

RMS vibration levels at the location with highest vibration are shown in Figure 20. It is evident that vibration levels in the 2023 data are low (< 0.25 in/s RMS) due to the detuning effect.

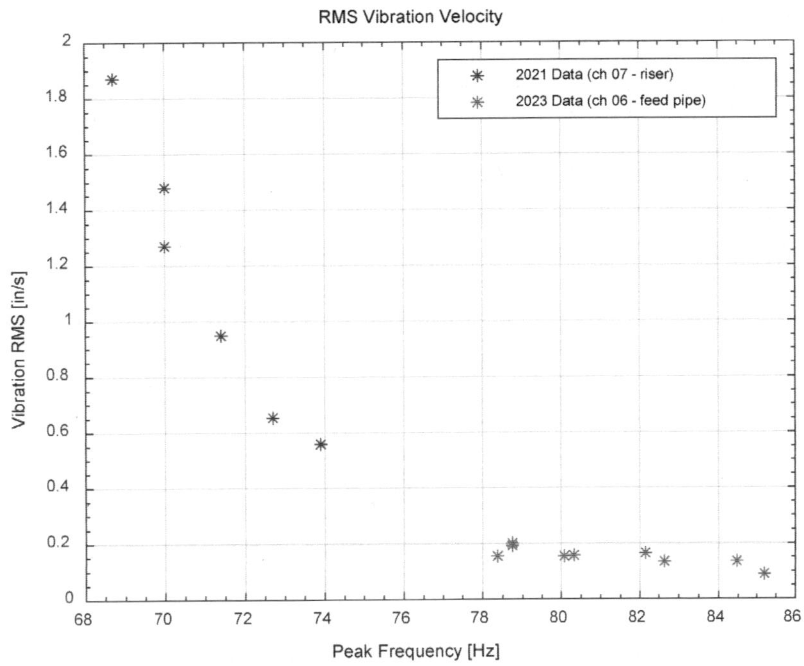

Figure 20: RMS Vibration Trend with Frequency – 2021 and 2023 Data

It is not anticipated that the feed temperature will exceed 535° for operations in the foreseeable future. However, it was desirable to conservatively estimate an upper allowable limit such that vibration levels will not be excessive. The peak frequency vs. temperature data was extrapolated in Figure 21 to determine the upper limit on feed temperature. At 550° the pulsation/vibration peak frequency is expected to be 77 Hz, indicating that the acoustic natural frequency is within 10% of the mechanical natural frequency. Low vibrations are expected with temperatures below 550°, though higher vibrations may result above this limit.

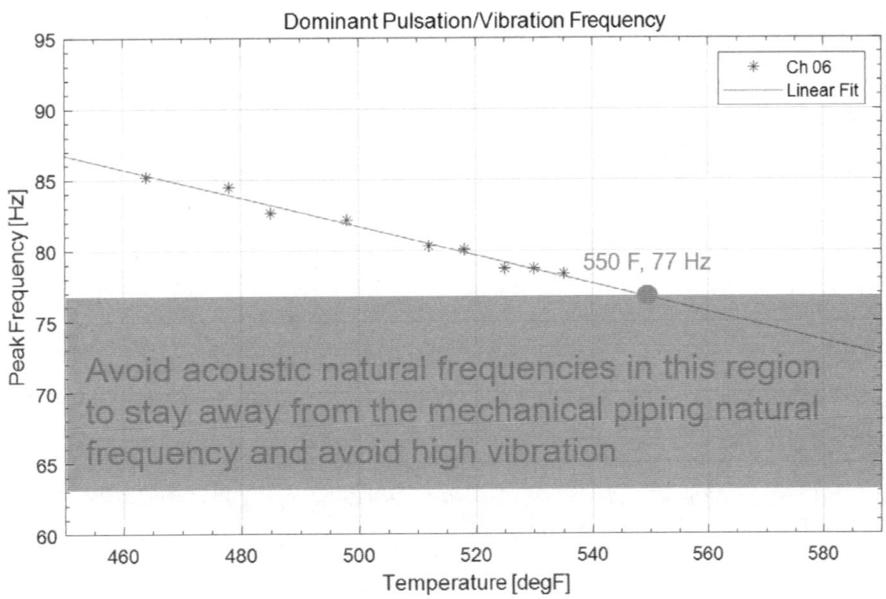

Figure 21: Pulsation/Vibration Peak Frequency vs. Temperature in 2023 - Extrapolated

SUMMARY AND CONCLUSIONS:

Excessive FCCU feed piping and nozzle vibrations were noted in 2019 through 2021. Initial pulsation/vibration measurements, along with mechanical FEA and acoustic natural frequency calculations indicated that excessive vibrations could be due to the alignment (resonance) of an acoustic natural frequency in the nozzle, which varies with temperature, with a mechanical piping natural frequency at 69.4 Hz.

More rigorous acoustic modeling was performed using properties of the feedstock obtained from the plant to determine the sonic velocity in the fluid. Using the calculated sonic velocity along with nozzle geometry and wall flexibility, the acoustic natural frequencies were calculated. A metric termed "margin to resonance" was computed as the normalized difference between the actual sonic velocity (given the operating conditions) and the sonic velocity required for resonance with the mechanical piping mode. This metric was used to limit operating temperature and predict the amount to shorten the nozzle to detune the nozzle acoustic natural frequency from the mechanical piping natural frequency.

Vibration and pulsation data were collected later in 2021, confirming the acoustic and mechanical resonance. The acoustic natural frequency aligned with the 69.4 Hz piping mode when the feed temperature was increased to 527° F, leading to high vibrations. When the feed temperature was 477° F, the acoustic natural frequency increased to 74 Hz, creating separation from the piping natural frequency, reducing the vibration levels.

Shortening the nozzle by 18" during a 2023 turnaround resulted in an increase in the acoustic natural frequency by about 10 Hz (from 69-74 Hz to 79-84 Hz, depending on feed temperature). This shift created separation between the acoustic and mechanical natural frequencies, even at feed temperatures above 535°, which is higher than the typical operating range. Measured vibrations levels were reduced significantly (< 0.25 in/s), which are below allowable vibration levels.

However, if feed temperatures exceed 535° F by a large enough amount, the potential still exists for higher vibration, as the acoustic natural frequency decreases, getting closer to the mechanical natural frequency around 69.4 Hz. By extrapolating the measured data results, it is predicted that resonance (high vibration) will not occur below 550° F, if the properties of the feed do not change drastically. In light of the new findings, it was recommended that the temperature limit be removed/modified accordingly.

AKNOWLEDGEMENTS:

The authors would like to thank the management of ExxonMobil for permission to publish this work.

REFERENCES:

[1] McNeill, S., *"Vibration Assessment of FCCU Piping"*, Proceedings of AFPM Cat Cracker Seminar, 2018.

[2] McKee, R. and Broerman, E., *"Acoustics in Pumping Systems"*, Proceedings of the Twenty-Fifth International Pump Users Symposium, 2009.

BIBLIOGRAPHY:

[1] Dashti, H.H. and Riazi, M.R., *"Acoustic velocities in petroleum fluids: Measurement and prediction"*, *Journal of Petroleum Science and Engineering*, Vol. 124, pp. 94-104, December 2014.

[2] Chadda, T.S., *"Effect of Refractory Lining on the Stiffness of Steel Pipe"*, by T. S. Chadda; *Proceedings of the Joint Conference of the Pressure Vessels and Piping (PVP)*, Vol. 53, pp. 79-88, 1981.

VIATC TRAINING CONFERENCE & EXPO
COVINGTON, KENTUCKY | AUGUST 7-9, 2024

Case Studies on Hydro Turbines with High Vibrations

Bernard F Boueri
Senior Mechanical Engineer
Ontario Power Generation
bernard.boueri@opg.com

Kevin Morrissette
Senior Mechanical Plant Engineer
Ontario Power Generation
kevin.morrissette@opg.com

Bernard Boueri is currently a Senior Vibration Engineer in the Machine Dynamics Section at Ontario Power Generation (OPG). He has 20 years experience supporting rotating equipment commissioning, diagnostic and balancing activities. In recent years he involved in the new commissioning and upgrades of the hydroelectric fleet at OPG. He was involved in the commissioning of the Upper and Lower Mattagami Projects which included the installation of 13 new units.
Bernard is a graduate from the University of Florida. He is a licensed Professional Engineer in the province of Ontario and is a certified Category IV from the Vibration Institute.

Kevin Morrissette is a Senior Mechanical Plant Engineer at Ontario Power Generation. He holds a bachelor's degree in mechanical Engineering from Carleton University. Kevin has experience in the Petro Chemical, Metallurgical industries with the last 14 years in the Power Generation industry working on turbines and generators.

ABSTRACT:

Ontario Power Generation (OPG) operates over 300 hydroelectric generators in its fleet. Some of these units have been in operation since the 1950s. These units are mostly vertical with a combined thrust and guide bearing below the generator and a turbine guide bearing. For very large hydroelectric generators there is a third guide bearing above the generator. The average rotational speed of these units varies between 100 to 200 RPM. Sources of vibrations like other machines are mechanical, electrical, and hydraulic in nature. This paper will present two cases on both old and newly refurbished units with high vibrations due to magnetic unbalance, and a rub.

KEY WORDS: Hydro, generator, magnetic, balancing, rub, stator, circularity, concentricity

Case 1: High Generator Shaft Vibrations due to Magnetic Unbalance
The unit is a 70 MW Kaplan (propeller) unit built in 1950 with a rated speed of 100 RPM. The unit has a combined thrust and guide generator bearing and a turbine guide bearing. Shaft relative displacement levels are monitored by two proximity transducers at 90° apart (Figure 1).

Figure 1: Proximity Transducers at Turbine Guide Bearing

The unit is known to have an uneven stator expansion causing a distortion in the air gap. The unit has high shaft relative displacement levels at the generator guide bearing in excess of 30 mils p-p (Table 1).

Table 1: Generator Shaft Relative Displacement Levels

Condition	Generator Y			Generator X		
WG Opening	Overall (mils p-p)	1x RPM (mils p-p)	1x Phase (deg)	Overall (mils p-p)	1x RPM (mils p-p)	1x Phase (deg)
SNL* Field Off	38	27	333	47	37	42
SNL Field On	37	33	273	36	29	35
70%	38	36	282	38	32	34
100%	35	34	282	37	31	31

***Speed No Load**

The data also showed a generator shaft centerline movement of about 20 mils between field off and field on (Figure 2), this sudden movement in the generator shaft centerline when the field is applied is an indication of an uneven air gap.

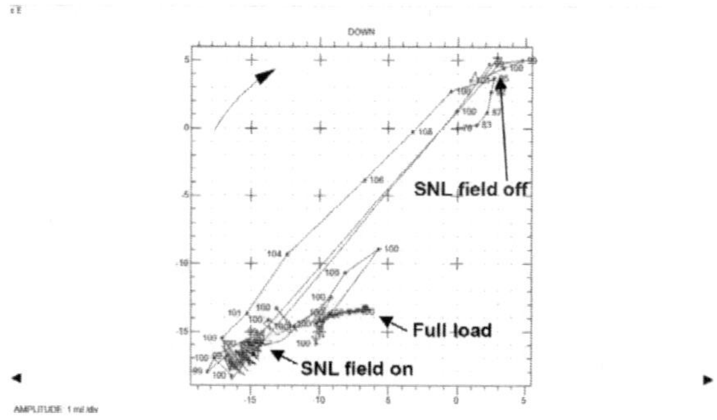

Figure 2: Generator Shaft Centerline Movement

The generator orbit (Figures 3) at full load showed a flat spot. No rub was suspected because the generator guide bearing temperatures were well below alarm and the shaft generator relative displacement trends were constant and not increasing over time as they would be in case of a rub (Figure 4). The conclusion was that the shaft centerline movement of 20 mils when the field was applied, was consuming a significant amount of the bearing clearance and causing a preload at the generator guide bearing.

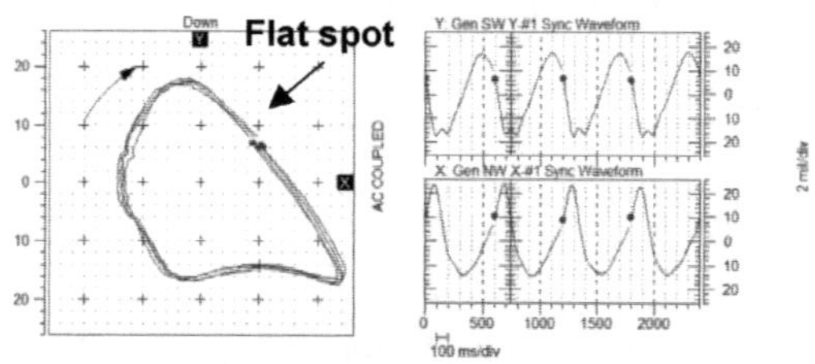

Figure 3: Generator Shaft Orbit at Full Load

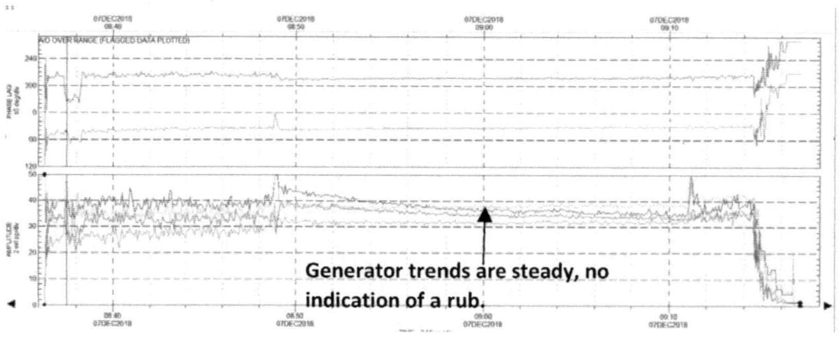

Figure 4: Generator Shaft Relative Displacement Trend

The generator shaft relative displacement spectra at full load (Figure 5) show the dominant vibration component at 1x RPM and its harmonics, the harmonics could be the result of the preload on the bearing and/or possible verticality issues with the unit due to the uneven air gap.

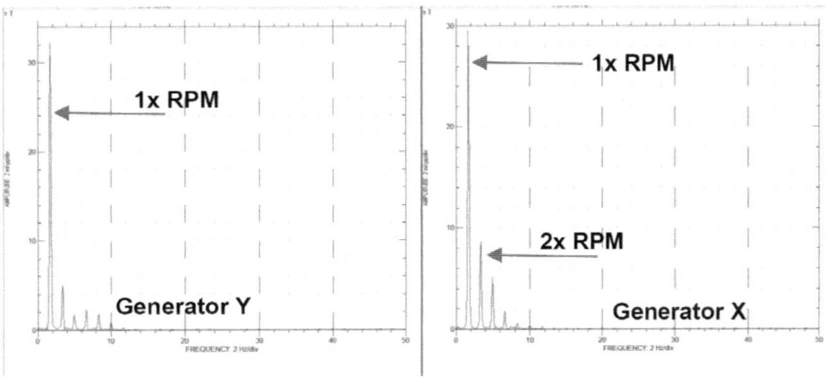

Figure 5: Generator Shaft Relative Displacement Spectra at Full Load

The operators requested that this unit be made available due to grid demand. The decision was made to do a compromise balancing of the machine in order to reduce the shaft relative displacement levels when the field is applied. A recommended weight of 400 lb. at 75° opposite rotation was installed inside the rotor rim (Figure 6).

Figure 6: Balancing Weights at the Inner Side of the Rotor

Table 2 are generator shaft relative displacement levels post balancing. The balancing reduced the generator shaft relative displacement levels once the field is applied. It was recommended that unit be synced to the grid as soon as unit reach full speed and operation be limited between 70% and 100% WG opening. Figure 7 is the generator spectra post balancing.

Table 2: Generator Shaft Relative Displacement Levels Post Balancing

Condition	Generator Y			Generator X		
WG Opening	Overall (mils p-p)	1x RPM (mils p-p)	1x Phase (deg)	Overall (mils p-p)	1x RPM (mils p-p)	1x Phase (deg)
SNL Field Off	44	38	100	48	41	174
SNL Field On	16	14	325	9	6	125
70%	17	14	352	9	5	141
100%	16	12	347	9	5	142

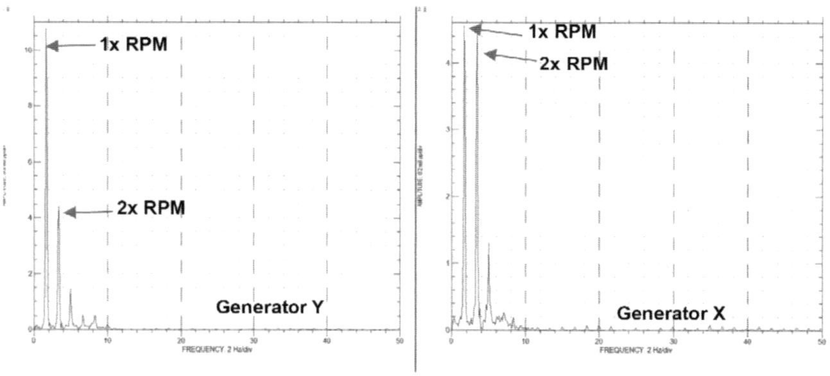

Figure 7: Generator Spectra at Full Load Post Balancing

After a bi-annual outage, site operators noted an increase in the vibrations at the generator guide bearing. A test was conducted to identify the reason for the increase in vibrations. A trend of the generator shaft relative displacement levels at all tested conditions is shown in Figure 8 (post outage).

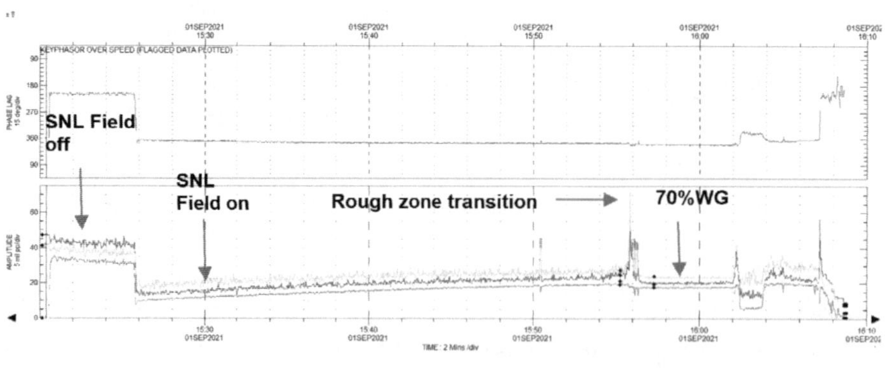

Figure 8: Generator Shaft Relative Displacement Trends Post Outage

The trend shows a high shaft relative displacement level at SNL field off, when the field is applied, the shaft relative displacement levels are reduced as expected. However, the data indicate an increase in the shaft relative displacement levels over time. Such a behavior is a possible indication of a rub. The generator shaft centerline movement when the field was applied (Figure 9) shows a movement of about 35 mils. This is an increase of 15 mils from the previous measured movement in Figure 3.

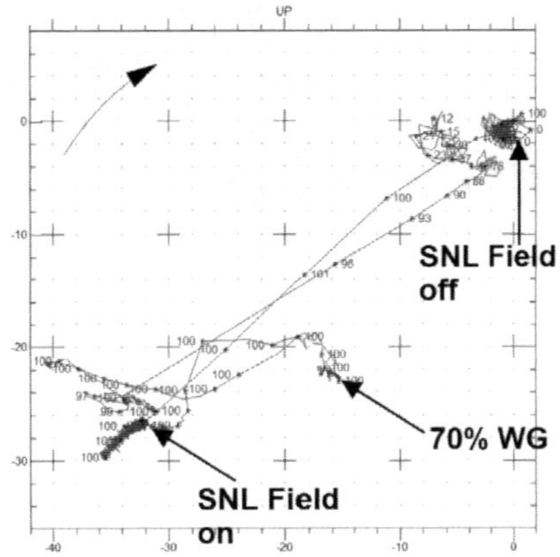

Figure 9: Generator Shaft Centerline Movement Post Outage

The generator orbits at SNL field on and off also show possibility of a rub (Figure 10). The generator guide bearing temperatures were constant and below alarm, as such a bearing rub was not suspected. However, a rub on the bearing seals is highly probable. It was recommended to declare the unit unavailable and inspect the generator guide bearing.

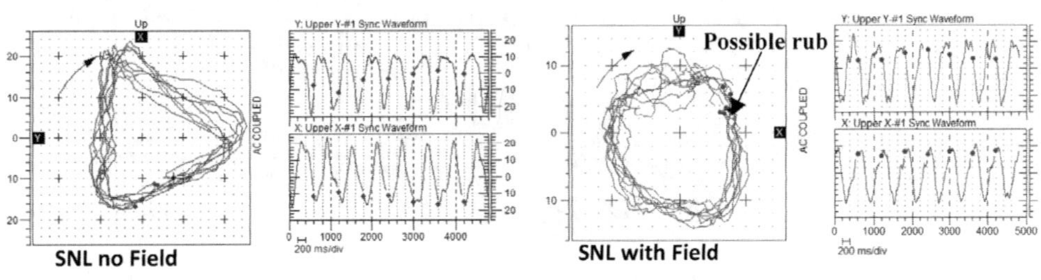

Figure 10: Generator Orbits at SNL Field off and on Post Outage

Site personnel performed inspection of the generator guide bearing (Figure 11), the diametrical guide bearing clearance was found to be 60 mils. The design clearance should be 30 mils.

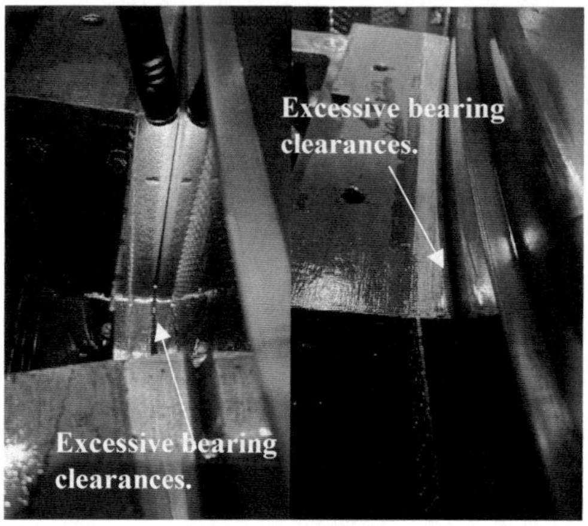

Figure 11: Generator Guide Bearing Pads Excessive Clearances

Further inspections identified several poles that rubbed with the stator (Figure 12) and caused damage to one of the air gap probes.

Figure 12: Rotor Poles with Contact Marks

During the outage it was recommended to adjust the circularity and concentricity of the stator and rotor to adjust the air gap. A pinning structure (Figure 13) was designed and welded on the splits of the stator, the structure will allow the expansion of the stator during operation and will maintain the stator original circularity when the unit is shut down and the stator contracts.

Figure 13: Designed Stator Pinning Structure

The 400 lb. that were installed prior to the air gap adjustment were removed before restarting the unit. The unit was operated at SNL and from 70% WG to 100% WG opening (Figure 14).

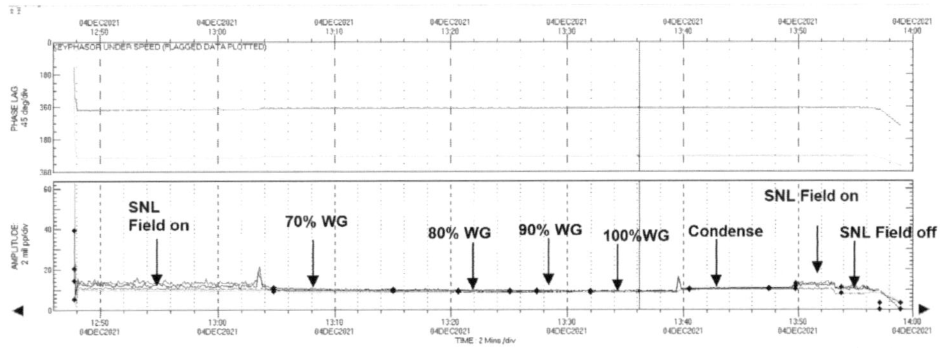

Figure 14: Generator Shaft Relative Displacement Levels after Air Gap Adjustment

Tables 3 is a summary of the shaft relative displacement levels at all tested conditions at the generator guide bearing after the air gap adjustment, the levels are now close to the recommended levels in ISO 20816-5 [1]. The overall shaft relative displacement levels are much improved. The generator shaft 1x RPM amplitude and phase are relatively constant when the field is applied. This would indicate that the magnetic unbalance has been corrected, by adjusting and re-centering the stator and rotor. Figure 15 is the generator orbits before and after air gap adjustment at full load. The generator shaft centerline movement

Table 3: Generator Relative Shaft Displacement Levels after Air

Condition	Generator Y			Generator X		
WG Opening	Overall (mils p-p)	1x RPM (mils p-p)	1x Phase (deg)	Overall (mils p-p)	1x RPM (mils p-p)	1x Phase (deg)
SNL Field off	10.9	8.0	272	10.5	8.0	3
SNL Field on	12.7	9.8	273	13.0	10.0	2
Condense	10.3	10.0	273	10.7	10.2	3
70%	9.2	8.9	276	10.3	9.4	5
80%	9.4	8.7	275	9.7	9.0	5
90%	9.2	8.6	274	9.4	8.7	5
100%	9.0	8.5	274	9.3	8.6	4

***Generator guide bearing clearance 30 mils diametrical**

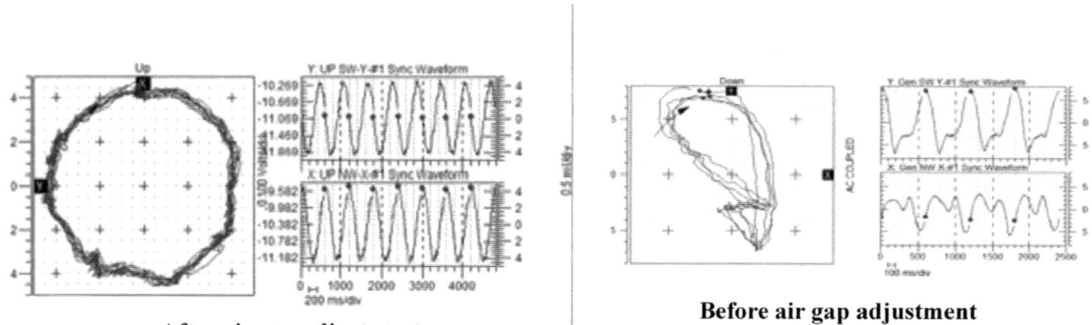

Figure 15: Generator Shaft Orbit Before and After Air Gap Adjustment at Full Load

Figure 16 is the generator shaft centerline movement after the air gap adjustment. The data indicate a shift of about 1 to 2 mils in the generator shaft movement when the field is applied on the unit. This is a significant improvement as compared to the generator shaft centerline movement of about 35 mils measured in Figure 9.

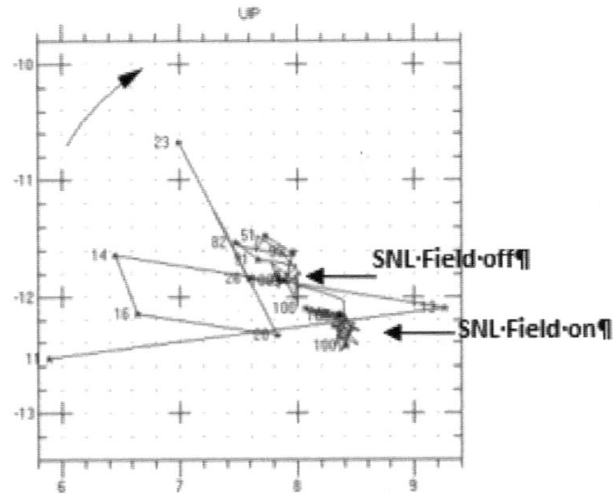

Figure 16: Generator Shaft Centerline Movement

Case 2: High Generator Guide Bearing Vibration.
The unit is part of a powerhouse consisting of 2 Westinghouse Francis turbines with rated power of approximately 85 MW at a speed of 100 RPM. The unit underwent a generator refurbishment and vibration support was required by plant personnel prior to return to service. Two proximity probes were temporarily installed at 90° apart at the generator and turbine guide bearings. Figure 17 shows the turbine transducers. A temporary optical tachometer was installed on the turbine cover in line with one of the proximity probes to measure rotational speed. A reflective tape was installed on the shaft in line with the rotor leads.

Figure 17: Proximity Probes on Turbine Guide Bearing

The unit was operated at speed no load (SNL) without field to investigate its mechanical behavior. Data indicated that generator shaft relative displacement levels are trending upward with time (Figures 18). This behavior is an indication of a possible rub. Both guide bearing temperatures were steady and below alarm, this indicated that the rub is most likely at the bearing seals.

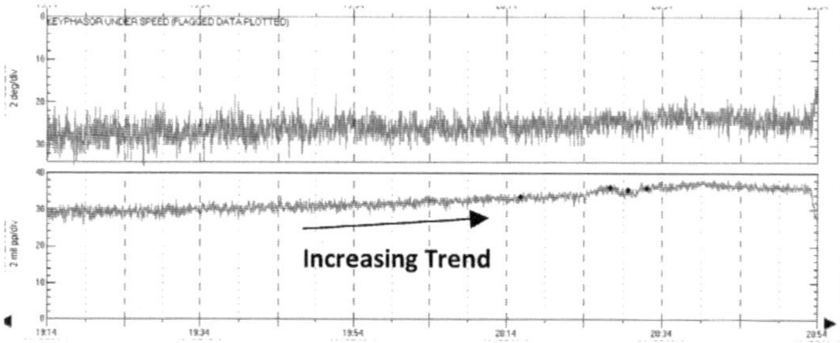

Figure 18: Generator Shaft Relative Displacement Levels SNL Field off

In addition, the generator shaft orbit (Figures 19) showed a restriction indicating a high possibility of a rub. It was recommended to inspect the generator guide bearing seals and verify that all clearances were set properly.

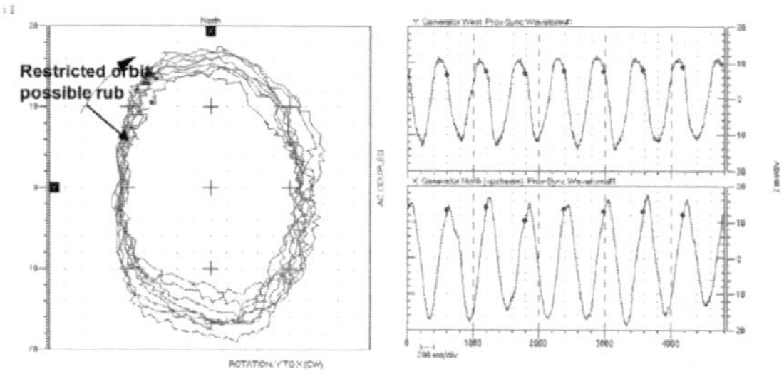

Figure 19: Orbit and Time Waveform of Generator Shaft at SNL Field off

Inspection of the upper generator bearing indicated a rub at the upper seals as shown in Figure 20. Clearances were found to be set incorrectly. The upper seals were readjusted, and additional testing done.

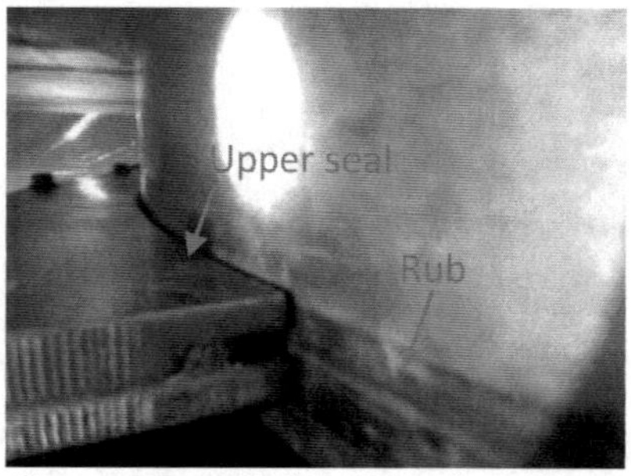

Figure 20: Generator Guide Bearing Upper Seal Rub

After adjusting the upper seal, the overall generator shaft relative displacement levels at SNL field were constant. However, when the field was applied and the unit was loaded, it was noted that the shaft relative displacement trend was still increasing over operating time (Figure 21). The generator shaft orbit at full load shows possible additional rub at a different location (Figure 22).

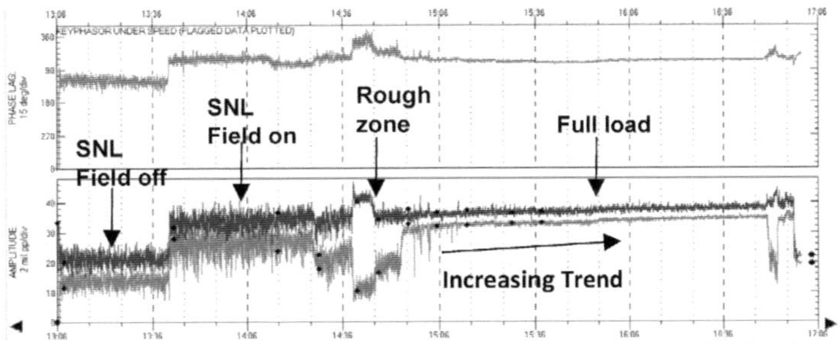

Figure 21: Generator Shaft Relative Displacement Levels during Loading

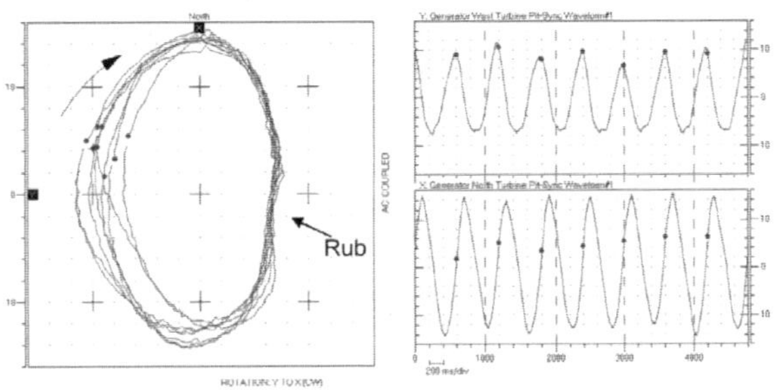

Figure 22: Generator Shaft Orbit at Full Load

This required the disassembly of the generator guide bearing for further investigation to identify the cause of the rub. Once the generator guide bearing was disassembled a severe rub was noted on the shaft at the location of the lower seals, the rub was severe enough to cause discoloration of the shaft.

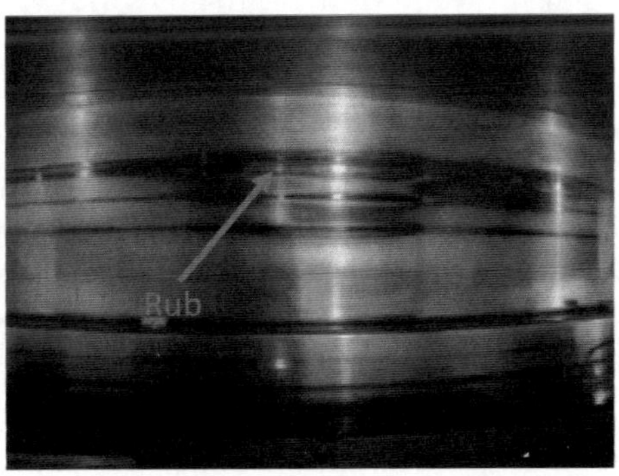

REFERENCES:

[1] ISO 20916-5 Mechanical vibration – Measurement and evaluation of machine vibration – Part 5: Machine sets in hydraulic power generating and pump-storage plants. 2018

VIATC TRAINING CONFERENCE & EXPO

COVINGTON, KENTUCKY AUGUST 7-9, 2024

INVESTIGATING A THRUST BEARING REPEAT FAILURE ON A BOILER FEED PUMP

James Chin Yut
Senior Technical Specialist - Bruce Power
Tiverton, Ontario, Canada

James Chin Yut, P Eng, is employed at Bruce Power as a Senior Technical Specialist. Current duties include the life cycle management of rotating equipment- large pumps and motors. His career experience spans various roles in maintenance, inspection and engineering in aviation, petrochemical and nuclear power generation. Mr. Chin Yut is a graduate of the University of the West Indies with a B Sc. – Mechanical Engineering (Hons) and holds a Cat III certificate with the Vibration Institute (VI) and Canadian Machinery Vibration Association (CMVA).

ABSTRACT:

This case study will focus on a repeat failure of the thrust bearings a Bruce Power Bruce A Main Boiler Feed water pump (MBFW) the investigation, findings and rebuild.
4-43230- P3 was refurbished under the asset management project to address the end-of-life issues and installed with a new motor in 2020. The new motor required a transition base between the motor base and original baseplate due to its smaller size. Post installation the pump set experience increased vibration attributed to flexibility of transition base.
In May 2022, during a pump start up, the pump experienced the first failure of the thrust bearings when the inboard thrust bearing temperature reached 220C, findings was inconclusive, and the pump returned to service for 50 days before duty swapped. The transition base was replaced in the next planned outage in 2022 and operated for 6 days before being duty swapped. On Jun 2023, the pump was started in recirculation and the inboard thrust bearing temperatures climbed to 107C before being shut down. Case study will highlight the findings after the 4-43230-P3 was disassembled, inspected and steps taken to prevent recurrence.

KEY WORDS: Large horizontal pump, boiler feed water, Kingsbury thrust bearings, Byron Jackson DVMX, cavitation

SECTION 1: Pump data

The boiler feed water system transfers feedwater from the deaerator to the preheater and boilers to produce steam for the turbine for electrical production. The boiler feed water system consists of 3 X 50% main boiler feed water pumps (MBFW) and 1X 3% auxiliary boiler feed pump (ABFP).

4-43230-P3- Main Boiler Feed Water Pump (MBFW) pump was removed from service due to high thrust bearing temperature, this resulted in reduction in redundancy of the number of U4 available MBFW pumps with the potential impact to production.

Pump is a Byron Jackson 10 X12 X 18 – 2 stage – DVMX, rated flow and pressure 8600 IGPM @ 703 psig driven originally by a Westinghouse, 5300 HP, 3580 rpm, 4000 V motor.
Casing material ASTM-351-CA15.
These pumps were originally commissioned in the 1970s, had a history 1st stage impeller cavitation damage which prompted a redesign in the 1990's to address the cavitation issues.

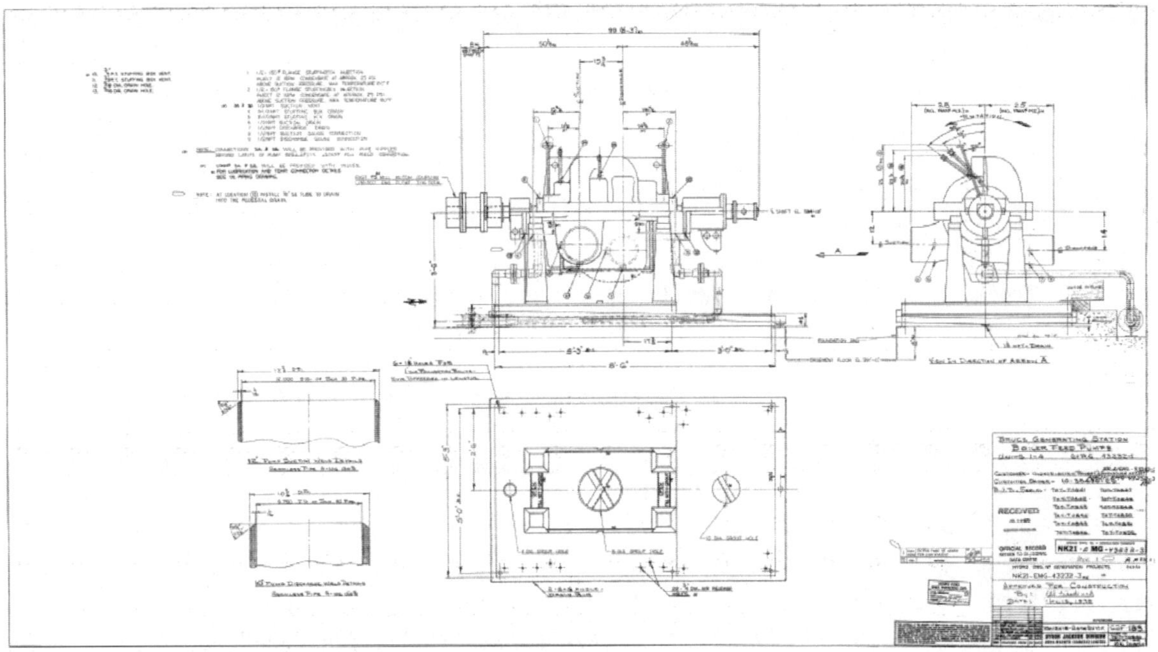

Figure 1: Pump Arrangement

SECTION 2: Failure history

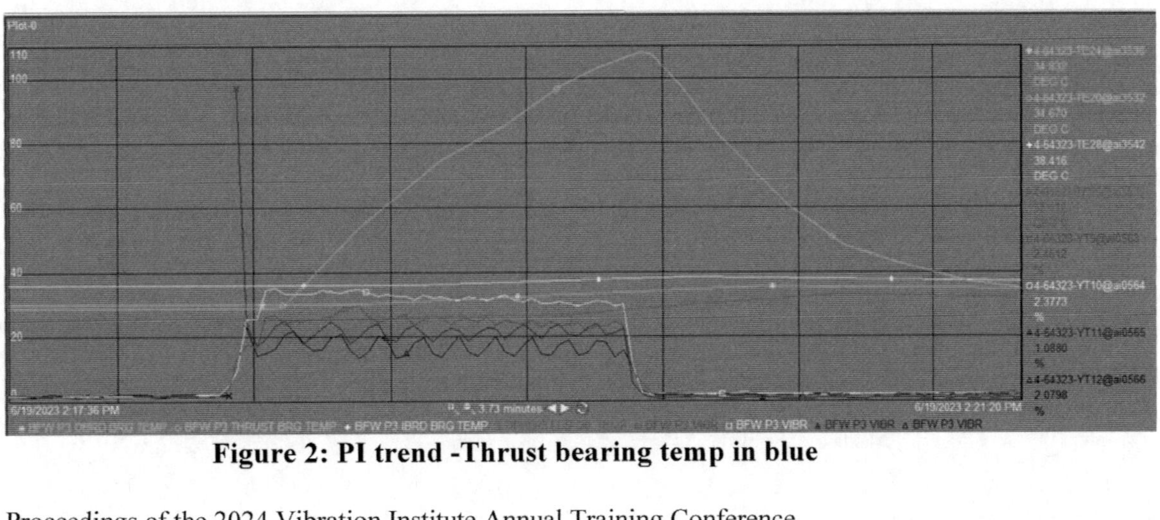

Figure 2: PI trend -Thrust bearing temp in blue

4-43230- P3 was refurbished under the asset management project to address the end-of-life issues and installed with a new motor in 2020. The new replacement Toshiba motor is physically smaller and requires a transition base between the motor and original baseplate. Post installation the pump set experience increased vibration attributed to flexibility of transition base.

By Mar 2021, the pump had operated for approximately 300 days before being duty swapped.

In May 2022, in the lead out from an outage, during a pump start up, the pump experienced the first failure of the thrust bearings when the inboard thrust bearing temperature reached 220C, findings was inconclusive, and the pump returned to service for 50 days before duty swapped.

The transition base was replaced in the next planned outage in 2022 and returned to service for 6 days before being duty swapped. In Jun 2023, the pump was started in recirculation and the inboard thrust bearing temperatures climbed to 107C before being shut down.

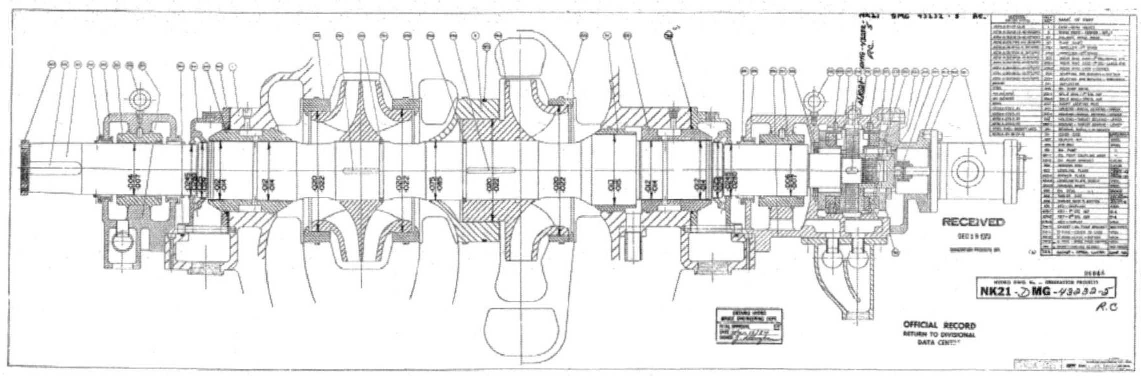

Figure 3: Pump cross section

Internal process for significant component failures requires the completion of an Equipment Failure Checklist (EFC). The EFC is an internal investigation tool that seeks to identifies gaps in process, procedures, design, supply chain and practice under subsections- prevention, detection and correction. Forensic analysis and 5 Whys method was used to complete the EFC.

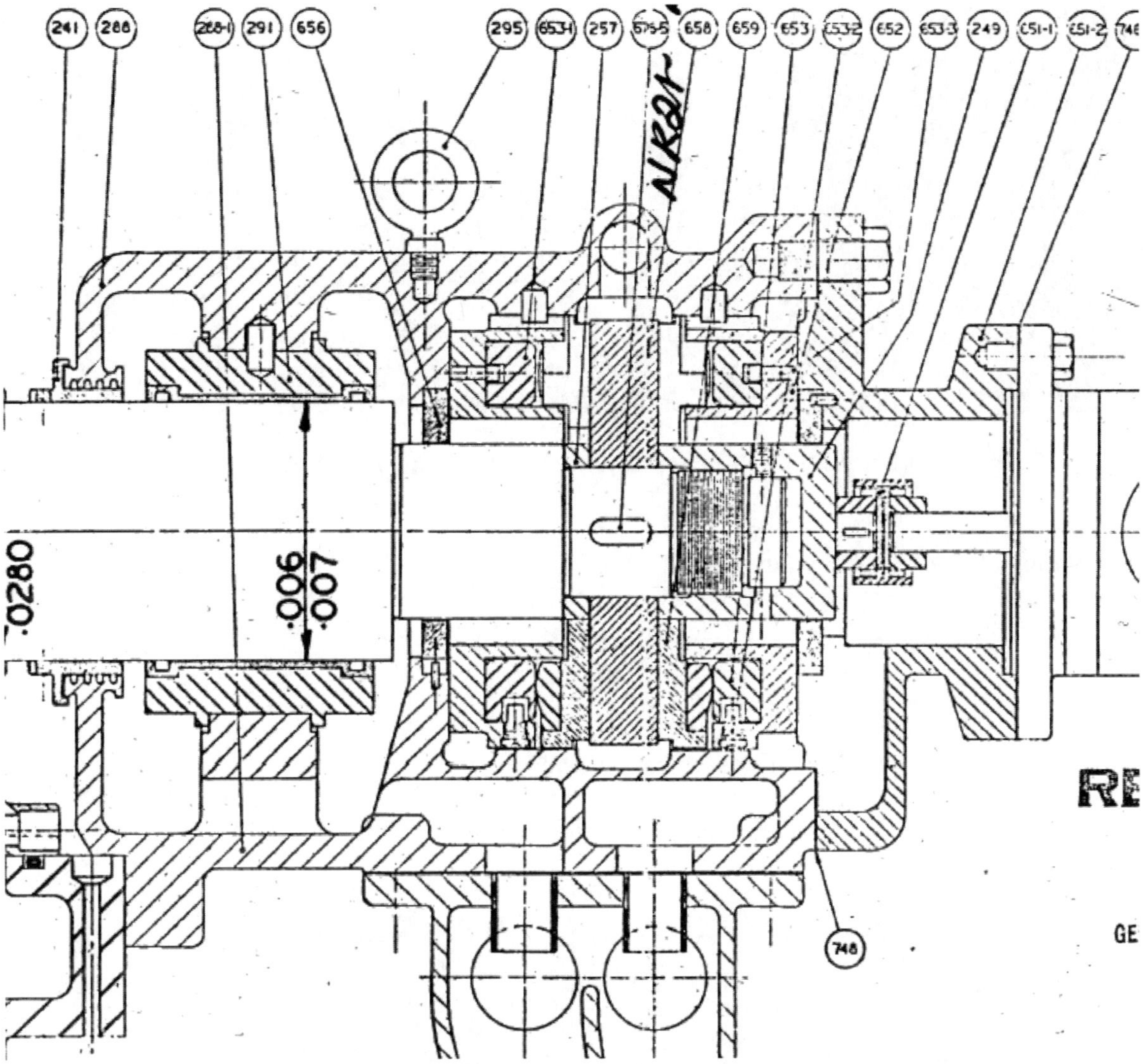

Figure 4: Thrust bearing detail

SECTION 3: Disassembly – initial findings

When the thrust end bearing cover was removed, the inboard thrust bearing was found wiped. Initial inspection results:

- Inboard thrust bearing shoes was wiped
- Thrust disc contact face incurred surface damage
- Outboard thrust bearing shoes were in acceptable condition

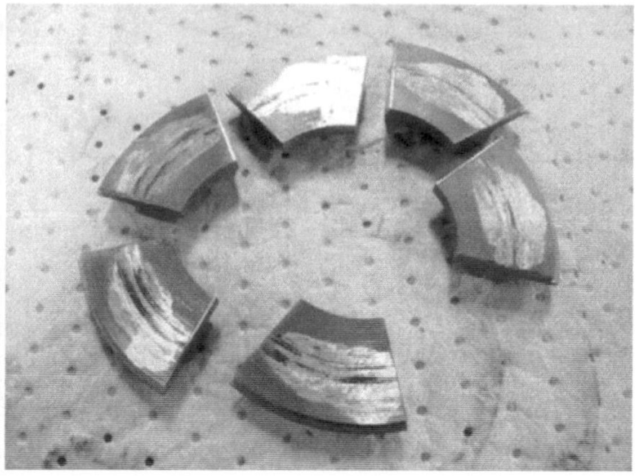

Figure 5 and 6: Thrust bearing location and failed bearing

The rotating element on this pump has a double suction 1st stage impeller and single suction 2nd stage impeller. To rule out any internal issues as the cause of the thrust bearing failure the pump cover was removed, and the rotating element and internals inspected.

Internal inspection results:

- 1st stage impeller had signs of cavitation damage
- On one inboard suction vane a small section of 1" x 1.5" was missing from leading edge
- Splitline and casing was in acceptable condition

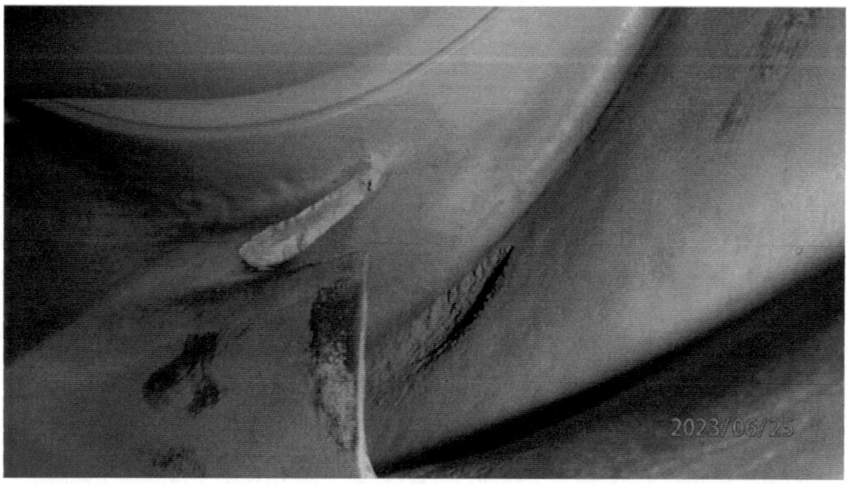

Figure 7: Vane missing section

SECTION 4: Review of findings

A general guide to the principles, operation and troubleshooting of hydrodynamic bearings by Kingsbury list the possible causes and conditions of overheating applicable to our circumstances:

- Inadequate lubrication supply
- Interrupted fluid film
- Poor thrust disc surface finish
- Insufficient bearing clearance
- Excessive load

From the list the two most probable causes are overload of the thrust bearing beyond its design capacity or lack of lubrication. These conditions can be independent or occur at the same time.

Initially, it was thought that the resultant pump thrust was away from the drive end of the rotating element, making the outboard thrust bearing shoes as the active contrary to the findings. Pump thrust calculation performed by a vendor supporting the repairs indicated that the thrust direction was flow dependent and resulting direction switched through the operating range. The thrust bearing capacity is 9000lbs in either direction.

The OEM and the third part vendor did not agree on positive thrust direction at full flow. However, the important take away is that the thrust bearing capacity was greater than the maximum calculated loads across the pump operating range.

Physical evidence indicated that the inboard thrust bearing pads and the mating thrust disc surface was wiped, because of overheating. On disassembly and inspection, it was found that one levelling plate was stuck and the oil seal ring was measured as 0.012" oversized on the failed thrust bearing side.

The stuck levelling plate can directly affect the supply of adequate lubrication or disruption of the oil film leading to the 4-43230-P3 thrust bearing failure. The oversized oil seal ring effect on the failure was thought to be minimal. Foreign material and interference with the base ring caused the levelling plate to stick, removing the foreign material and the small interference restored the functionality of the levelling plate. There is no documentation of a stuck levelling plate at previous overhaul, so it is unknown whether this anomaly existed or grew worse since the last overhaul.

This Kingsbury model thrust bearing design utilizes a base ring and levelling plates to support and distribute loads from the thrust pads. The thrust pads transfers load to an upper levelling plate. The upper levelling plate transfers the load to the two adjacent lower levelling plates which articulates and distributes to the adjacent pads. A sticking levelling plate inhibits the ability of the thrust bearing functioning normally.

Figure 6: Thrust bearing housing and levelling plate shoes

Visual inspection of the failed thrust pads indicates material loss of the babbitt mostly in the center of the pads but there is some conflicting evidence on two pads with respect to direction of babbitt smear. The pump rotates clockwise from the driver end so the inboard thrust shoes should have the babbitt smeared in the counterclockwise direction and the outboard in the clockwise direction. The damaged pads were cleaned, so not in as found state when inspected. The buttons at the back of the pads wear spot did not indicate excessive wear.

This pump in position 4-43230-P3 was refurbished and had a history of higher audible noise during operation but the vibration during full flow operation was acceptable to operate.

Since 2020, this pump had limited operating hours from its status as a non preferred pump due to high vibration stemming from an error in construction of the transition base. This transition base was replaced in 2022 which allowed the pump to operate without the restriction.

To facilitate internal inspection, the pump top cover was removed to determine whether excessive wear contributed to a high load on the thrust bearing. Internal clearances was verified within tolerance and dispositioned as acceptable by the OEM. The rotating element the 1st stage impeller had a missing section approximately 1" x 1.5" on the leading edge of one of the vanes. The damaged vane is located on inboard suction side of the double suction impeller, cavitation damage was also observed on one vane on the outboard suction. The pump split line was in good condition with no gasket loss and a clean machined surface finish leading to the conclusion that there was no interstage recirculation.

An anomaly was discovered with the running location of the pump in the casing. The location ring used to centralize the rotating element on 4-43230-P3 was 0.341" on existing vs 0.502" using new rotating element. For comparison as sister pump, 4-43230-P1- location ring thickness from rebuild records 0.489" was thick. This suggests that in the original rebuild that the rotating element was operating at 0.160" off the centerline. The rotating element position is dictated by the 2nd stage impeller discharge and if not centralized can result in variable thrust loading especially in low flow conditions.

The pump has low run time since installation and the damaged vane and other cavitation damage suggest that the audible noise was most likely 1st stage cavitation. With the double suction design, cavitation can interrupt the flow at the suction and if severe result in thrust reversal and shuttling of the shaft. Shuttling would also contribute to the observed failure.

There was an anomaly with the as found alignment between the pump and motor compared to the record as left data on the previous overhaul which was rechecked during reassembly.

The stuck levelling plate on the inboard thrust bearing, off centered position of rotating element, pump operating in cavitation, and pump / motor alignment are all credible contributing causes to the failure.

The failed impeller and thrust bearings have been sent for further analysis, at the time of the writing of this paper the no results of those investigations were available.

The EFC was completed, and the following actions created to close the identified gaps:

- Revise the Mechanical Maintenance Procedure (MMP) to include to include explicit instructions for inspection of thrust bearing shoes, levelling plates and supporting ring and criteria for replacement. Also, clearer instructions for the centralization of the rotating element in the casing.
- Revise operating procedure on routine pump swaps to operate auxiliary oil pump for minimum 5 minutes before initiating start.
- Review the failure analysis report for thrust shoes and 1 stage impeller, revisit the EFC corrective actions and create new actions if gap identified.

SECTION 5: Installation new rotating element and post maintenance test

Pump was reassembled with new rotating element and commissioned. Attached are the vibration spectrum at the pump drive end (DE) and non drive end (NDE)post maintenance.

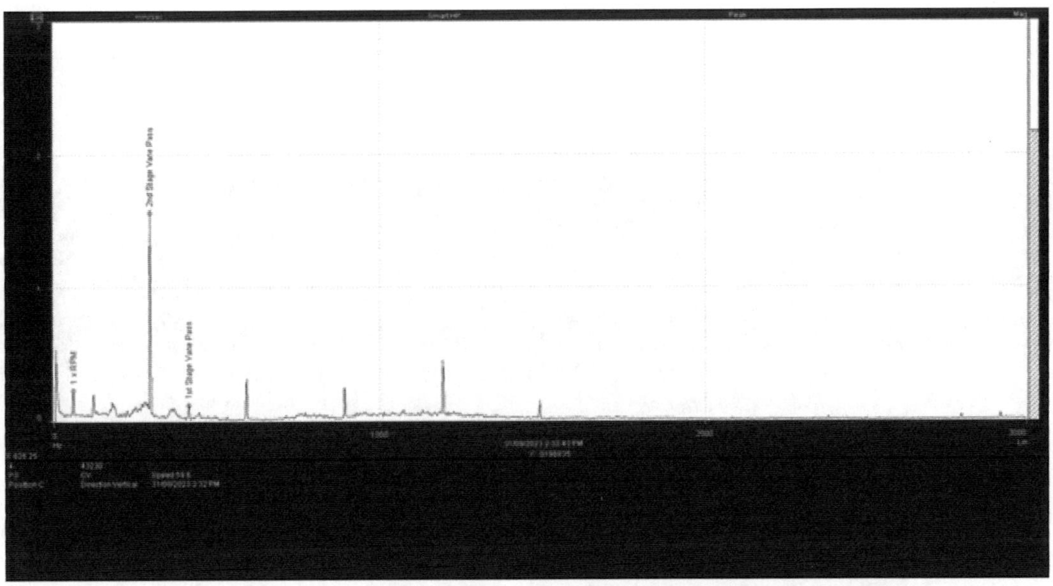

Figure 7: CV- Pump DE- Vertical mm/s

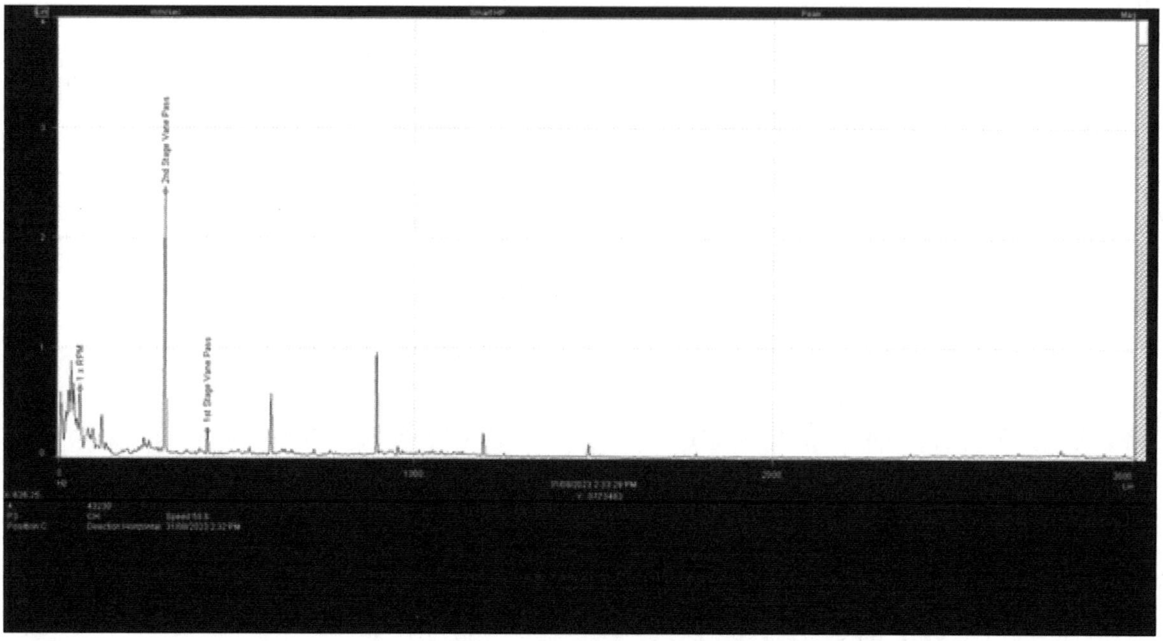

Figure 8: CH- Pump DE horizontal- mm/s

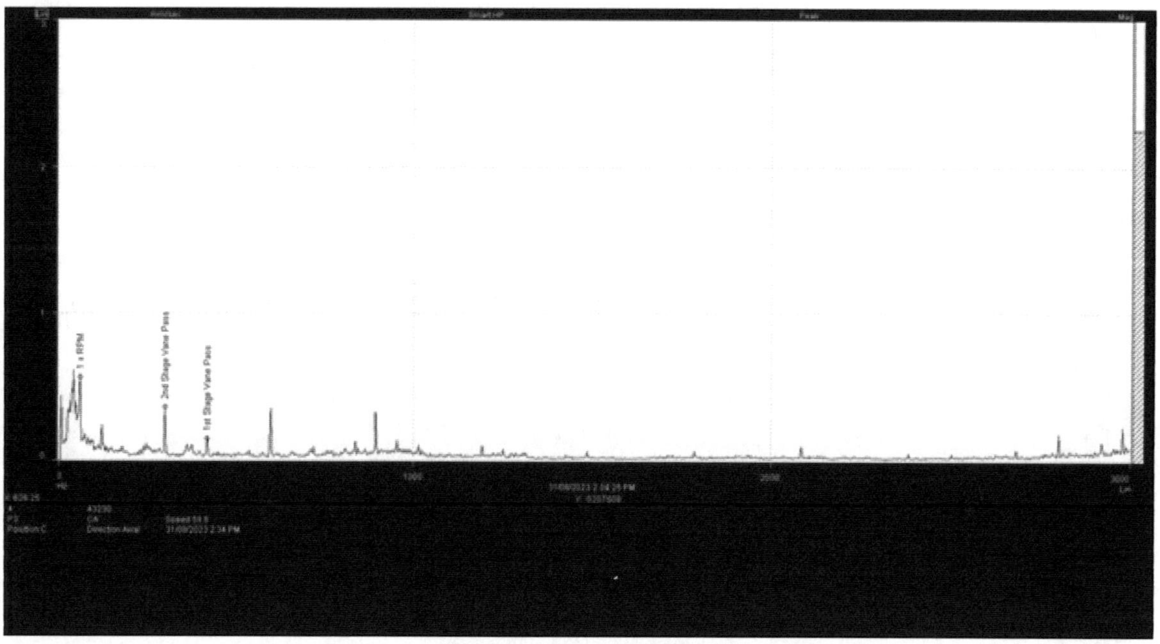

Figure 9: CA- Pump DE axial- mm/s

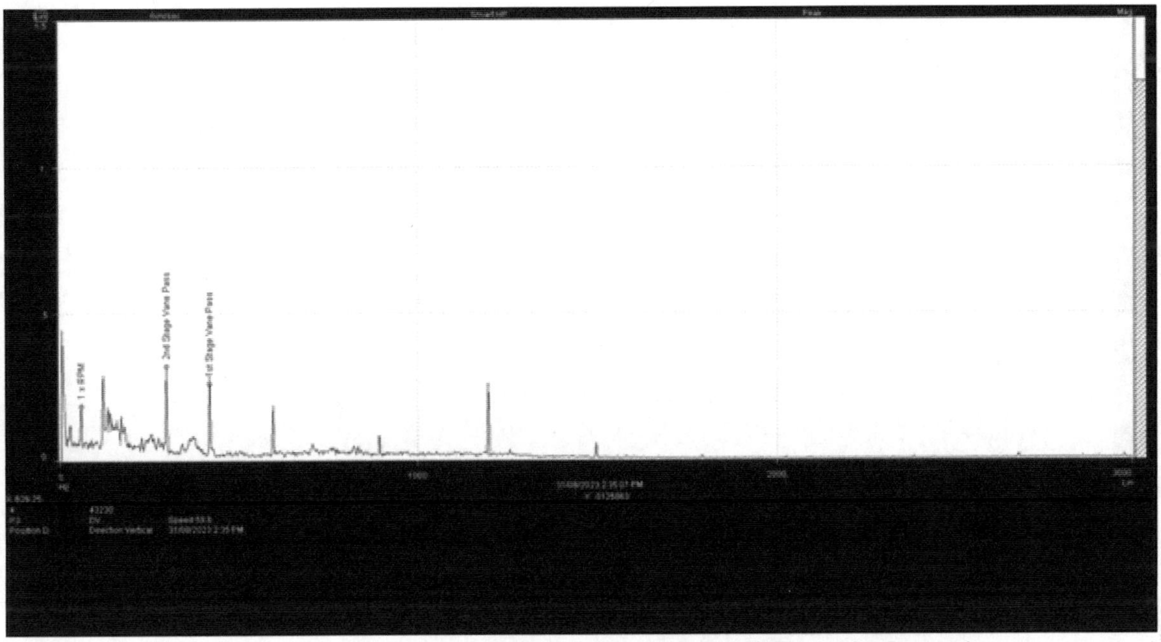

Figure 10: DV- Pump NDE vertical- mm/s

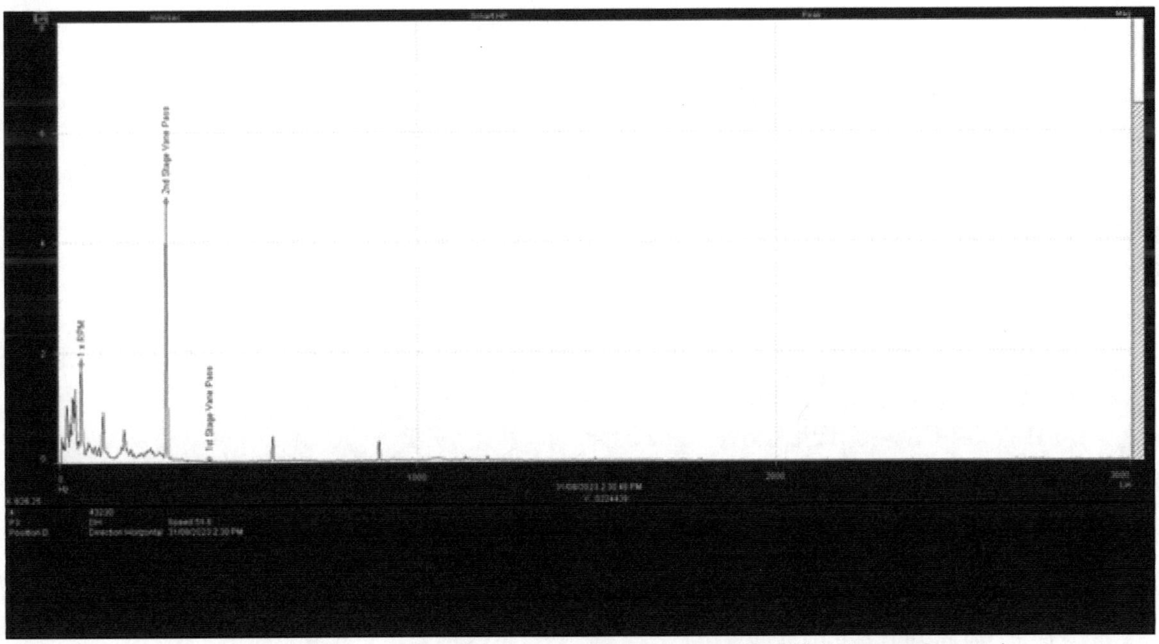

Figure 11: DH- Pump NDE horizontal- mm/s

Figure 12: DA- Pump NDE axial- mm/s

AKNOWLEDGEMENTS:
Hydroaire, Bruce A maintenance, Umar Afzaal, Sean Zhen, Jason Dales, Geoffrey Lawson, David Durance, Zbigniew Komosa

REFERENCES:

- A general guide to the principles, operation and troubleshooting of hydrodynamic bearings- Kingsbury
- Equalizing Thrust bearings- comprehensive design guide- Kingsbury
- Engineering evaluation – Riley Murnaghan – 2Jun 2022
- FMEA 4-43230-P3 Thrust Bearing Failure
- NK21-OM-43230-04.09
- NK21-OPM-20-4821A
- NK21-MMP-43230-00001
- NK21-DM-43230
- NK21-DMG-43232-00005
- NK21-FS-43000-00001-U4
- NK21-FS-43000-00003-U3
- NK21-DE-64323-000009-U4
- NK21-EMG-43232-00007-U4

BIBLIOGRAPHY:
NK21-EMG-43232-00003- Byron Jackson- C2F-183
NK21-DMG-43232-00005- Byron Jackson- C1F-306

VIATC TRAINING CONFERENCE & EXPO
COVINGTON, KENTUCKY | AUGUST 7-9, 2024

ADVANCED ANALYTICS SUPPORT MALFUNCTION DETECTION FOR RECIPROCATING COMPRESSORS

Fayyaz Karim Qureshi
Senior Technical Leader for Reciprocating Compressors and Analytics
Bently Nevada, a Baker Hughes business
fayyaz.qureshi@bakerhughes.com

Fayyaz Karim Qureshi is the Senior Technical Leader for Reciprocating Compressors and Analytics for Bently Nevada looking after Middle East, India, Africa and Asia. He is responsible for commissioning asset health monitoring and advanced analytics solutions for reciprocating machines around the world. In addition, Fayyaz performs diagnostics and advises customers on proactive maintenance of reciprocating compressors. As part of his job, Fayyaz retrofits existing rotating and reciprocating assets with sensors and protection systems.

ABSTRACT:

Amid of industries' concentration towards cost optimization to have an improved bottom line, the optimum resource utilization is inevitable. This requirement has prompted the imminent need for automated systems that can help identify developing issues autonomously and resources can concentrate on resolving the issues rather than spending hours in finding problems. Today, there is a shortage of in-house skilled industrial expertise, especially in narrow-but-deep areas of knowledge, such as machine condition monitoring. During the next 5-10 years, an estimated 50% of the experienced workforce will be retiring, taking their knowledge and expertise with them unless codified in a capable digital platform. Integration of human, and artificial intelligence can form a consolidated approach towards effective machine condition monitoring.

Large amounts of data from field instrumentation installed on machine, its auxiliaries and process streams are flowing into asset health monitoring platform's historian. Powerful logic engines are designed to process the data in real-time through pre-defined diagnostic rules (analytics) and pinpoint towards the failing components on machines providing actionable insights. In this paper, techniques will be presented on how diagnostic strategies are embedded within automated diagnostics platform using physics-based calculation models which provide actionable information about failing components to operate assets more efficiently. Use cases will be presented demonstrating the end-to-end workflow of Asset Health Management and value realization.

KEY WORDS: Reciprocating Compressors, Condition Monitoring, Analysis and Diagnostics, Analytics, Automated Diagnostics.

INTRODUCTION:

Medium to large scale industries are experiencing great competition to be more productive while minimizing expenses and maximizing resource utilization, which means they need to increase reliability and availability of their assets while reducing operational and maintenance costs. There is no one-fit-for-all solution to achieve this target but at least on enabling asset health monitoring, there is consensus to have 2-tier monitoring approach. Large amounts of IIoT (Industrial Internet of Things) -related data from field instrumentation installed on machines, their auxiliaries and process streams are flowing into data storage and visualization systems for asset health monitoring platform's historian.

The first tier monitoring is to identify any anomalies within the data for different operating conditions; systems are typically designed to baseline the data trends for various load conditions and define statistical alarm boundaries. While these anomaly detection systems scan through large streams of continuous trend data and raise the hand wherever certain parameter(s) start deviating from baseline values; yet these tools are still presenting annotations about a certain measurement presenting abnormal behavior from typical trends (which can sometimes be also due to a sensor malfunction). Mostly these are not able to pinpoint towards the root cause of this anomaly. Moreover, the fine tuning of these models is a continuous process; following every major overhaul and minor maintenance activities, the models must be re-trained or at least fine-tuned to get valid results.

The end users are mostly interested in knowing the failing component on the machine or abnormality in the process conditions. This is where a second tier monitoring comes into play wherein the static and dynamic data are reviewed in detail for that machine, auxiliary or process and compared against defined data signatures pertaining to particular failure modes. This condition monitoring scope has historically been human resource intensive, requires subject matter knowledge and primarily time consuming. With the advancements in the field of automation and AI/ML applications, attempts have been made to minimize the manual intervention so the experts can invest their time on more vital tasks e.g. root cause analysis (RCAs), recurring events' mitigation strategies and defect elimination. Some automated diagnostics tools are designed which have the capability to do first principle physics-based analytics [1] on static and dynamic data where the subject matter knowledge is programmed into logics and systems are able to pinpoint towards failing components (rather than just highlighting an anomalous measurement).

In this paper, techniques will be presented on how diagnostic strategies are embedded within such automated diagnostics platforms (Decision Support Analytics from Bently Nevada will be taken as an example) using physics-based calculation models which provide actionable information about failing components to operate assets more efficiently.

The following use cases will be presented to showcase the functionality of these tools, followed by validation of alarms using conventional diagnostic methodology of reviewing static data and dynamic waveforms, to demonstrate the end-to-end workflow of Asset Health Management:

1. Valve leakage detection in early stages using autonomous triggers.
2. Valve stiction was detected as soon as one chamber of a cylinder was unloaded which prompted the need to re-evaluate the optimal cylinder lubrication rate with OEM.
3. The effect on productivity using these tools will be demonstrated whereby now the Reliability Engineer spends just10 mins on reviewing the health status of components on a 6-throw machine rather than 4 hours spent previously looking through the data to identify problems.

MOST COMMON FAILURE MODES ON RECIPROCATING COMPRESSORS

Mechanical Issues

Valve Failure: Valves are generally responsible for the highest maintenance costs [2] for a reciprocating compressor. To detect this failure, analytics can be created using the temperature measurement of each valve and the flow balance (ratio of suction capacity to discharge capacity) for each acting chamber of the cylinder. The flow balance will be compared with its baseline and generate a notification of suction and discharge valve leakages if the flow balance exceeds or falls below its baseline respectively. In addition, analytics compare the temperature of each valve with the average (suction/discharge) valve temperature for that compression stage. If the temperature of the valve is higher than the average temperature by the predefined threshold, that valve would be identified as malfunctioning. *Note*: while calculating the average (suction/discharge) valve temperature for each compression stage, the chamber loading condition must be checked; valve temperatures belonging to unloaded chambers must not be accounted for while calculating the average valve temperature.

Valve Stiction: Valve stiction contributes to the indicated horsepower loss which becomes aggravated on lubricated cylinders in the events of over lubrication and in cases when the process gas is not clean. Due to over lubrication of cylinders, the lube oil gets accumulated in the valves which results in the valve plate sticking to the seat. A higher differential pressure is required to open the valve which causes higher impulse impacts resulting in valve degradation. To detect this malfunction, the dynamic cylinder chamber pressure measurements (suction, discharge, minimum and maximum pressure) are used as inputs to the analytics. Stiction in suction valves (or discharge valves) can be detected by identifying the undershoot (or overshoot) in a measured dynamic pressure curve and by comparing the suction pressure and the minimum pressure (or maximum pressure and discharge pressure) throughout the crank rotation. As these differences increase beyond predefined thresholds, a valve stiction notification is generated.

Crosshead Pin Overloading: It is important to monitor the crosshead pin as this component is the weakest link in the running gear which can experience excessive/overload forces during severe cylinder trim leakages. Automated analytics continuously compare the peak rod compression and peak rod tension (parameters calculated from online cylinder chamber pressure measurement) with the maximum allowable combined rod load in compression and tension as per the compressor datasheet. Simultaneously, the degrees of rod load reversal (calculated from online cylinder chamber pressures) is also compared with the minimum degrees of rod reversal for each load condition. In cases where crosshead pin temperature is available, it is also used in this diagnostic rule providing supplementary indication of a crosshead pin in distress.

Pressure Packing Leak: Packing rings at the stuffing box tend to deteriorate over time and may start to leak. To detect a pressure packing leak, analytics use the packing case and/or vent line temperature as inputs and generate notification as the temperature exceeds the predefined setpoint. In addition, the flow balance in the crank end chamber is also considered. If both measurements are available, analytics are enhanced by ANDing the two logics to make the malfunction detection more concrete.

Piston Ring Leak: Piston rings provide the sealing between the crank end and head end chambers. These can get cracked or broken, resulting in a gas flow between both chambers and increased discharge gas temperature. Analytics are designed to generate a notification of piston ring leak if discharge gas temperature exceeds its pre-defined threshold level AND the flow balance in both crank end (CE) and head end (HE) chambers exceed the baseline or predefined limit simultaneously.

Liquid/Debris Ingestion: It is vital to detect liquid/debris ingestion as soon as accumulation starts in reciprocating compressors since continued operation of a compressor with considerable accumulated quantities of liquid can cause expensive running gear damage. Crosshead and/or cylinder acceleration measurement is banded referenced to crank angle within the monitoring system. Bands near top dead center (TDC) and bottom dead center (BDC) can be used to identify liquid ingestion in the HE and CE chambers respectively. Analytics use crosshead and/or cylinder acceleration as an input and compares these bands with predefined limits to detect liquid or debris ingestion in the cylinder chambers.

Machine Power Overloading: To detect machine power overloading, analytics are devised to sum up the indicated horsepower of all compression chambers and compare it with the lowest of motor rating or compressor frame rating. A notification will be generated if the value exceeds a certain percentage of the design limit. The total machine indicated horsepower is also displayed on Health and Performance HMI (Human Machine Interface) for statistical information and to estimate the energy cost per unit volume throughput.

Foundation, Misalignment or Structural Issue: Frame vibration amplitudes are used to detect foundation, misalignment, or structural issue. Analytics compare frame velocity with pre-defined limits and generate a notification in case the value exceeds the setpoint.

Piston Rod Failure: Piston rod peak-to-peak displacement is the only measurement that can provide an early indication of an impending piston rod failure [3,4]. The rule logic is written where the piston rod vibrations in both vertical and horizontal directions are compared against predefined limits and a notification is triggered if the value exceeds the setpoint.

Worn Rider Bands: One of the primary monitoring parameters for maintenance planning is the current piston-to-liner bottom clearance/standout. Advanced analytics use gap voltage change, scale factor of the vertical probe installed on the pressure packing flange and mechanical dimensions to detect the piston drop from its reference/zero position and calculate the remaining piston-to-liner bottom clearance. It generates an alarm if the value falls below the OEM recommended minimum clearance.

Defective Stepless Unloader: In case of a malfunctioning stepless unloader, the difference in the volumetric efficiencies in both CE and HE chambers will not be consistent; rather, the efficiencies will start to diverge [5]. The difference in volumetric efficiencies in HE and CE chambers is compared against the baseline difference in analytics; as it exceeds the predefined limits, a notification is triggered.

Damaged Crankshaft Main Bearings: Analytics are built to compare main bearings' temperatures with the predefined limits and generate an alarm if the values exceed the alarm setpoint. This can be either because of lubrication problems or due to excessive misalignment resulting in rubbing.

Instrumentation Issues

Pressure Transducer Drift: Overheating or overcooling of a pressure transducer can result in a bias shift in the measured pressure waveform which can eventually lead to incorrect performance calculations and diagnosis. Analytics are devised to compare the suction pressures from both CE and HE pressure transducers. As the difference between the two starts to increase, the transducer with a higher suction pressure is identified and a notification is generated as the likely faulty transducer.

Process and Auxiliaries Issues

Advanced analytics can also be used to detect process and auxiliary related issues. Rule logics are developed to detect malfunctions such as:

- Choked lube oil filter: using the differential pressure across the filter
- Frame lube oil system failure: using the frame lube oil level, the frame lube oil pressure, and the lube oil temperature
- Choked interstage cooler: using the differential pressure across the intercooler
- Choked suction filter: using the differential pressure across the suction filter
- High condensate level: using the knockout drum / scrubber level
- Cylinder lubrication system failure: using the mechanical lubricator pressure and the cylinder lube oil tank level
- Cylinder cooling system failure: using the tempered water (main header) supply temperature, individual cylinder's return line temperature and surge tank level
- Packing buffer gas system failure: using the buffer gas supply pressure

CASE STUDIES:

Case studies will be presented here which will show data from real machines collected across System 1 Asset Health Monitoring platform and the analytics suite of Decision Support (DS) is explained here as example of an automated diagnostics tool.

Cylinder Valve Leakage Detection:
As stated earlier, valves are the most troublesome components on a reciprocating compressor and account for the highest maintenance costs [2]. The valve health is generally monitored by dynamic pressure measurement per chamber and by measuring the valve temperature. With the former measurement, the flow balance is compared with its baseline. In case of suction valve leakages, the flow balance exceeds 1.0 (or baseline); discharge valve leakages cause the flow balance to decrease below 1.0 (or baseline). Using the valve temperatures, DS analytics compare the temperature of a suction (or discharge valve) with the average suction valve (or discharge valve) temperature per compression stage. By doing so, small deviations during early stages of valve failure can be detected.

In this case, the customer operates three 4-throw Recycle and Makeup gas compressors in a Diesel Hydrotreater (DHT) Unit with the following operating conditions:

- Makeup gas pressure is increased from 25 barg to 81 barg in two stages (1 throw per stage)
- Recycle gas pressure is increased from 52 barg to 83 barg in a single stage (2 throws per stage)
- 2 of the 3 units must be operational to fulfill downstream flow requirements

Figure 1 shows the compressor layout.

Discharge Valve Leakage Detection
After deploying the Decision Support analytics suite and integrating the results into the Health and Performance HMI as shown in Figure 1, the customer reliability team noticed severity 2 alarms on all of the three discharge valves at the HE chamber of Cylinder 3 on one of the compressors as highlighted below.

Machine Running		☐
Machine Power Overloading		☐
Discharge Capacity Recycle Stage 1	217856 Sm³/h	
Discharge Capacity Makeup Stage 1	53430 Sm³/h	
Discharge Capacity Makeup Stage 2	63203 Sm³/h	
Total Indicated Horsepower	4692 kW	
Total Power Losses	306.1 kW	

MOTOR CONDITION	
High Motor Bearing Vibration	☐
Damaged Motor NDE Bearing	☐

CRANKSHAFT MAIN BEARING CONDITION	
Foundation, Misalignment, Structural Issue	☐
Damaged Crankshaft Main Bearing 1	☐
Damaged Crankshaft Main Bearing 2	☐
Damaged Crankshaft Main Bearing 3	☐
Damaged Crankshaft Main Bearing 4	☐

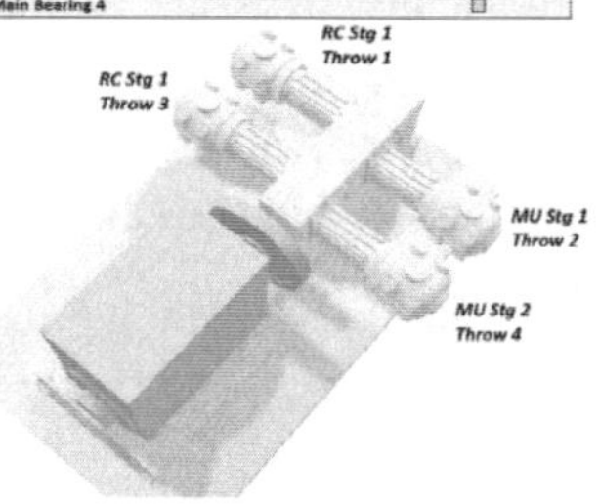

COMPRESSOR PROCESS & AUXILIARIES	Throw 1, Recycle Stage 1	Throw 2, Makeup Stage 1	Throw 3, Recycle Stage 1	Throw 4, Makeup Stage 2
Frame Lube Oil System Failure ☐				
Choked Suction Filter	☐	☐	☐	☐
Cylinder Cooling System Failure	☐	☐	☐	☐

RUNNING GEAR CONDITION	Throw 1, Recycle Stage 1	Throw 2, Makeup Stage 1	Throw 3, Recycle Stage 1	Throw 4, Makeup Stage 2
Crosshead Vibration Overall	☐	☐	☐	☐
Crosshead Pin Overloading	☐	☐	☐	☐
Debris or Liquid Ingestion	☐	☐	☐	☐
Piston Rod Failure	☐	☐	☐	☐
Worn Rider Bands	☐	☐	☐	☐
Piston-to-Liner Bottom Clearance	598.7 µm	1745 µm	1594 µm	1847 µm

CYLINDER TRIM	Throw 1, Recycle Stage 1		Throw 2, Makeup Stage 1		Throw 3, Recycle Stage 1		Throw 4, Makeup Stage 2	
Flow Balance	HE 1.359	CE 1.047	HE 1.040	CE 1.022	HE 0.871	CE 1.041	HE 1.032	CE 1.026
Suction Volumetric Efficiency	HE 97.09 %	CE 93.01 %	HE 34.81 %	CE 80.67 %	HE 78.34 %	CE 92.91 %	HE 55.36 %	CE 87.77 %
Discharge Volumetric Efficiency	HE 79.44 %	CE 68.66 %	HE 20.57 %	CE 47.70 %	HE 70.72 %	CE 69.68 %	HE 37.47 %	CE 59.39 %
Suction Power Loss	HE 21.93 kW	CE 27.20 kW	HE 4.799 kW	CE 9.324 kW	HE 31.82 kW	CE 30.40 kW	HE 10.23 kW	CE 12.28 kW
Discharge Power Loss	HE 16.05 kW	CE 41.33 kW	HE 8.319 kW	CE 5.920 kW	HE 43.27 kW	CE 31.25 kW	HE 5.424 kW	CE 14.04 kW
Degrees Rod Load Reversal	101.0 °		159.0 °		167.0 °		153.0 °	
Suction Valve Failure	HE SV 1 ☐	CE SV 1 ☐	HE SV 1 ☐	CE SV 1 ☐	HE SV 1 ☐	CE SV 1 ☐	HE SV 1 ☐	CE SV 1 ☐
	HE SV 2 ☐	CE SV 2 ☐	HE SV 2 ☐	CE SV 2 ☐	HE SV 2 ☐	CE SV 2 ☐	HE SV 2 ☐	CE SV 2 ☐
	HE SV 3 ☐	CE SV 3 ☐	HE SV 3 ☐	CE SV 3 ☐	HE SV 3 ☐	CE SV 3 ☐		
Discharge Valve Failure	HE DV 1 ☐	CE DV 1 ☐	HE DV 1 ☐	CE DV 1 ☐	HE DV 1 ■	CE DV 1 ☐	HE DV 1 ☐	CE DV 1 ☐
	HE DV 2 ☐	CE DV 2 ☐	HE DV 2 ☐	CE DV 2 ☐	HE DV 2 ■	CE DV 2 ☐	HE DV 2 ☐	CE DV 2 ☐
	HE DV 3 ☐	CE DV 3 ☐	HE DV 3 ☐	CE DV 3 ☐	HE DV 3 ■	CE DV 3 ☐		
Suction Valve Stiction	HE ☐	CE ■	HE ☐	CE ☐	HE ☐	CE ☐	HE ☐	CE ☐
Discharge Valve Stiction	HE ☐	CE ☐	HE ☐	CE ☐	HE ☐	CE ☐	HE ☐	CE ☐
Pressure Transducer Drift	☐		☐		☐		☐	
Piston Ring Leak	☐		☐		☐		☐	
Pressure Packing Leak	☐		☐		☐		☐	

Figure 1: Machine Health and Performance HMI

The Reliability Engineer referred to valve temperature and by reviewing the bar graphs, as shown in Figure 2, it was noted that the individual valve temperatures were higher than the threshold limit of 105 deg C defined in the tunable configured properties in DS analytics. Individual valve temperatures were around 12 deg C higher than the average discharge valve temperature for this compression stage. The offset threshold defined for this trigger was 5 deg C in the tunable configured properties whereas the actual difference reached ~12 deg C. The Flow balance decreased from 1.0 down to 0.87 which is below the defined low limit of 0.9 in the tunable configured property in DS. This triggered discharge valve failure notifications across all discharge valves in the HE chamber as both the flow balance and individual valve temperatures violated limits for all three valves.

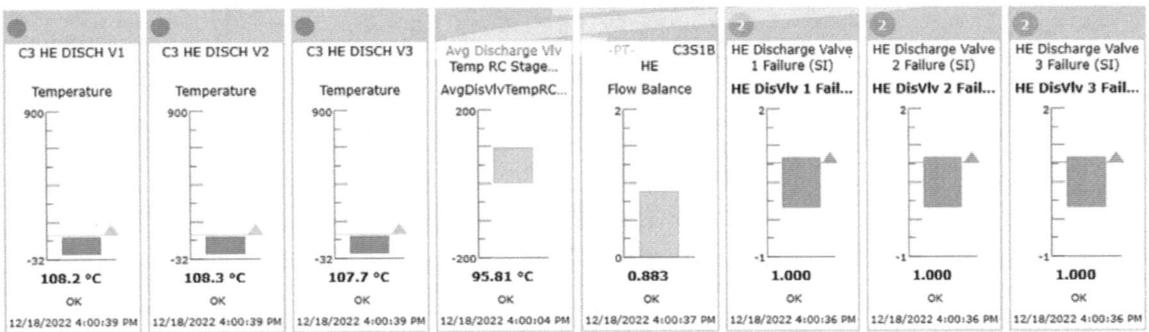

Figure 2: Discharge valve leakage detection – Valve Temp. & Flow balance Bar graphs

The site team also reviewed the trends of last four months, as shown in Figure 3, and noticed that the valves started to deteriorate around mid October and the leakage became substantial by early November. If DS analytics was in place at that time, this issue would have been highlighted in its early stages and the maintenance team could have planned the replacement in a timely manner during the first two weeks of November while the machine was already shut down for maintenance. This could have saved machine operation with additional energy consumption, extended downtime and extra maintenance cost associated with re-assigning manpower.

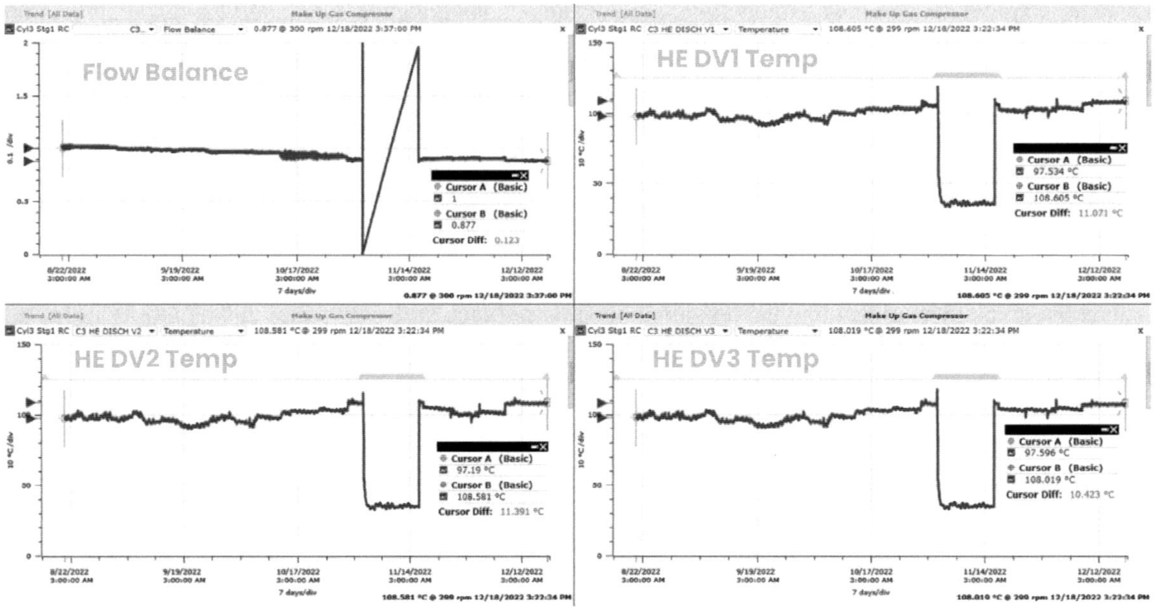

Figure 3: Discharge valve leakage detection – Valve temperature and flow balance trends

For further analysis, dynamic plots were consulted as shown in Figure 4. Pressure vs crank angle plot (lower left): measured pressure (red) in HE chamber is rising quicker than expected (orange): indication of discharge valve leak. In the unfiltered crosshead vibration waveform (blue) in upper right plot, there is a high frequency energy (hiss leakage sound) content picked up by the accelerometer during the time when the HE discharge valves are closed, which also indicates leaking discharge valves.

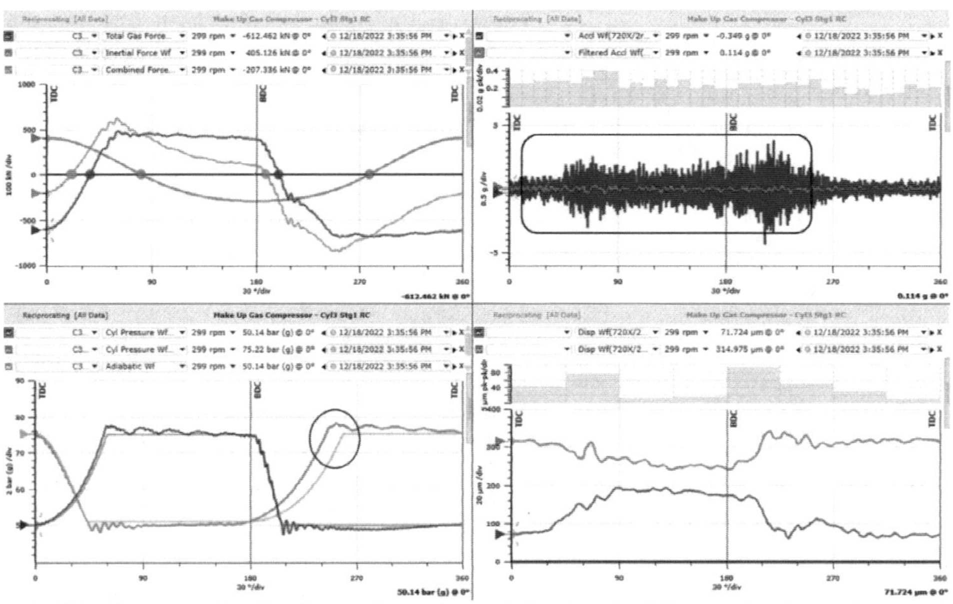

Figure 4: Discharge valve leakage detection – Mechanical Integrity (crank angle) plots

Four month trends of indicated horse-power consumption in HE chamber is shown in Figure 5, which depicts that discharge valves' leakage resulted in increased horsepower consumption of around **28 kWh**.

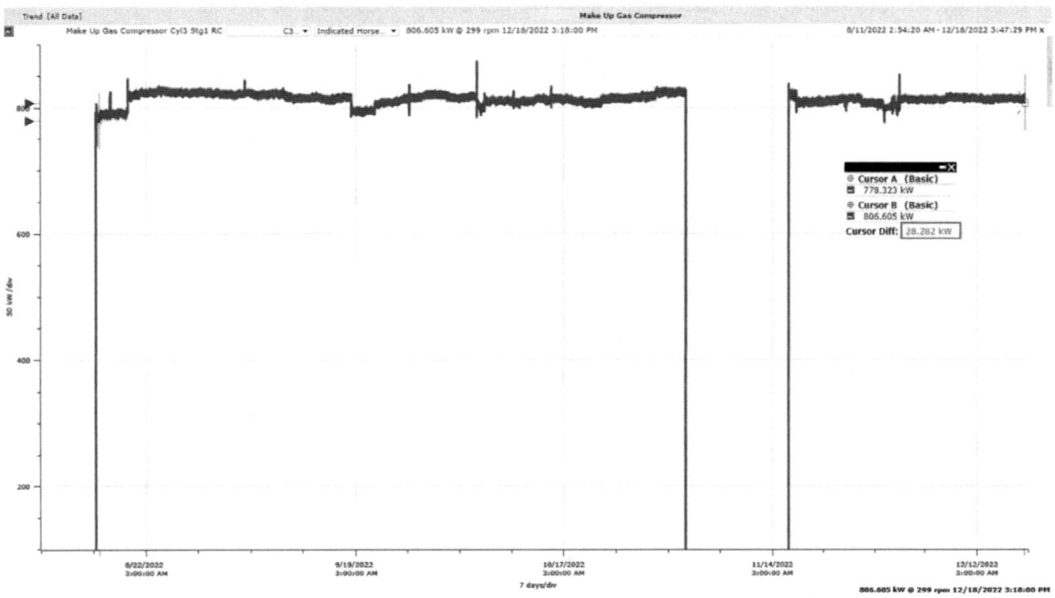

Figure 5: Discharge valve leakage detection – HE Indicated Horsepower trend

The customer further acknowledged the positive impact of Decision Support Analytics on their productivity whereby now the Reliability Engineer would spend just a few minutes in reviewing the health status of 40 valves on this 4-throw compressor at a glance rather than spending hours looking through the data to locate problems.

Suction Valve Leakage Detection

Customer's reliability staff noticed a severity 2 software alarm triggered in System 1 for one of the reciprocating compressors in the same DHT Unit. The alarm was related to Suction Valve failure in the CE chamber of Cylinder 1 triggered by DS Analytics software. To investigate this alarm, the staff quickly reviewed bar graphs of related parameters as shows in the Figure 6. It was noted that valve temperature was higher than the alarm setpoint of 75 degC and flow balance increased from 1.0 (baseline) to 1.25 which is above the defined high limit of 1.1 in tunable configured properties within DS. This triggered the Suction Valve failure rule pinpointing directly towards the failed valve number 4.

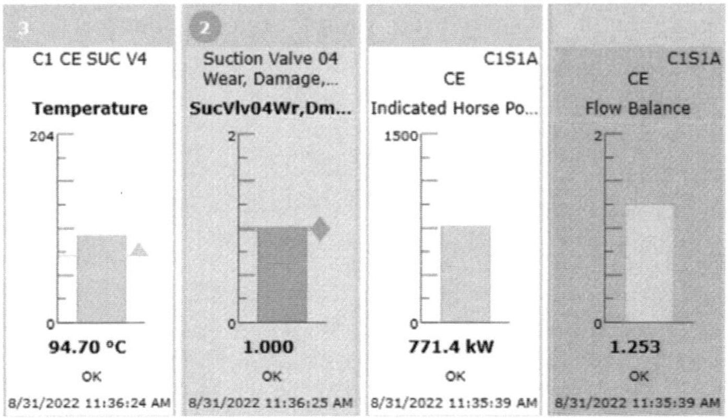

Figure 6: Suction Valve Leakage detection - Valve Temp. & Flow balance Bar graphs

One month trend data was also reviewed as shown in the figure 7, where the flow balance (ratio of suction to discharge capacity) increased from 1.06 to 1.20 at the CE chamber which is an indication of suction valve or packing or piston ring leak. Suction valve no. 4 temperature in CE chamber showed a clear increase in trend over time which confirmed the suction valve leak in the CE chamber. Suction valve leakage resulted in an increased horsepower consumption of around 57 kWh.

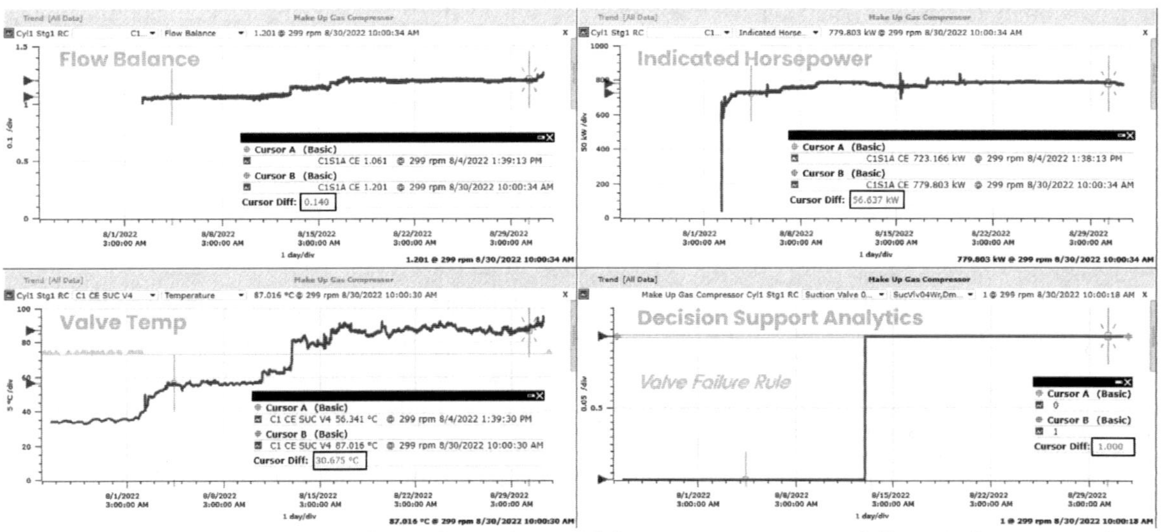

Figure 7: Suction valve leakage detection – Valve temperature and flow balance trends

To further validate this finding, dynamic plots were reviewed by comparing dynamic samples before and after the Suction Valve failure as shown in Figure 8:

- Pressure vs crank angle plot (upper left) during normal valve operation: measured pressure (blue) in CE chamber is rising as expected (green) during compression. Note, the HE chamber is unloaded during that time

- Pressure vs crank angle plot (lower left) during distress: measured pressure (blue) in CE chamber is rising slower than expected (green) during compression which can be attributed to possible suction valve or pressure packing leak. Since the pressure packing temperatures was constant, a leak flow via the packing was excluded.
- In crosshead (lower right) unfiltered vibration waveform (blue), there is high frequency energy ('hiss' leakage sound) content picked up by the accelerometer during the time when the CE suction valve is closed which is an indication of a leaking suction valve in CE chamber.

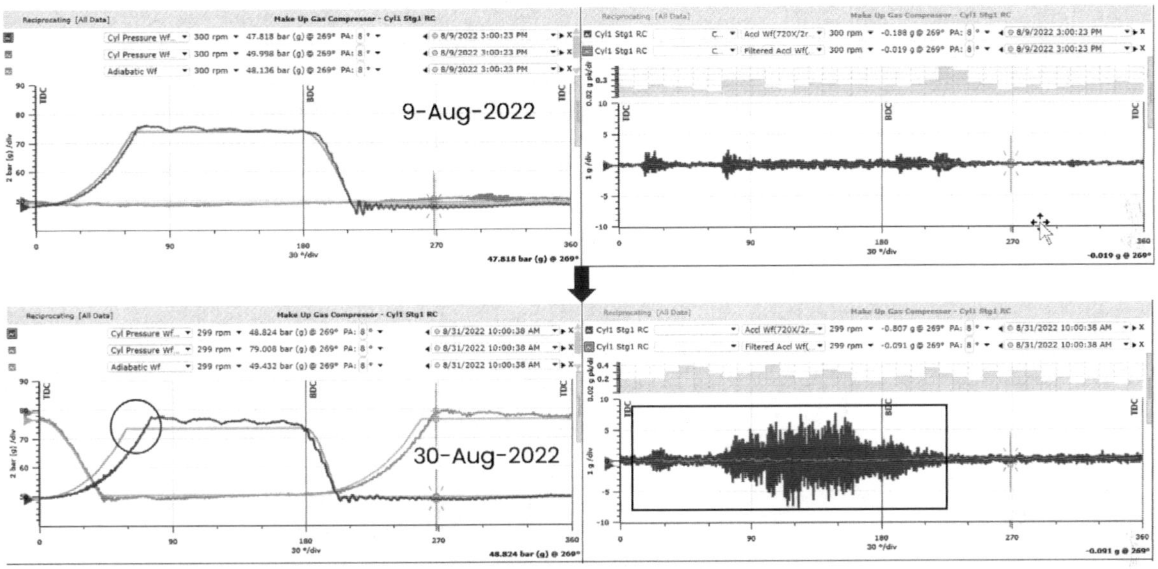

Figure 8: Suction valve leakage detection – Dynamic Cylinder Pressure and Crosshead filtered & unfiltered vibration vs crank angle plots

<u>Valve Stiction Detection</u>:
Valve stiction is one of the most common causes of valve life degradation. It occurs mainly due to improper lubrication or due to ingress of foreign debris when the process gas is not clean or during decomposition of non-metallic components (rider bands, piston rings) [1]. Stiction in suction valve (or discharge valve) can be detected by identifying the undershoot (or overshoot) in the dynamic pressure curve and by comparing the suction pressure and the minimum pressure (or maximum pressure and discharge pressure) throughout the crank rotation. Severity of suction valve (or discharge valve) stiction can be quantified by trending the difference between suction pressure and minimum pressure (or maximum pressure and discharge pressure) and generating a notification if the difference exceeds a pre-defined threshold level.

Customer's reliability team noticed a severity 2 software alarm related to suction valve stiction at the CE chamber for cylinder 1 as highlighted in Figure 9. This centered attention towards cylinder pressure measurements and by reviewing the bar graphs, as shown in Figure 9, it was noted that the difference between suction and minimum pressure was exceeding 5 bar (the offset threshold defined for this trigger was 5 bar in the tunable configured properties within DS). This triggered a suction valve stiction notification in the CE chamber of cylinder 1.

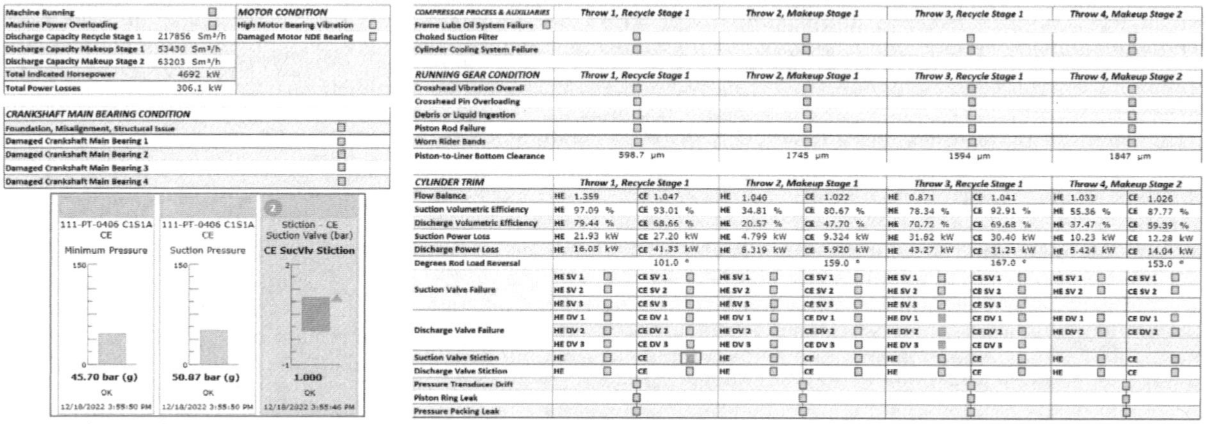

Figure 9: Valve Stiction Detection – Health & Performance HMI and Bar graphs

For further analysis, the mechanical integrity plot was referred to for dynamic waveform analysis as shown in Figure 10. The pressure vs crank angle plot (lower left) – the measured pressure (blue) in the CE chamber exhibited an undershoot and pulsations during suction valve opening instant which is indicating suction valve stiction.

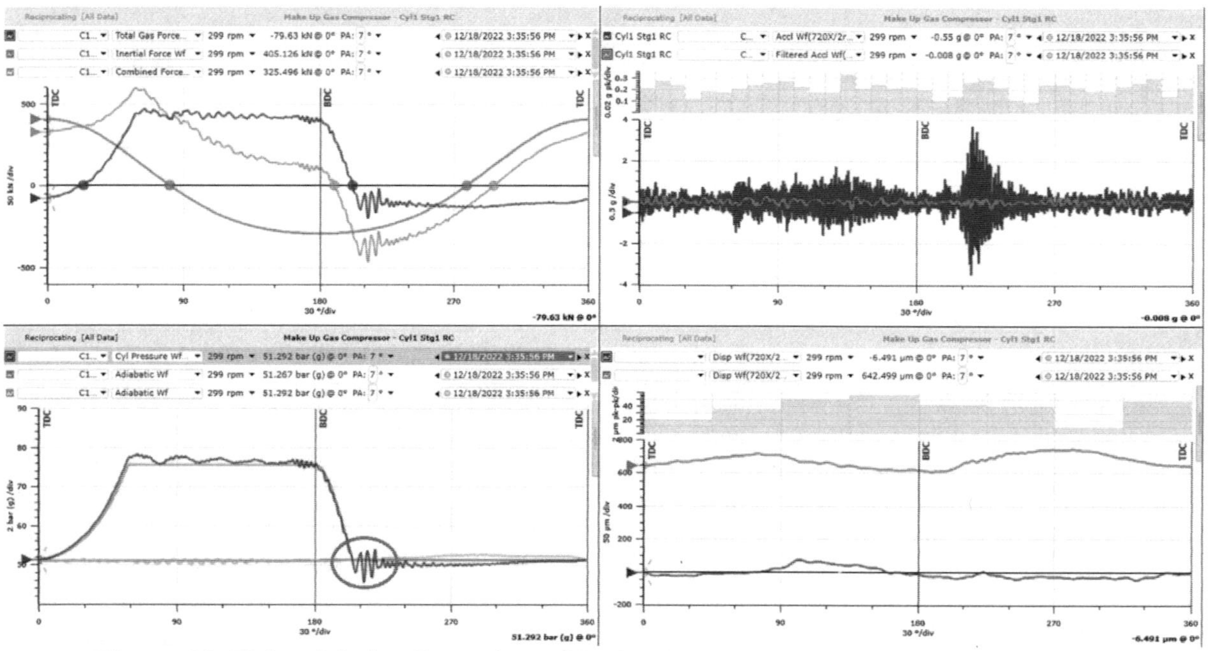

Figure 10: Valve Stiction Detection – Mechanical Integrity (crank angle) plot

Three-month trend data of the CE cylinder pressure (left), as shown in Figure 11, depicts that the difference between suction and minimum pressure increased from 3 barg to 6 barg as the stiction started to get worse over time. By comparing CE and HE cylinder pressures, it was observed that stiction started getting aggravated as soon as the HE chamber was unloaded. The customer noted this observation and decided to compare the lifetime of valves operating in partially loaded cylinders with the valves' life operating in fully loaded cylinders. If the stiction is found to be affecting the mean time between failures (MTBF), then the OEM shall be consulted to fine-tune the cylinder lubrication rate for different load conditions.

While the cylinder lubrication rate is not changing when a chamber is unloaded, less lubrication oil is discharged and remained in the cylinder, causing valve stiction in the loaded chamber.

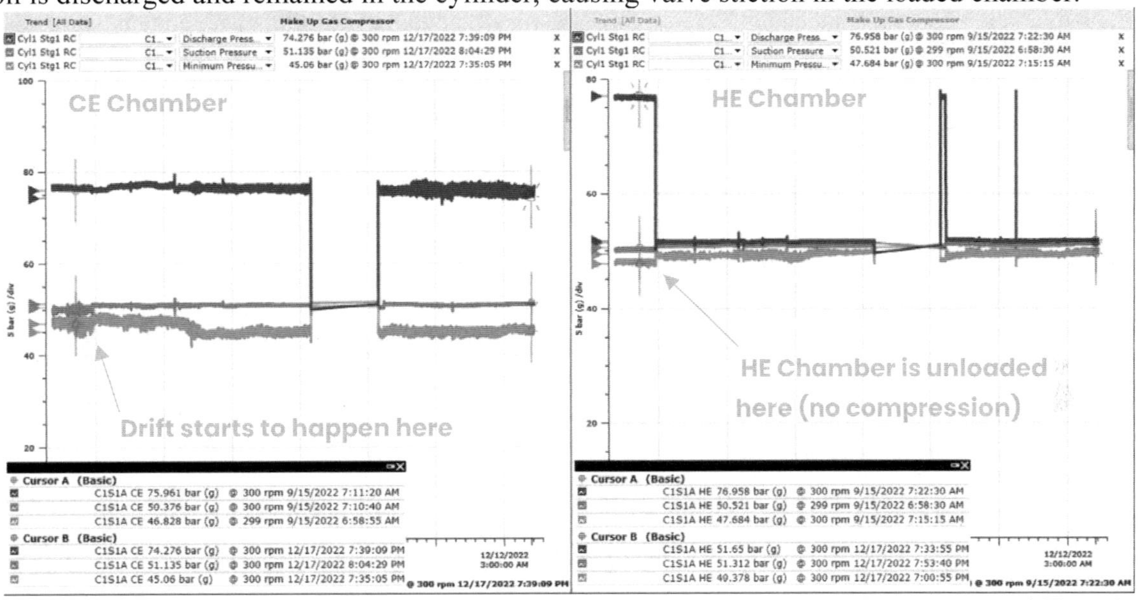

Figure 11: Valve Stiction Detection – Cylinder Chamber Pressure Measurements' Trends

<u>Pressure transducer drift detection</u>:
Cylinder pressure measurement is vital for reciprocating compressors' condition monitoring and proper functioning of these pressure sensors is critical to avoid erroneous readings. Overheating or overcooling of pressure transducers can result in a bias shift which can eventually lead to incorrect performance calculations and diagnosis. DS analytics can be used to generate a notification if the difference between the suction pressure from both the crank end and head end pressure transducers starts to increase and exceeds a specific threshold level defined in the tunable configured properties within DS. The transducer with higher suction pressure would indicate a probable faulty transducer. The discharge pressure is not used to trigger this malfunction since the chamber can be unloaded for controlling capacity during compressor operation. Hence only the suction pressure is used as a defining criterion for this malfunction [1].

After deploying DS Analytics, customer noticed a severity 2 software alarm related to pressure transducer drift at the HE chamber of cylinder 5 as highlighted in Figure 12. When the bar graphs were reviewed, it was noted that the suction pressure in the HE chamber was reading 85 barg compared to 75 barg in CE chamber (10 barg difference). Since, the threshold offset defined in DS tunable configured properties was 5 barg, the transducer drift notification was triggered. With this observation, it was evident that the HE chamber pressure transducer had become overheated over time. A corrective action was taken to review the isolation valve design (port length and diameter) for checking temperature profile across the valve length while also ensuring that acoustic resonance was avoided.

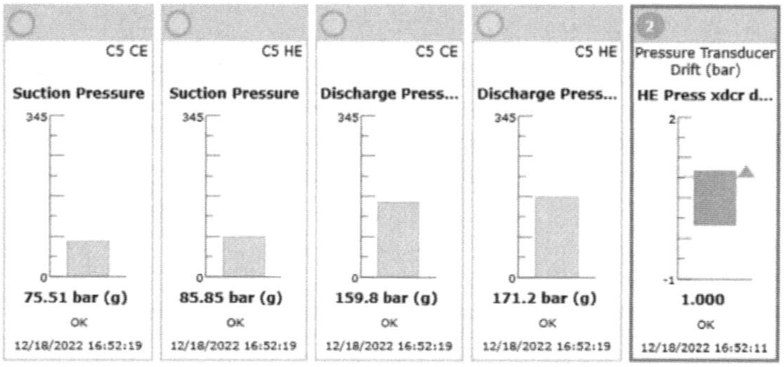

Figure 12: Pressure transducer drift detection - Bar graph

A one-month trend for suction and discharge pressure in the CE and HE chambers was reviewed as shown in Figure 13. Note that both suction and discharge pressures in HE chamber are around 10 barg higher than the ones in CE chamber which is an indication of the HE pressure transducer overheating over time. Since both chambers belong to the same cylinder and share common suction and discharge lines, unless there is a drastic valve or packing leakage, the suction and discharge pressure in both chambers should be similar in the pressure waveform.

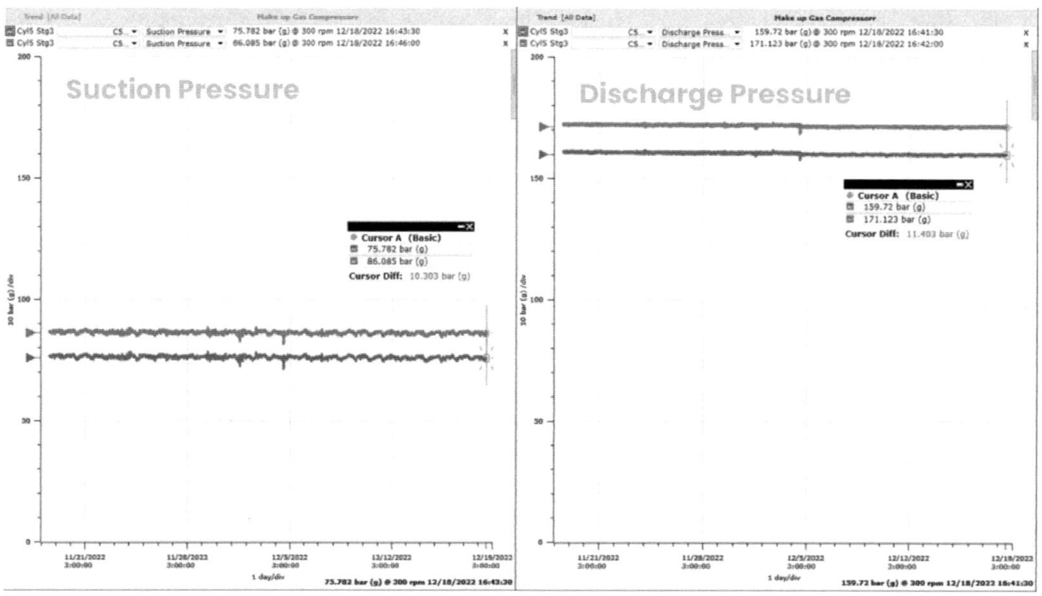

Figure 13: Pressure transducer drift detection - Suction & discharge pressure Trend plots

The mechanical integrity plot was also reviewed as shown in Figure 14. Pressure vs crank angle waveform (Lower Left) showed a bias shift in the HE pressure waveform (red) compared to CE pressure waveform (blue).

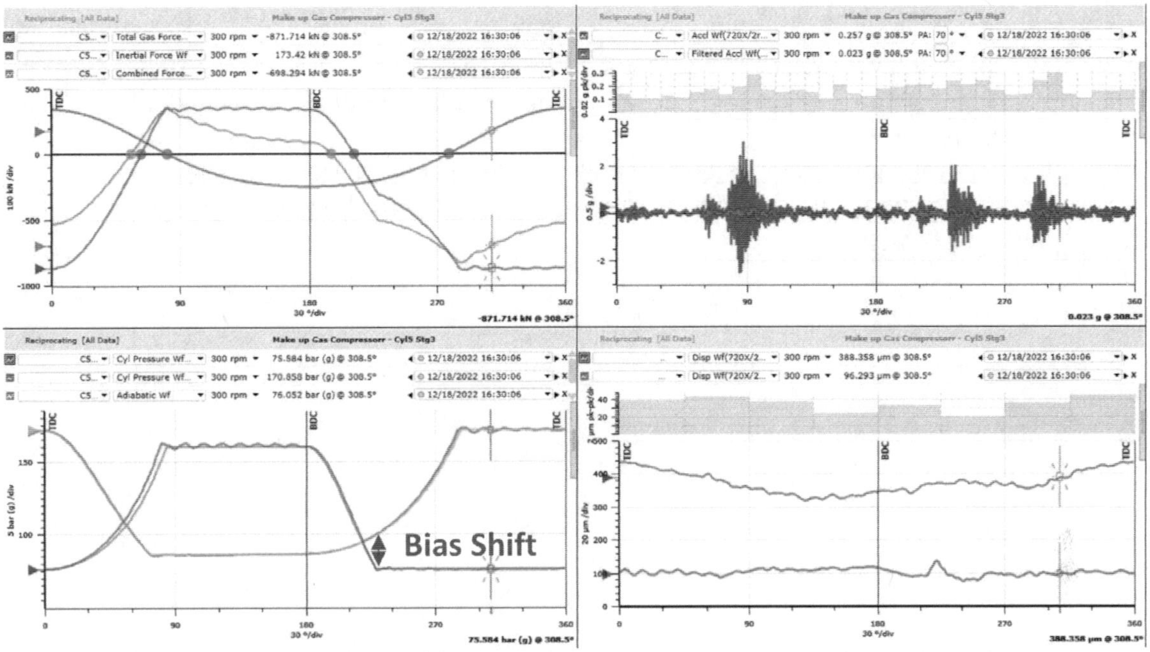

Figure 14: Pressure Transducer Drift Detection - Mechanical Integrity (crank angle) plot

CONCLUSION

With carefully configured advanced analytics software, data-driven insights are generated for machine condition monitoring through advanced rule-building and data-mapping capabilities. In addition to packaged algorithms of known failure signatures, users have the flexibility to create custom rules that help to capture, disseminate, and leverage knowledge of equipment, processes, and business solutions. For reciprocating compressors, automated diagnosis of typical malfunctions such as pressure packing leaks, valve failures, crosshead pin / frame overloading, debris/liquid ingestion, auxiliary systems (lube oil, cylinder cooling system, unloader etc.) failures and several process related issues can be realized. Further, the improvement in productivity is unprecedented whereby now the Reliability Engineer spends 10 mins on reviewing the health status of components on a 6-throw compressor at a glance of health and performance HMI rather than 4 hours spent previously looking through complete dataset to locate problems. Advanced analytics and auto-diagnosis software are the need of the hour to maximize the utilization of workforce and systems as they help to eliminate excess data searches and focus attention towards pertinent machine issues and their source.

REFERENCES

1. “Advanced Analytics And Diagnostic Rules Automatically Notify Operators About Developing Failures In Rotating And Reciprocating Machines”, SPE-211244-MS, ADIPEC 2022.
2. “Monitoring Reciprocating Compressors”, Presented to the NPRA, 1997.

3. "Online condition monitoring system identifies developing faults in reciprocating compressor at the very outset", ADIPEC SPE, 2018
4. "What's going on with my recip? How uncommon failure modes get detected", 12th EFRC Conference, 2021
5. https://my.bakerhughes.com/sites/ds-site/News/343036/natref-achieves-optimal-operation-through-bently-nevada-s-comprehensive-reciprocating-compressors-condition-monitoring-system

VIATC TRAINING CONFERENCE & EXPO
COVINGTON, KENTUCKY | AUGUST 7-9, 2024

RECIPROCATING COMPRESSOR CROSSHEAD GUIDE VIBRATION

Brian Howard
Principal Engineer, Bently Nevada
brian.howard@bakerhughes.com

Brian Howard, P.E. is a Principal Mechanical Engineer at Bently Nevada in Minden, Nevada. In this role, he focuses on sensor development, machinery analytics, diagnostic algorithm development, and research. Prior roles included Signal Processing Engineer and Global Technical Leader for Reciprocating Machinery Diagnostics, where he provided diagnostic support to customers to solve problems with rotating and reciprocating machinery. Mr. Howard received a B.S. in Mechanical Engineering from the University of Houston and an M.S. in Electrical Engineering from the University of Washington. He is a member of ASME and IEEE.

ABSTRACT:
This paper focuses on reciprocating compressors found in hydrocarbon processing applications. These machines' running gear (piston assembly, connecting rod assembly, and crosshead assembly) includes a force-fed lubricated crosshead that drives a double-acting piston via a piston rod. Looseness in the running gear results in knocks (impulse events), typically detected by an accelerometer mounted to the crosshead guide.

Crosshead guide accelerometers have proven successful in detecting potentially serious failures. For this reason, crosshead guide accelerometers provide machinery protection in many installations. Over the years, protection system suppliers have developed measurements for the signal produced by the crosshead accelerometer. These measurements include direct, rms, segmental bands (pattern recognition), and tracking bands.

What do these measurements mean, and how do they relate to running gear and cylinder trim faults? This paper presents a crosshead system model that estimates the crosshead guide's vibration response to the forces and impulses in the running gear. Next, the paper provides an overview of the signal processing for each measurement. The model provides a conceptual framework that relates the measurements to specific compressor faults. Finally, the paper uses this conceptual framework to examine the measurement responses for reciprocating compressors experiencing a variety of faults in a series of case histories.

To encourage reproducibility, the Python simulation code, model input parameters, model results, and case history data will be published in a public GitHub repository under the MIT software license.

KEYWORDS: acceleration, crosshead guide, crosshead, crosshead shoes, piston, piston nut, reciprocating compressor, vibration

INTRODUCTION:

Reciprocating compressors provide a variety of compression services in the hydrocarbon processing industry. These designs range from high-pressure ("hyper") compressors operating at less than 200 RPM to reciprocating engine-driven compressors operating at 1800 RPM. This paper focuses on electric motor-driven heavy-duty units built to the American Petroleum Institute (API) 618 standard, typically operating at 300 to 600 RPM [1]. In compliance with this standard and plant safety requirements, these compressors include instrumentation to provide protection and condition monitoring. In particular, accelerometers mounted to the crosshead guide have proven valuable in providing protection and diagnostic information.

The signal produced by these accelerometers contains information about mechanical events on the running gear and gas flow events within the compressor cylinder. Differentiating the possible source of an event in the waveform enables diagnostics and reliable protection. For this reason, system suppliers have developed several measurements for this signal. These measurements include direct, rms, segmental bands (pattern recognition), and tracking bands.

This paper presents a model for the crosshead movement in response to forces in the running gear and cylinder to improve diagnostic capabilities. It then covers the measurements commonly made on these signals and provides a framework relating them to specific faults. Finally, the paper uses this conceptual framework in case histories of machine faults.

CROSSHEAD GUIDE ACCELERATION MEASUREMENTS

Sensor Mounting

Figure 1 shows the terminology for the crank-slider mechanism in a typical API-618 process compressor. Several mechanical elements (piston, piston rings, piston rod, crosshead, and connecting rod) work together to turn the rotating torque at the crankshaft to linear force to compress the gas. In addition to these mechanisms, pressure or hydraulically actuated valves in the cylinder open and close each cycle (many times a second). With more parts than a turbocompressor, reciprocating compressors benefit from condition monitoring to detect developing problems and failures.

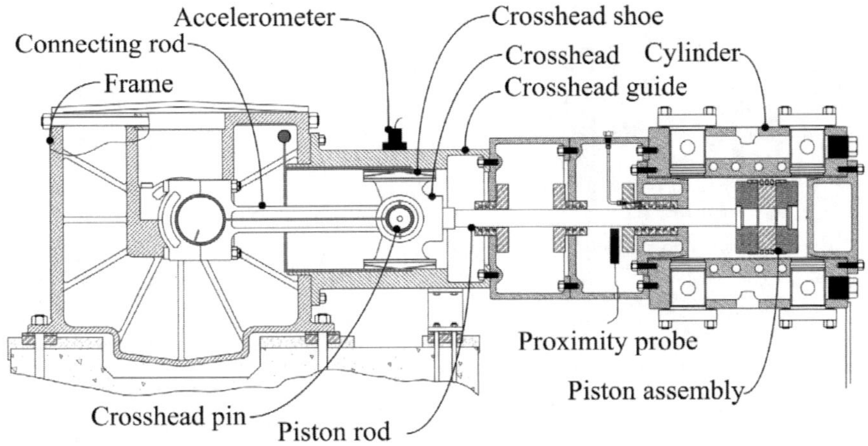

Figure 1: Reciprocating compressor nomenclature and throw vibration instrumentation. Red highlights key running gear lube oil passages.

Techniques for instrumenting and detecting these failures have evolved over the years. Initially, mechanical switches mounted to the frame provided a form of protection. Roughly fifty years ago, end users added proximity probes to the pressure packing case [2]. The introduction of industrial piezo-electric accelerometers made additional measurements possible. Nearly thirty years ago, engineers installed these sensors to the crosshead guide acceleration to detect faults [3].

The accelerometers mount flush to the crosshead guide to measure vibration [4], as shown in Figure 2. This arrangement ensures that the accelerometer senses the full range of vibration available on the crosshead guide. This location provides sensitivity to mechanical events and cylinder valve opening and closing events, which have high-frequency content (more than 5 kHz). For this reason, the mounting surface must have a good surface finish, and the accelerometer must attach to the crosshead guide with a stud.

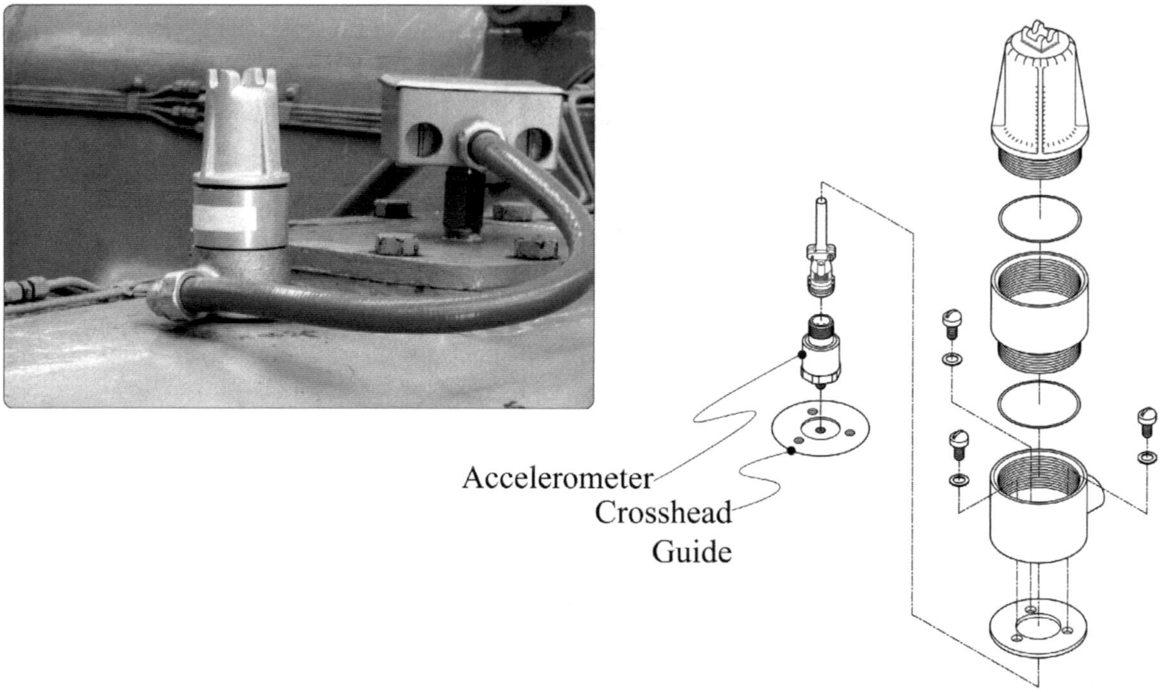

Figure 2: Typical API-670 mounting arrangement for crosshead accelerometer.

As the later topic on case histories shows, the crosshead guide accelerometer signal ("crosshead acceleration") varies in frequency and amplitude over the revolution, making conventional vibration analysis techniques difficult. Interpretation of the signal content proved to be one of the challenges with early installations. Physics-based models provide insight into dynamical system behavior that can aid in signal interpretation. The next section introduces a model that has proven helpful.

MODEL DEVELOPMENT

Referring to Figure 1, the red highlights key oil passages. The pressurized oil, fed through the crankshaft and connecting rod or from separate lines from the oil header, lubricates the crosshead. This lubrication produces an oil film between the crosshead shoes and the guide. This oil film provides stiffness and damping in the vertical direction. The oil film stiffness and damping depend on the relative velocity of the crosshead assembly to the crosshead guide and the clearance between the crosshead assembly and the crosshead guide. The crosshead forces acting on this oil film largely determine the operating clearance between the crosshead shoes and guide during operation.

Mass-elastic diagrams

The model estimates the oil film stiffness as a non-linear spring supported by the stiffness and damping of the crosshead guide structure [5], as shown in Figure 3. The model switches between the system on the left and right depending on the vertical crosshead forces. The need to switch the model introduces non-linear and time-varying behavior. Assumptions of a linear-time invariant (LTI) and signal stationarity may not hold for this system. As explained later in the paper, this behavior informs much of the diagnostic strategy for crosshead accelerometer signals.

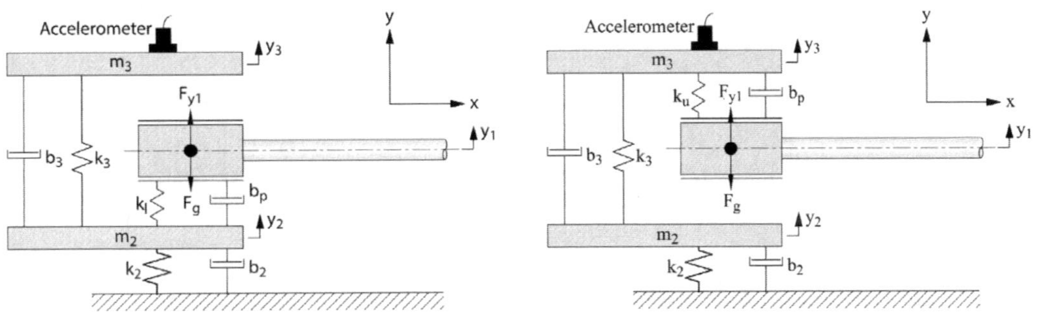

Figure 3: Model mass-elastic diagrams.

The model switches to the up-running condition (right side of Figure 3) when the vertical force exceeds the crosshead weight plus the connecting rod and piston rod weight supported by the crosshead.

Combined Rod Load

The angle of the connecting rod and the combined rod load create this vertical force. The combined rod load force includes loading from cylinder pressures inside the cylinder and inertia forces resulting from the piston changing direction twice per revolution [5],[6]. Appendix A explains how the cylinder pressure and compressor kinematics produce this rod load in a typical API-618 compressor cylinder.

Crosshead Vertical Movement

A thought experiment might help illustrate how the forces and kinematics interact to produce changes in crosshead vertical position. For this experiment, assume that the horizontal rod load force at the crosshead pin has a constant value for the entire stroke. The top of Figure 4 shows the forces at the crosshead pin for constant compression. The bottom of Figure 4 shows crosshead pin forces for constant tension.

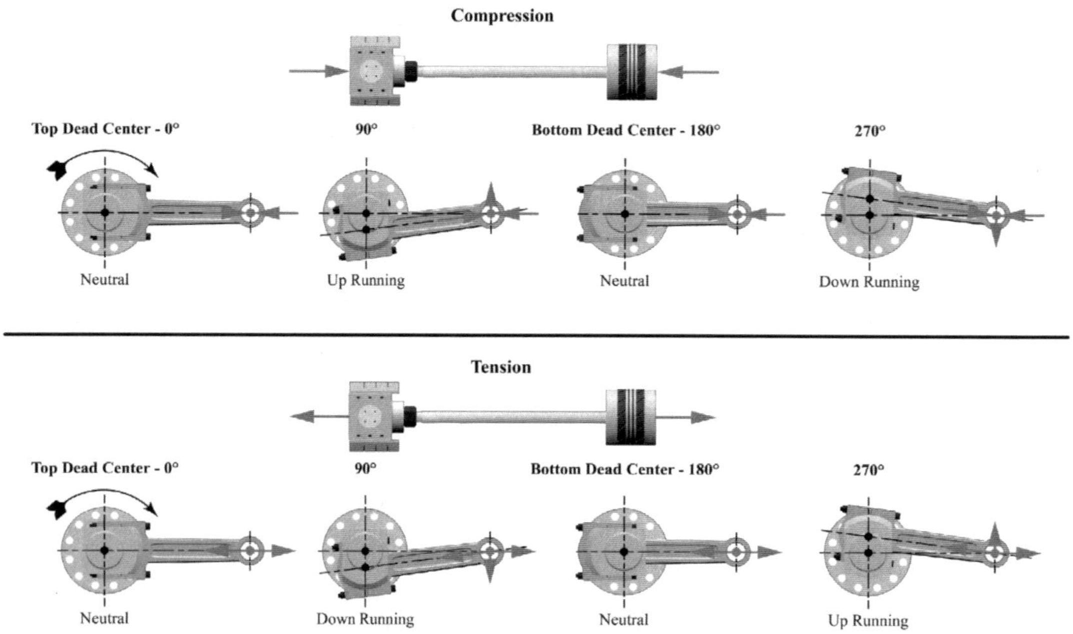

Figure 4: Crosshead pin reaction forces.

In both cases, the crosshead vertical force changes direction during the stroke even though the horizontal rod load force remains constant. For example, the top pane shows the force upwards ("Up Running Crosshead") at 90°, but at 270°, this force points downward ("Down Running Crosshead") even though the rod load force remains in compression for both cases. In either case, the system does not produce a vertical force when the piston reaches either the top dead center (TDC) or bottom dead center (BDC). These locations' lack of vertical force makes them helpful reference points when analyzing crosshead acceleration and piston rod vibration signals. This illustrates how the combination of rod load and connecting rod angle creates different vertical crosshead positions.

Industry practice often labels throws as up-running or down-running. Where do these labels come from? For many API-618 compressors, the cylinder conditions and kinematics create combined rod load curves with predominantly tensile loads during the first half of the stroke and predominantly compressive loads during the second half of the stroke (see Figure 28 for an example). A throw on the machine's right side, as shown in Figure 5, creates loads that place the crosshead in the down-running position for most of the stroke.

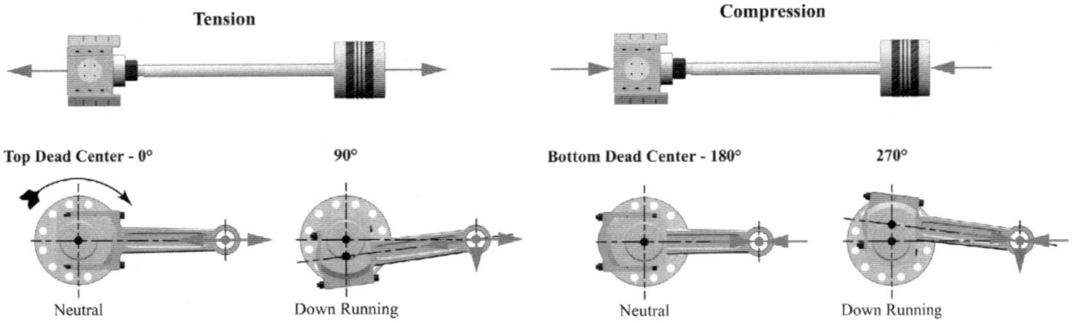

Figure 5: Crosshead pin reaction forces for a typical API-618 compressor, throw on the right side.

For a throw on the left side, the same typical rod loading conditions create forces that place the crosshead in the up-running position for most of the stroke. Figure 6 shows opposing throws with the same operating conditions and kinematics. When in tension (top half of Figure 6), the right throw operates in the down-running condition, while the left throw operates in the up-running condition. When the piston direction reverses and the rod load changes to compression, the right-side throw returns to the down-running condition and the left throw to the up-running condition.

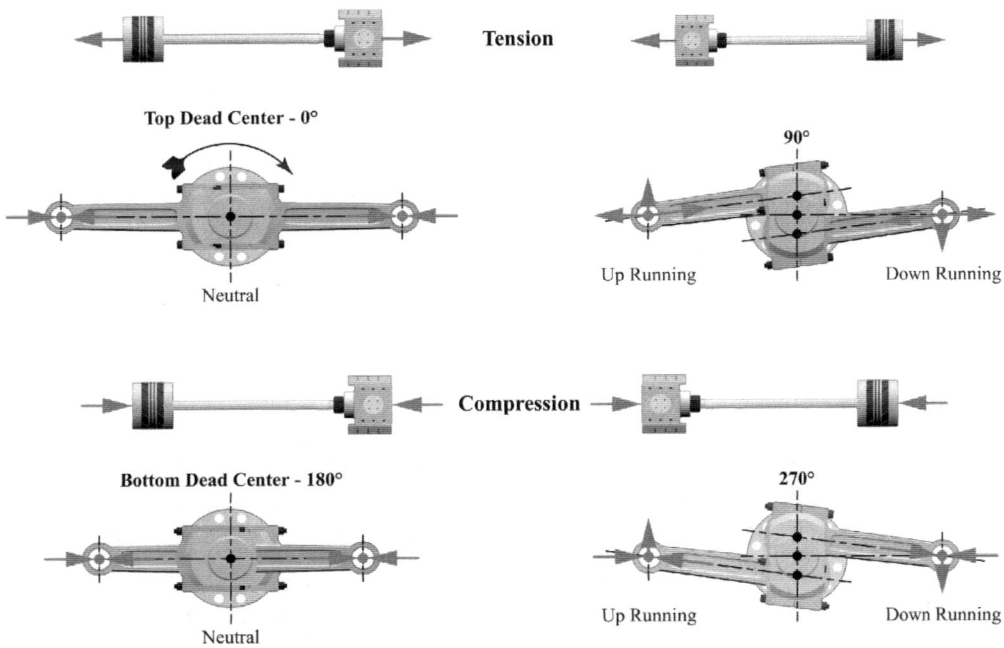

Figure 6: Crosshead pin reaction forces for a typical API-618 compressor, opposing throws.

Not all machines operate with this typical behavior. For example, a cylinder with a discharge pressure of 200 bar(g) (2900 psig) has a smaller crank-end piston area than the head-end piston area and will operate primarily in compression. A cylinder with head-end unloaders will likely operate mainly in tension. Standard practice often labels the physical throws as up-running or down-running; however, diagnostics improve when these labels apply to portions of the load throughout the stroke.

The preceding discussion gives a qualitative sense of crosshead behavior. To better understand crosshead behavior and crosshead acceleration signals, examine the model response to a pure inertia load. Such a condition occurs during a mechanical run-in (no cylinder valves installed). Figure 7 shows the load curves, crosshead displacement, and crosshead acceleration signal for this condition.

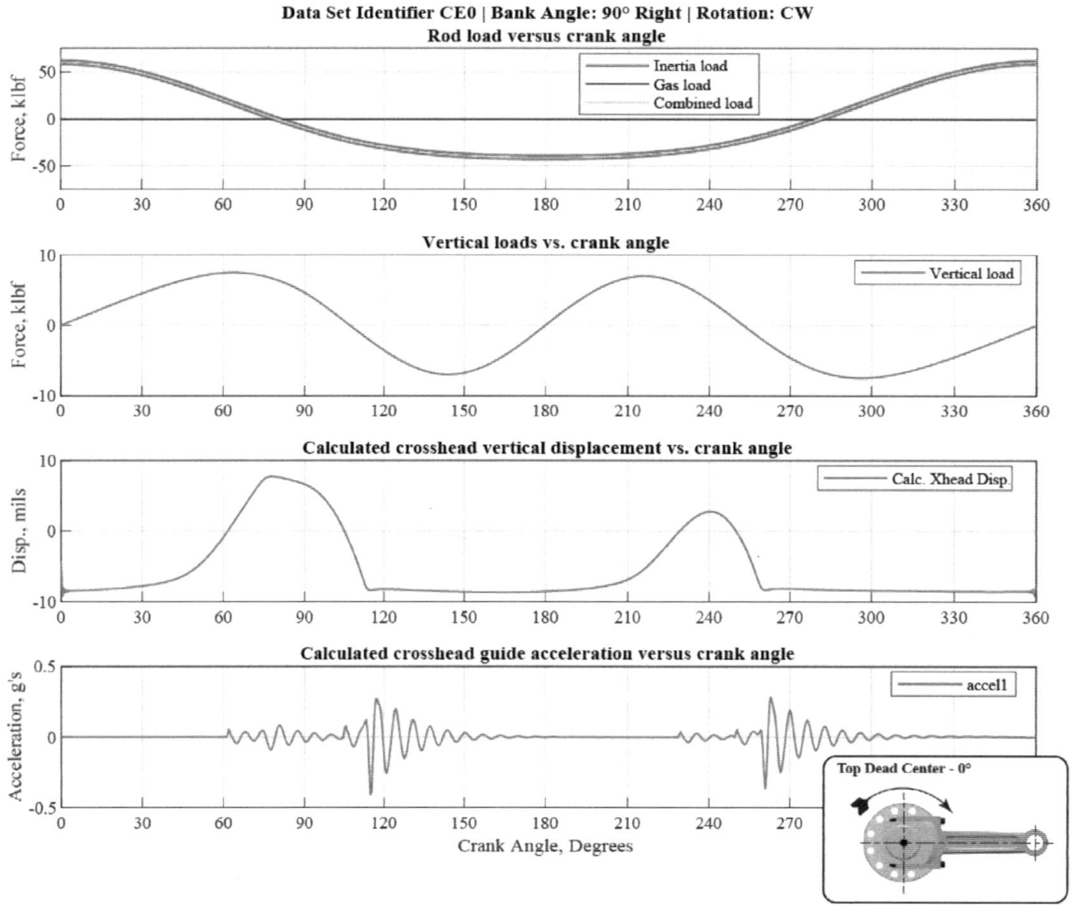

Figure 7: Rod load (top pane), calculated vertical crosshead load, crosshead displacement, and crosshead guide acceleration (bottom pane).

The crosshead acceleration signal shows a complex response. The amplitude changes throughout the revolution. The signal has no amplitude when the vertical forces lift the crosshead off the bottom rail (from 0° to about 60°). In contrast, when the crosshead contacts the upper rail (70°) or lower rail (112° and 260°), the amplitude increases as the guide responds to the impact. In addition, inspection of the impact and ring-down events also reveals changing frequency. For example, the upper rail impact at 70° has a lower frequency than the impact with the lower rail.

The frequency and amplitude variation over the revolution indicates a non-stationary process. Stationarity describes a process with time-invariant statistical properties. In other words, the statistical values for a sensor observing a stationary system remain the same no matter when sampling occurs. Turbomachinery often vibrates in response to rotating unbalance. When the machinery operates at steady state conditions, this approximates a stationary process.

For this reason, turbomachinery direct vibration measurements do not have to relate to the rotor angular position. In contrast, crosshead acceleration does not meet conditions of stationarity. For this reason, signal processing and measurements designed for turbomachinery generally do not produce good results when applied to reciprocating machinery.

If not turbomachinery measurements, what measurements work on crosshead guide acceleration signals? Reciprocating machinery protection system (MPS) companies have developed several measurements, including direct (0-peak), rms, segmental bands (pattern recognition), and tracking bands.

Direct (pk)

Amplitude measurements for turbomachinery signals typically use an attack-decay stage with a ripple filter to smooth the signal. This approach produces reasonably accurate measurements since these signals usually have a sinusoidal shape.

Early crosshead acceleration monitoring systems used a similar approach. Since the rise time of crosshead acceleration signal events can be much greater than the ripple filter rise time, the turbomachinery peak detector does not always provide a good indication of the actual vibration level. Figure 8 shows an example of an attack-decay peak detector struggling to respond to the crosshead acceleration signal.

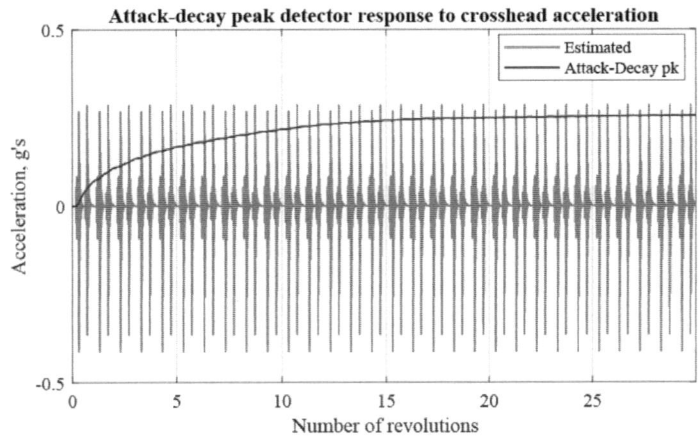

Figure 8: Attack-decay-ripple filter response to crosshead guide acceleration.

Modern systems calculate the peak level for each revolution by subtracting the minimum value from the maximum value over each revolution to provide a more accurate indication of the vibration levels. This approach builds on the observation that, for steady-state conditions, the crosshead acceleration signal for each revolution should have similar statistical properties (i.e., the crosshead acceleration signal follows a cyclostationary process). This approach accurately captures the fast-rise events shown in Figure 9.

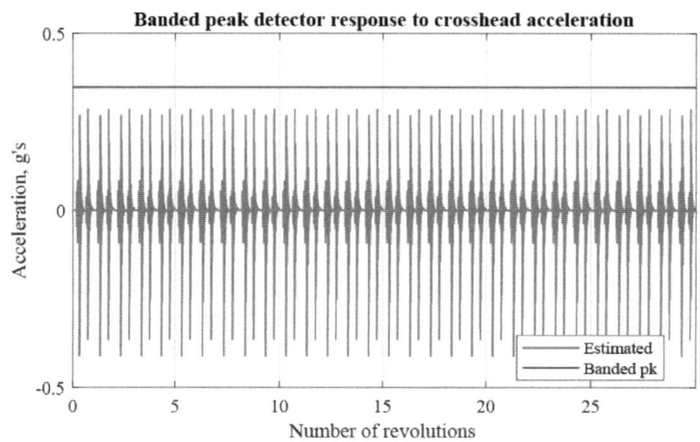

Figure 9: Banded response to crosshead guide acceleration.

Root-mean-square (rms)

The direct measurement above characterized a signal by the amplitude. In contrast, the root-mean-square (rms) measurement represents the energy in a signal. As the name suggests, this measurement takes the square root of the mean of the squares. Discrete-time systems perform the calculation over a fixed number of samples ("kernel"), as shown in Eq. (1).

$$Value_{RMS,DT} = \sqrt{\left(\frac{1}{n}\right)\sum_{i=1}^{n} x_i^2} \qquad \textbf{Eq. (1)}$$

Where:

n Number of samples in the calculation

x_i The ith sample in the calculation

While calculation has a well-behaved form, determining the number of samples requires additional information. Turbomachinery MPSs usually determine the sample length based on the channel highpass corner frequency. For example, suppose the highpass corner frequency sits at 10 Hz. The kernel should be long enough to capture at least one cycle of a signal at this frequency. The 10 Hz corner works out to a minimum sampling period of $1/10\ Hz = 100\ ms$. For a system sampling 100,000 times per second (100 ksps):

$$Min.\,kernel\ length = 100\ ms\left(\frac{1\ second}{1000\ ms}\right)\left(\frac{100{,}000\ samples}{second}\right) = 10{,}000\ samples \qquad \textbf{Eq. (2)}$$

As mentioned above, this approach works for turbomachinery because the unbalance force behaves as a stationary process. This approach causes problems for reciprocating compressor instrumentation. For example, a 300 RPM compressor takes 200 ms to complete a revolution. The rms measurement described above would indicate a value for only half the revolution. Applying this kernel design to the crosshead signal from Figure 7 results in cases where the kernel will have one, two, or no events, as shown in Figure 10.

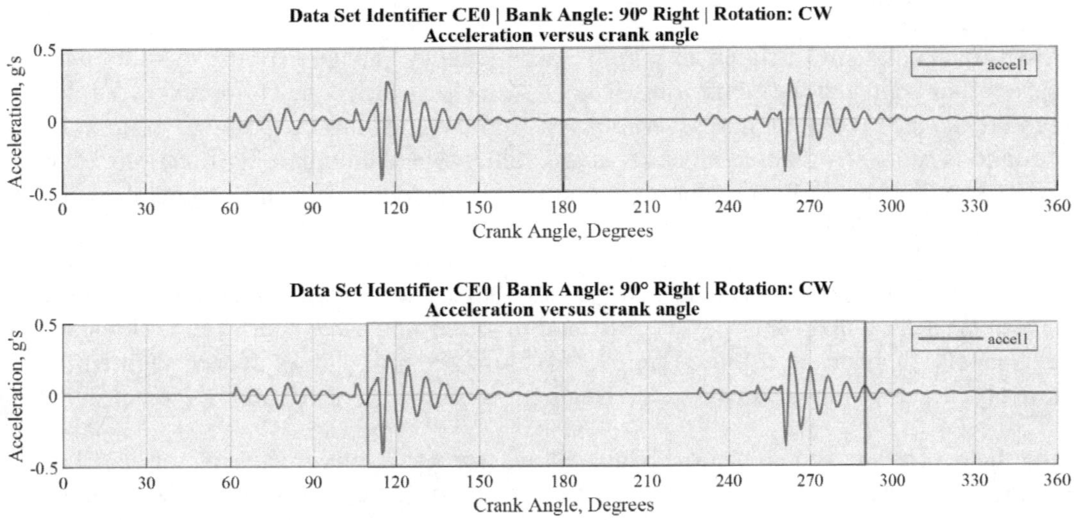

Figure 10: Kernel position at different times for a 300 RPM machine configured with a 10 Hz highpass corner.

Many MPS manufacturers add a lowpass ripple filter after the rms calculation to work around this response on slow-speed equipment. This filter reduces variance in the rms reading but adds a time delay in system response.

To avoid adding delay and improve performance, today's reciprocating compressor MPS manufacturers set the kernel to equal the number of samples in each revolution. This approach takes advantage of the cyclostationary nature of the crosshead acceleration signal and provides a stable reading that responds quickly to changes in machinery conditions.

Before moving on, it might help to compare pk and rms. The rms measurement responds to the signal energy, not the amplitudes. For crosshead acceleration signals containing rapid rise-time events with short duration, the energy content does not change as rapidly as the peak values. Table 1 summarizes the performance of these two measurements to changes in the system [7]. The pk measurement consistently changed to indicate either higher forces or reduced stiffness.

In contrast, the rms measurement did not always provide the same strength. Further, when damping changed, the rms measurement responded. Since damping has little diagnostic value, this change is a contrary indicator (the rms measurement changed even though stiffness and forces did not).

Table 1 – Summary of measurement response to changes in vibration

Condition	pk	rms
Change in Force	Strong Indicator	Indicator
Change in Damping	Strong Indicator	Contrary indicator
Change in Crosshead Guide Stiffness	Indicator	Strong Indicator
Change in Oil Film Stiffness	Strong Indicator	Indicator

Segmental bands

The crosshead accelerometer signal amplitude and frequency change over the revolution suggests another approach for monitoring. Rather than characterize the entire revolution, this approach breaks the revolution into several bands. Some companies break the revolution into non-continuous segments focused on crank angle values where common malfunctions appear [1]. Other MPS suppliers break the revolution in continuous segments, typically 36 bands with a 10° span for acceleration signals [8].

Figure 7 shows segmental bands applied to the calculated crosshead guide acceleration predicted by the model. Each segment spans 10°, confirmed by observing three bands fall within each of the x-axis labels. In this case, the amplitude levels show the peak values in each segment. Alternatively, the user may characterize the segment using rms to give rms segmental vibration.

Breaking the revolution into 36 distinct bands allows users to configure alarms for each band. More importantly, higher-level (more critical alarms) alarm logic uses logic AND to include adjacent bands to provide pattern recognition. In this method, a single band might trigger a low-level alarm, but the next-level alarm cannot be driven until adjacent bands cross their alarm levels.

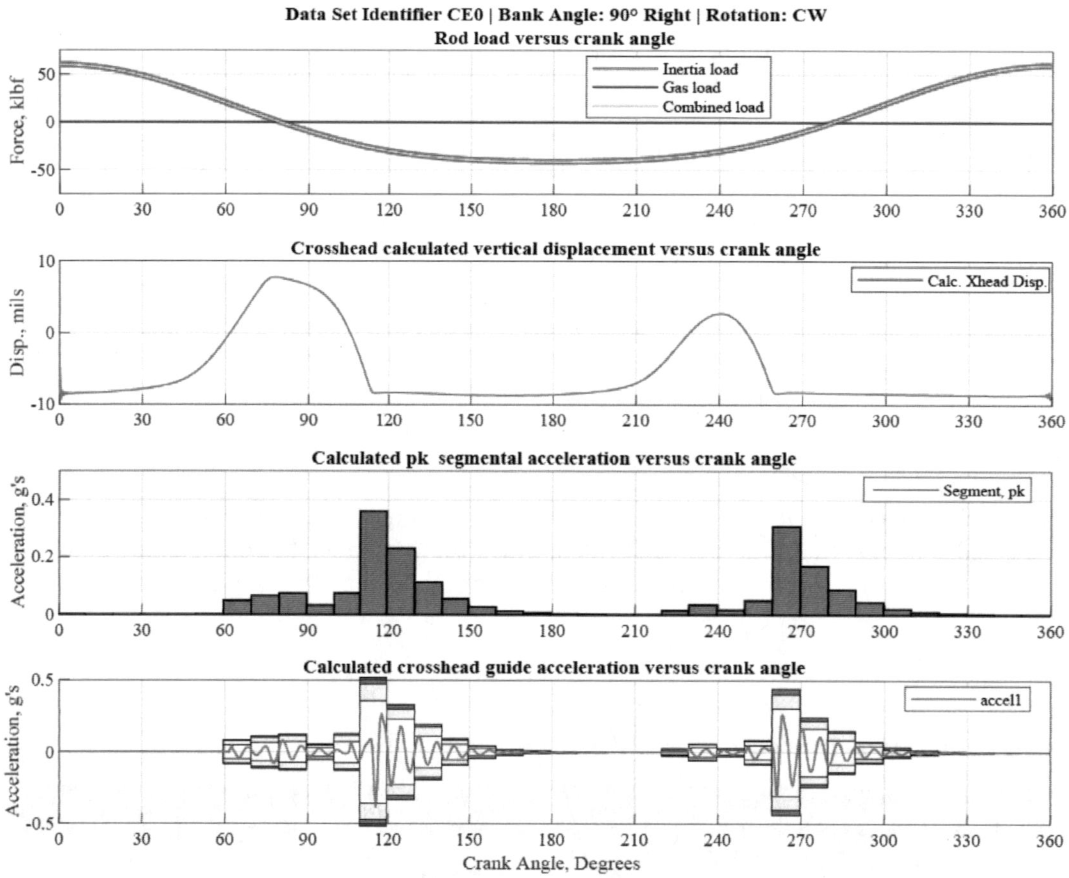

Figure 11: Rod load (top pane), calculated crosshead displacement, peak segmental acceleration, and crosshead guide acceleration (bottom pane).

Tracking bands

Tracking bands analyze portions of the crosshead acceleration signal most likely to contain knock or impact signals. Knocks occur when these force curves cross zero for gas and combined rod load curves. For pressure curves, knocks happen when the valves open or close (the forces acting on the valve assembly change direction). Tracking bands follow these points. In contrast to segmental bands, these tracking bands change angular position as load or pressure reversals shift due to load changes, process changes, or cylinder trim leaks.

Figure 12 shows an example of tracking bands applied to the model with only inertia load. For pure inertia, the rod load curve has two crossings. The location of these crossings, shown with solid green circles, locates a tracking band symmetrically in the crosshead acceleration signal (indicated with green dashed lines). The middle pane shows the vertical load at the crosshead pin. Two solid red circles indicate the crossings for this curve. Two lines show how these crossings connect to the tracking bands. Like the static segmental bands, the user may calculate the amplitude in pk or rms.

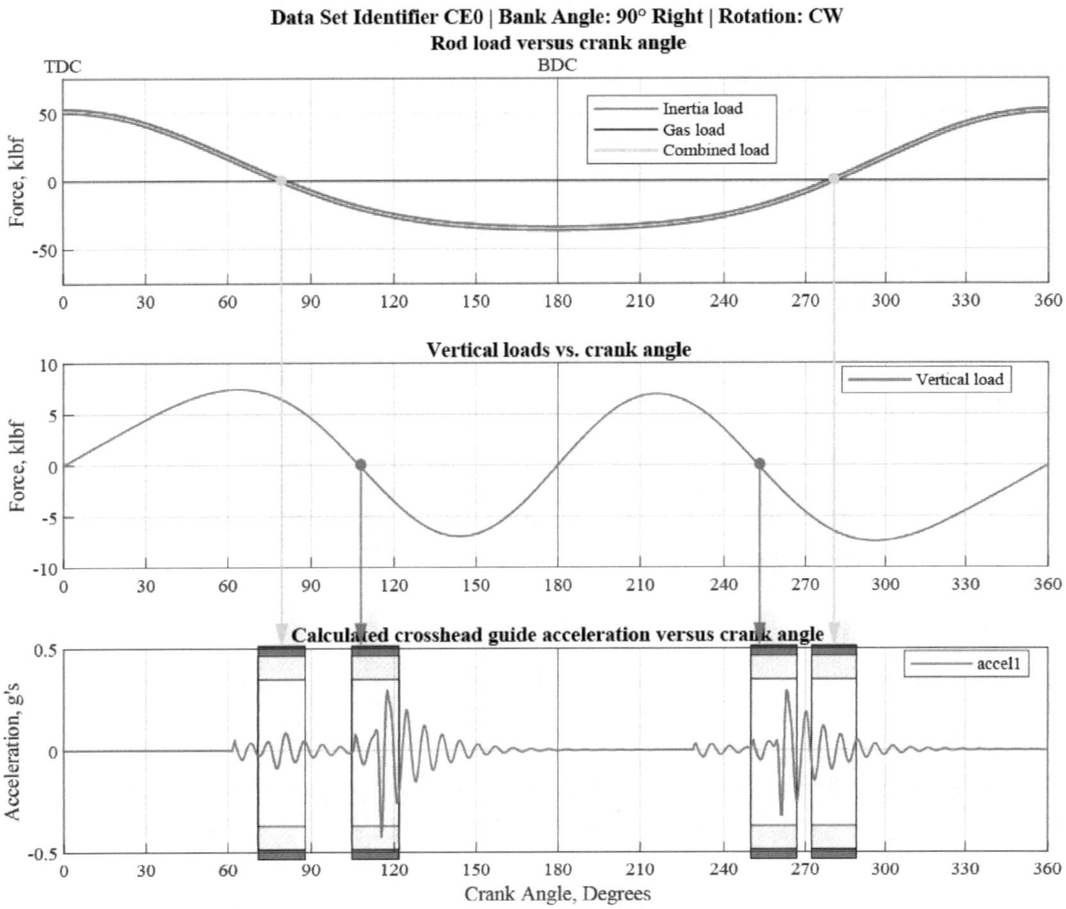

Figure 12: Rod load (top pane), calculated crosshead pin vertical force (middle pane), and crosshead guide acceleration with tracking bands (bottom pane).

When conditions change inside the cylinder, the position of the bands shifts in the crank angle domain. For example, the bands have moved with the valves re-installed, and the cylinder operating normally, as shown in Figure 13. For example, the bands associated with combined rod load (green lines) have shifted by about 30°. In addition, the cylinder pressures create gas loads with their reversal points. The gas force reversals (blue lines) have their tracking bands. Loose pistons impact the rod collar or piston nut, creating impulse events within the tracking bands associated with the gas force reversals. In contrast, crosshead pin looseness creates impulse events inside the tracking bands following the combined load reversals.

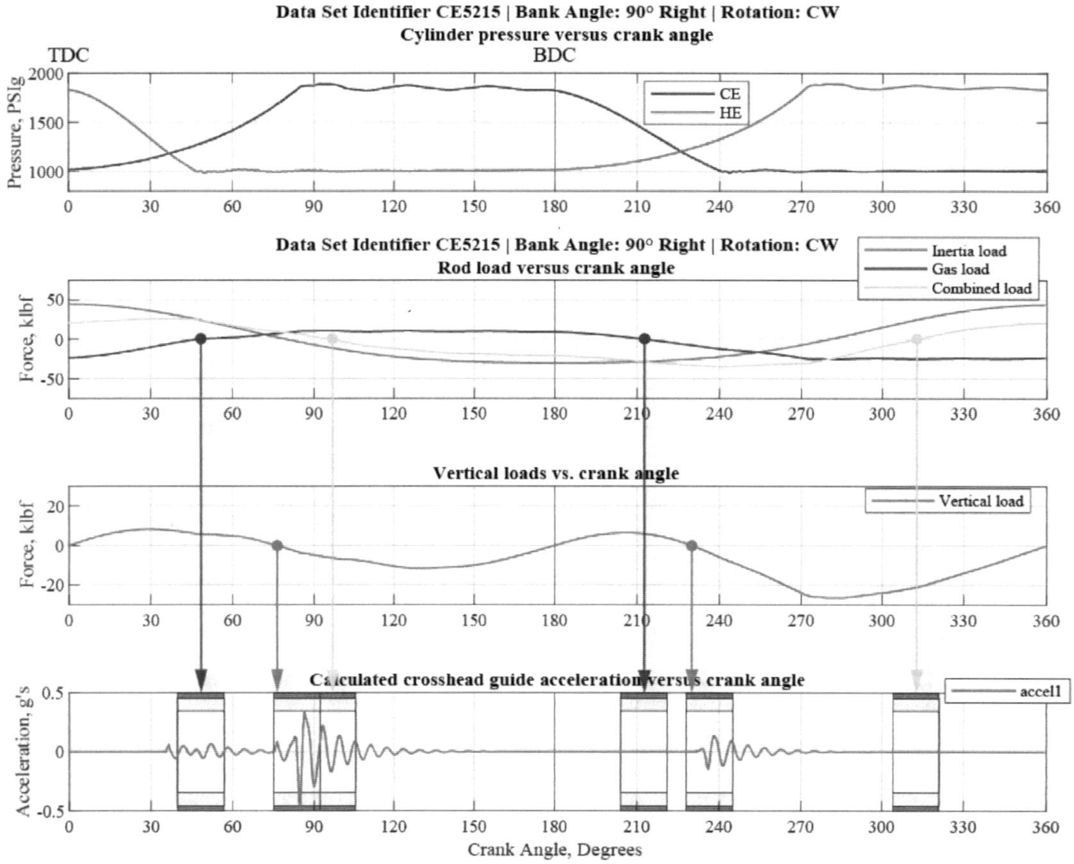

Figure 13: Indicated cylinder pressure (top pane), rod load, calculated crosshead pin vertical force (middle pane), and crosshead guide acceleration with tracking bands (bottom pane).

This section used the output of a simple model to explain how various measurements characterize the crosshead guide accelerometer signal. The following section provides an overview of crosshead accelerometer diagnostics.

CROSSHEAD GUIDE ACCELEROMETER DIAGNOSTICS

Crosshead guide accelerometer signals provide the most diagnostic value when plotted alongside other signals in the crank angle domain. The signals with the most value include the internal cylinder chamber pressure (or indicated pressure) values and the rod loads. Figure 14 shows these signals acquired from an unloaded cylinder and plotted in the crank angle domain where 0° indicates the piston has reached TDC.

The pressure curves in the top pane of Figure 14 represent typical curves found on an API-618 process compressor. The blue line indicates the crank end (CE) chamber pressure. The red line indicates the head end (HE) chamber pressure. Since the first point in the curve (0°) marks TDC, the head end chamber will have higher pressure here than the crank end. As the piston moves away from the head end head, the HE chamber pressure drops and the CE chamber pressure rises. At about 35° after TDC, each chamber has approximately the same pressure values.

If the piston had the same area on the HE and CE face, the gas force would fall to 0. The blue line in the middle pane of Figure 14 shows the load calculated from pressures acting against the actual HE and CE piston areas. This reversal occurs later than pressure equalization, at about 45°, because the piston rod area reduces the CE area compared to the HE area.

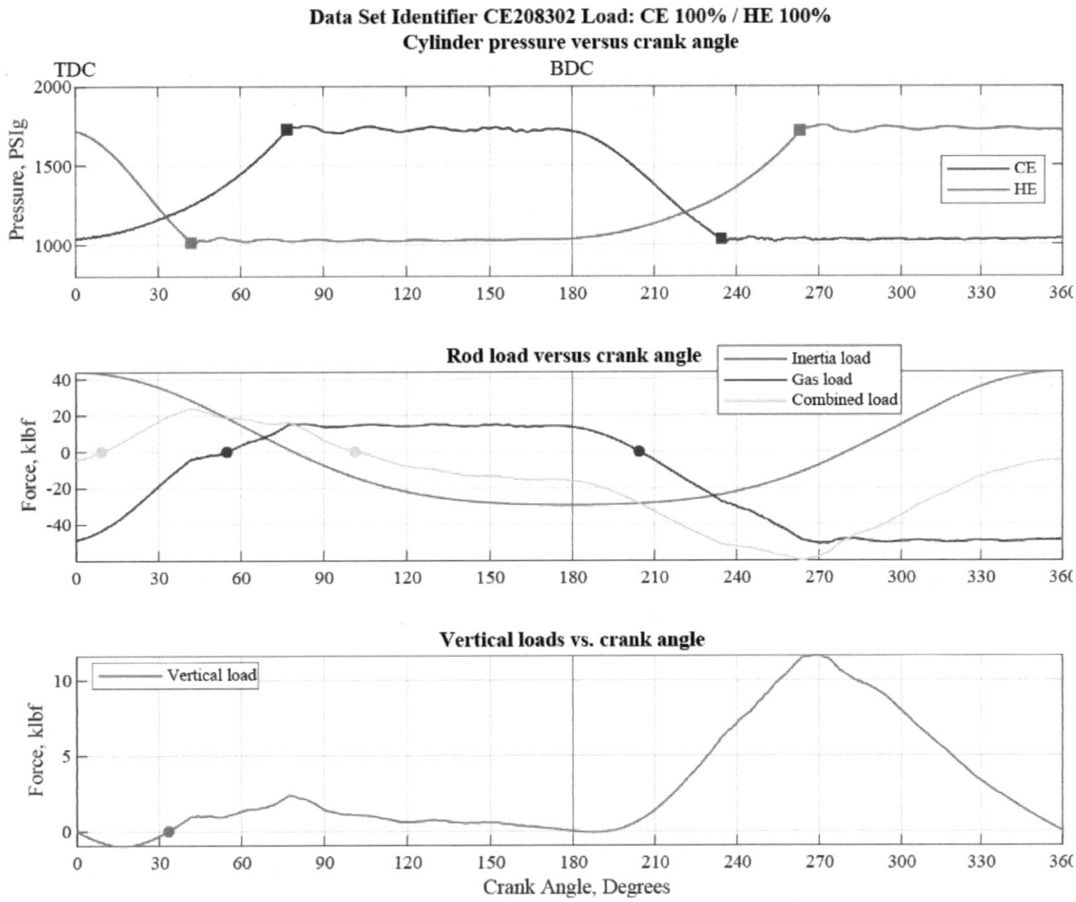

Figure 14: The top pane shows the indicated pressure for a loaded cylinder, the middle pane shows the rod loads, and the bottom pane shows the vertical loads at the crosshead pin.

How do diagnostic engineers use rod load data with crosshead guide accelerometer signals to detect problems? Crosshead guide accelerometers have value because they detect mechanical looseness and impacts. A mechanical joint with looseness or a crack cannot create an impact unless the forces acting on that joint change direction. For this reason, locations in the stroke with a change in force direction often provide locations to look for faults. The list below describes some of the more reversals:

- **Pressure reversals.** Compressor cylinder valves open when pressures inside the cylinder equal the manifold pressure (the blue and red rectangles in Figure 14). Accelerometers mounted to the cylinder provide the best diagnostic signals for cylinder valves; however, severe knocks (such as a loose valve cage) can produce impact events in the crosshead guide accelerometer signal.
- **Gas load reversals**. When the gas forces acting on the piston change direction (indicated by the blue dots in Figure 14), a loose piston moves, creating an impulse event.
- **Combined rod load reversals.** For API-618 compressors, the combined load indicates the forces acting on the crosshead pin. When these forces reverse (shown by the green dots in Figure 14), the crosshead pin moves from one side of the bushing to the other. Excessive crosshead-pin-to-bushing clearance results in impacts and impulse events at these locations. Note that if excessive clearance exists but conditions inside the cylinder do not produce a reversal, then no impact occurs. For this reason, low crosshead acceleration signals alone do not always indicate healthy mechanical joints.
- **Crosshead vertical force reversals.** When the vertical forces acting on the crosshead change direction (rod dots in Figure 14), the crosshead may change position. Loose crosshead shoes, inadequate lubrication, or excessive crosshead-to-crosshead guide clearance issues can appear near these locations.

With the explanation of measurements and diagnostic strategy complete, the following section examines real-world crosshead guide accelerometer signals.

CASE HISTORIES

Crosshead Slipper Damage

The compressor in question had recently returned to operation from service. As part of the start-up process, the plant maintenance team removed the valves and ran the machine to validate mechanical integrity and lube oil flow. Machine start-up occurred at 8:39:52. As part of the mechanical run-in process, the plant vibration analysts monitored the vibration trends from their workstations. For the first hour, the frame velocity transducers reported normal vibration levels. As shown in Figure 15, at 9:51:49, the "Frame Vib. N.W." and "Frame Vib. N.E." direct amplitudes began to increase.

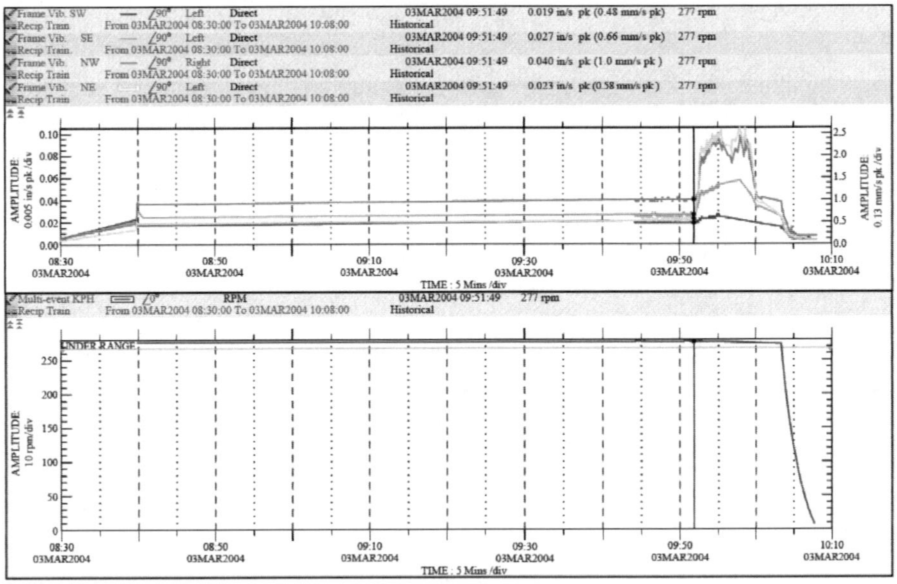

Figure 15: Frame Velocity Direct Value Trend (Top) and Compressor RPM (Bottom).

To better understand the nature of the events reflected in the change in the direct readings, the plant plotted the time base waveform from each of the frame velocity transducers, as shown in Figure 16.

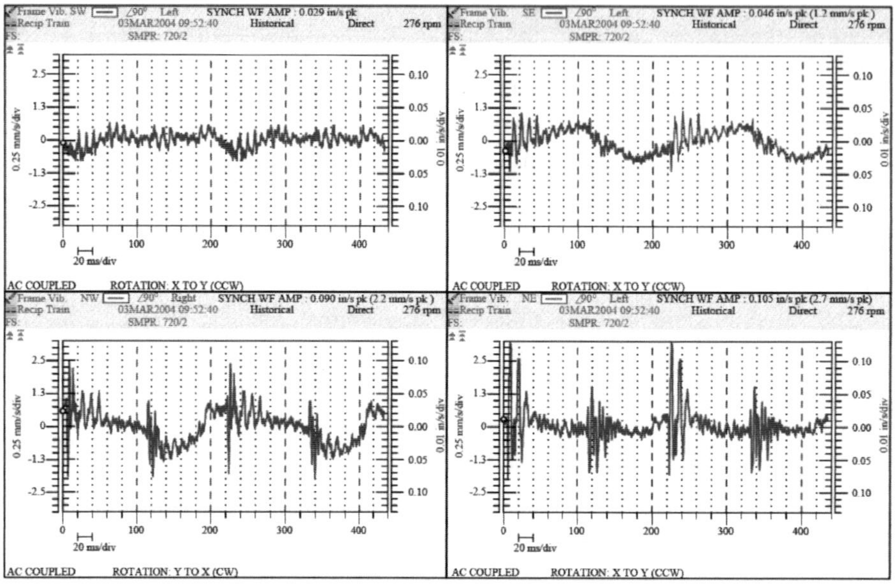

Figure 16: Frame Velocity Waveforms at 9:52:40 am.

The timebase waveforms show two complete crankshaft revolutions. These waveforms' four events (two per revolution) strongly suggest a mechanical knock event. The "Frame Vib. N.E." transducer (lower right) shows the highest amplitude, indicating that the cylinder nearest this transducer, L.P. Stage 1, should be examined further to determine the cause.

As the valves were out for the air run, diagnostics began with the rod load curves, frame velocity transducer, and crosshead accelerometer. Figure 17 shows this data. Two impulse events in the crosshead accelerometer align with the two events in the frame velocity signal, suggesting a common cause. Interestingly, the unfiltered crosshead acceleration signal shows a third event at approximately 300 degrees. This event provides essential insight into the cause of the high vibration amplitude.

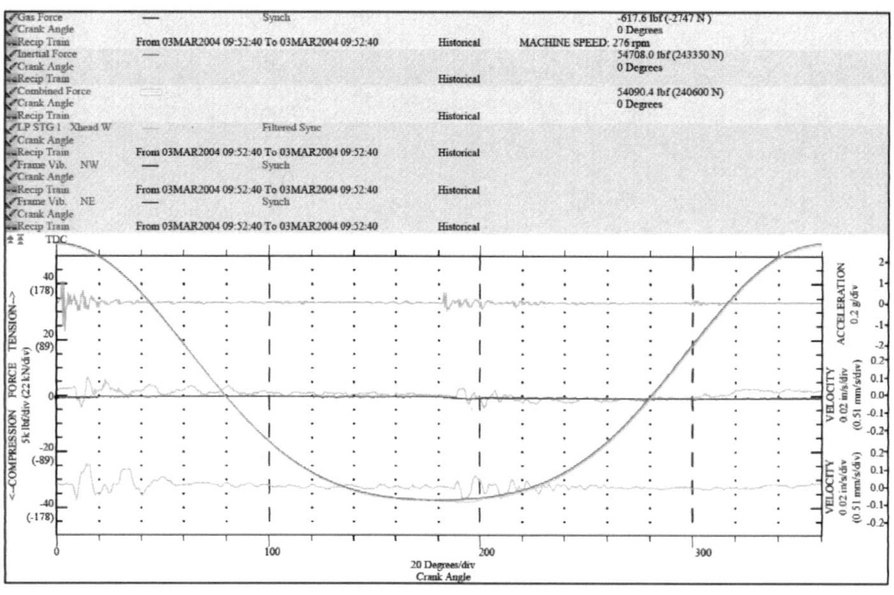

Figure 17: Rod Load, L.P. Stage 1 Crosshead Acceleration and Frame Vibration Versus Crank Angle.

The two impulse events at top dead center (TDC) and bottom dead center (BDC) would normally indicate liquid/debris ingestion; however, the valves had been removed from the compressor so that no process gas could pass through the machine. The lack of valves eliminates liquid ingestion as a possible cause. Although debris ingestion could be a cause, it would not explain the appearance of the third event at 300 degrees.

In addition to piston-to-cylinder head clearances reaching their minimum values at TDC/BDC, another interesting change occurs at these locations. At TDC/BDC, the connecting rod lies parallel to the true horizontal plane. In this orientation, the connecting rod does not exert vertical forces on the crosshead. In such a situation, the crosshead could transition from the upper guide to the lower guide, resulting in an impulse event.

The orientation of this throw with respect to the crankshaft rotation results in this being labeled a "down-running" crosshead. This term indicates that the crosshead would spend most of its time on the lower crosshead guide under normal load conditions. Interestingly, under no load conditions, this label no longer accurately describes the location of the crosshead within the guide.

Figure 18 shows the estimated crosshead vertical position derived from the first principle of kinematic analysis [5]. The filtered signal has a banded pk value of 1.1 g's pk, less than typical alarm thresholds [5], but warrants attention since the machine operates without any valves or gas load.

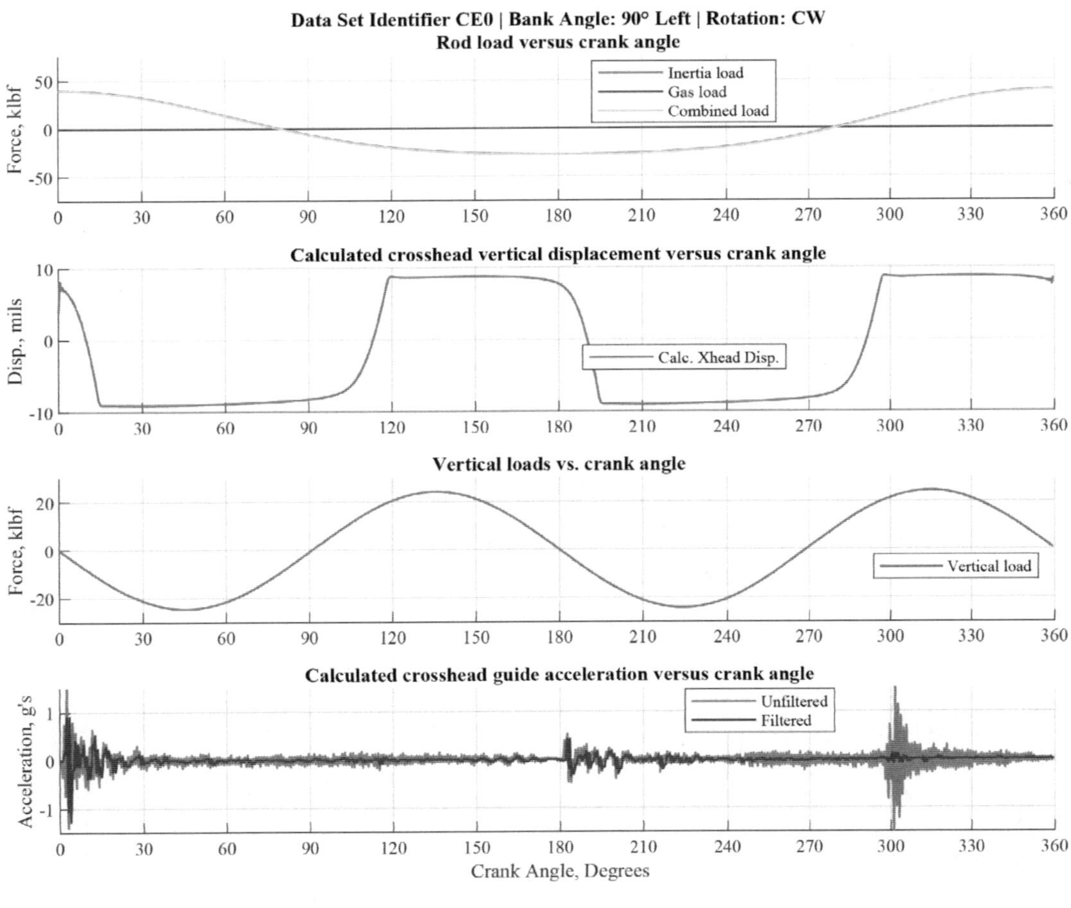

Figure 18: Rod load (top pane), estimated crosshead motion, estimated crosshead pin vertical force, and crosshead acceleration. 9:52:40.

The comparison of this motion to the impulse events of the crosshead accelerometer waveform yields some helpful information. The first two events align with a change in the direction of the vertical force. The third event lags the zero crossing because the upward force must overcome the crosshead weight. All three events align with a crosshead transition from one rail to another, including the previously unexplained third event, which aligns with the second transition to the top rail.

The model predicts a fourth event that should happen at about 110°. In the waveform in Figure 18, at 9:52:40, no impulse event appears in the crosshead acceleration; however, just a few minutes later, as shown in Figure 19, this event does appear as predicted by the model.

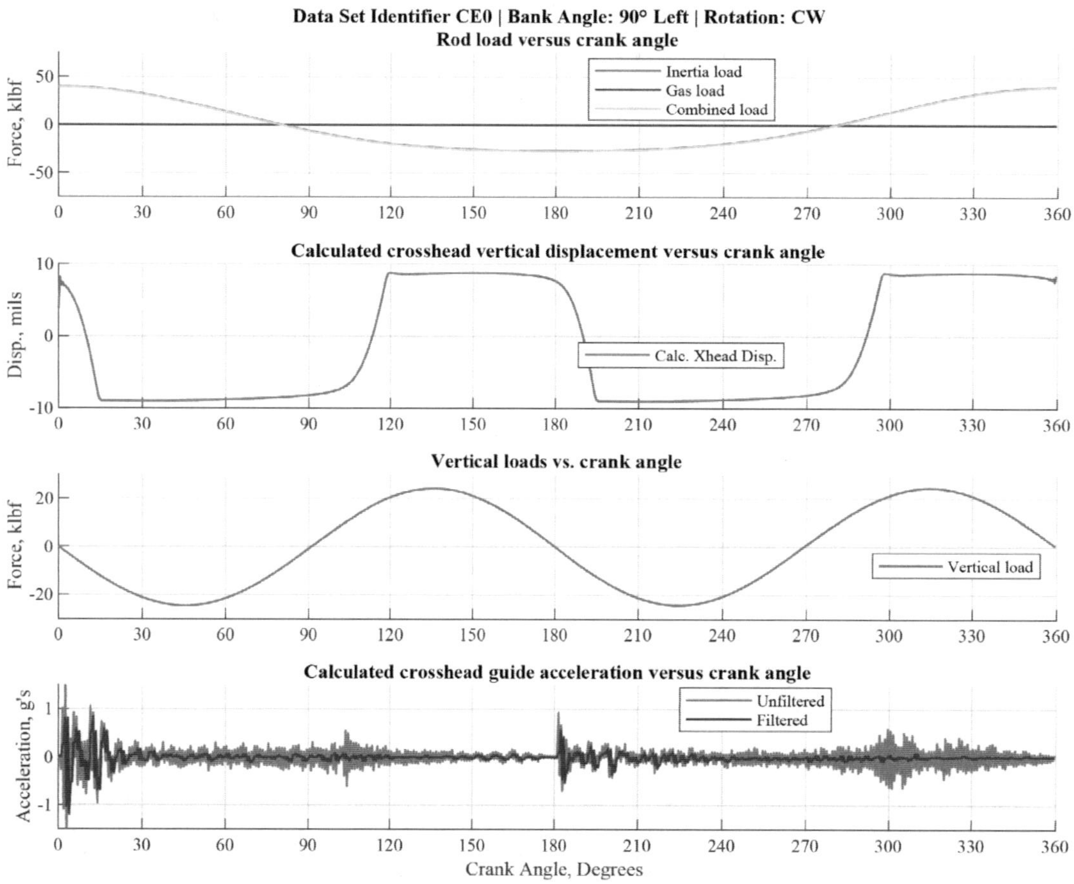

Figure 19: Rod load (top pane), estimated crosshead motion, estimated crosshead pin vertical force, and crosshead acceleration. 9:55:11.

This correlation of impulse events to crosshead motion suggests a problem at the crosshead. The equipment engineer terminated the mechanical run and inspected the crosshead assemblies.

Based on the sudden change in velocity direct values, the plant decided to shut the compressor down and inspect the running gear. The inspection found pieces of babbitt floating in the oil film around the crosshead rails. The plant removed the crosshead and found significant wear on the top and bottom shoe, as shown in Figure 20. The shiny, mirror-like surfaces indicate babbitt worn away from the crosshead shoe.

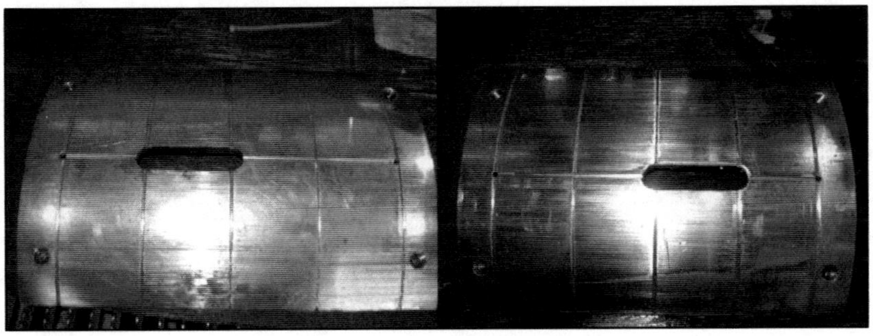

Figure 20: Crosshead Top and Bottom Shoe

Loose Piston

Analytics using tracking bands alerted the plant operators to a potential loose piston on the first stage cylinder. To confirm the findings, the plant engineers reviewed the historical data starting with rod load curves and crosshead acceleration signals several days before the alert (shown in Figure 21). This plot shows the rod load curves in the foreground and crosshead acceleration curves in the foreground. Two blue lines indicate the gas load reversal points since loose piston knocks often show up close to these points. The throw has redundant crosshead accelerometers (A and B). The top two signals show the unfiltered synchronous waveform. In this case valve opening events have significant amplitude. To separate the mechanical knocks from the valve noise the signal passes through a 2 kHz lowpass filter, creating the bottom two traces. These filtered synchronous waveforms have content related to mechanical knocks and looseness. At this time these signals show no distinct events.

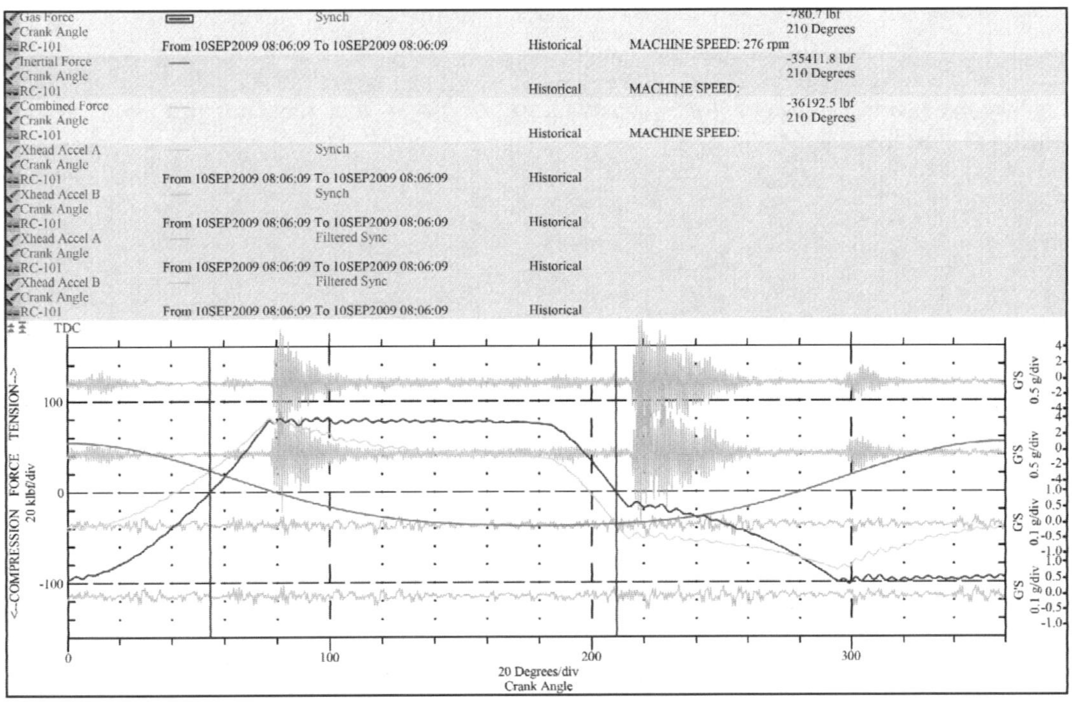

Figure 21: Stage 1 rod load curves (foreground) and crosshead acceleration signals (background) several days before the loose piston alert.

Figure 22 shows waveform data around the time of the alert. These filtered synchronous waveforms have changed compared to those shown in Figure 21. Just after the gas load reversal (indicated by the blue lines), both accelerometers showed increased acceleration levels compared to Figure 21. These events indicate looseness in the piston assembly or between the piston assembly and the piston rod.

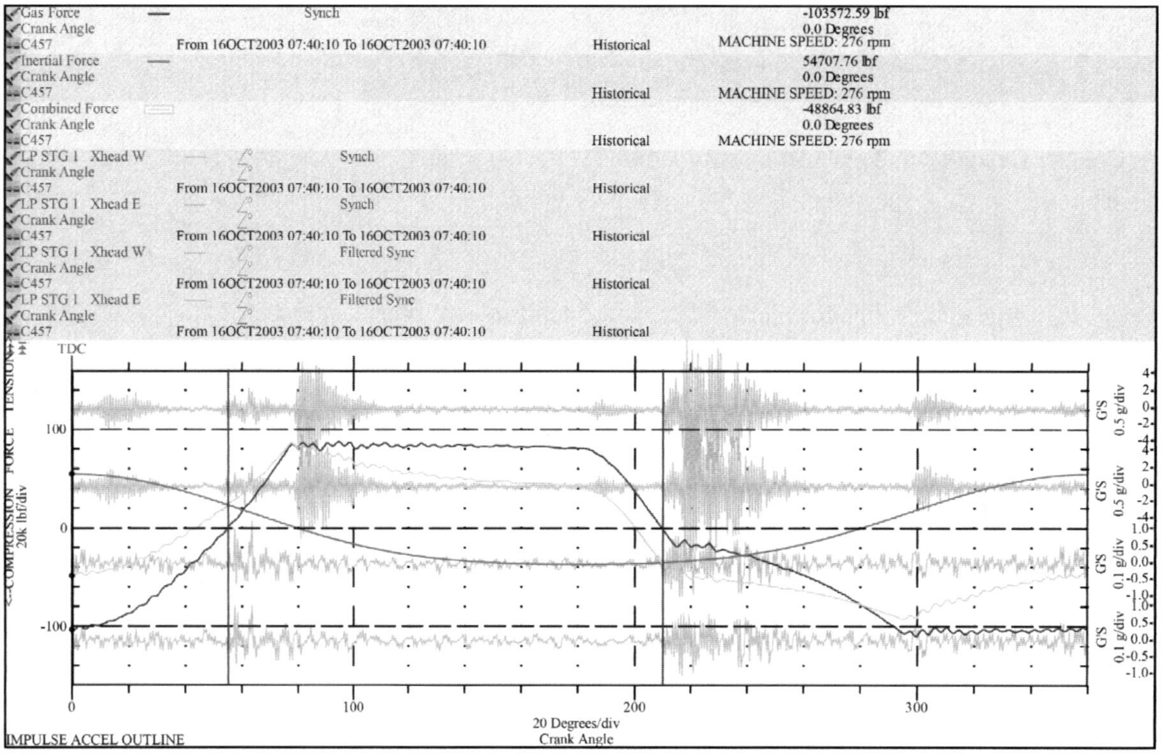

Figure 22: Stage 1 rod load curves (foreground) and crosshead acceleration signals (background) at time of alert.

After running the compressor for another two months, other equipment brought the unit down, so the plant engineer decided to pull the head end head and inspect the piston. Figure 23 shows the inspection results. The locking pin had fractured, allowing the nut to turn. A portion of the pin, highlighted by the red rectangle, had become trapped between the nut and rod.

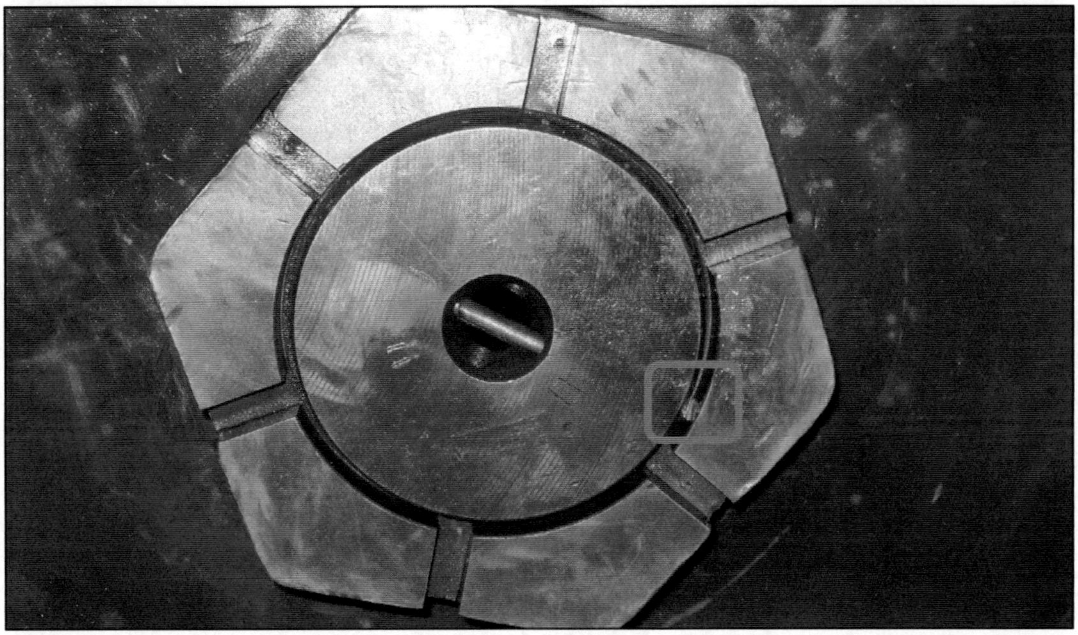

Figure 23: Murphy was an optimist (picture showing broken retaining pin on the piston nut).

CONCLUSION

The paper explained how loads and compressor kinematics interact to create forces at the crosshead pin. These forces have a horizontal component (parallel to the piston rod centerline) and a vertical component. The horizontal rod load forces have good definitions in industry documentation and API standards. In contrast, the vertical forces receive little attention. Combining these forces with a simple model allows the prediction of crosshead motion and impact events. Various measurements characterize the crosshead guide signal, including banded pk, rms, segmental bands, and tracking bands. Using these measurements and the model with well-established crosshead guide acceleration diagnostic principles allows for diagnosing faults under various conditions.

APPENDIX A – RECIPROCATING COMPRESSOR ROD LOAD

The term "rod load" has been used for decades to describe the maximum forces a reciprocating compressor assembly can withstand. API-618 uses the terms combined rod load and gas load to describe these forces.

Gas Load

The pressures inside the cylinder act on the piston and cylinder faces. API-618 (paragraph 6.6.2) [9] calls the sum of these forces the gas load. Convention from statics describes tensile forces as positive and compressive forces as negative. For a double-acting cylinder, the gas force equals the crank end piston area times crank end chamber pressure minus the head end piston area times the head end chamber pressure:

$$F_{\mathrm{GasLoad}} = A_{\mathrm{CE}} P_{\mathrm{CE,Gauge}} - A_{\mathrm{HE}} P_{\mathrm{HE,Gauge}} \quad \textbf{Eq. (3)}$$

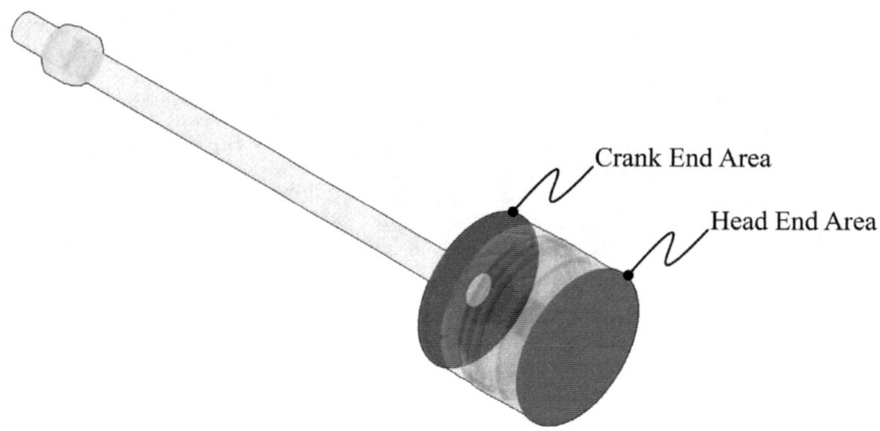

Figure 24: Piston Areas Used for Gas Rod Load Calculation

For reciprocating compressor diagnostics, the plot usually displays the pressure and load curves together. Figure 25 shows an example of this format. The x-axis labels indicate the degrees of revolution, with 0° indicating top dead center (TDC). At this angular position, the head-end discharge valves have just closed, as have the crank-end suction valves. The head end chamber has a higher pressure than the crank end chamber. For this cylinder, see Table 2, the head end does not have a tail rod, so it also has more surface area. The combination of high pressure and more area places the rod in compression (negative sign). The bottom pane of Figure 25 shows the gas load curve starting as a negative value. As the chamber pressures change, the gas load changes as well. It becomes positive (tension) just after the chamber pressures equalize and reaches a maximum value when the crank end discharge pressure reaches a maximum.

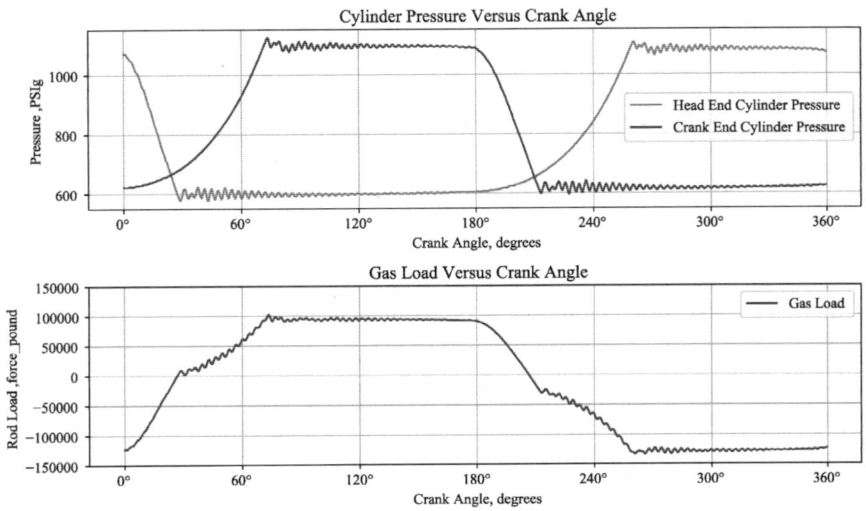

Figure 25: Plot showing indicated cylinder pressure curves (top pane) and the calculated gas load curve (bottom pane).

Table 2 – Throw dimensions.

Chamber	Bore Diameter (inches)	Rod Diameter (inches)	C' Rod Length (inches)	Stroke (inches)	RPM (-)
Crank End	17.52	5.51	50.59	14.17	327
Head End	17.52	0	50.59	14.17	327

Inertia load

Starting and stopping the piston and crosshead assembly twice per revolution creates a force acting on the crosshead pin and connecting rod. This force can be calculated from Newton's Second Law of Motion (force equals the mass times acceleration):

$$F_{\text{InertiaLoad}} = (Acc * \ Recip\ Mass) \qquad \textbf{Eq. (4)}$$

the reciprocating mass must include the piston and crosshead assembly masses to calculate the forces at the crosshead pin.

The reciprocating acceleration has a more complicated formula. [5] provides the derivation for the acceleration formula using the nomenclature shown in Figure 26.

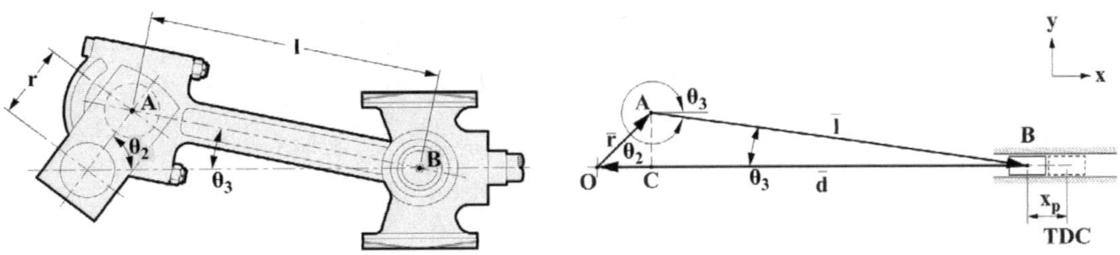

Figure 26: Kinematic nomenclature used for reciprocating acceleration calculation.

Where:

$\bar{r}$	Crank radius (1/2 of the stroke)
θ_2	Crank angle (0° at TDC), equals ω*t
ω	Crankshaft speed in radians/second.
l	Connecting rod length
θ_3	Connecting rod angle
$\bar{d}$	Crankshaft centerline to crosshead centerline
θ_4	Angle of vector $\bar{\mathrm{d}}$, fixed at 180°
x_p	Distance from top dead center (TDC)

Eq. (5) shows the expression for calculating the reciprocating acceleration term. From a vibration analysis perspective, this equation shows that the vibration components in the inertia load depend on the kinematics.

$$Acc = -r\omega^2\left(\mathrm{Cos}(\theta_2]) + \frac{r\,\mathrm{Cos}(2\theta_2)}{l\left(1-\frac{r^2\,\mathrm{Sin}\,(\theta_2)^2}{l^2}\right)^{3/2}} + \frac{r^3\,\mathrm{Sin}(\theta_2)^4}{l^3\left(1-\frac{r^2\,\mathrm{Sin}(\theta_2)^2}{l^2}\right)^{3/2}}\right)$$

Eq. (5)

As with gas load, reciprocating compressor diagnostic plots include the inertia curve along with the gas load and indicted cylinder pressure curves. Figure 27 shows the plots, with the inertia curve added to the bottom pane. In contrast to the gas curve above, the crosshead pin must slow the crosshead, piston rod, and piston assembly as the piston approaches the head end head. This places the crosshead pin in tension, opposite of the gas force curve. Another way to look at this: the gas force acts to slow the piston assembly. Since the gas force and inertia force act in opposite directions, lowering the pressure can result in higher combined rod load, especially for machines operating above 600 RPM.

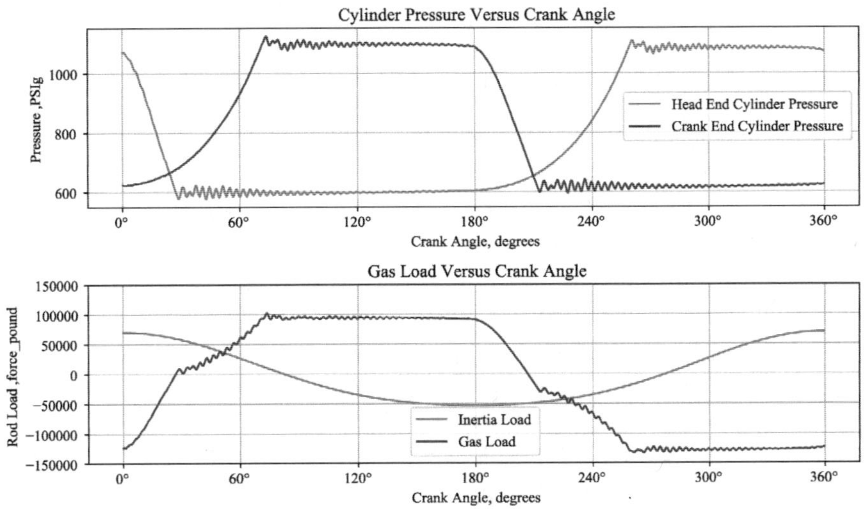

Figure 27: Plot showing indicated cylinder pressure curves (top pane), the calculated gas load curve (bottom pane, blue line), and the inertia load curve (bottom pane, orange line).

Combined Rod Load

Adding the gas and inertia load curves together gives the total combined rod load curve, shown in Eq. (6). It represents the force acting on the crosshead pin to which the connecting rod reacts.

$$F_{\text{CombinedLoad}} = (F_{\text{GasLoad}} + F_{\text{InertiaLoad}}) \quad \textbf{Eq. (6)}$$

The reciprocating compressor diagnostics use this combined rod load to detect looseness at the crosshead.

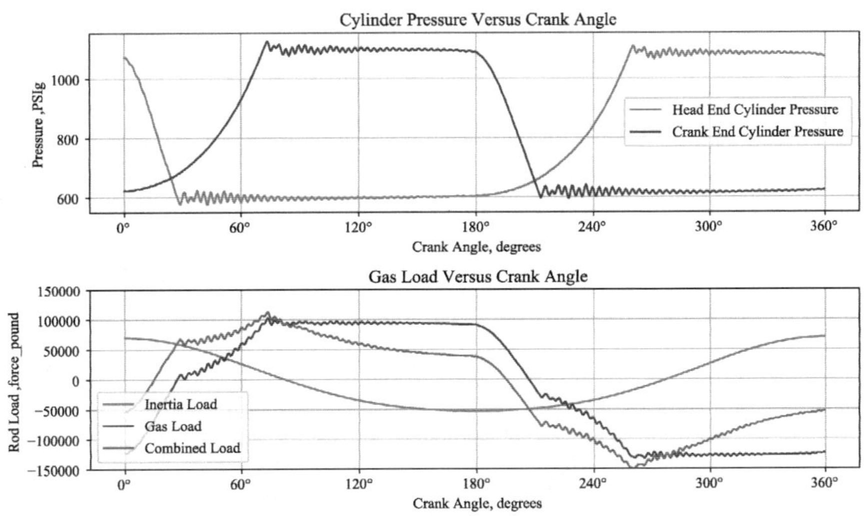

Figure 28: Plot showing indicated cylinder pressure curves (top pane), the calculated gas load curve (bottom pane, blue line), inertia load curve (bottom pane, orange line), and combined rod load curve (bottom pane, green line).

REFERENCES

[1] Bently Nevada. Datasheet. "3500/70M Impulse Velocity Monitor." Document No. 166766 Rev. M.

[2] De Waal, Cornelis G. Apparatus for Measuring the Thickness of Piston Rider Rings. 19 Jan. 1991. United States Patent No. 4,987,774.

[3] Buquoi, Dennis. "Detecting Internal Looseness in a Reciprocating Compressor." Orbit, pp. 8-9, March 1995.

[4] API-670, 2014, "Machinery Protection Systems." Fifth Edition, American Petroleum Institute, Washington D.C.

[5] Howard, B. F., Hatch, C., Yonker, G., and Philips, K. Measuring and Modelling Vertical Crosshead Vibration in API-618 Reciprocating Compressors, Proceedings of the 44th Turbomachinery Symposium, Texas A&M University, Houston, TX, 2015.

[6] Den, Hartog J. P. Mechanical Vibrations, New York: Dover Publications, 1985. Print.

[7] Bently Nevada. Application Note. "Reciprocating Compressor Vibration Measurements." Document No. 173M7207.

[8] Dietermann, Alex and Jetalina, David. Proven Methods for Evaluating the Health Condition of Piston Compressors, Proceeding of the Thirtieth Turbomachinery, pp. 219-225, 2001.

[9] API 618, 2007, "Reciprocating Compressors for Petroleum, Chemical, and Gas Industry Services," Fifth Edition, American Petroleum Institute, Washington, D.C.

[10] Horowitz, Paul, and Thomas C. Hayes. The Art of Electronics. Cambridge Univ. Press, 2001.

Gap in pagination due to formatting issues.

Page 357

VIATC TRAINING CONFERENCE & EXPO
COVINGTON, KENTUCKY | AUGUST 7-9, 2024

BELT-DRIVEN FAN 1X & 2X VIBRATION CASE STUDY: THE ROTOR RESONANCE THAT WASN'T…OR WAS IT?

Neil Tugby, P. Eng
Senior Vibration Engineer
Ontario Power Generation
neil.tugby@opg.com

Neil Tugby is presently a Senior Vibration Engineer at Ontario Power Generation in Toronto, Ontario, Canada. He has 15 years of experience supporting rotating equipment commissioning, diagnostic & balancing activities, as well as numerous M&D and reliability initiatives, at a variety of hydroelectric, nuclear and fossil power generation facilities.

Neil holds a B. Eng degree (Mechanical Engineering) from Carleton University and an M. Eng degree (Mechanical & Industrial Engineering) from the University of Toronto. He is a licensed Professional Engineer in the province of Ontario and is certified as a Category IV Vibration Analyst with the Vibration Institute.

ABSTRACT:
A belt-driven, double-wide, double-inlet, between rolling element bearings, squirrel cage fan was experiencing excessive fan 1x & 2x running speed vibrations. Multi-channel steady state and coast down vibration data, along with motion amplification videos were collected to inform a diagnostics effort.

Based on the resulting data, two independent machine faults were suspected: a fan rotor imbalance responsible for the fan 1x response and a second, independent machine fault responsible for the fan 2x response. Maintenance checks were performed, and a series of diagnostics tests undertaken including impact testing of the fan rotor and its support structure.

Single plane balancing was successfully carried out to correct the suspected quasi-static imbalance. The fan 2x issue proved elusive so several theories were entertained and tested. The responsible machine fault was ultimately identified and successfully corrected but identification was possible only through indirect means. The exact cause of the negative diagnostic test results en route to the indirect diagnosis is yet to be fully understood.

KEY WORDS: Fan, belt, quasi-static, balance, resonance, motion amplification, rolling element, nuclear, spring isolator, coast down, impact test

INTRODUCTION & MACHINE DESCRIPTION:

Diagnosis and correction of a rotating equipment vibration issue can at times take the analyst on a meandering path. This is especially true when the machinery suffers from multiple underlying issues and so called red herrings are present. Navigating these kinds of vibration issues can be a challenge - especially in a production-focused, yet highly change-resistant environment like a nuclear power plant - but can also provide some of the best opportunities for growth as an analyst. The following details an experience of this type that makes for an interesting case study.

The common services mechanical equipment room of a 4 Unit nuclear power plant (4 x 878MW) is provided with ventilating air flow by a redundant set of 950rpm fans (2x100% duty, F1 & F2). The fans function to dissipate the heat gain generated by a large collection of brine chillers, pumps, fans and electrical equipment within the common services room. The fans are classified as safety-related, critical equipment due to the operational importance of the collection of equipment they dissipate heat from.

Each fan is belt-driven by a 15hp, 575V, 1780rpm constant speed, 3-phase induction motor with rolling element bearings. The belt drive consists of 3 V-belts and fixed ratio sheaves, with taper-lock bushings that are also keyed to their respective shafts.

Each fan rotor is housed in a barrel/casing within a plenum and consists of a pair of back-to-back squirrel cage impellers with backward-inclined airfoil blades. The impellers are keyed to a steel shaft. The pair of impellers intake air axially from opposite ends of the fan, through inlet bells with stationary vanes. Air discharge is radially upwards to ductwork connected to the plenum with a flexible joint. Each fan rotor is supported by 3 self-aligning ball bearings in pillow blocks, in the configuration shown in Figure 1. The plenum, fan barrel, pillow blocks support structures and motor support structure are all fastened to a common support frame. The frame is then spring-isolated to a plinth via 4 vertical & 2 horizontal springs.

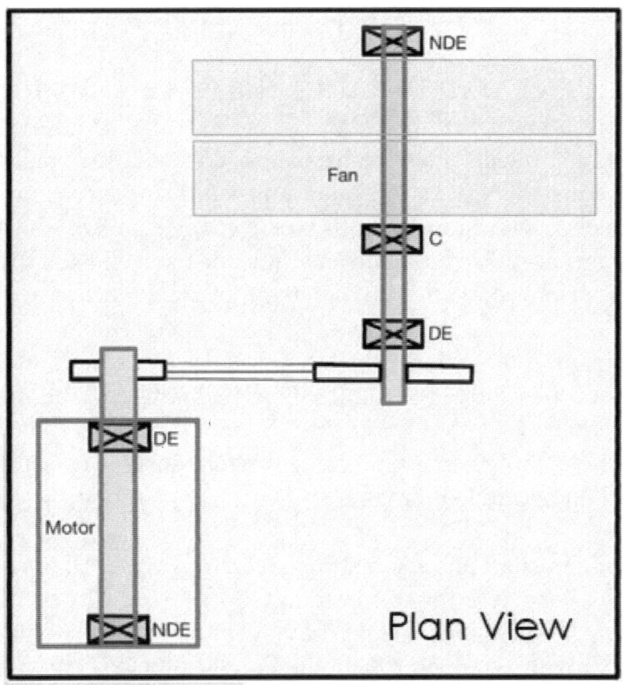

Figure 1: Fan Overview (left - view from south, right – machine overview, not to scale)

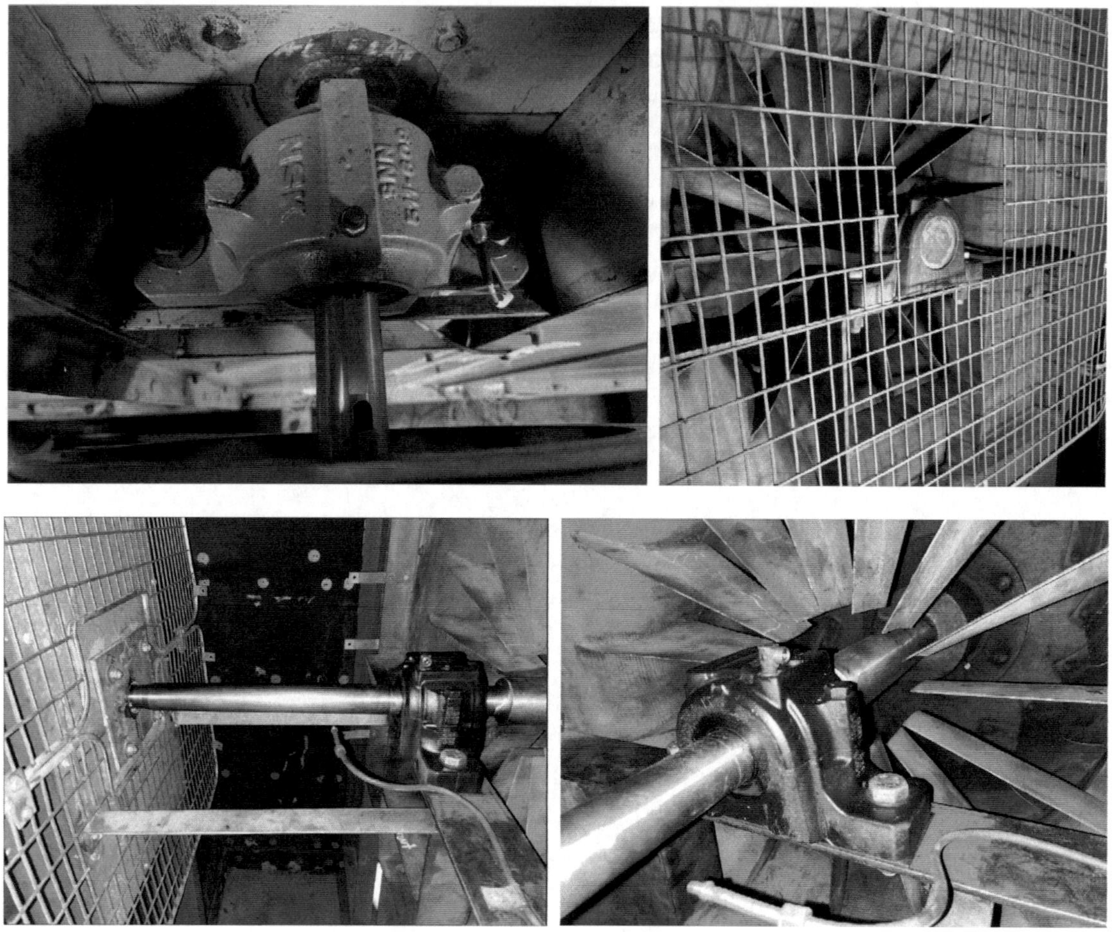

Figure 2: Fan Bearings
(Upper L - DE, Upper R - NDE, Lower L - Center, Lower R - Center & Impeller)

RECENT MAINTENANCE HISTORY AND PROBLEM DESCRIPTION:

Recent history of the subject fan (F2) starts with installation of a new motor. The existing motor was removed for a time-based rewind under the station's Motor Program. There were no notable motor performance or condition monitoring issues prior to replacement. Post-maintenance baseline vibration readings are typically taken by the station Predictive Maintenance (PdM) Program following major maintenance activities. However, in this instance baseline readings were not recorded following the motor replacement.

Approximately 6 months after motor replacement, the fan shaft severed between the DE and center bearings. Fan shaft failure had not occurred on either F1 or F2 in the past. No PdM data had been recorded leading up to the failure. The failed material was unfortunately not retained but a verbal description of the fracture surface (discussions with maintenance) suggested failure in bending due to fatigue.

The fan shaft and fan bearings were replaced but only through significant effort. The fan impellers were found seized to the shaft and substantial force was required to pull them free. This maintenance insight proved valuable for the balancing task, which will be detailed below. Removal of the rotor from the fan barrel and plenum was also difficult. The F1 & F2 plenums are in close proximity axially and there is no radial access to the fan barrel to facilitate removal. This poor maintainability also limited inspection scope and acted as a deterrent for a subset of the diagnostic & corrective options that would have required rotor removal.

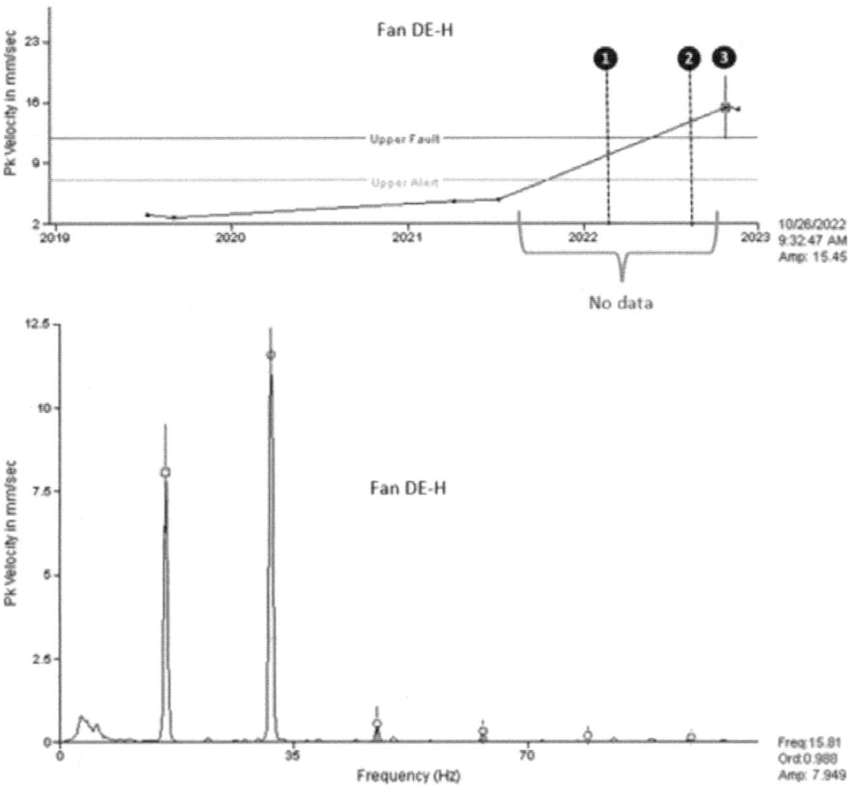

Figure 3: PdM Program Vibrations (Fan DE-H)
(❶: Motor Replacement ❷: Fan Shaft Failure ❸: Post Fan Shaft & Bearings Replacement)

Table 1: PdM Post-Mtce Vibes

Position	Velocity Amplitude (mm/s-pk overall)
Mtr NDE-H	11.3
Mtr NDE-V	7.4
Mtr NDE-A	4.0
Mtr DE-H	11.8
Mtr DE-V	11.1
Mtr DE-A	4.3
Fan DE-H	15.5
Fan DE-V	7.2
Fan DE-A	2.0
Fan NDE-H	8.7
Fan NDE-V	11.6
Fan NDE-A	3.4

1st alarm: 7mm/s-pk overall
2nd alarm: 12mm/s-pk overall

Vibrations were recorded by the PdM Program following fan shaft and bearings replacement. This was the first set of readings since prior to the motor replacement, as indicated in the trend plot of Figure 3. Overall casing vibration amplitudes at several positions now exceeded alarm limits as shown in Table 1. The velocity spectra were dominated by fan 1x & 2x running speed at most locations, as shown in Figure 3 for the Fan DE-H position.

Machine Dynamics was contacted to perform vibration diagnostics.

INITIAL DIAGNOSTICS – MULTI-CHANNEL STEADY STATE VIBRATIONS:
The machine was instrumented with tri-axial accelerometers (5 in total, 1 for each bearing location) and a laser tachometer viewing reflective tape at the fan DE. During instrumentation setup, scaffolding was found clamped to the support structure beneath the motor NDE. The scaffolding was modified to eliminate contact prior to recording data.

After clearing the scaffold interference, steady-state amplitudes worsened significantly relative to the PdM levels above, as indicated in Table 2. The scaffold had been constraining motion of the support frame and adding stiffness to the system. Table 2 is therefore a more accurate quantification of the true post-maintenance vibration levels.

Table 2: Multi-Channel Post-Mtce Vibrations

	Position	1X Amp	1X ϕ*	2X Amp	2X ϕ	Overall (mm/s-pk)
Mtr	NDE-H	4.7	55	7.2	23	15.1
	NDE-V	5.7	60	7.2	214	15.5
	NDE-A	3.4	351	**11.4**	164	17.2
	DE-H	4.3	47	7.5	19	13.8
	DE-V	5.6	55	5.6	268	10.8
	DE-A	2.7	356	**11.7**	158	16.1
Fan	DE-H	6.8	42	**9.3**	2	17.4
	DE-V	5.7	34	2.8	255	7.6
	DE-A	1.7	146	0.7	109	2.7
	C-H	0.7	20	**14.4**	198	21.6
	C-V	5.0	18	0.4	187	9.2
	C-A	2.0	191	0.8	281	7.6
	NDE-H	6.7	33	1.7	322	10.3
	NDE-V	7.6	31	3.0	233	9.5
	NDE-A	2.2	148	1.0	175	4.4

*true phase

Several noteworthy observations were made at this point:
- Axials amplitudes now high on motor (not just radials)
- Fan 1x in-ϕ across the fan
- Fan 2x high at motor, fan DE & fan center bearings but not at fan NDE bearing
- Fan 2x highest in axial direction on motor and in horizontal direction on fan
- Large H:V fan 2x amplitude ratio at fan DE & center bearings
- Fan 2x out-of-ϕ across adjacent fan bearings

INITIAL DIAGNOSTICS – MOTION AMPLIFICATION:
Several motion amplification videos were recorded during the initial run. Various camera positions were selected to view accessible areas of the motor, the fan and their support structures. No standout issues were indicated in motion amplification videos specific to the motor, the fan DE, the fan sheave or the accessible support structures.

The fan center bearing and its support structure were largely inaccessible for video recording during operation but we were able to precariously shoot through a small window just above the fan sheave, by removing a plate on the fan plenum that housed the fan shaft seal. This camera position and corresponding perspective is shown in Figure 4.

The resulting fan 2x (~31.6Hz) band-pass filtered video clearly illustrated highly directional vibration that was out-of-phase between the fan DE and center bearings (indicated by blue arrows in Figure 4). This did not reveal anything that wasn't already in the vibration data from the accelerometers. However, it proved very useful as a communication tool when describing the vibration issue to maintenance and operations. Those that have spent several years regularly analyzing vibrations may underappreciate or forget how weakly-grasped conventional vibration data is by the novice. In contrast, you can see faces light up when observing the same dynamics via motion amplification videos. There is a nearly immediate comprehension and a better appreciation for the severity of the vibration symptoms than is possible with conventional vibration tables and plots alone. This acts to renew maintenance and operations enthusiasm towards solving the problem.

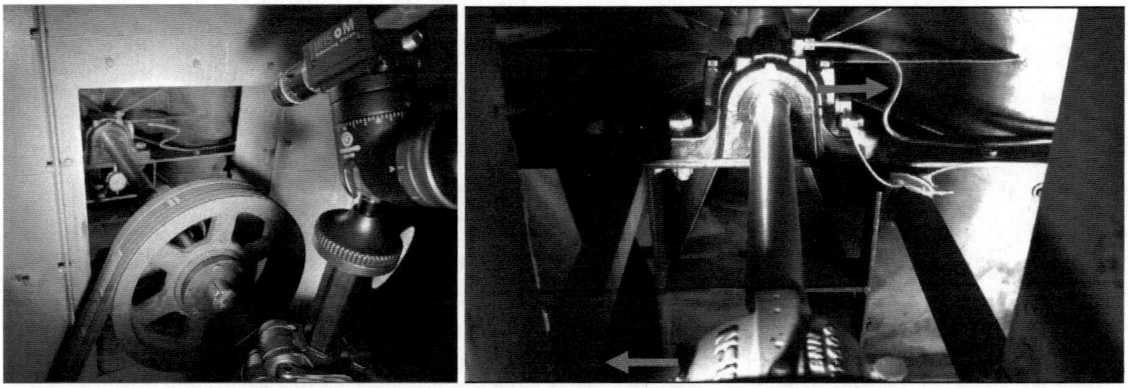

Figure 4: Fan Center Bearing Motion Amplification

The fan NDE bearing was shot using the camera position and corresponding perspective shown in Figure 5. The photo also illustrates the poor maintainability due to proximity of F1 & F2 plenums, described above in the Recent Maintenance History section.

Figure 5: Fan NDE Bearing Motion Amplification

The resulting fan 1x (~15.8Hz) band-pass filtered video revealed a rotating arc of the impeller shroud that was proud of the inlet bell. The visible arc of the shroud was relatively reflective and lit up by the DC flood lighting as shown in Figure 5. Bearing vibration appeared to lag the motion of the shroud. More accurately, bearing vibration was lagging the motion of the impeller 180° opposite of the visible shroud, as the impeller's center of gravity would be offset to the axis of rotation along that vector.

Following coast down, the fan rotor was slowly turned by hand and the outboard impeller shroud still appeared radially offset (~½") from the inlet bell. The same observation was not present at the inboard impeller. Recalling that the impellers were found seized on the failed shaft and significant force was required to remove them, it is likely that the outboard impeller was distorted during removal. Taken together with the vibration data, these observations indicated a rotor imbalance, due to an eccentric impeller, exciting fan 1x radial vibration especially at the fan NDE.

INITIAL DIAGNOSTICS – COAST DOWN:

A fan center-H coast down Bode plot is shown in Figure 6. Overall amplitude, fan 1x and fan 2x are plotted. A global natural frequency of the entire isolated machine in the horizontal direction is indicated at ~3Hz. More importantly, the fan 2x amplitude is seen to rapidly decay after the power is cut and speed starts to decrease.

This same behaviour is shown in a different manner in Figure 7, which contains the fan DE-H and fan center-H TWFs captured at 950rpm, ~900rpm and ~850rpm. Each fan shaft revolution is shown by a blank-dot indication. Visually, the fan 2x component is barely present at the fan DE-H by 900rpm and non-existent at both positions by 850rpm.

The sharp amplitude drop-off with speed change, along with the highly directional steady-state vibes, suggests a resonance at fan 2x. However, one reservation with that diagnosis is the absence of a fan 2x phase shift in the Bode corresponding to the drop-off in amplitude.

Another possibility could be that the phenomenon is due to some change in mechanics that occurs when transitioning between the energized (motor on) and de-energized (motor off) states. Coast down is fairly rapid (< 2 secs to 850rpm) and there is very little damping in the rotor or support structure, so the fan 2x excitation force may be immediately cut at the moment of de-energization but then require a few seconds for the bearing vibration response to decay to zero.

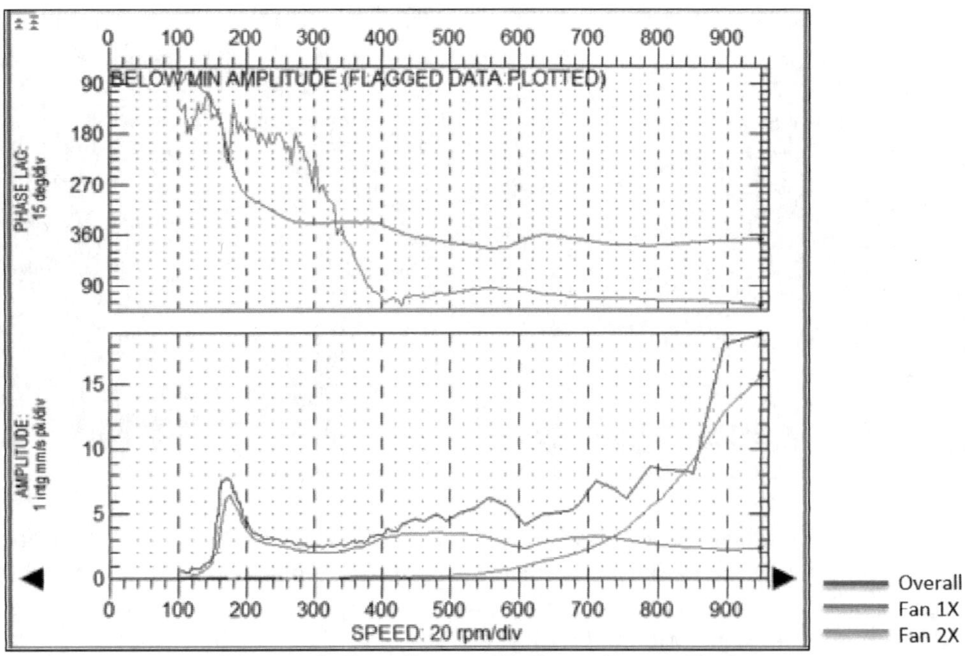

Figure 6: Fan Center-H Coast Down Bode

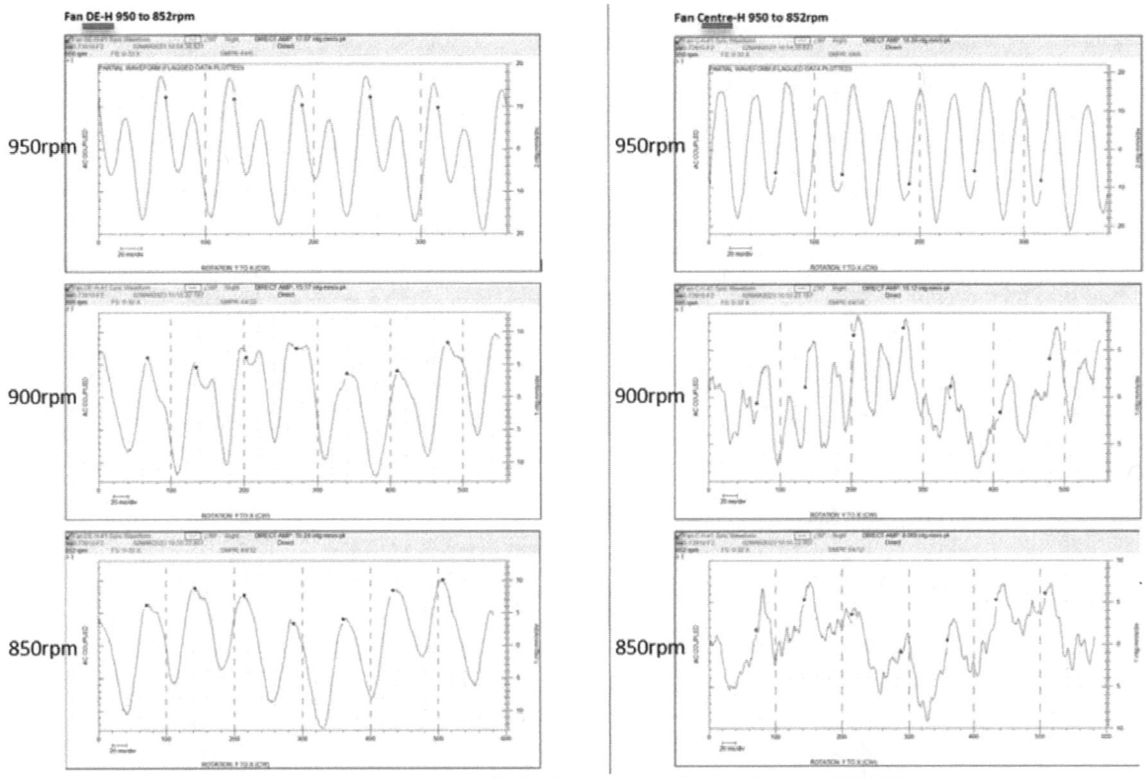

Figure 7: Fan DE-H & Center-H Coast Down TWFs

PRELIMINARY DIAGNOSIS:

Based on the information obtained to this point, the preliminary diagnosis was two-fold:

1. Fan 1x: due to imbalance, and
2. Fan 2x: most likely due to structural or rotor resonance, more testing required

The presence of a 2x excitation force is somewhat atypical for an otherwise healthy machine of this design. So regardless of whether there was a natural frequency near fan 2x, the source of fan 2x excitation had to be considered. It was theorized that the fan 2x excitation force was due to:

i. Asymmetric shaft stiffness (due to long impellers & sheave keyways) "seeing" a static radial load (belt tension) twice per revolution, or
ii. Shaft crack (another potential source of radially asymmetric stiffness), or
iii. Fan bearing misalignment, or
iv. Some currently unknown phenomenon related to energization (motor on vs motor off)

IMPACT TESTING:

Impact testing of the fan structure and fan shaft was undertaken next to confirm or refute resonance at or near fan 2x (~31.6Hz). Impact and response positions included the fan center bearing, fan DE & center bearing support structures, motor casing, fan shaft near the center bearing and fan shaft inboard and outboard of the DE bearing. The fan shaft impacts were first performed with the keyways in the same plane as the impact direction and then with the keyways rotated 90° to the impact plane. Impacts and response measurements were in horizontal plane.

Impact test results were quite similar for the various impact-response pairs. A representative result is shown in Figure 8 for the fan center pillow block in the horizontal. Recall that this position & direction had the highest fan 2x amplitude and the largest H:V amplitude ratio, so if a mode was present, it would be reasonable to expect high modal participation at this location.

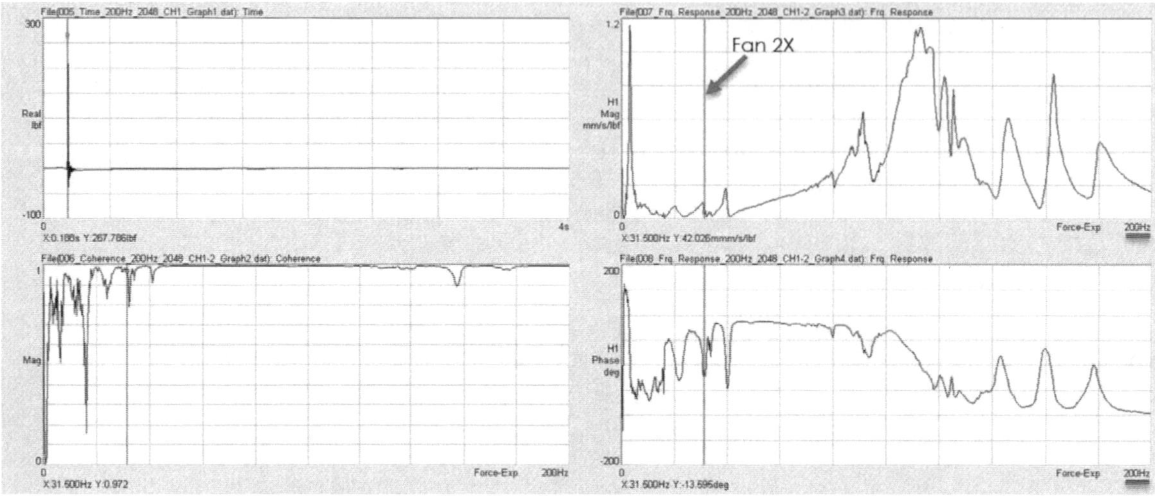

Figure 8: Fan Center-H Impact Test Results (linear, 200Hz)
(Upper L - Force TWF, Upper R - Mobility FRF Mag, Lower L - Coh, Lower R - Mobility FRF ϕ)

Coherence was acceptable at and around fan 2x and the global ~3Hz mode seen in the coast down Bodes was clearly indicated, lending confidence in the results. The FRF ϕ plot showed something of a phase shift through 30Hz but there was not a corresponding high mobility magnitude, at least when compared to the mobility at the 3Hz mode and at several frequencies above 80Hz. Switching the FRF Mag y-scale to decibels and zooming in to a 60Hz span (Figure 9), we could possibly make the case for a natural frequency just above 30Hz but it was still a hard sell. In theory, a lower frequency lateral rotor mode at rest could be driven-up in frequency through some rotor dynamic phenomenon, akin to gyroscopic or centrifugal stress stiffening. However, that was a leap in logic with the information available at this point.

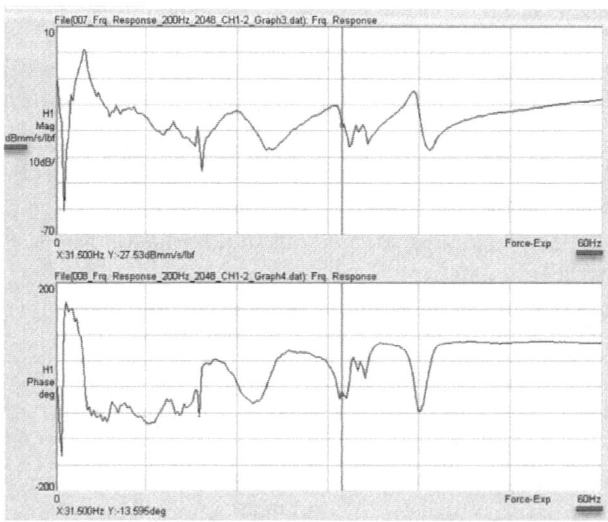

Figure 9: Fan Center-H Impact Test Results (dB, 60Hz)
(Upper - Mobility FRF Mag, Lower - Mobility FRF ϕ)

Overall, the set of impact tests did not produce strong evidence for a fan 2x resonance.

This was a somewhat surprising outcome given the coast down behaviour and high H:V fan 2x amplitude ratio in the steady state vibrations, so some self-doubt was exercised and the setup and test procedure scrutinized. Impact energy was confirmed to be sufficient across the frequency range of interest, using a -10dB decay rule of thumb [1][2] in the impact force spectra and per coherence results. In addition, it was audibly evident that we were exciting the structure, as it would energetically ring out for a long duration upon impact. Clearly we were not impacting a fan 2x modal node, given that the operating deflection shape is very active at the fan center bearing in the horizontal and multiple impact-response pairs were tested. Lastly, the test was repeated with a different analyzer and velomitor make & model and the results were repeatable.

MISCELLANEOUS FAN 2X CONSIDERATIONS:
With the negative impact test results, it was time for a different approach. We considered several potential issues that were related to the recent maintenance scope, easy enough to quickly check but not a perfect match with the vibe data. The aim of discussing these items is to provide a window into the meandering path that diagnostics can take while developing familiarity with the design, operation and maintenance of a particular machine. Not all paths are fruitful. The report that follows a diagnostics campaign often shows a relatively straight line from problem statement to success. Those familiar with diagnostics know this is often not the case.

Station engineering expressed uncertainty about the materials and geometry of the new shaft relative to that used historically. This raised concern about a potential unknown change in stiffness of the shaft. No drawings of the original or replacement shaft were available. Visual comparison of dimensions with the redundant fan (F1) was not possible as Operations declared F1 as Protected Equipment. As a result, it could not be inspected or tested (unless you wanted to lose your job in the process).

So we were suspicious of an issue with the new shaft - whether it be with the design or due to degradation - but wanted to avoid removing it. The fan shaft was inspected visually for any evidence of cracking. In-situ inspection scope in the assembled state was of course limited and the visual inspection method is less than ideal. That said, no issues were identified.

The fan support structures were inspected at rest and with a strobe while in-service. No degradation was noted, bolt tightness was confirmed throughout and axial & radial play of the rotor was minimal.

Pillow blocks are usually supplied with bearing fixing rings. The end user decides if an individual bearing position should have the fixing rings installed (fixed/thrust bearing) or left out (free/floating bearing). Detrimental effects can occur if the wrong bearing or multiple bearings are axially-fixed, especially for designs that must accommodate thermal expansion/contraction. F2 had fixing rings installed only the fan center bearing, per design.

Sheave and shaft runouts were measured. No issues were indicated: 0.006" face & 0.002" rim on the fan sheave and 0.001" on the shaft adjacent to the fan sheave. Alignment of the belt-drive was also checked and found acceptable.

Concerning alignment of the fan bearings, the NDE is not visible from the DE or center bearing once the rotor is installed, so alignment has to be indirectly verified. An alignment method was proposed and presented to the maintenance crew. It was evident from the resulting discussion that little scrutiny had been applied to alignment during installation.

Static deflection of the long and slender fan shaft was also an alignment consideration. The fan shaft is subject to static radial loads from the belt drive and weight of the impellers. Figure 10 offers a set of rough fan shaft static deflection curves to illustrate this thought.

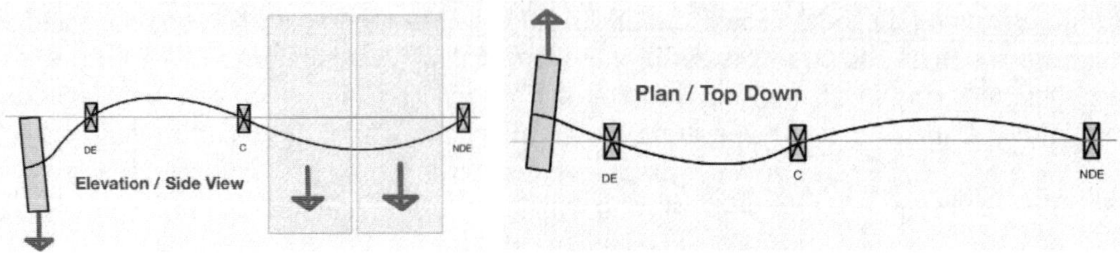

Figure 10: Fan Shaft Static Deflection

Field checks were performed for fan alignment and came back acceptable for the most part, despite the aforementioned concerns. Checks included vertical and horizontal cold-springing measurements when loosening hold down bolts, bearing face ½ sweeps indicated from the shaft with the bearing caps off and feeler gauge measurements side-to-side relative to the grease seals.

This result suggests that a precision setup simply isn't required for this bearing configuration. These particular self-aligning ball bearings (Figure 11) have 3° permissible angular alignment so the shaft static deflection is accommodated. Inducing a significant vertical or horizontal offset misalignment would require applying a static force sufficient to elastically deform the shaft. For most tool choices, this is unlikely to occur unnoticed by the mechanic. Lastly, the same level of scrutiny had been applied on numerous other fans at the station without the vibrations that resulted on F2.

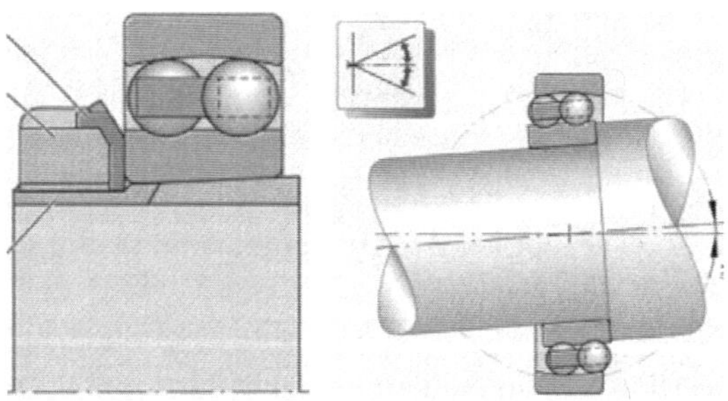

Figure 11: Self-Aligning Ball Bearings - Permissible Angular Misalignment [3]

All that said, a 0.010" cold spring in the vertical direction was found at the center bearing, likely due to fan bearings alignment being performed prior to tensioning the belts. A shim was installed at the center bearing to correct. Following shimming, steady state vibes were recorded and there was no significant change.

A significant static radial deflection of the fan sheave (0.020") was noted at fan DE as belt tension was applied. With reference to Figure 12, this sparked further thought about the shaft deflection curve and whether that deflection was changing with changes in static belt tension vs full speed energized belt tension (motor driving the fan) vs full speed de-energized belt tension (fan's inertia driving the motor).

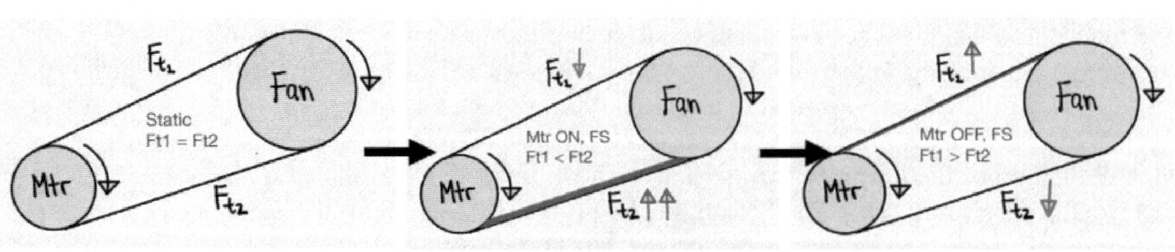

Figure 12: Belt Tension

To test, we shifted the fan sheave closer to the fan DE bearing and reduced the static belt tension. This minimized radial deflection at the fan sheave (static deflection reduced from 0.020" to 0.006"). A minor improvement to fan 2x amplitudes resulted. While this gave some credibility to the static radial load + asymmetric shaft stiffness theory for fan 2x excitation, the result was not sufficient to attribute fan shaft deflection as the culprit overall.

Dynamic offset misalignment of fan bearings was also considered. The idea was that a bearing relative position shift could be occurring only once energized, resulting in a horizontal offset misalignment. We recorded a motion amplification video at moment the power was cut and did not see a shift in the average relative position of DE and center bearings.

Lastly, we found that the NW corner vertical spring isolator was compressed relative to the other three vertical isolators and relative to the same spring on F1 (Figure 13). It was sitting so low that the common support frame was contacting the plinth.

Figure 13: NW Spring Isolator

Inspection did not reveal any difference in spring model vs F1. Ultimately, this isolator will have to be replaced at first maintenance opportunity but for the moment, adjustments were made to clear the plinth. A very minor improvement to amplitudes resulted. Both the fan 1x and 2x issues remained excessive.

IN-SITU BALANCING:

Best practice order of operations is to resolve any non-imbalance machine faults prior to in-situ balancing. This is done to avoid balancing rework, as the presence of other faults may impact the rotor's balance sensitivity. The benefit here is greater for fluid-film bearing machines, where changes in shaft centerline position affect bearing stiffness and damping coefficients, but it is still good practice on rolling element bearing machines.

Whether it be due to the pressure to produce quantifiable progress or simply impatience, we elected at this point to flip the script and attempt to balance the fan. Recall that the fan 1x response had been attributed to imbalance.

The machine configuration, vibration characteristics and recent maintenance history should all be considered to determine the nature of an imbalance, which in turn dictates the set of suitable balancing methods. Being a rolling element bearing machine, the fan operating speed could be approaching, but is unlikely to be above the rotor's 1st critical speed. Traversal through a critical speed with very low damping (as is the case for rolling element bearings) could result in rapid degradation or failure. Indication of such behaviour was not present during run-up or coast down. Therefore, the rotor could be considered rigid as opposed to flexible and modal balancing was not required. The general types of rigid rotor imbalance are shown in Figure 14.

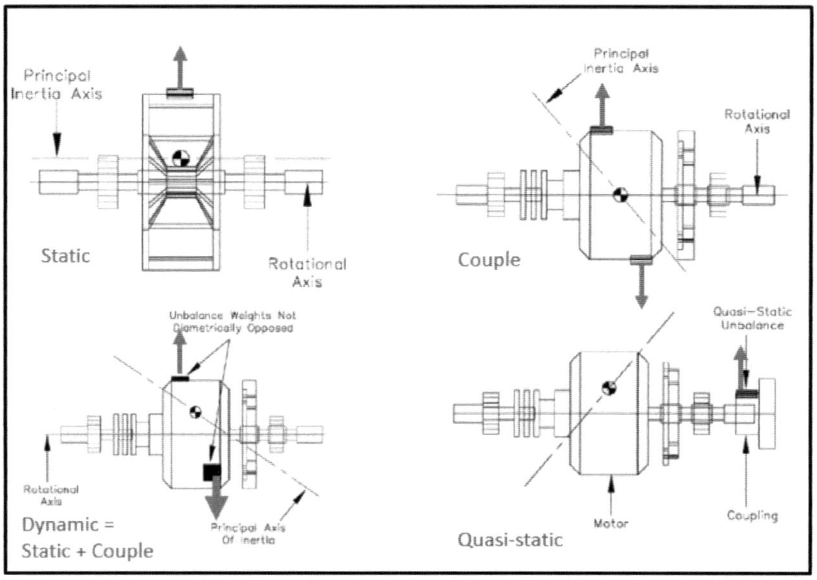

Figure 14: General Types of Rigid Rotor Imbalance [4]

Concerning the number of required balance planes, this rotor has a fairly large length to diameter ratio (1.6:1.0) due to its double-wide impellers. One rule-of-thumb [5] is that for any simply-supported rotor above 150rpm, a 2-plane correction is required when the rotor's discs have an L:D > 0.5. So normally a 2-plane balance would be required for this fan. However, as detailed in the Motion Amplification section the outboard impeller shroud was visibly eccentric while the inboard impeller appeared concentric.

Given the above and an in-phase fan 1x response across the fan bearings (Table 2), it seemed reasonable to treat the imbalance as quasi-static in nature and correct at the seemingly known plane of imbalance, which would lie somewhere axially along the outboard impeller.

The general case of a quasi-static imbalance is shown on the lower right of Figure 14. For a static imbalance the principal inertia axis is offset but parallel to the axis of rotation. In contrast, for a quasi-static imbalance the principal inertia axis is offset and at an angle to the axis of rotation. This is similar to a dynamic imbalance but unlike a dynamic imbalance, a quasi-static imbalance can be corrected in a single plane.

$245lb_f$ per impeller was obtained from documentation and the shaft weight was approximated based on geometry, for a total fan rotor weight of 746lbf. From this and a 9" weight radius, a suitable trial weight was calculated by equating the radial force generated by the trial weight at full speed to 10% of the rotor weight and solving for mass per Equation (1). An approximately 5.2oz (147gram) trial weight was the result.

$$|\overrightarrow{F_W}| = mr\omega^2 = 10\% * \text{rotor weight} \quad (1)$$

Weights were installed near the back plate of the outboard impeller to reduce stress on the short welds joining each fan blade to the impeller shroud. The type of weights that were used are shown in Figure 15. These are barbed to better hold their position and were installed at the inner edge of the fan blades, such that they would be flung out tighter onto the blades during operation, further reducing the chance of throwing a weight.

Figure 145: Balancing Weights

Balance sensitivity vectors ($\overrightarrow{S_{11}}, \overrightarrow{S_{21}}$ *and* $\overrightarrow{S_{31}}$, where plane 1 is the fan NDE) were calculated with the trial weight installed using Equation (2). The phase lag of $\overrightarrow{S_{11}}, \overrightarrow{S_{21}}$ *and* $\overrightarrow{S_{31}}$ were similar. This suggested that the NDE plane could be used to improve fan 1x response at all 3 planes, as the original fan 1x vectors were also in phase across the fan bearings.

$$\overrightarrow{S_{ij}} = \frac{\overrightarrow{TW_j}}{\overrightarrow{T_i}} \qquad (2)$$

where i = response plane, j = balance plane, $\overrightarrow{TW}$ = trial weight vector, $\vec{T}$ = response due to trial weight [6]

Using the influence coefficient method, an effective correction weight of 220 grams < 99° was installed at the fan NDE. As expected, the correction weights were in line with the eccentric shroud as shown in Figure 16.

Figure 16: Balancing Correction Weights

The quasi-static balancing significantly reduced the fan 1x response in all planes, including at the motor. The outboard impeller visual eccentricity remained, of course, but the dynamic force at fan 1x imposed on the bearings due to the eccentricity had been counteracted. Table 3 provides post-balancing vibration readings and percent change in fan 1x, fan 2x and overall amplitudes.

Table 3: Post-Balancing Vibrations

	Position	1X Amp	1X ϕ*	2X Amp	2X ϕ	Overall (mm/s-pk)	% Change 1X	2X	OA
Mtr	NDE-H	2.8	43	7.8	23	15.1	-40%	8%	-2%
	NDE-V	2.2	59	7.6	197	15.5	-61%	6%	-10%
	NDE-A	0.5	3	**11.9**	153	17.2	-85%	4%	-9%
	DE-H	1.9	34	7.8	17	13.8	-56%	4%	-29%
	DE-V	1.9	44	5.4	258	10.8	-66%	-4%	-25%
	DE-A	0.6	29	**12.3**	150	16.1	-78%	5%	-7%
Fan	DE-H	1.4	27	**9.6**	356	17.4	-79%	3%	-17%
	DE-V	1.3	42	2.8	255	7.6	-77%	0%	-34%
	DE-A	0.3	309	0.7	104	2.7	-82%	0%	-22%
	C-H	1.5	215	**14.4**	198	21.6	-	0%	0%
	C-V	0.4	290	0.4	187	9.2	-92%	0%	-2%
	C-A	0.3	270	0.4	137	7.6	-85%	-	5%
	NDE-H	0.7	27	1.7	322	10.3	-90%	0%	-34%
	NDE-V	0.7	296	3	233	9.5	-91%	0%	-45%
	NDE-A	0.3	340	0.9	175	4.4	-86%	-10%	-30%

*true phase

SPEED CHANGE:
As expected, the fan 2x response was unaffected by balancing. Despite the negative impact test results encountered earlier, the idea of a structural or rotor resonance still seemed to be the best fan 2x theory in play.

With this stubbornly in mind, the decision was made to test run with a larger sheave to slightly reduce the full speed of the fan. The speed reduction would also reduce the frequency of fan 2x excitation. If resonant at full speed, a sufficient change in the frequency of excitation should cause amplitudes to drop off. A compatible fan sheave was selected to reduce fan speed by 18% from 950rpm to 775rpm. The corresponding fan 2x reduction was from 31.6Hz to 25.8Hz.

As shown in Table 4, fan 2x amplitudes were nearly completely eliminated as a result of the change in full speed and overall levels dropped below alarm limits. Now we're getting somewhere!

This outcome ruled out a cracked shaft, a fan internal misalignment or some unknown energized vs de-energized phenomenon, as those machine faults, if present at 950rpm would remain present at 775rpm steady state.

Table 4: Vibrations with Fan Speed = 775rpm

	Position	1X Amp	1X ϕ*	2X Amp	2X ϕ	Overall (mm/s-pk)	% Change 2X	% Change OA
Mtr	NDE-H	2.2	27	0.8	280	4.6	-90%	-69%
	NDE-V	2.1	36	1	262	4.6	-87%	-67%
	NDE-A	0.1	337	0.8	276	3.1	-93%	-80%
	DE-H	1.8	30	0.8	276	4.3	-90%	-56%
	DE-V	1.8	41	0.6	255	4.3	-89%	-47%
	DE-A	0.1	300	0.7	276	3.1	-94%	-79%
Fan	DE-H	1.5	26	1.2	259	3.5	-88%	-76%
	DE-V	1.5	24	0.2	130	2.4	-93%	-52%
	DE-A	0.4	337	0.2	73	1.4	-71%	-33%
	C-H	0.3	207	0.8	105	3.4	-94%	-84%
	C-V	0.5	3	0.4	109	1.8	0%	-80%
	C-A	0.5	316	0.3	83	1.8	-25%	-78%
	NDE-H	0.3	66	0.3	288	2.4	-82%	-65%
	NDE-V	0.8	271	0.6	146	2.1	-80%	-60%
	NDE-A	0.5	6	0.3	156	1.7	-67%	-45%

*true phase

Revisiting the impact test results (Figure 17), we can compare fan 2x at 775rpm (blue dashed line) and 950rpm (red line). The mobility FRF magnitude (mm/s/lb$_f$) is similar at these two frequencies, so this does not offer an explanation for the large change in velocity amplitudes.

There is a difference of ~75° in the FRF phase and a similar phase shift was evident in the steady state running vibes, comparing 950rpm (198°) to 775rpm (105°) full speed operation. However, a phase shift only partially satisfies the criteria to declare a natural frequency. Interestingly, the change in 2x phase from 950rpm to 775rpm is not present in the coast down from 950rpm as shown previously in Figure 6. Perhaps this is related to the fan 2x filter settling time. Some uncertainty exists here.

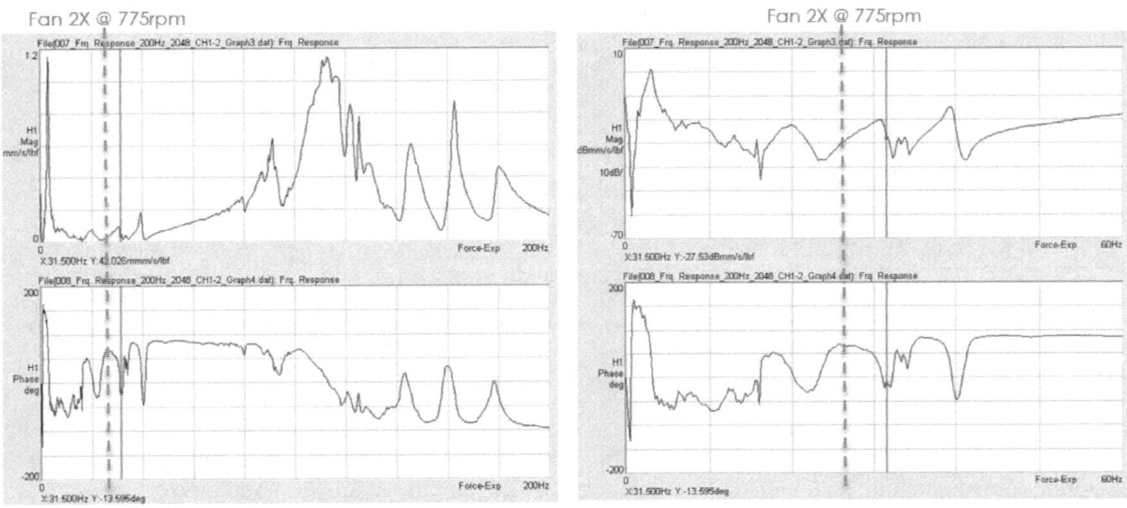

Figure 17: Fan Center-H FRFs with Fan 2x @ 950rpm and 775rpm
(Left – linear, 200Hz, Right – dBs, 60Hz)

High level, the fault behaviour smells like a resonance and quacks like a resonance. There is a very high H:V amplitude ratio and the amplitude & phase drop off significantly with a small change in forcing frequency. Based on this, I've rested on the conclusion that this is a fan 2x lateral rotor resonance excited due to asymmetric shaft stiffness (long keyways for impellers & fan sheave) "seeing" a static radial load (belt tension) twice per revolution. Unfortunately, resonance could not be confirmed irrefutably via impact testing.

A natural question arises: if resonant, why would we not also see the issue on F1, or on the previous installations of F2 prior to motor replacement? This could be due to:

i. the inherently narrow bandwidth of a lightly-damped mode, together with small differences from one fan to the next (e.g. weld lengths & penetration, angle iron dimensions, precise bearing positions along the shaft, etc.) causing a shift of modal frequencies sufficient to tune or detune a resonance [7], and/or
ii. suspected material or geometric differences in the replacement fan shafts (lack of documentation and F1 inaccessible for comparison)

PERMANENT FIX:

At this point, it was decided to proceed towards a permanent fix. Nuclear plants are notoriously change-resistant and bureaucratic processes are in place encourage this culture. The result is that a subset of potential modifications would not be feasible or at least subject to delays, on the order of months or years. This must be front-of-mind when evaluating alternatives. In descending order of practicality and confidence, the following options were discussed with the client:

1. **Speed Change**: Permanently reduce the fan speed. The lower the speed, the better, while still maintaining the documented design basis and an acceptable air flow rate. Without strong evidence on the FRF, it was unclear whether 31.6Hz was slightly below, slightly above or on the natural frequency for operation at 950rpm. Also, exact bandwidth of the mode was not known. Given these uncertainties, a larger speed decrease would increase probability of success from a vibrations perspective.

2. **Bearing Span Adjustment**: Difficult to implement as the footprint of the each pillow block base has essentially the same axial length as the corresponding horizontal beam that supports it. A wider beam would have to be installed to facilitate a shift in position. This would likely be classified as a modification and result in significant delays.

3. **Shaft Modification:** Could work, but is relatively expensive, time-consuming and is onerous to replace the shaft due to the aforementioned maintainability issues.

4. **Add Horizontal Stiffness to Center Bearing Support:** The softest spring in a series of springs largely dictates the equivalent stiffness of the system. So, adding stiffness to the bearing support structure may not work as the stiffness "seen" by the rotor in the horizontal direction may still be dictated by the very soft spring isolators. To increase confidence, modelling and/or an Operating Deflection Shape field survey would be required prior to design and implementation. May be classified as a modification.

5. **Bearings and/or Pillow Blocks of Higher Stiffness**: Same drawbacks as Option 4.

In reviewing these alternatives with the station, it was discovered that the existing design basis allowed for nominal fan speed +/- 5% without modifying documentation. Adding to that, our starting point of 950rpm was above the nominal fan speed in the design basis, providing room to reduce speed by slightly greater than 5%. It was decided to proceed with Option 1.

A compatible fan sheave was procured that would reduce fan speed by 9% from 950rpm to 864rpm. The corresponding fan 2x reduction was from 31.6Hz to 28.8Hz.

Vibration results remained acceptable at 864rpm and the change was made permanent. Table 5 provides the final set of vibration readings including % change relative to the original post-maintenance levels from Table 2.

Table 5: As-Left Vibrations with Fan Speed = 864rpm

	Position	1X Amp	1X ϕ*	2X Amp	2X ϕ	Overall (mm/s-pk)	% Change		
							1X	2X	OA
Mtr	NDE-H	2.2	33	2.1	286	5.7	-53%	-71%	-62%
	NDE-V	1.8	41	1.2	212	4.5	-68%	-83%	-71%
	NDE-A	0.2	10	1.6	285	3.6	-94%	-86%	-79%
	DE-H	1.7	31	1.9	280	5.2	-60%	-75%	-62%
	DE-V	1.5	49	1.1	174	5.1	-73%	-80%	-53%
	DE-A	0.2	355	1.5	283	3.5	-93%	-87%	-78%
Fan	DE-H	1.1	26	2.6	264	5.4	-84%	-72%	-69%
	DE-V	1.2	39	1.3	127	3.8	-79%	-54%	-50%
	DE-A	0.1	270	0.2	325	1.4	-94%	-71%	-48%
	C-H	1	213	2	116	5.4	43%	-86%	-75%
	C-V	0.2	315	0.6	126	2.9	-96%	50%	-68%
	C-A	0.2	275	0.4	316	2.0	-90%	-50%	-74%
	NDE-H	0.7	10	0.2	121	3.2	-90%	-88%	-69%
	NDE-V	0.6	305	1.5	157	3.8	-92%	-50%	-60%
	NDE-A	0.1	11	0.8	206	2.3	-95%	-20%	-48%

*true phase

1st alarm: 7mm/s-pk overall

2nd alarm: 12mm/s-pk overall

REFERENCES:

[1] "*Frequency Response Measurement by Impact Excitation by Using Impulse Hammer*", 2015. onosokki.co.jp/English/hp_e/c_support/ope_manual/pdf/cf9000_impact_analysis_e.pdf.

[2] "*Simcenter Testlab Impact Testing*", 2020. community.sw.siemens.com/s/article/simcenter-testlab-impact-testing.

[3] "*1309 EKTN9 Self-aligning ball bearing with tapered bore",* 2024. https://www.skf.com/my/products/rolling-bearings/ball-bearings/self-aligning-ball-bearings/productid-1309%20EKTN9.

[4] Feese, T.D., Grazier, P. E. "*Balance This - Case Histories from Difficult Jobs*", 2004.

[5] Mobius Institute, "*Vibration Training Course Book, Category III*", 2013.

[6] Eisenmann, R. "*Machinery Malfunction, Diagnosis and Correction*", 1998.

[7] Sayer, Robert J., Eshleman, Ronald L., "Cat IV: Advanced Vibration Control",

VIATC TRAINING CONFERENCE & EXPO

COVINGTON, KENTUCKY | AUGUST 7-9, 2024

STANDARDIZED FAN DIAGNOSTICS ACCORDING TO ISO 13373-5

Aly El-Shafei
RITEC, Cairo, Egypt
Email address elshafei@ritec-eg.com

Dr. Aly El-Shafei is a Certified Vibration Analyst, Category IV. He holds a Ph.D. from the Massachusetts Institute of Technology (MIT). In addition to serving as Managing Director at RITEC, Dr. El-Shafei is also a Professor of Vibration Engineering at Cairo University, Egypt and is the Vice President of the Vibration Institute, USA. Prof. El-Shafei chairs the ISO working group ISO/TC108/SC2/WG10 responsible for developing international standards on machinery vibration diagnostics ISO 13373. Dr. El-Shafei is known as a seasoned diagnostics engineer and has solved many chronic problems worldwide. He is also known worldwide for his work on Squeeze Film Dampers, and for developing new technology for the control of rotating machinery vibration. Prof. El-Shafei is the recipient of the distinguished Grand African Innovation Award in 2017. Dr. El-Shafei has received 5 US patents and has published more than 80 papers on rotordynamics and machinery diagnostics.

ABSTRACT:

There is a new push to standardize vibration diagnostics of machines. The ISO 13373 series of standards describe the vibration diagnosis of machines. ISO 13373-3 describes the basic diagnosis of machines including Fault Tables and Flow Charts, including initial analysis and safety precautions. It also describes the diagnosis of common machine faults like bearings and installation faults. ISO 13373-9 describes electric motor diagnosis, while ISO 13373-7 describes the diagnosis of hydraulic machines. Other standards (under preparation) include ISO DIS 13373-4 for diagnosis of steam and gas turbines, ISO NWI 13373-10 for diagnosis of generators and ISO PWI 13373-8 for the diagnosis of pumps.

ISO 13373-5 for the diagnosis of fans and blowers became an international in 2020. In addition to the Flow Charts and Fault Tables, the standard use Symptom Tables. This paper describes the vibration diagnosis of fans according to ISO 13373-5. An abundance of case studies are used to explain the diagnostic process.

KEYWORDS: Standardized, Diagnosis, Flow Charts

INTRODUCTION:

It has been debated for a long time whether the diagnosis of rotating machinery could be standardized. In fact the debate always centered on whether the knowledge of a human being in diagnosing machine faults, is a personal process that relies on the knowledge and experience of an analyst and is more of an art rather than a science.

However, ISO through the 13373 series has adopted an approach that relies on Fault Tables to document available knowledge on machine faults, and the use of step-by-step flow charts or process tables to document the structured procedure that is used by many analysts in reaching an appropriate diagnosis of a machine. This process, particularly the use of flow charts for diagnosis, was pioneered by the author in his papers on the diagnosis of installation faults [1] and the diagnosis of fans [2].

The author of this paper has participated in developing the ISO 13373 series as the Convenor of ISO TC108/SC2/WG10 and as the project leader for many of its standards. This paper summarizes the author's opinion and many of the approaches that are accepted now as international standards.

In particular, this paper focuses on the newly released ISO 13373-5 for the diagnosis of fans and blowers [3]. The standard describes the diagnosis of directly coupled fans and blowers, as well as belt-driven fans and blowers. The standard uses Fault Tables and Flow Charts, as well as Symptom Tables to help in the diagnosis process. This paper describes the ISO standards for the vibration diagnosis of machines [4], and focuses on the diagnosis process used in ISO 13373-5 [3]. Examples of the application of the diagnosis of fans and blowers are then given [2].

STANDARDS OVERVIEW [4]:

The ISO 13373 series consists of the following parts, under the general title *Condition monitoring and diagnosis of machines — Vibration condition monitoring:*

- *Part 1: General procedures*
- *Part 2: Processing, analysis and presentation of vibration data*
- *Part 3: Guidelines for vibration diagnosis*
- *Part 4: Diagnostic techniques for gas turbines* and steam turbines on fluid film bearings
- *Part 5: Diagnostic techniques for fans and blowers*
- *Part 7: Diagnostic techniques for machine sets in hydraulic power generating and pump-storage plants*
- *Part 9: Diagnostic techniques for electric motors*

The following parts are under development:

- *Part 6: Diagnostic techniques for gas turbines on rolling element bearings*

- *Part 8: Diagnostic techniques for industrial pumps*
- *Part 10: Diagnostic techniques for generators*
- *Part 11: Diagnostic techniques for gearboxes*

ISO 13373-3 [5] is the base document for the diagnosis process. It provides guidelines for the diagnosis of machines. ISO 13373-1 [6] describes the measurement procedures on machines for diagnosis purposes, while ISO 13373-2 [7] describes the processing and analysis of data for diagnosis purposes also. While the other parts [3], [8], and [9] are machine specific standards that describe the unique diagnostic procedures for each machine type.

ISO 13373-3 [5] is described in the standard as: "has been developed as a set of guidelines for the general procedures to be considered when carrying out vibration diagnostics of machines. It is intended to be used by vibration practitioners, engineers and technicians and it provides them with useful diagnostic tools. These tools include diagnostic flowcharts, process tables and fault tables. The material contained herein presents a structured approach of the most basic, logical, and intelligent steps to diagnose vibration problems associated with machines. However, this does not preclude the use of other diagnostic techniques." [5]

The diagnosis process is described in ISO 13373-3 [5] as consisting of flowcharts, process tables and fault tables. The flowcharts and the process tables are essentially a step-by-step question and answer procedure that guides the user in the diagnosis process. The flowcharts are used for an overview of the vibration events and characterize the features, while the process tables are used for more in-depth analysis. The fault tables are used to illustrate common machinery events and how they manifest themselves [5].

Annex A in 13373-3 [5] specifies the systematic approach to the vibration analysis of machines:

a) Annex A.1 is used to gather background information regarding the machine, nature and severity of the vibration.

b) Annex A.2 is used to answer a set of questions aimed at arriving at a probable diagnosis of such common faults as unbalance, misalignment and rubs.

c) Annex A.3 is used to set out certain considerations when recommending actions following a probable diagnosis.

The diagnosis of faults common to a wide range of machines are shown in other annexes [5]:

— Installation faults and examples are described in Annex B.

— Radial hydrodynamic fluid-film bearing faults and examples are described in Annex C.

— Rolling element bearing faults and examples are described in Annex D.

The use of ISO 13373-3 [5] is necessary with all machine specific diagnostics standards, including ISO 13373-5 [3] and should be used in conjunction with the machine specific standards, as it describes the initial analysis that must be applied to the diagnosis of all machines [5] as well as common faults to all machines including installation faults, fluid film bearing faults and rolling element bearing faults [4].

FAULT TABLE FOR FANS AND BLOWERS ACCORDING TO ISO 13373-5 [3]

The systematic approach to vibration analysis of fans and blowers is given by the fault table in Table 1. The fault table includes mainly installation faults. For faults regarding fan or blower bearings, see ISO 13373-3:2015 [5], Annexes C and D. Several faults can give similar indications and further investigation would be necessary to distinguish between them.

Table 1 — Fault table for fans and blowers

Fault	Vibration characteristics	Other descriptors	Comment
Shaft misalignment/ concentricity errors	1x, or 1x and 2x, sometimes 1x and 2x and 3x.	Directional force 180° phase shift across coupling. Offset misalignment tends to produce phase shift across the coupling in the radial direction, while angular misalignment tends to produce the phase shift in the axial direction.	There are two types of misalignment: parallel and angular, and in most cases there would be a combination of the two.
Looseness	Usually a series of peaks at rotational speed and integer harmonics of rotational speed, generally the amplitude of these peaks decreasing with higher harmonic numbers.	Looseness can be at bearings or skid, or anchor bolts. Check for difference in amplitude and/or phase at the interface to discern position of looseness.	Looseness can be at the bearing housing (sometimes due to the bearing installation), and/or at the pedestal or the skid.
Excessive bearing clearance	1x. With low amplitude harmonics.	Directional.	Can be due to wear, in both fluid film and rolling element bearings.
Piping strain	1x	Directional, wave clipping in time waveform.	Piping flanges should match without jacking.

Soft foot	1x, plus 2x line frequency in the electric motor	Soft foot test.	Soft foot is the condition that exists when all feet are not correctly supporting the machine. See also ISO 13373-9.
Shaft rubbing	Clipping in time waveform, with 1x and multiple harmonics in spectrum. Light rubbing can cause rotating vectors (spiral vibration).		Not commonly observed on fans.
Unbalance	1x	Phase shift across coupling depends on the mode. Cylindrical modes tend to have 0° phase shift across the coupling, while conical modes tend to have 180° phase shift. Usually, 90° phase shift between the horizontal and vertical measurements at the same bearing location.	Unbalance is often due to erosion, or deposits on blades. Overhung fans may require a couple balance, while centre-hung fans can generally be balanced in a single-plane.
Bent shaft	1x similar to unbalance, manifests itself at slow roll speed.	Can cancel with unbalance at particular rotational speeds.	Rarely seen on fans.
Casing distortion	1x, sometimes 2x.	180° phase shift from end to end.	Only important where bearings are integral with the casings.
Resonance	High vibration at a particular frequency.	Resonance testing indicates natural frequency.	Avoid operating close to a resonant frequency e.g. by changing speed, or by changing resonant frequency, e.g. by stiffening machine or adding mass. Sometimes damping may be needed.
Tilting foundation	High 1x vibration levels that cannot be explained by unbalance, misalignment, bent shaft or eccentricity.	Rocking motion in 1x ODS.	ODS study to analyse problem in more depth.

Aerodynamic forces	Blade passing frequency.	Can have high noise.	Usually caused when fan is operating off best efficiency point.
Belt faults	Belt Passing frequency.	Less than 1x.	Typically due to belt wear, misalignment and/or incorrect tension.
Belt resonance	Belt resonance frequency.	Usually less than 1x.	Usually due to lack of belt tension.
Excessive belt tension	1x	Directional.	Similar symptoms to misalignment
Belt pulley eccentricity	Usually directional 1x, sometimes 1x and 2x.	Sometimes visually observed as wobbly motion.	
NOTE ODS stands for operational deflection shape.			

The Fault table (Table 1) above includes the most common faults in fans and blowers including the descriptors of their vibration characteristics, as well as other fault descriptors. To complement the Fault table, ISO 13373-5 introduces the concept of a Symptom table (Table 2 below). The Symptom table includes the observable symptoms of typical faults, in the form of elevated vibration signals (subharmonic, 1x, 2X, ..,,, etc.) as well as vibration phase and other diagnostic discriminating factors. The Symptom table distinguishes between symptoms almost certain to be seen if fault occurs .and symptoms that may or may not be seen.

As a foot note Table 2 corrects some of the errors in the Symptom table in ISO 13373-5 [3]. These errors have been reported to ISO, and corrective action is going to be taken soon.

Table 2 — Observable symptoms of typical faults

Fault type	Elevated vibration signals					Vibration phase, etc.				Other diagnostic discriminating factors		Typical corrective action to be considered
	Sub 1x	1x	2x shaft speed	2x electrical supply	>2x	90° V - H	180° shift at coupling	ODS findings	Time wave clipping	Soft foot test	Other observations	
Shaft misalignment/concentricity errors		●	○		○		●				Phase shift across coupling in axial or radial direction	Align machine
Looseness	○	●	●		●						Locate looseness by vibration – test. See A.2.2	Check source of looseness and remedy – e.g. tighten bolts
Excessive bearing clearance		●	○		○							Repair or replace bearing
Piping strain		●							●		Disconnect pipes - does gap appear? Vibration almost always is directional	Adjust pipes to fit without jacking
Soft foot		●		○						Yes, see A.2.1	2x electrical supply frequency indicates distortion of motor stator or frame. See ISO 13373-9 for motor driven fans.	Measure clearance and fit shims or packing or as appropriate
Rubbing	○	●	●		●				●			Identify root cause and rectify it.

Unbalance		●				○					Usually 90 deg phase shift between vertical and horizontal measurements at same bearing location	Check cause, remedy e.g. by cleaning fan blades, re-balance if still required
Bent shaft		●									Axial vibration, phase shift between the bearings	Difficult to balance an overhung rotor with bent shaft
Casing distortion		●	○								180 degrees phase shift end-to-end	Relieve distortion
Resonance	○	●			○			●			High vibration at a frequency that is NOT an obvious multiple of shaft speed	Change speed, or change natural frequency by changing stiffness or mass
Tilting foundation		●						●			Usually result of uneven isolation and uneven mass distribution or poor soil conditions	Requires civil engineering solution
Skid not level		●						●				Level up skid
Aerodynamic forces					●						Usually at blade pass frequency	Correct flow, e.g., by changing damper angle

Belt faults	●	○									Belt pass frequency	Replace belt, align, and tension correctly
Belt resonance	●										Belt resonance frequency	Usually resolved by setting to correct tension
Belts too tight		●	○								1x directional	Tension correctly
Belt sheave eccentricity		●	○								1x directional	Correct sheave eccentricity and align
Belt misalignment		●									Axial	Align sheaves

This table is not exhaustive but contains the most prevalent faults associated with fans and blowers.

● indicates symptom almost certain to be seen if fault occurs.

○ indicates symptom may or may not be seen.

DESCRIPTIONS OF TESTS RELATED TO TABLE 2

SOFT FOOT TEST

Loosen holding down bolts one at a time, checking for gaps opening up under the base and measuring with a feeler gauge if appropriate. Retighten the loose bolt before loosening another bolt. If one or more feet show a gap, it implies that the machine casing has been distorted when tightened down. Filling the gap with the appropriate thickness of shims and then re-tightening is a typical route to correcting this problem.

Alternatively, follow the same procedure by loosening one bolt at a time, while measuring live vibration at the bearings. If the vibration decreases immediately upon loosening one bolt, then the soft foot is at this bolt location.

LOOSENESS LOCATION VIBRATION TEST

Measurements of vibration magnitude and phase are taken at a range of locations, from the bearing housings and machine casing, through the bearing holding down bolts, machine holding down bolts, support frame holding down bolts and skid holding down bolts, looking for the place where the vibration magnitude and/or phase are significantly different at the two sides of the interface. This indicates the location of the looseness.

ODS Test

Measure amplitudes and phase angles at several points to find the operation deflection shapes of the system. Investigate the possible causes for high mobility or high flexibility of the system.

EXCESSIVE BEARING CLEARANCE

This refers to several possible situations.

— Journal bearing, where the clearance between the shaft and the bearing is too large (oil gap).

— Rolling element bearing, where the bearing race is loose in the bearing housing.

— Rolling element bearing, where the clearances between the rolling elements and the races is too large, e.g. the wrong specification of bearing was selected, or serious wear has occurred to the races and rolling elements.

FLOW CHART OF VIBRATION DIAGNOSIS OF FAULTS IN FANS AND BLOWERS [3]

Fans and blowers are amongst the simplest machines. Most fans and blowers are driven either through a coupling or a belt, whilst other fans and blowers are driven by a gearbox.

The most common faults for fans and blowers can be categorized into four main types:

a) installation faults,

b) bearing faults,

c) unbalance, and

d) aerodynamic forces.

The methodology shown in the flow chart of Figure 1 guides the user through the steps taken in the diagnosis process. For an installation faults step-by-step methodology, see ISO 13373-3:2015, Annex B, and for a bearing faults step-by-step approach, see ISO 13373-3:2015, Annexes C and D.

In many industries, particularly the cement industry, fans require frequent balancing. Fan unbalance is depicted by high 1x radial vibration and 0° phase across the coupling. Usually, 90° phase shift between the horizontal and vertical measurements at the same bearing location confirms the unbalance condition. Sometimes 1x axial vibration appears in overhung fans due to unbalance. Cleaning fan blades is recommended before balancing. Users should eliminate all other sources of 1x vibration e.g., misalignment, looseness, bearing clearance, before attempting to balance fans and blowers.

In some machines, a high vibration component at the blade passing frequency (*n*x vibration, where *n* is the number of blades) occurs. This is usually due to aerodynamic forces and is usually attributed to an incorrect air damper positioning. This can be corrected by using a suitable damper angle.

In belt driven fans (see Figure 2), particular faults occur. Belt faults appear at frequencies less than 1x, and in this case, the belt should be replaced. Belt tension problems can have adverse effects. A low belt tension can cause a belt resonance, usually below 1x, and may be corrected by increasing the belt tension. Excessive belt tension, on the other hand, results in directional 1x vibration, and in this case, belt tension may be corrected to resolve the problem.

In some cases, with belt driven machines, sheave problems occur. Usually, sheave problems appear as a misalignment, and either 1x or 1x and 2x vibration appear in the spectrum. If possible, visual observation can confirm sheave problems.

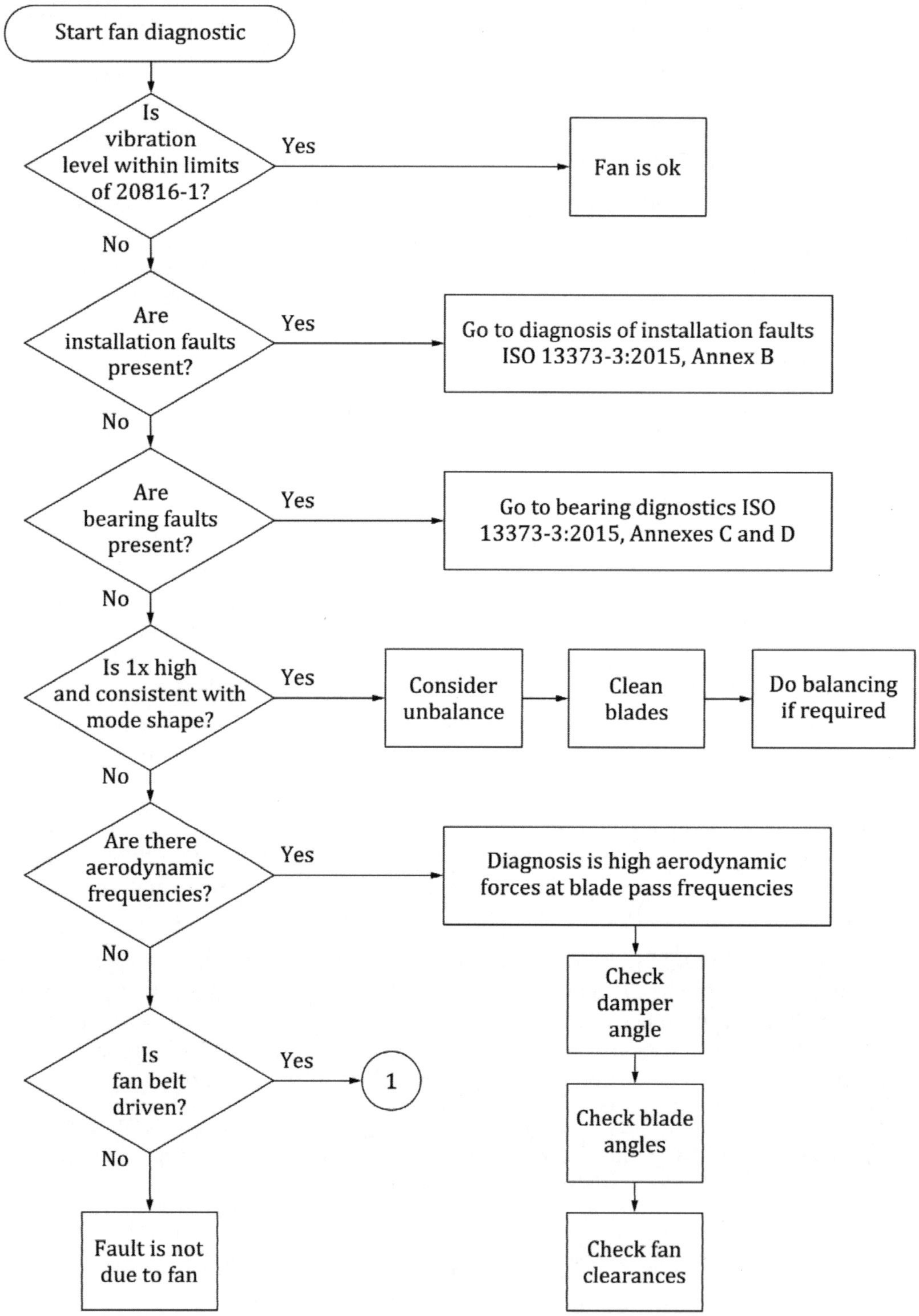

Figure 1 — Flow chart for the diagnosis of faults in fans and blowers

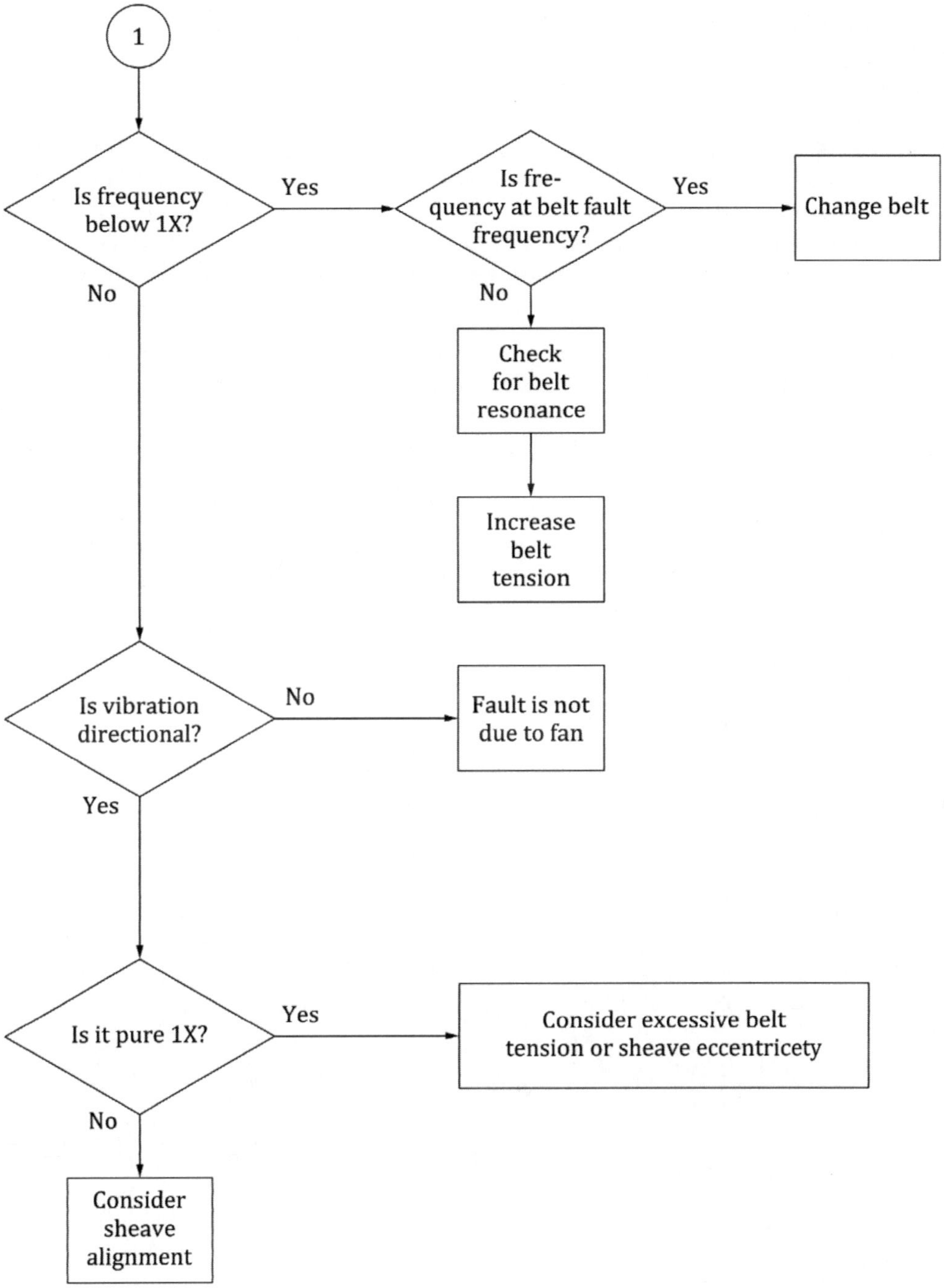

Figure 2 — Flow chart for the diagnosis of faults in fans and blowers with belt drive

EXAMPLE APPLICATIONS OF THE ISO 13373-5 METHODOLOGY [2]

EXAMPLE OF SPECTRAL ANALYSIS [2]

The main air blower in a sulphur plant had been experiencing frequent motor bearing failure for the previous two years. The 600 kW AC motor operating at 1 500 r/min was mounted on two rolling element bearings. The axial load was taken on the non-drive end angular contact bearing, while the drive end roller bearing was the one experiencing the frequent failures. Most failures were in the form of a shearing cage. The motor was torn apart several times, but no cause was found for this problem.

Upon investigating this machine, the bearing failure rate was at a rate of once every two to three weeks. Spectral measurement on the machine confirmed that the motor had no malfunction, but the axial vibration was relatively high at the speed of the blower. Figure 3 shows the spectrum of the motor drive end bearing with high vibration at the speed of the blower. The motor was driving through a flexible coupling and a speed increaser gearbox mounted on journal bearings. The gearbox was driving the blower at 4 000 r/min through a flexible coupling. The blower was overhung and mounted on two journal bearings. A survey of the blower and the gearbox showed a high axial component at both the gearbox and blower bearings at the speed of the blower.

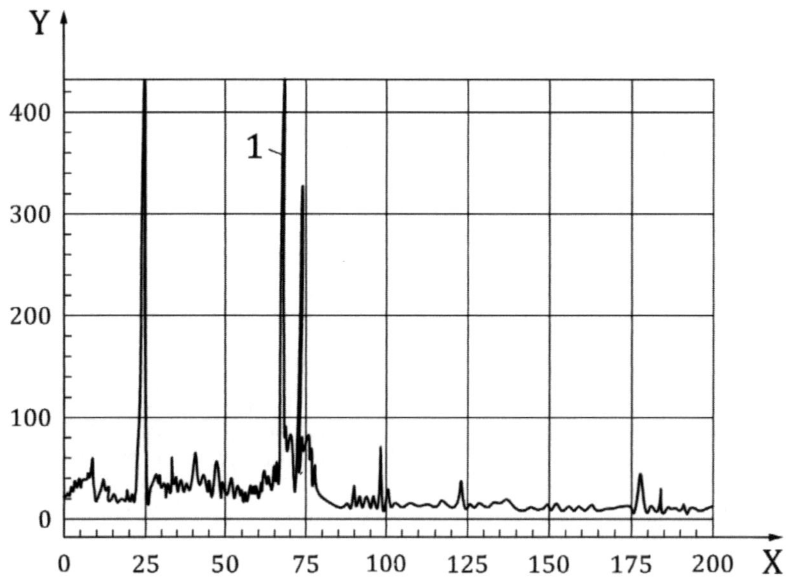

Key

X frequency in Hz

Y axial vibration velocity in µm/s

1 high blower speed component

Figure 3 — Motor drive end spectrum

Figure 4 shows a typical spectrum from this blower gearbox. The gearbox showed a high gear mesh frequency of 1 966 Hz and a high twice gear mesh frequency. There is a pronounced series of peaks centred around 2 365 Hz, with sidebands. These have been interpreted as indicating some form of impact, triggering a natural frequency at 2 365 Hz. The highest vibration values were at the vertical and axial directions of the blower bearings with an RMS reading of 5.2 mm/s.

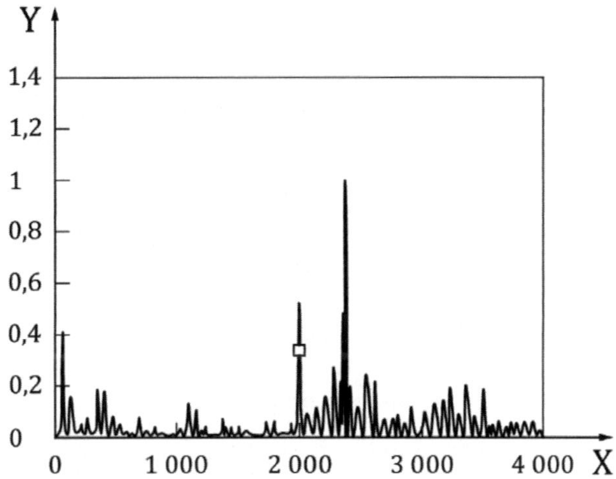

Key

X frequency in Hz

Y axial RMS vibration velocity in mm/s

Figure 4— Gearbox spectrum

The diagnosis of this sensitive machine was as follows: the blower is experiencing a couple unbalance producing an axial force that is creating an impact on the gearbox and deteriorating the gear meshing. This axial force is then transmitted to the electric motor causing the bearing failure. The recommendation was to inspect the gearbox, correct any problems and balance the blower.

The gearbox was dismantled; the drive end journal bearing on the gearbox was completely destroyed. The axial thrust collar on the gearbox and on the blower bearings showed clear signs of impact. The blower bearings showed excessive clearance and rubbing on the rotor and bearings.

The gearbox was replaced, the blower bearings and shaft were replaced, and the motor bearings were replaced. The machine showed reduced overall vibration, but balancing was still recommended. The vibration vectors showed the need for a couple correction weight. A trial mass of 10 g was used with this sensitive machine. The final trim balance weight was 3 g. The highest RMS vibration value was reduced to 1.2 mm/s. No motor bearing failures have been reported since.

This case study illustrates the need to clearly identify the cause of vibration from spectral analysis. These problems should be corrected before attempting to balance the machine. The diagnostic procedure in Figure 1 was followed. No bearing frequencies were identified; however, spectral analysis indicated deterioration of the gearbox, and a clear 1x indicated unbalance in the blower. 1x axial vibration due to unbalance occurs only in overhung machines. This axial vibration was transmitted to the motor, and its presence at the blower speed on the motor drive end was key to the diagnosis of this machine train.

EXAMPLE OF BALANCING AND ODS ANALYSIS [2]

A cement factory had a kiln fan that was experiencing high vibration. The fan had been installed for 12 years, during which time it had experienced one catastrophic failure and a major design modification to overcome its problems. In recent years, the fan had been experiencing high vibration resulting in the need to reduce its speed to reduce the vibration. This adversely affected the production output.

On inspection of the machine, it was found to be a huge 10.5 t overhung fan on rolling element bearings with an impeller of a diameter over 5 m and driven by a 1 600 kW DC motor. The maximum fan speed was 490 r/min. Production required the machine to be operating at 95 % of its maximum speed, but, because of the high vibration, the machine was operating at 82 % of its maximum speed only. The 4 m high foundation showed visible cracks. Vibration measurements on the non-drive end bearing were 290 mm/s, mainly at 1x.

Applying the diagnostic procedure of Figure 1, no bearing faults were found. Clearly the machine foundation required attention; however, due to production requirements, it was determined to balance the machine. This is unusual, and it was therefore recommended that other problems be resolved before balancing, but in this case the plant had requested a temporary solution, pending the resolution of the problem of the skid and foundation.

Before balancing, it was requested to clean the fan blades, by which the vibration values were reduced to 50 mm/s. A run-up test was performed to try to determine if any resonance conditions occurred and to mitigate any undesirable effects from the skid and foundation. Table 3 illustrates the results of the run-up test.

Table 3 — Run-up test on kiln fan

Speed	Drive end bearing		Non-drive end bearing	
r/min	RMS magnitude mm/s	Phase °	RMS magnitude mm/s	Phase °
284	6,74	–160	7,27	–159
309	3,36	–149	5,67	–144
355	3,52	173	2,72	172
380	3,74	162	5,2	155
415	3,73	140	10,9	135
425	5,8	126	16,5	125
440	10,6	112	31,8	111

From Table 3, it is clear that the machine is approaching resonance as its speed increases. A computer model was developed and it was determined that resonance occurs at the maximum speed of the machine. The mode shape is shown in Figure 5. Yet, it was decided to balance the machine. Inspection of the results of the run-up test showed that the machine exhibits vibration that is in phase at both bearings at nearly all speeds, even though the computer model indicates that it exhibits a conical mode at resonance.

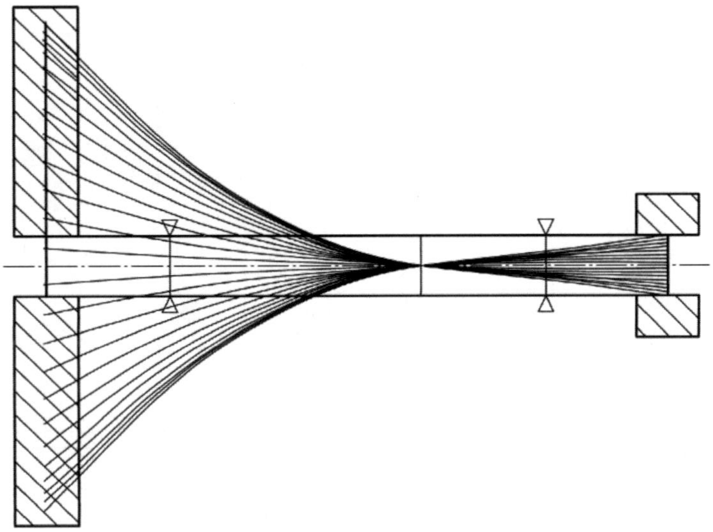

Figure 5 — Mode shape of the fan

It was decided that the machine only needed static balancing because of the in-phase measurement. A trial mass of 2 kg barely made the machine respond but was used to obtain the balance sensitivity. The machine was finally balanced by a 6.1 kg correction mass, as shown in Figure 6, after removal of the trial mass. The vibration was reduced to 2.7 mm/s and has been performing satisfactorily since then. With the balanced rotor, production can achieve maximum throughput with no excessive fan vibration. Figure 7 shows the vibration spectra before and after balancing.

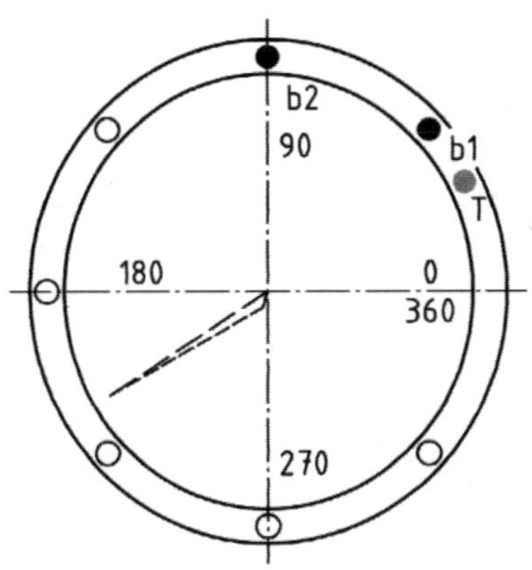

Key

b1, b2 correction masses

T trial mass

Figure 6 — Static balancing of the fan

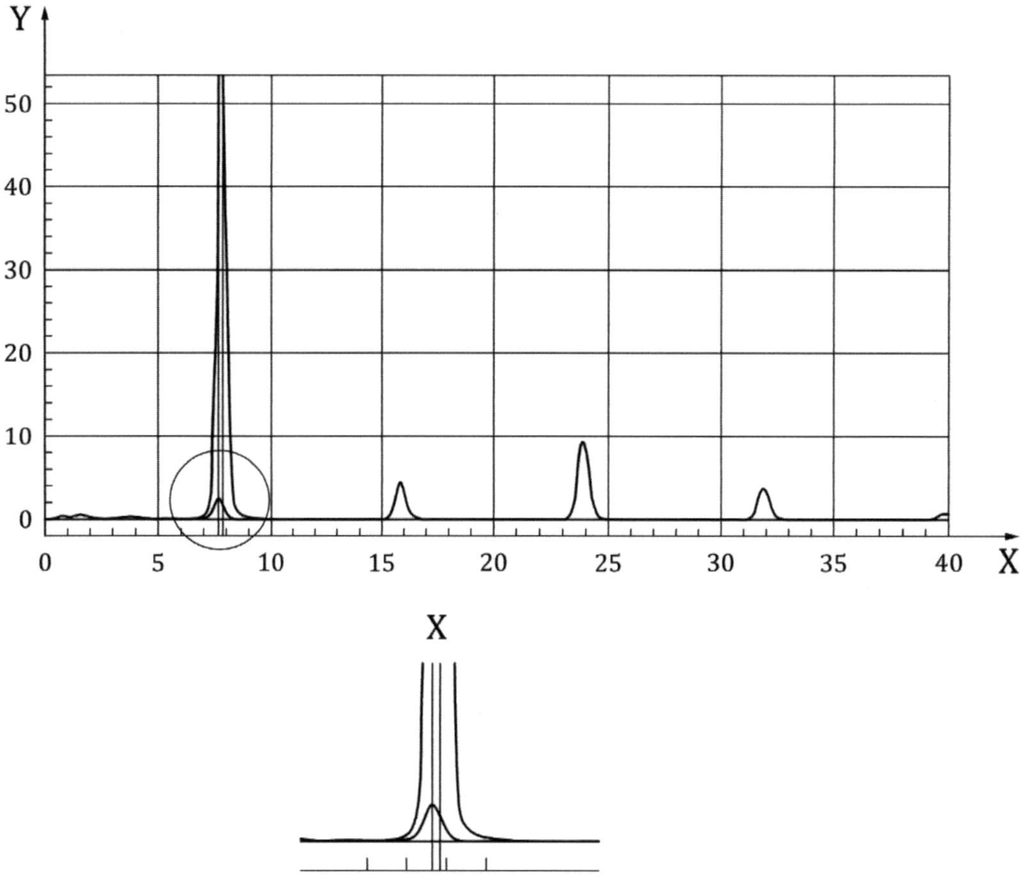

Key

X frequency in Hz

Y vibration velocity in mm/s

Figure 7 — Spectra before and after balancing showing significant reduction

Upon successful balancing, it was required to investigate the skid resonance problem. An operational deflection shape (ODS) of the skid was measured. Figure 8 shows a clear rocking motion for the skid in the vertical direction, while Figure 9 shows a clear swaying in the skid in the horizontal direction. These measurements increase in amplitude as the machine speed increases and approaches resonance. Apparently, the long period at which the fan was operating with severe unbalance resulted in the I-beams, embedded 0.8 m in the foundation and carrying the skid, separating from the concrete. A finite element model of the skid and foundation was developed and a suitable supporting structure was designed and implemented to prevent the rocking and swaying and to increase the natural frequency well beyond the operating speed of the fan.

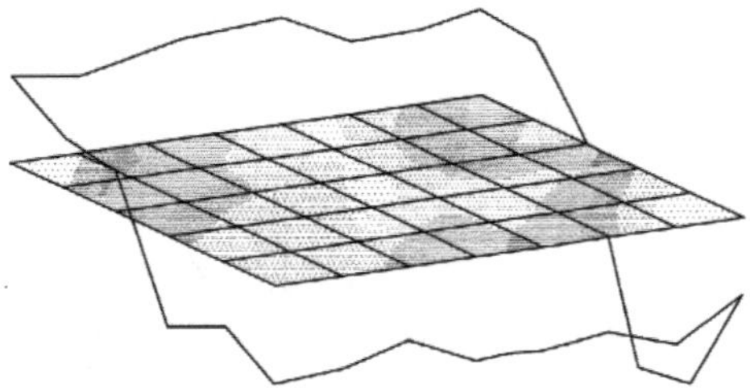

Figure 8— Vertical ODS showing rocking of skid

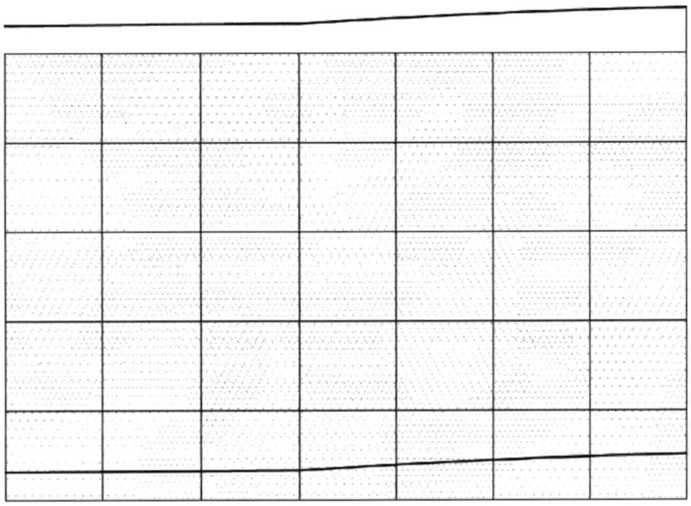

Figure 9 — Horizontal ODS showing swaying of the skid

EXAMPLE OF ODS ANALYSIS [2]

This case study is from the petrochemical industry and is for the supporting frame of a vertical air cooling fan supported on a horizontal steel frame with a horizontal motor and a bevel gearbox (see Figure 10). The unit is one of 12 units supported on an elevated steel frame, used to cool air for three gas turbines 9 MW each. The units were experiencing frequent shear failures of the shaft connecting the motor and the gearbox at the input shaft of the gearbox. The manufacturer suggested the use of a larger diameter shaft. However, this resulted in accelerated wear in the gearbox.

Upon investigation of the problem, and applying the diagnostic procedure shown in Figure 1, a spectral analysis indicated that no bearing faults were present, and no unbalance, but the main exciting frequency was at 6x, i.e. the blade passing frequency. However, investigating the operating conditions, it was clear that the blade angles were correct and the blade clearance was also set correctly. Therefore, aerodynamic excitation was eliminated as a cause of the problem. A bump test was conducted but was not conclusive because of transmitted vibration and it was not possible to shut down all coolers at the same time. An ODS was conducted at 6x the exciting frequency, and the results are shown in Figures 11 and 12.

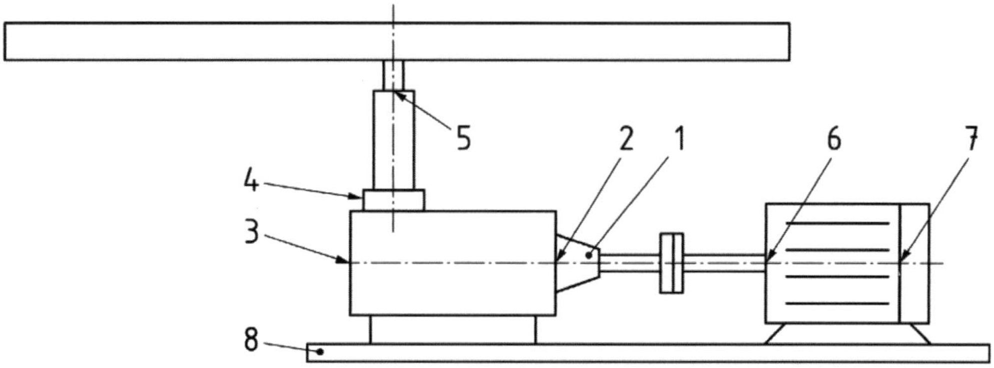

Key

1 gear box bearing 1
2 gear box bearing 2
3 gear box bearing 3
4 gear box bearing 4
5 fan bearing
6 motor drive end bearing
7 motor non-drive end bearing
8 base plate

Figure 10— Schematic layout of the machine

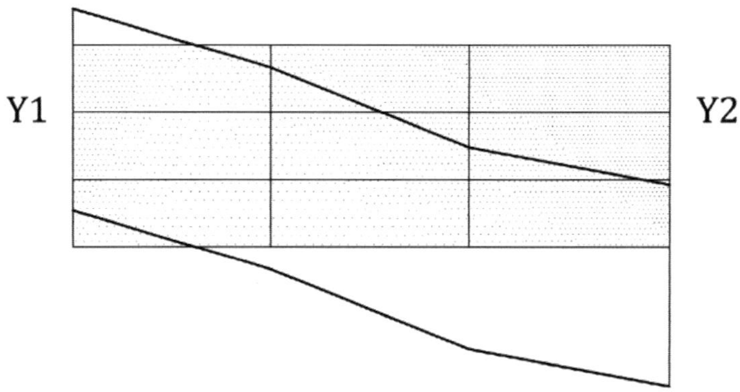

Key

Y1 fan
Y2 motor

Figure 11 — Base plate ODS — Horizontal direction

It is seen quite clearly that the main deflection occurs in the middle of the frame, right where the repeated failure occurs, both in the horizontal and vertical directions. The reason of the failures is quite clear: the flexibility of the supporting structure at the gearbox input shaft position results in shaft failure. Increasing the shaft diameter only transmitted the problem to the gearbox, resulting in the accelerated wear. The root cause of the problem is the flexibility of the supporting structure. This supporting structure should be stiffened.

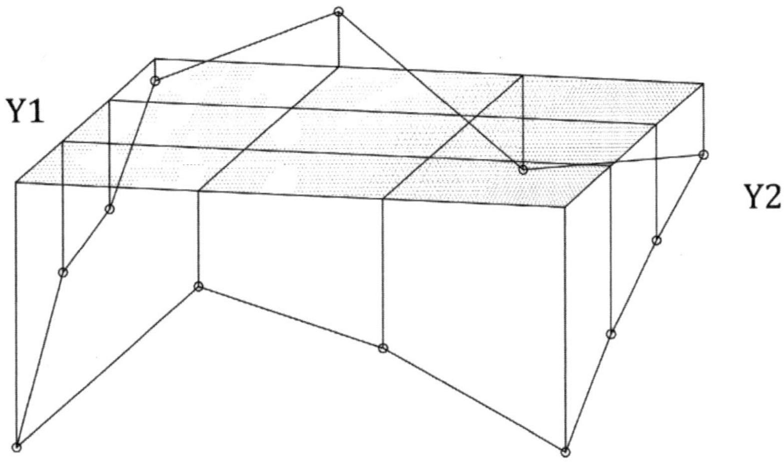

Key

Y1 fan

Y2 motor

Figure 12— Base plate ODS — Vertical direction

EXAMPLE OF BEARING PROBLEM [2]

This is an unusual case. During acceptance testing of a cement plant, the vibration values were relatively high for a new fan operating on rolling element bearings. The spectra are shown in Figures 13 and 14.

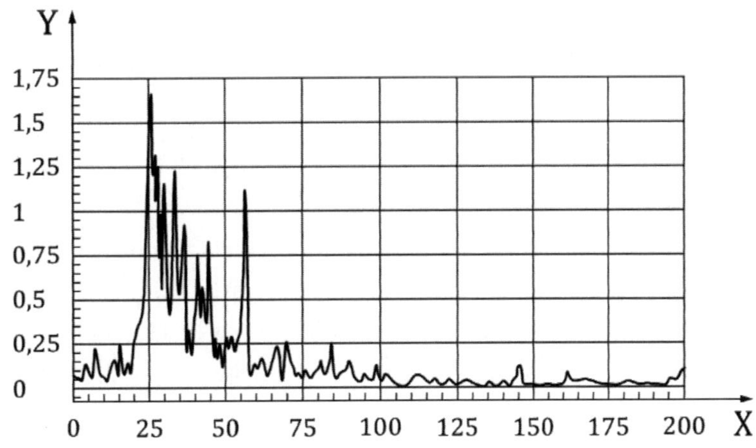

Key

X frequency in hz

Y vibration velocity in mm/s

Figure 13 — Drive end bearing spectrum

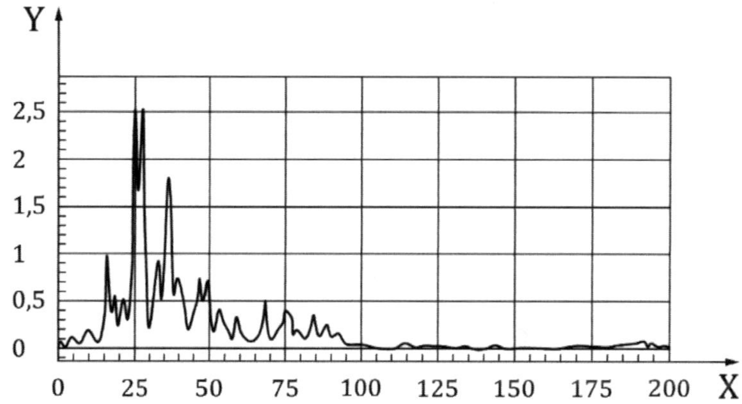

Key

X frequency in hz

Y vibration velocity in mm/s

Figure 14 — Non-drive end bearing spectrum

Applying the diagnosis procedure shown in Figure 1, no bearing frequencies were found, no installation faults were found and the fan was well balanced. Also, no vibration at the blade passing frequency was seen. However, the fan bearings were noisy. The vibration pattern shown in Figures 13 and 14 is not usually associated with bearing faults. Although the spectrum is noisy with significant energy and has many frequencies, none of the frequencies matches with the bearing frequencies, namely ball pass frequency of the outer race (BPFO), ball pass frequency of the inner race (BPFI), ball spin frequency (BSF) and fundamental train frequency (FTF).

However, discussion with plant personnel revealed that welding was performed near the fan. This immediately caused a reconsideration of the diagnosis. The pattern shown in Figures 13 and 14 is an indication of electric discharge on the bearings and occurs when welding is performed without proper electrical insolation. The fan was stopped and the bearings were removed, and the discharge marks were clearly evident. The fan was ran smoothly after the bearings were changed, allowing its acceptance by the customer.

CONCLUSION:

In this paper, the standardized methodology for the vibration diagnosis of machinery was discussed. It is a logical step-by-step approach based on the use of Flow charts, Fault tables and Symptom. In particular, this paper focused on the newly released ISO 13373-5 for the diagnosis of fans and blowers [3]. The standard describes the diagnosis of directly coupled fans and blowers, as well as belt-driven fans and blowers. Examples of the application of the diagnosis of fans and blowers are then given [2].

It is the opinion of this author, and contributor to the international standards, as well as a field user of this methodology, that the standardized methodology does lead to accurate diagnosis. Application of the methodology in the field [1-2] has proved to be quite successful. Many examples of the successful application of this methodology are given in the standards.

REFERENCES:

[1] El-Shafei, A., "Diagnosis of Installation Problems of Turbomachinery", Proceedings of ASME Turbo Expo, Amsterdam, The Netherlands, June 2002, ASME paper GT-2002-30284

[2] El-Shafei, A., "Fan Diagnosis in the Field", I. Mech. E. International Conference on Vibrations in Rotating Machines, Exeter, UK, 2008.

[3] ISO 13373-5 Condition monitoring and diagnosis of machines — Vibration condition monitoring: Part 5: Diagnostic techniques for fans and blowers, 2020.

[4] El-Shafei, A., "Machine Diagnosis as a Standardized Approach", IFToMM International Conference on Rotor Dynamics, pp. 207-221, Springer, 2018

[5] ISO 13373-3 Condition monitoring and diagnosis of machines — Vibration condition monitoring: Part 3: Guidelines for vibration diagnosis, 2015.

[6] ISO 13373-1 Condition monitoring and diagnosis of machines — Vibration condition monitoring: Part 1: General procedures, 2002.

[7] ISO 13373-2 Condition monitoring and diagnosis of machines — Vibration condition monitoring: Part 2: Processing, analysis and presentation of vibration data, 2005.

[8] ISO 13373-7 Condition monitoring and diagnosis of machines — Vibration condition monitoring: Part 7: Diagnostic techniques for machine sets in hydraulic power generating and pump-storage plants, 2017.

[9] ISO 13373-9 Condition monitoring and diagnosis of machines — Vibration condition monitoring: Part 9: Diagnostic techniques for electric motors, 2017.

VIATC TRAINING CONFERENCE & EXPO
COVINGTON, KENTUCKY | AUGUST 7-9, 2024

HIGH-FREQUENCY VIBRATION MITIGATION THROUGH CYLINDER DESIGN FOR RECIPROCATING COMPRESSORS

Tom Stephens P.E., VI certified category 4
Sr. Equipment Specialist
Ariel Corporation – Mt. Vernon, OH
tstephens@arielcorp.com

Tom Stephens is a Senior Equipment Specialist within the Research & Development group at Ariel Corporation in Mount Vernon, Ohio. He has over 30 years of experience with high-speed reciprocating compressor and has been with Ariel Corporation since 1998, predominately working to reduce lateral and torsional vibration levels and improve the performance efficiency & reliability of high-speed reciprocating compressors. Mr. Stephens received an Engineering degree from the Colorado School of Mines in 1993, a Professional Engineer license from the State of Ohio in 2004, and a Category IV vibration certification from the Vibration Institute in 2005. He has participated in several standard workgroups, including API 688, ISO 20816-8, and the GMRC torsional sub-committee. He has published multiple papers at the Gas Machinery Conference, Vibration Institute, and Torsional Vibration Symposium.

ABSTRACT:
Some horizontally-opposed reciprocating compressor cylinders can exhibit a mode of vibration where opposing cylinders oscillate out-of-phase from each other. This mode of vibration is relatively high in frequency (usually in the 150-300 Hz or 9000-18000 CPM range) and not typically analyzed in dynamic studies for reciprocating compressor packages. The gas forces driving the vibration are intermittent and depend on valve timing, plate dynamics, and gas passage reflections inside cylinders. On single stage units, the forces from opposing cylinders can align, which compounds the problem.

Large, efficient cylinder designs (i.e., pipeline cylinders) can be prone to this problem due to the heavy cylinder construction and large valving needed for the application. Mitigation methods have been limited to asymmetric loading using additional clearance, or speed restrictions; neither of which are robust or desired solutions.

Recent efforts have led to the development of new internal cylinder features which attenuates and cancels the forces that excite this resonant vibration mode. The new cylinder design, termed diverter cylinder, will be described. The paper will conclude with three short case studies that show significant reductions in vibration achieved with the new design features.

KEY WORDS: Vibration, High-Frequency, Acoustics, Compressor, Cylinders, Natural Gas, Diverter Feature,

INTRODUCTION:

The term "Stretch Resonance" has sometimes been used to describe high-frequency cylinder vibration on horizontal-opposed, reciprocating compressors. The "Stretch" term is the direction of the vibration which is the direction of piston motion, and therefore rod load force. The "Resonance" term indicates that there is a structural, mechanical resonance involved. These vibration modes are relatively high in frequency (usually in the 150-300 Hz range) and are not pre-determined as problematic by traditional design studies.

Symptoms of high frequency vibration can include failures of small-bore attachments on bottles and pipe, damage to control devices and wiring, tubing failures, and other nuisance problems [1,2]. When levels are excessive, there are only a few limited options available for correction, such as asymmetric loading using clearance and/or speed restrictions, both of which can be undesirable.

Section 1 of this paper examines the vibration modes and the excitation forces that occur during the valve events and in the gas passages. Section 2 describes a new cylinder design feature that reduces high frequency shaking forces with case studies that document vibration reduction. Appendix 1 contains a terminology section to aid the reader with the unique terms and abbreviations for reciprocating compressors. Appendix 2 describes the units and measures used for the vibration analysis in this paper.

SECTION 1 – RESONANCE MODES AND FEATURES:

Mechanical Natural Frequency Mode Description

Understanding of the stretch resonance mode and excitation has greatly improved from field testing and data analysis. The stretch mode is a horizontal full compressor mode that acts in the same direction of piston motion. The highest level of displacement occurs at the end of each cylinder. The even throws tend to move out of phase with the odd side throws. As the compressor vibrates, the cylinders act as masses while the crosshead guide and crank case act as springs with the base of the crank case being grounded (Figure 1). We have seen two significant stretch modes on four-throw compressors. The first stretch mode on a four-throw unit has all cylinders vibrating in/out together (Figure 2). The second stretch mode on a four-throw unit has the drive end (DE) cylinders vibrating 180 degrees out-of-phase with the non-drive end (NDE) cylinders (Figure 3). The second mode can be several harmonics higher in frequency than the first mode and tends to cause higher frame vibration levels than the first mode. The mode plot deflection amplitudes in these figures are greatly exaggerated, considering

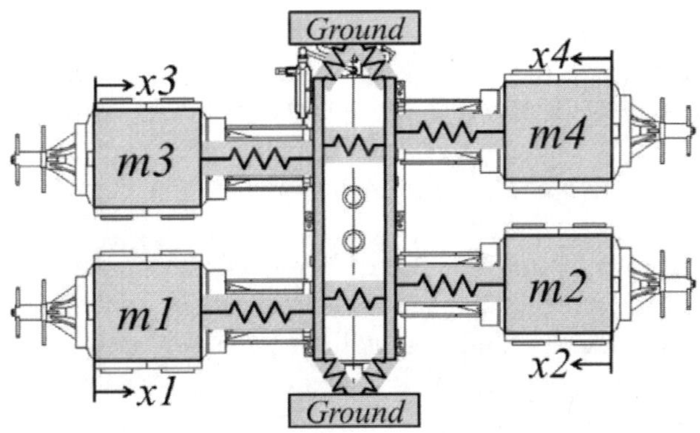

Figure 1 - Modal Spring/Mass Model

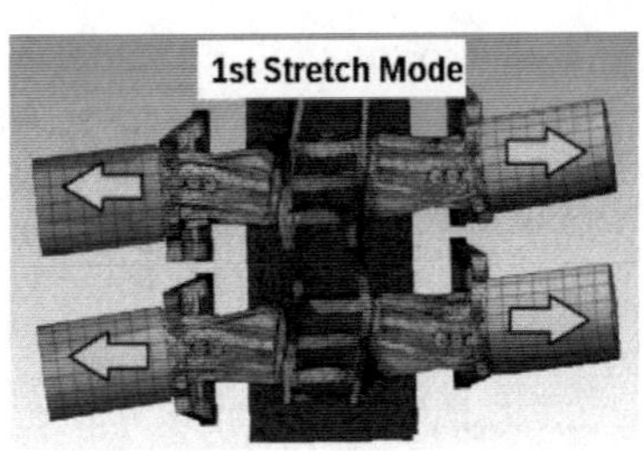

Figure 2 – Lower Stretch Mode

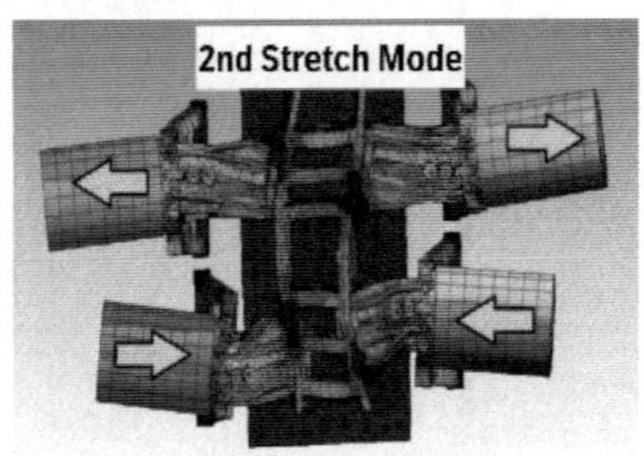

Figure 3 – Upper Stretch Mode

the deflection at 1 ips peak (25.4 mm/s) at 150 Hz is only about 0.002 inches P-P (0.05 mm), and less at higher frequencies.

Shaker test data shows excitation on just one throw can excite the mode and cause all of the cylinders to respond. The peak hold FFT plot in Figure 4 shows the amplitude of each cylinder during a shaker test. A 100 lbf shaker was attached to the cylinder on throw three (T3) in the horizontal direction and swept from 100 Hz to 200 Hz. The results showed that every cylinder shook at almost the same amplitude with the cylinder on throw two (T2) having the greatest amplitude, although it was the cylinder furthest from the shaker. This same trend was followed during run testing, with cylinder on throw two (T2) having the highest vibration.

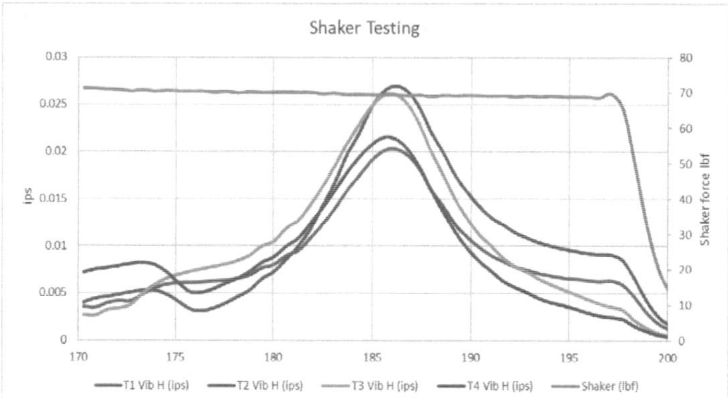

Figure 4 – Shaker Test

Excitation Sources

The cylinder resonance can be excited by gas forces inside the cylinder, which are dependent on valve timing and gas passage pulsation. Suction and discharge pulsation can reflect and amplify from one end of the gas passage to the other. High-frequency pressure fluctuations from the gas passages and valve dynamics get transmitted to the bore pressures during the valve events. This causes high-frequency gas rod load forces, which is the dominant excitation force. The force is intermittent, in that the high frequency components are experienced only during valve events. These force harmonics are speed dependent, and when they align with the cylinder resonance, the vibration can be excessive. On single stage units, the forces from opposing cylinders can align, which compounds the problem.

Extensive testing was done on a KBZ4 compressor with high horizontal cylinder vibration due to the stretch mode resonance. It was noticed that changing the variable volume clearance pocket (VVCP) setting and/or installing valve spacers would change the amplitude of vibration. The following observations were concluded from this testing.

Gas Force at Valve Events

The best way to visualize the gas rod load excitation is to band-pass filter the vibration and gas rod load (calculated from the HE and CE bore pressures), and overlay them in the time domain. In Figure 5, the bore pressures measurements were converted to force by multiplying the HE/CE pressure times the piston/head area for each end and subtracting the two. The resulting gas rod load force and vibration were then band-pass filtered from 100 Hz to 350 Hz. This figure shows the comparison with the final filtered rod load (red) in lbf and filtered vibration (green) in ips. The vibration mode was being intermittently excited during the valve events and decaying during expansion and compression events. The pressure fluctuations during the valve events (from gas passages pulsation and valve plate dynamics) cause the high frequency forces shown in the plot.

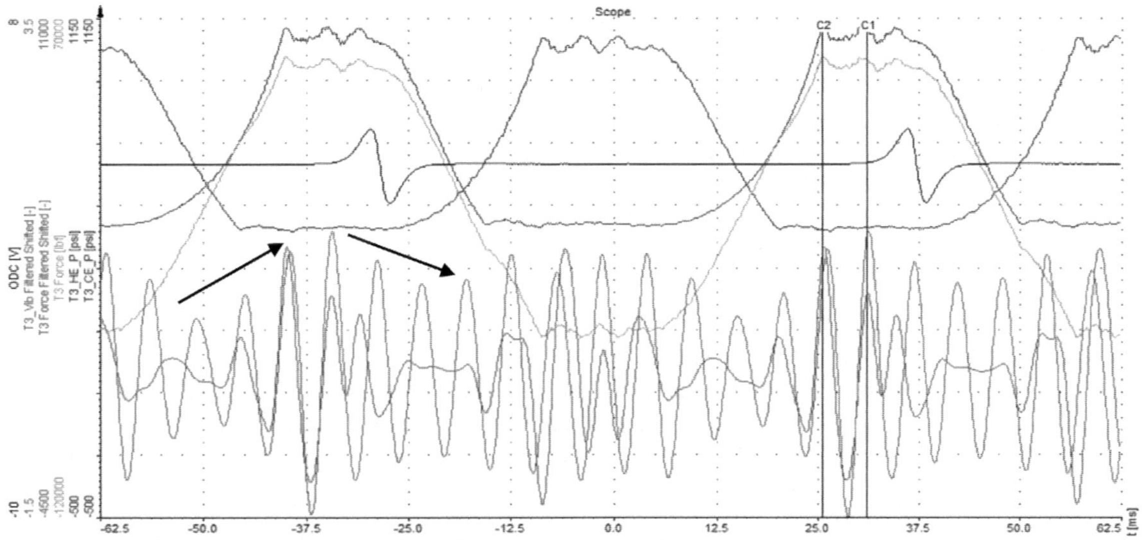

Figure 5 – Filtered Excitation (Gas Rod Load) and Response (Vibration)

Force Distribution

The cylinder gas passage pulsation contributes to the overall high-frequency shaking force, but the gas rod load force (from the bore pressures) was the largest component of the combined excitation force. In Figure 6, it can be seen that the bore gas rod load force is the dominant excitation. The "VVCP=5in" configuration has 70% of the excitation from the bore and only about 15% coming directly from the suction and discharge gas passages. This plot also shows the different 12× force levels at three different clearance configurations resulting in 1400, 950, and 600 combined shaking force (lbf). Force reduction by changing clearance is addressed later in the paper.

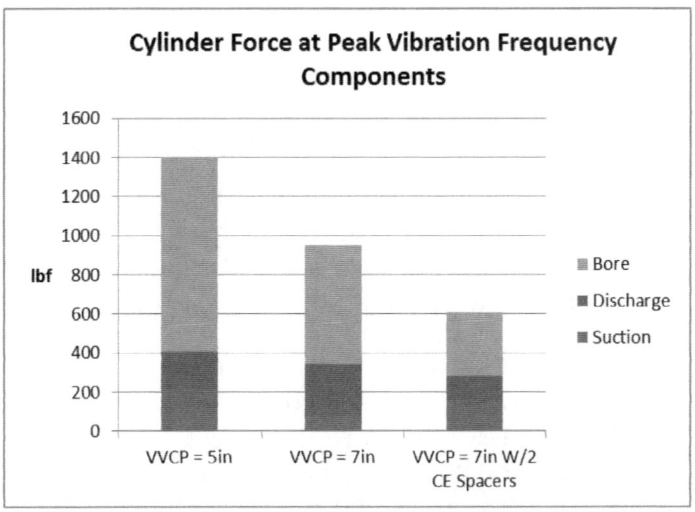

Figure 6 – 12× Shaking Force Distribution

The Influence of Valve Timing on Excitation

Valve timing has a direct impact on the high-frequency excitation, which is a function of pressure ratio, clearance, speed, gas composition, and other parameters. Changing clearance is one of the more practical ways to change valve timing and affect the excitation. Three main factors were found to affect the overall force excitation by changing the valve timing in a compressor. These factors are the length of the valve flow event, the HE to CE force phase within the cylinder, and the opposing cylinder force phase including the CE-CE, and HE-HE phase.

- **Length of Flow Valve Event**:

Test data showed that cylinder vibration reduces with shorter valve events, which are a result of high-pressure ratio or increased clearance. Since the shaking forces are only present during the valve events, a shorter valve event will yield less excitation Adding clearance can reduce excitation, but it also reduces flow. Low ratio applications can have high excitation due to inherently long valve events.

- **Force Phasing between the HE-CE in a Single Cylinder:**

The HE and CE valve timing can affect the excitation force within a single cylinder. By simply opening or closing a HE clearance pocket, the excitation can change. This method of force reduction is usually marginal at best, and can be unreliable as the excitation forces may fluctuate with operating conditions.

- **Force Phasing Between Opposed Cylinders:**

The cylinder clearance for a single stage is often identical for every cylinder, which means the valve events for any two opposed cylinders will occur at the same time. When opposed cylinders are symmetrically loaded, the excitation forces align. By changing clearance in one cylinder (asymmetric loading), the valve events misalign, which reduces the excitation forces. This occurs from both HE to HE and CE to CE events between two opposed cylinders. The CE usually has less clearance than the HE, so the CE valve events tend to be longer and generate more force. Adding CE clearance (i.e., valve spacers) to one cylinder and not the opposing cylinder is an example of asymmetric loading that can have a large effect on the overall shaking force. Figure 7 shows the CE force component from two opposed throws with the same clearance (blue and black) along with both traces summed. The data has been converted from pressure to force and band pass filtered from 100 Hz to 350 Hz. Figure 8 shows the same data from a different set of throws but one cylinder has two CE clearance spacers installed. This resulted in a ~33% reduction in the peak-to-peak CE force. In this case, we were also able to reduce the HE clearance on the same cylinder by closing the VVCP, maintaining the same overall power of the compressor with a further decrease in shaking force.

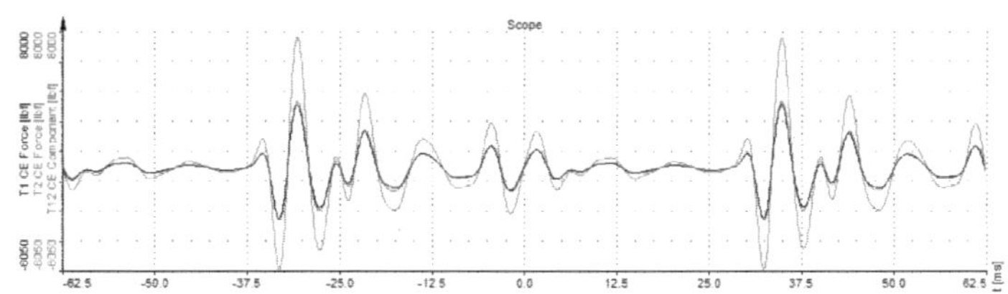

Figure 7 – Symetrical Crank End Force

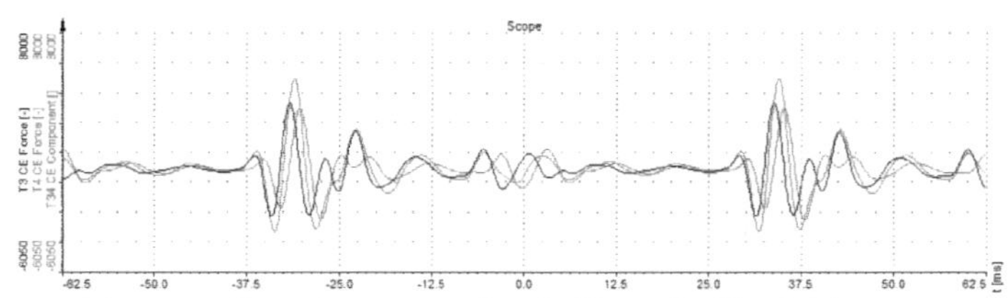

Figure 8 – Asymetrical Crank End Force

Cylinder Vibration and Gas Force

The summation of the filtered rod load force from every cylinder contributes to the overall vibration of the compressor. Figure 9 shows the 12× vibration plotted against the force amplitudes on a Z/4 at resonance for each cylinder individually and averaged for the compressor. Four different clearance configurations are displayed. The per-cylinder vibration responses appear semi-random while the averaged vibration response shows an almost linear relationship with the excitation force.

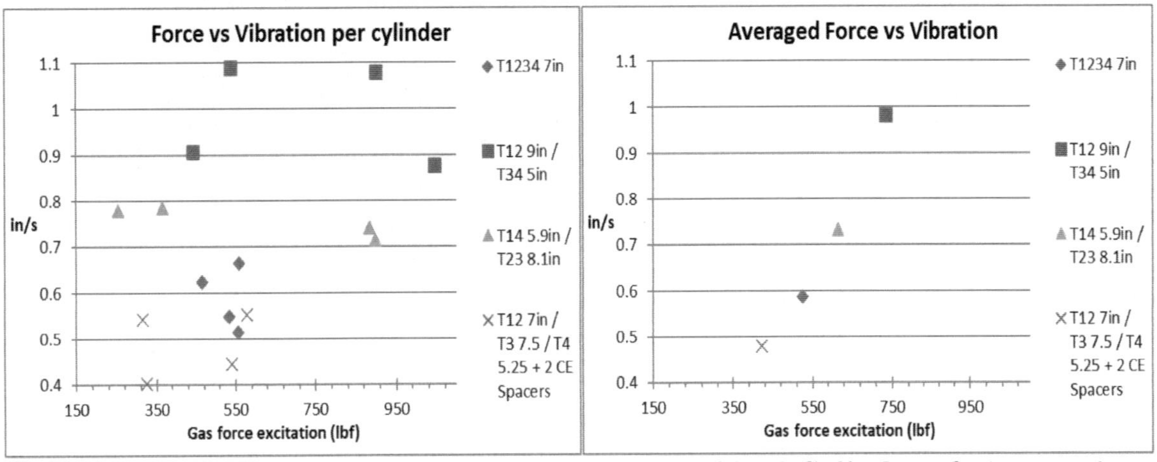

Figure 9 – Excitation vs Response plots (Individual Cylinders & Average)

Title Figures as **Figure 1: Figure Title** in 12 pt New Roman Bold. Place Figure title directly below the Figure. Wide tables should be centered on the page. Narrow tables may be placed to one side with text wrapping around them. Unlike figures, table numbers and titles should be place above the table. Equations should be numbered consecutively, with the number appearing in parenthesis, flush-right to the margin. In the text, refer to equations by number, and keep the number in the parenthesis. For example: see Eq. (4). Please note that an abbreviation, if it is the first word of the sentence, should be spelled out as well as for its first use.

SECTION 2 DIVERTER CYLINDER:

New Gas Passage Feature

The implementation of a gas passage diverter (Figure 10) has proven to significantly reduce high-frequency excitations that drive the stretch mode mechanical resonances. By partially isolating the head end and crank end gas passages in a double acting cylinder, pulsations associated with the valve events tend to cancel, rather than reflect and amplify inside the cylinder. The reduced pulsation in the gas passages has a beneficial effect on the bore pressures and gas rod load. The diverter feature also adds some acoustic damping to the system.

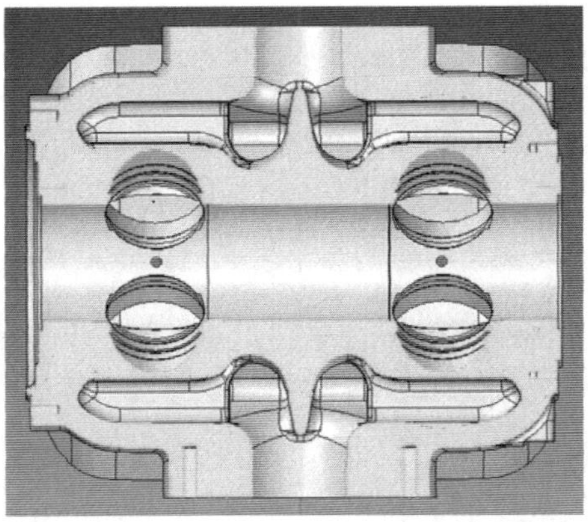

Figure 10 – Diverter Cylinder Cross Section

The feature has been optimized to reduce flow losses and aids in turning the flow from the suction nozzle into the suction valves and out of the discharge valves in to the discharge nozzle. Extensive performance testing has been conducted which has indicated negligible performance differences between the open and diverter gas passage designs.

Figure 11 shows the flow volume from one of these cylinders with the diverter rib in the middle void area below the nozzle.

A lower-frequency acoustic mode is introduced by the diverter feature that can be predicted in 1D pulsation studies. The pulsation mode has been measured inside the cylinder gas passages, but not in the bottles. The mode is typically less than 150 Hz (depending of temperature) which is lower in frequency than the stretch modes. This mode does not appear to generate significant shaking force and has not resulted in any vibration responses of concern on units tested to date.

Figure 11 – Flow Volume with Diverter Feature

Gas passages volumes are discretized for the pulsation analysis. For the Diverter cylinder, the model diagrams have changed from Figure 12 to Figure 13 and are detailed in reference [3]. The new model for the diverter feature incorporates a reduced nozzle area for each end (element F). By using an effective length of this new nozzle area equal to 0.6 × Diameter (typical for this geometry), the new acoustic mode mentioned above can be accurately predicted.

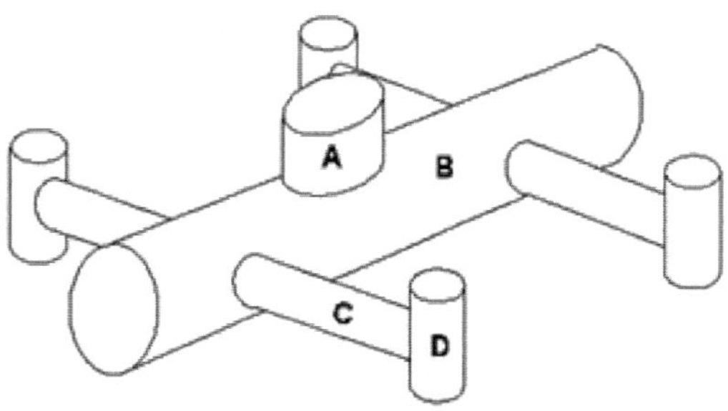

Figure 12 – Open Gas Passage Model

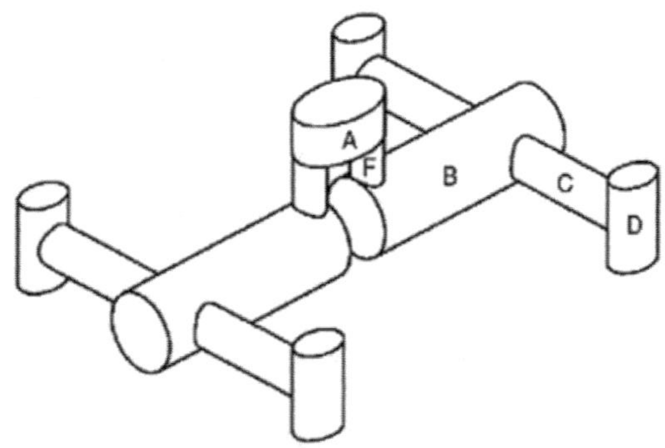

Figure 13 – Diverter Gas Passage Model

The photos below show an open gas passage (Figure 14) and a diverter gas passage (Figure 15) as viewed through the cylinder nozzle.

Figure 14 - Open Gas Passage

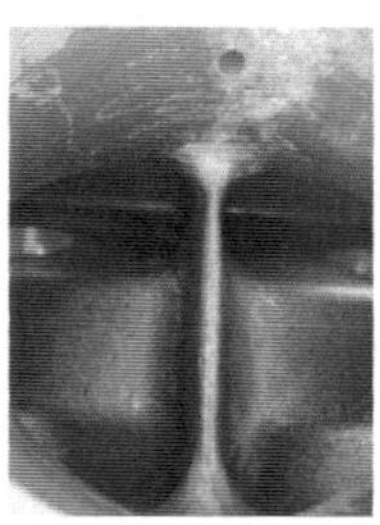

Figure 15 – Diverter Gas Passage

Case Study #1

The first installation of diverter cylinders was done on a KBZ4 driven by a Caterpillar G3612LE from 900 to 1000 rpm. Two diverter cylinders were installed on the non-drive end (NDE) throws of the compressor, and the original two cylinders remained on the drive-end (DE) throws. Vibration responses before and after the cylinder installation are displayed in the following peak-order plots (Figure 16). These plots show harmonic/order amplitudes over the frequency range on the top, and the same amplitudes (with an added over-all amplitude trace) over the speed range on the bottom. Pre-diverter test data is on the left, and post diverter data is on the right. Notable reductions occurred with the diverter cylinder at 11×, 12×, and 18× harmonics. The peak vibration at 915 rpm (12×/183 Hz) reduced from 0.8 ips (20.3 mm/s) to 0.2 ips (5.1 mm/s) for a 75% reduction. It was interesting to note on this test that the force reduction caused by only two installed diverter cylinders out of four was enough to reduce the non-diverter cylinder vibration responses by about 65% as well.

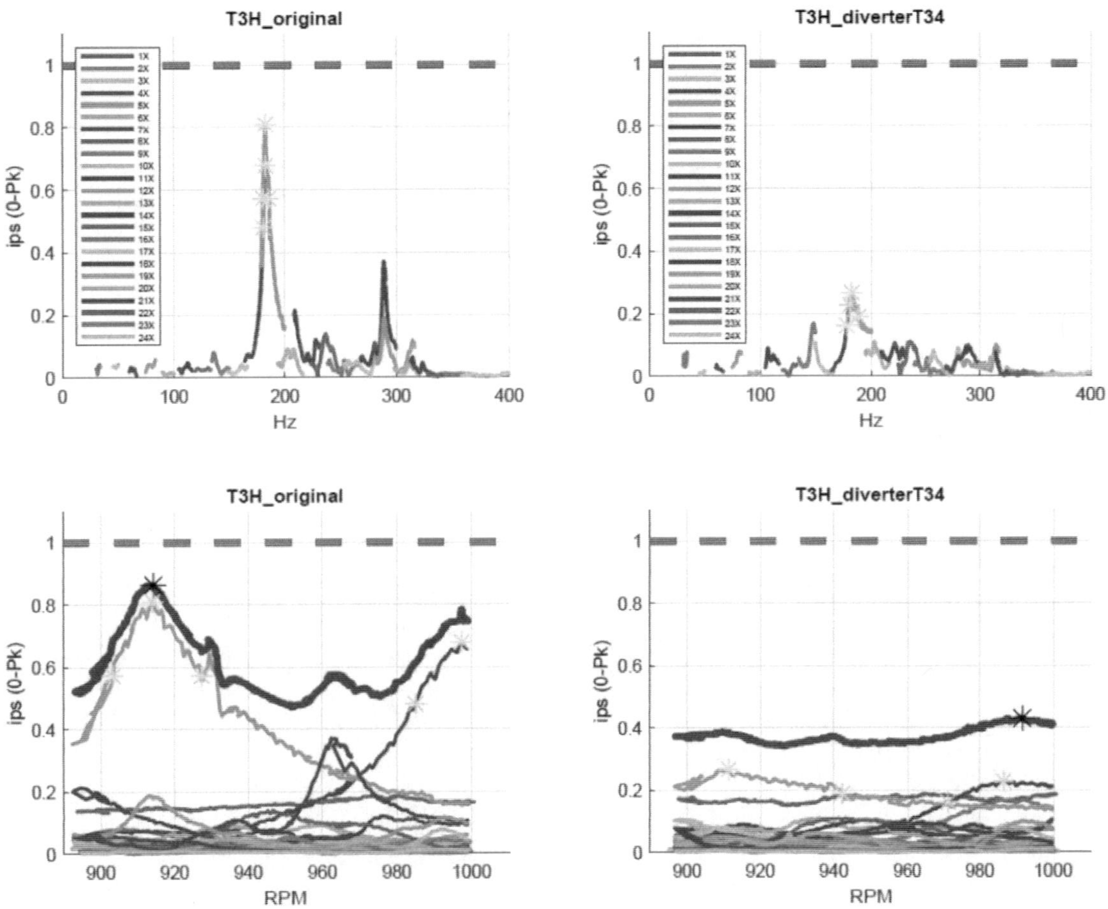

Figure 16 – Case Study 1: Cylinder Vibration Comparison

Case Study #2

The second installation of diverter cylinders were installed on a KBT4 compressor driven by a Caterpillar G3516B from 1000 to 1400 rpm. Diverter cylinders were installed on all four throws on the compressor. The vibration was initially causing a frame vibration shutdown event near 1350 rpm, which turned out to be from the 2nd stretch mode (10×/225 Hz). The frame vibration at 10× reduced from 0.6 ips (15.2 mm/s) to 0.1 ips (2.5 mm/s) for an 83% reduction after the cylinder change was complete (Figure 17).

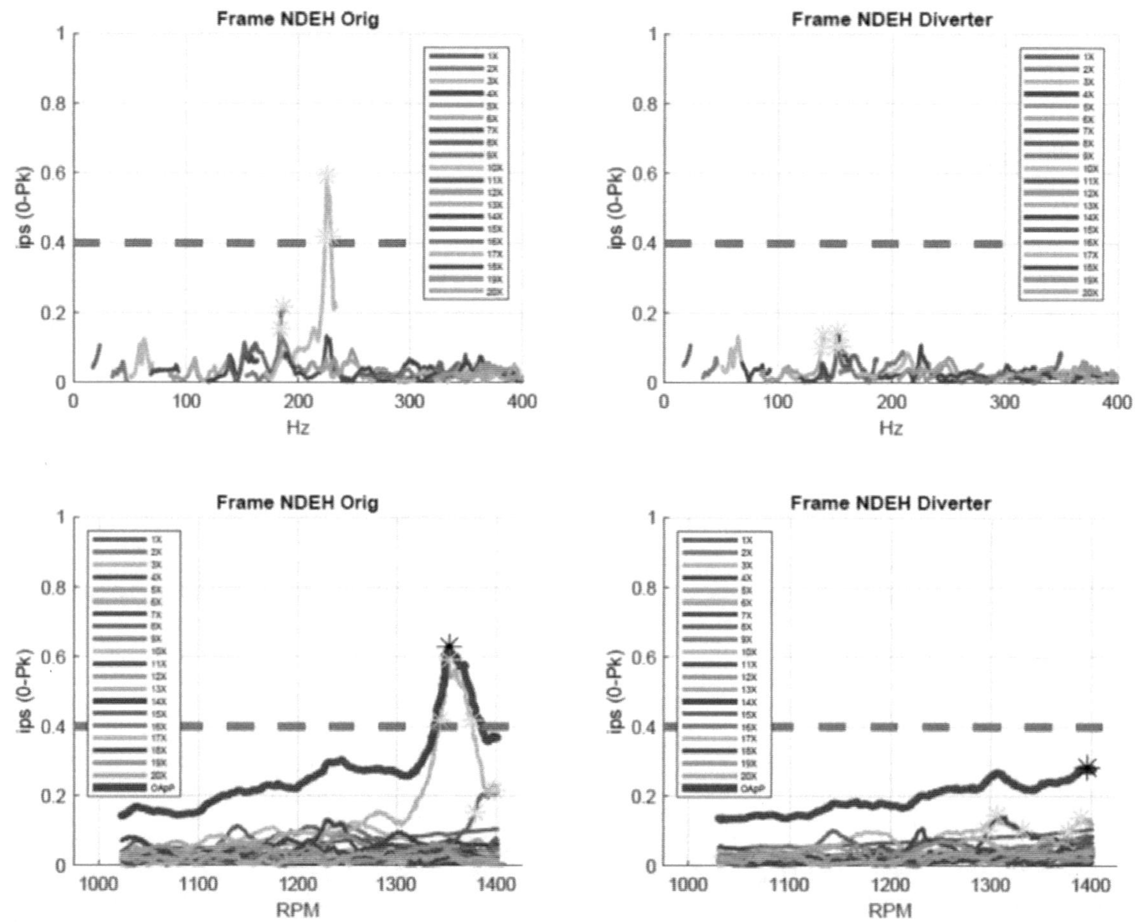

Figure 17 - Case Study 2: Frame Vibration Comparison

The cylinder vibration was found to be even more aggressive with responses at both stretch modes. The 1st mode was dominant with a high 1.3 ips (33 mm/s) response at 1400 rpm (8×/187 Hz). After the retrofit with diverter cylinders, the vibration was reduced to 0.2 ips (5.1 mm/s) for an 85% reduction. Responses at 1250 rpm (9×/187 Hz) and 1350 rpm (10×/225Hz) were also reduced from about 0.5 ips (12.7 mm/s) to less than 0.2 ips (5.1 mm/s) as shown in Figure 18.

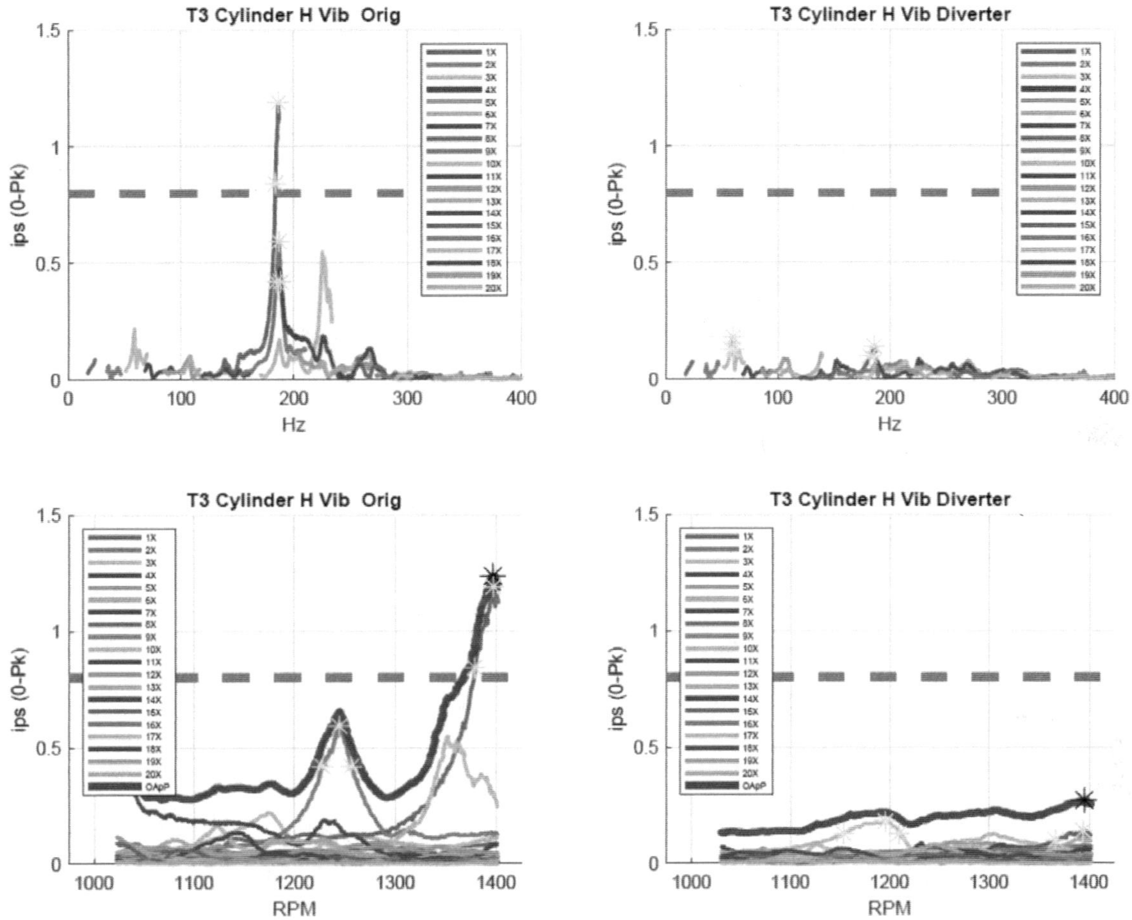

Figure 18 - Case Study 2: Cylinder Vibration Comparison

For this test, the pulsation consultant was contacted and asked to run the model with the new gas passage geometry. An additional shaking force was added to the model in the throw 2 discharge gas passage between nodes 5 & 13. The 1D analysis showed a shaking force reduction from 2703 lbf peak (12 kN) down to 265 lbf (1.2 kN) with the new geometry for a 90% force reduction in the gas passage alone (Figure 19). This was an interesting exercise and the authors are optimistic that this pulsation response may someday be incorporated in design studies [1,2] with guideline limits in order to prevent high-frequency vibration problems from occurring.

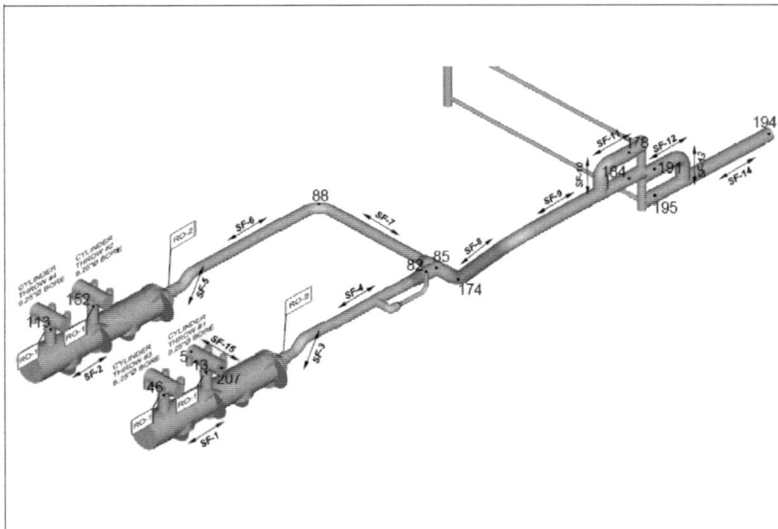

Case -1 (Original Muff)

Node #	Maximum Amplitude (psi pk-pk)	Maximum Amplitude (bar pk-pk)	Harmonic No. X RPM	% API 618 Allowable
Pulsation Amplitudes - Compressor Flange				
13	19.43	1.3399	14	80%
46	19.31	1.3317	14	80%
113	19.31	1.3317	14	80%
152	19.43	1.3399	14	80%
Pulsation Amplitudes - Muff*				
5	25.16	1.7346	9	**104%**
207	24.28	1.674	9	**101%**
Pulsation Amplitudes - Piping				
82	2.13	0.1465	4	35%
85	0.13	0.0091	12	8%
88	0.20	0.0139	12	10%
174	0.12	0.0085	12	7%
178	0.11	0.0072	12	6%
184	0.30	0.0204	1	5%
191	0.02	0.0014	1	0%
194	0.02	0.0014	1	0%
195	0.01	0.0006	6	0%
SF #	Maximum Amplitude lbf (pk)	Maximum Amplitude N (pk)	Harmonic No. X RPM	% API 618 Allowable
Shaking Force Amplitudes - Vessels				
SF-1	90.39	402.08	12	26%
SF-2	91.69	407.84	12	26%
Shaking Force Amplitudes - Piping				
SF-3	4.18	18.596	12	6%
SF-4	3.06	13.597	12	4%
SF-5	4.16	18.5	12	6%
SF-6	3.01	13.394	12	4%
SF-7	8.02	35.696	12	7%
SF-8	7.31	32.538	12	6%
SF-9	3.88	17.243	12	3%
SF-10	5.92	26.338	12	5%
SF-11	6.57	29.214	12	5%
SF-12	0.22	0.988	12	0%
SF-13	0.21	0.9446	12	0%
SF-14	0.27	1.2032	6	0%
SF-15	2703.19	12024	9	N/A

Case -1 (Split Muff)

Node #	Maximum Amplitude (psi pk-pk)	Maximum Amplitude (bar pk-pk)	Harmonic No. X RPM	% API 618 Allowable
Pulsation Amplitudes - Compressor Flange				
13	17.99	1.2404	14	74%
46	17.96	1.238	14	74%
113	17.96	1.238	14	74%
152	17.99	1.2404	14	74%
Pulsation Amplitudes - Muff*				
5	21.07	1.4526	2	87%
207	21.17	1.4596	2	88%
Pulsation Amplitudes - Piping				
82	2.31	0.1595	4	39%
85	0.34	0.0236	1	6%
88	0.14	0.0099	12	7%
174	0.09	0.0061	12	5%
178	0.31	0.0212	1	5%
184	0.30	0.0204	1	5%
191	0.02	0.0014	1	0%
194	0.02	0.0014	1	0%
195	0.01	0.0007	6	0%
SF #	Maximum Amplitude lbf (pk)	Maximum Amplitude N (pk)	Harmonic No. X RPM	% API 618 Allowable
Shaking Force Amplitudes - Vessels				
SF-1	89.67	398.86	10	18%
SF-2	105.73	470.32	10	21%
Shaking Force Amplitudes - Piping				
SF-3	2.98	13.258	12	4%
SF-4	6.99	31.073	4	3%
SF-5	2.97	13.189	12	4%
SF-6	1.68	7.4779	12	2%
SF-7	5.72	25.448	12	5%
SF-8	5.22	23.197	12	4%
SF-9	2.39	10.633	12	2%
SF-10	4.22	18.777	12	3%
SF-11	4.68	20.827	12	4%
SF-12	0.16	0.7046	12	0%
SF-13	0.15	0.6735	12	0%
SF-14	0.28	1.2597	6	0%
SF-15	264.74	1177.6	13	N/A

* API 618 Compressor Flange Pulsation limits were utilized for the cylinder Muff.

Figure 19 – Case Study 2: Discharge Gas Passage Force Prediction Comparison

Case Study #3

The 3rd case study involves a KBK4 driven by an electric motor at a constant speed of 1188 rpm. Both stretch modes were resonance at 8×/160 Hz, and 10×/200 Hz. Frame and cylinder vibration levels were over two times the OEM guideline at 0.82 ips (20.8 mm/s) and 1.4 ips (36.7 mm/s) respectively. Diverter cylinders brought the vibration down to ~1/2 guideline levels or .23 ips (5.8 mm/s) on the frame and .34 ips (8.6 mm/s) on the cylinders. The 72% reduction on frame vibration and 76% reduction on cylinder vibration (on average) was significant. Figure 20 shows the overall frame and cylinder peak vibration levels relative to guidelines for both the standard and diverter cylinder configurations.

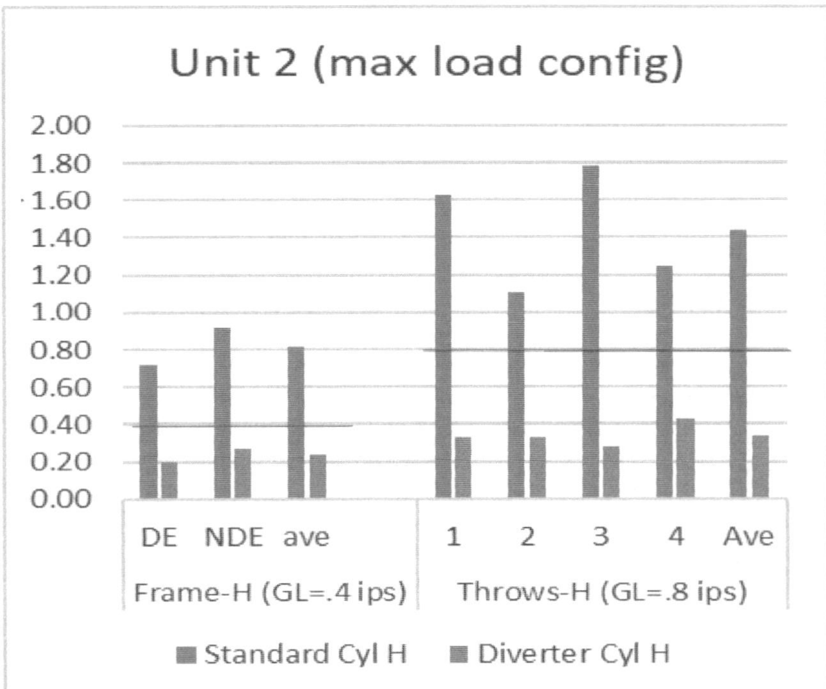

Figure 20 – Case Study 3: Frame & cylinder Vibration Comparison

Figures 21 & 22 below show the high 8× and 10× cylinder vibration levels (T3H maximum load configuration) in frequency & time domains. Levels before the diverter were 0.78 ips (19.1 mm/s) at 8×/159 Hz and 1.58 ips (40.1 mm/s) at 10×/199 Hz. After the diverter, the levels were less than 0.2 ips (5 mm/s) for all harmonics.

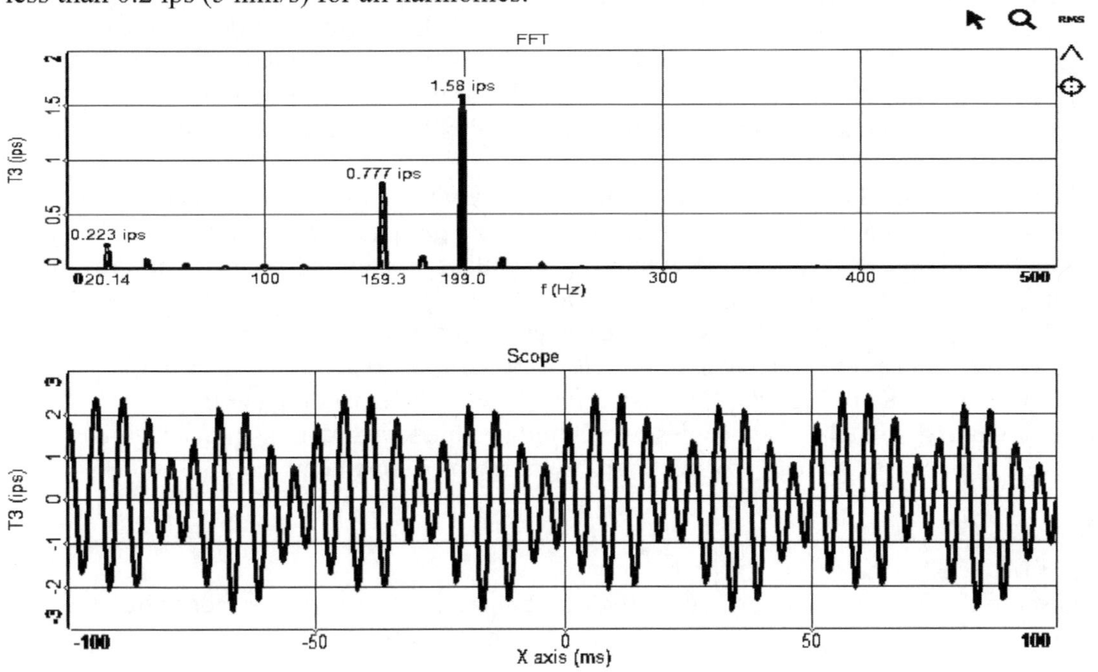

Figure 21– Case Study 3: Vibration T3H Before Diverter Cylinders Installed

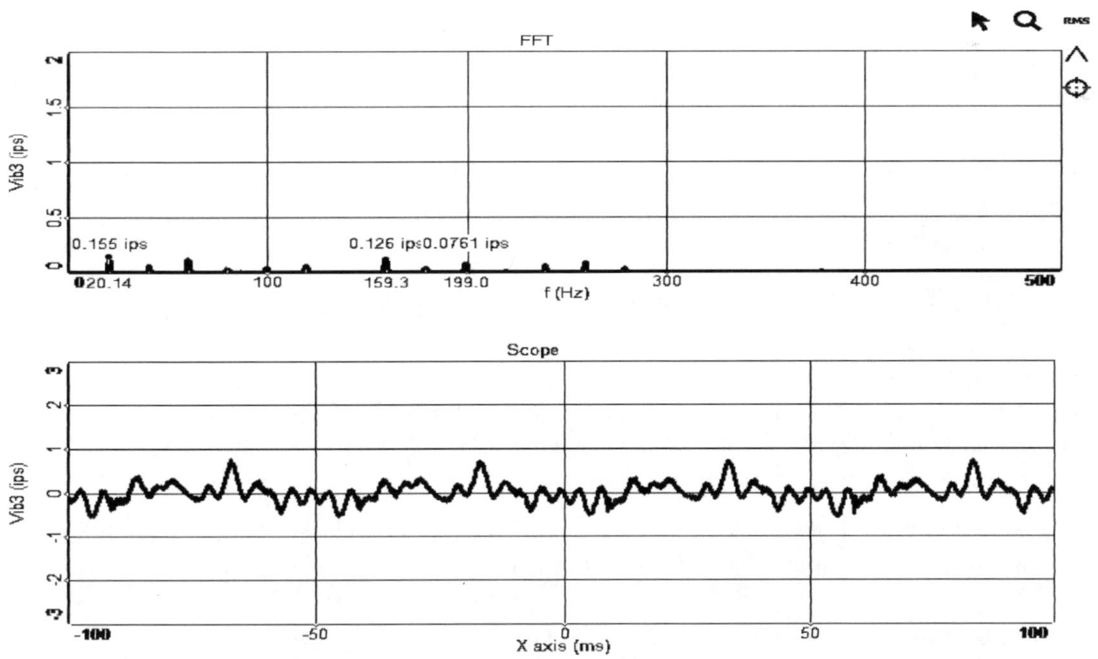

Figure 22 – Case Study 3: Vibration T3H After Diverters Cylinders Installed

Summary

The reciprocating compressor industry has struggled with high-frequency vibration for many years and most efforts have gone into the mitigation of individual components and trying to design more robust packages that would accommodate the vibration [1, 2]. While there are many causes of vibration, we have identified a major source of high-frequency vibration and have undertaken significant design changes to reduce the excitation forces moving forward. Vibration reductions over 75% have been demonstrated in multiple case studies. We now share our findings and solution details to enhance compressor reliability throughout the industry.

Several major providers of acoustic design solutions were involved in our research and testing. During this interaction, the predictability of internal cylinder gas passage forces was demonstrated. It would be of benefit to the industry to collaborate through a GMRC project to define design procedures for high frequency vibration. One improvement may be to provide guideline limits for internal cylinder gas passage forces. Another improvement may be to specify how a mechanical forced response analysis could be applied at these higher frequencies, which are currently limited to about 6 times operating speed. By expanding our design capabilities up into these higher frequencies, we should be able to avoid high-frequency vibration issues in the design phase and increase the reliability of reciprocating compressors.

ACKNOWLEDGEMENTS:

Many thanks to the following companies for their participation in our development efforts.

- Dearing Compressor and Pump Co.
- Dominion Energy
- Enerflex
- Engineering Dynamics, Inc
- MLB Engineers, Inc.
- Peerless Dynamics LLC.
- Southwest Research, Inc.
- Targa Resources Corp.
- Tech Transfer Inc.
- The Wood Group PLC.
- Williams

.

REFERENCES:

[1] "*GMRC Guideline for High-Speed Reciprocating Compressor Packages for Natural Gas Transmission and Storage Application, Official Release - Rev. 0*", October 7, 2013, Gas Machinery Research Council & ACI Services Inc.

[2] "*API Standard 618, Reciprocating Compressors for Petroleum*", Chemical, and Gas Industry Services, American Petroleum Institute, Washington D.C.

[3] "*Ariel Corporation, Engineering Reference (ER-111.6). Diverter Cylinders*", www.arielcorp.com

Appendix I: Compressor Terminology

To aid the reader in following this paper, the following terms and abbreviations are discussed regarding reciprocating compressors (see Figure 23).

Compressor Terms

1. Drive End (DE): The end of the frame that couples to the Driver.
2. Non-Drive End (NDE): The end of the fame opposite of the Driver
3. Throw (T1, T2, …): A compressor throw. Ariel numbers the throws with the first throw closest to the driver, which is the left side when facing the compressor drive end. Subsequent throws are incrementing for each opposed throw moving away from the driver with the last throw on the far-right side. Not all manufactures use the same numbering scheme.
4. Crank End (CE): In a double acting cylinder, this is the bore on the frame side of the cylinder.
5. Head End (HE): In a double acting cylinder, this is the bore on the outboard side of the cylinder.
6. Variable Volume Clearance Pocket (VVCP): An adjustable piston on the head end side of a cylinder that can increase or decrease the bore volume for clearance capacity control. More volume increases the gas expansion event duration and reduces capacity.
7. Fixed Volume Clearance Pocket (FVCP): A clearance capacity device mounted on a cylinder head that can be actuated to expose the bore to a fixed amount of increased volume.
8. Valve Spacer: A clearance capacity device in which the bore volume is increased by installing a short cylinder between the valve and the cylinder.

Figure 23 - Compressor Terms and Vibration Direction

Appendix 2: Vibration Terminology

1. Amplitude:
    a. Vibration motion can be described in terms of displacement, velocity, or acceleration.
    b. These measures can be in Peak (P), Peak-Peak (P-P) or RMS and depicted in Figure 24. All values in this paper are Peak (P), unless otherwise noted.
    c. Ariel prefers velocity due to the operating frequency of the equipment. In comparison, slow-speed compressors (below 500 rpm) typically use displacement, and high-speed equipment (over 5000 rpm) use acceleration.
    d. Units referenced in this paper include:
        i. Lateral vibration is in velocity inches/second Peak (ips P)
        ii. Torsional vibration is in velocity Peak (rpm P).
        iii. Pulsation is in pounds/inch^2 Peak–Peak (psi P-P)
        iv. Shaking Force is in pounds force Peak (lbf P)
2. Frequency
    a. Vibration frequency can be reported in Hertz (Hz) or cycles/second. Some analysts prefer cycles/minute (CPM) or Hz × 60.
    b. Rotating equipment responses typically occur at harmonics of run-speed, so another common method of referencing frequency is by the harmonic number or order. Dynamic variables at the first harmonic (1×) are at the operating frequency of the compressor (run-speed) or one cycle per revolution. Dynamic variables at the 2nd harmonic (2×) have a frequency of two times the operating frequency or two cycles per revolution, and so on.
    c. Complex wave forms are the summation of all harmonics as seen in Figure 25.
3. Direction
    a. Ariel uses Horizontal (H), Axial (A), and Vertical (V) to describe the direction of structural vibration (Figure 23). Structural vibration can also be termed lateral vibration to differentiate it from torsional vibration.
    b. The horizontal direction for this paper is perpendicular to the crankshaft. This is sometimes termed the "Stretch" direction for compressor cylinders. The axial direction is parallel to the crankshaft and vertical is up/down.
    c. Some analysts prefer the axial direction to be parallel with the piston rod, so it is important to define the intended direction for the data being reported.
4. Modes & Resonance
    a. Modes are resonant frequencies of a system. A resonance will amplify responses when excitation frequencies align. System damping limits the amplification. Complex systems can have multiple modes.
    b. Modes can be predicted in discretized models and verified through impact testing or loaded speed sweeps.
    c. For compressors, we deal with three types of resonance
        i. Mechanical resonance involves structures such as cylinders, bottles, and piping.
        ii. Torsional resonance is angular vibration a rotating drive train.
        iii. Acoustic resonance is the reflection and amplification of acoustic waves in the gas piping system, which cause pressure fluctuations or pulsation.

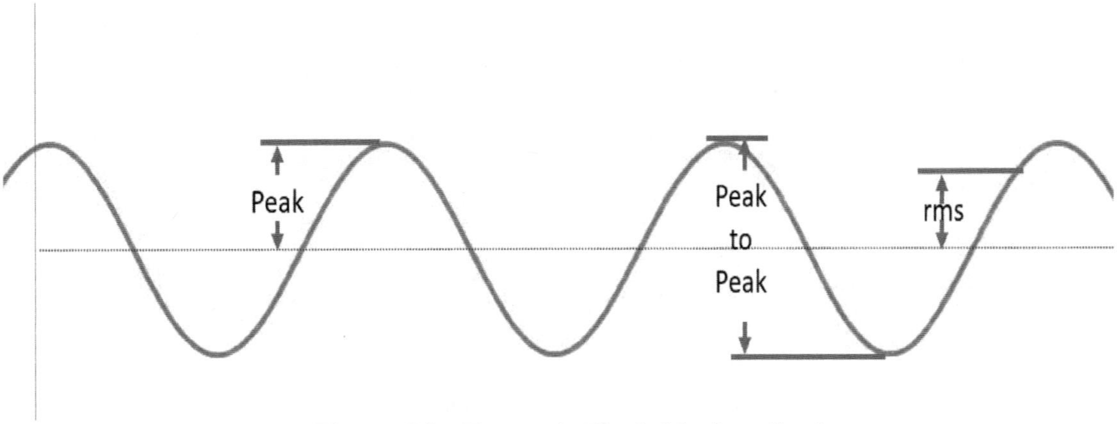

Figure 24 – Dynamic Variable Amplitudes

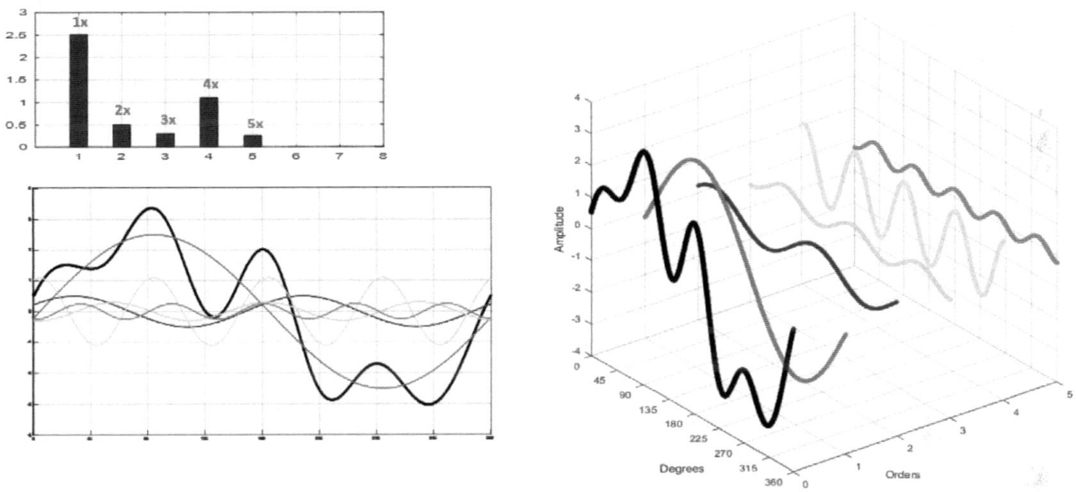

Figure 25 – Complex Waveforms and Harmonics (Orders)

BATTERIES AND IOT – ARE THEY MADE FOR EACH OTHER?

Ed Spence
Founder and Managing Director of The Machine Instrumentation Group
ed.spence@machineinstrumentation.com

Ed Spence is the Founder and Managing Director of The Machine Instrumentation Group, representing a network of consulting and contract engineering service providers helping clients define and develop new sensors for machine health instrumentation.

Ed has published or presented on subjects related to CBM / PdM on numerous occasions, maintaining an ongoing dialogue with the PdM eco-system regarding technology trends on subjects such as MEMS accelerometers, wireless sensor networks (IIoT) and data engineering for Predictive Maintenance.

Abstract:

There is extensive literature regarding estimation of battery life for the energy storage technologies installed in Electric Vehicles (EVs), in particular regarding methodologies for estimating state of charge. But there is relatively little written about choices of battery technology for IoT applications, how the IoT application influences power consumption, how to estimate battery life in the application or even interpret battery life specifications.

This talk will review a wireless condition monitoring sensor energy consumption profile, survey choices of primary battery technologies commonly found in wireless vibration sensors, and how this information might help us better understand what to expect from potential battery life.

Keywords: **Battery: Condition monitoring: Energy: IoT: Sensor: Vibration: Wireless:**

Introduction:

The cost of battery replacement in a typical industrial environment due to battery life not achieving expected longevity may be as high as 90% of the ongoing maintenance costs of a wireless vibration system[1]. With hundreds of thousands of sensors deployed today, maintenance managers and service providers are paying more attention to battery life, often finding that specified ratings don't seem to reflect reality once deployed.

---

[1] Allied Reliability

IoT Example:

A wireless sensor will have several operational modes managed by the radio SOC (microcontroller) firmware embedded by the SOC vendor. User configurable variables provided by the vendor include measurement time duration (sampling time) and measurement interval. When not measuring, connecting or transmitting the device microcontroller will instruct all functions to enter a 'sleep' state to conserve energy.

Typical operating modes include:

- System wake up
- Data acquisition (DAQ)
- Data formatting for transmission
- Radio wake up
- Transmission connection established
- Data transmission
- Shutdown

The total charge consumed by the sensor will be a composite of the energy consumed in each operational state and the time duration of the operation.

Wireless Condition Monitoring:

Wireless vibration sensors available today use a variety of radio protocols, both standards based and proprietary. Advanced Bluetooth Low Energy (BLE) is often used for the sensor to gateway connection, presumably due to the low cost, low power, high data transmission rates and SOC availability from multiple vendors.

The current consumption profile for a wireless vibration sensor equipped with BLE 5.0 is shown in Figure 1. The time duration of the data transmission is 6.6 seconds at 25C and 10m transmission distance. Once transmission is complete, the sensor drops back into a Sleep state punctuated by periodic 'Advertise' bursts every 2 seconds[2]. Even at an interval this frequent, the Advertise energy is low and typically accounts for around 10 % of the total power. The average Sleep mode current consumption is less than 5uA, as specified by the vendor (but only at 25C). Assuming infrequent measurement and waveform transmission cycles, Sleep mode will dominate the daily power consumption during the 86,400 seconds of a 24 hour period. More discussion of this when we talk about battery capacity.

---

[2] Frequent Advertise connection enables the user to trigger a waveform transmission on demand or change the measurement interval.

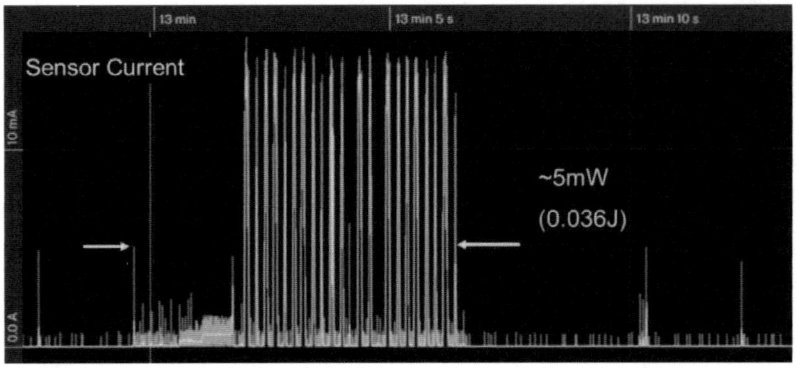

Figure 1. The current consumption profile of a wireless vibration sensor for Condition Monitoring applications provisioned with BLE 5.0. The average power during a typical 6.6 second transmission @ 25C and 10m transmission distance can be approximately 5mW. Once data transmission is complete, the radio SOC drops back into Sleep mode, punctuation by periodic transmission of low power Advertise signals.

Battery Technologies:

Numerous battery models are used in the the wireless vibration sensors on the market today. Form factor, chemical make up, capacity, open circuit voltages and of course, cost are all variables to be considered by the sensor designer. Given the higher demands for energy by the transmission of vibration waveforms in industrial environments, vendors tend to use higher capacity batteries than found in many IoT applications.

Virtually all the battery chemistries designed into wireless vibration sensors today are Lithium based, with Lithium Thionyl Chloride and Lithium Manganese Oxide the most common. One possible reason is that the technologies provide a relatively flat load voltage response of at least 3V over the useable life of the battery, easing power management requirements.

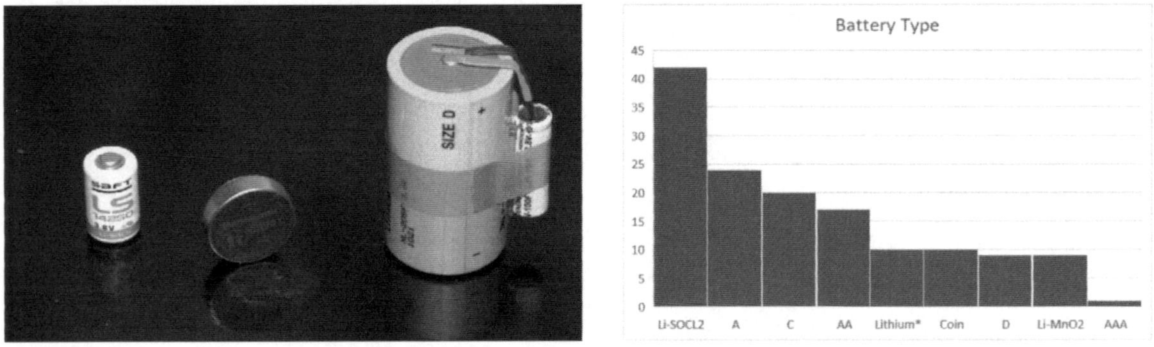

Figure 2. Primary cell batteries designed into wireless vibration sensors span the range from D to coin cells, with capacities from as high as 19Ah to as low as 470mAh. A histogram of battery types (RIGHT) from a database of 76 wireless vibration vendors with 158 sensor models.

Battery Life:

Operational life of a wireless sensor in an industrial environment is a complex function with many independent variables, most outside the control of the end user. At the heart of this are the technology choices – the battery, the communications protocol and the IoT SOC – all complex functions with complicated environmental sensitivities. Temperature, data transmission duration, transmit distances and interferers in the environment will all impact energy consumption.

For simple high / low current draw profiles with a given duty cycle where the average current can easily be calculated, the datasheet specifications can provide first order guidance regarding the energy the storage device will provide.

For example, if a constant current load for the Saft LSH 14 battery profile shown in Figure 3 is 24mA for 1 second out of a period of 10 seconds, the average current is 2.4mA. From the graph, we may interpolate between 2000 to 3000 hours of operation.

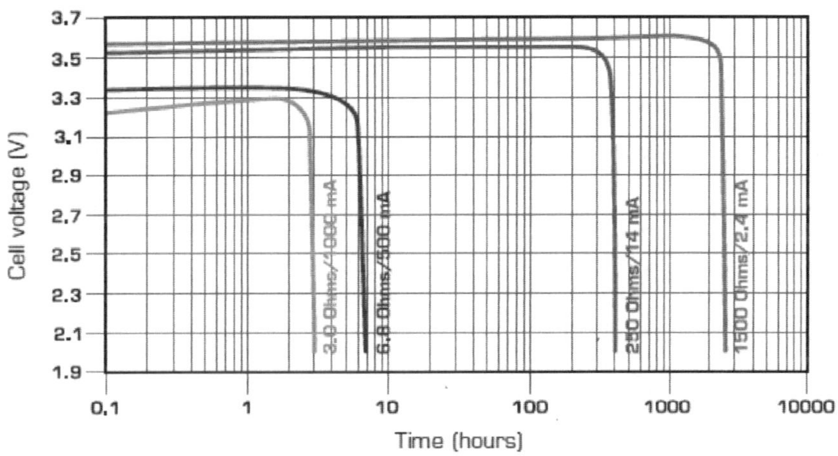

Typical discharge profiles at +20°C

Figure 3. Loaded voltage duration under constant current (CC) discharge. Saft LSH 14 3.6V 1300mAh rated capacity.

In reality, in addition to the specific battery capacity dependencies temperature and average current drawn, there are physical characteristics of the various components in a wireless sensor that act as parasitic consumers of power. We briefly discuss a few of them below.

Electronics Leakage and Sleep Current:

Semiconductor electronic devices are built from micro structures (transistors and gates) with dissimilar doped materials laid out adjacent to each other in such a way that current flow can be controlled by an applied voltage, forming a switch (or amplifier). As a practical matter, no silicon transistor turns completely off and instead suffers from 'leakage' current that is a function of temperature, chemistry and geometry. This means that radio micro-controllers with only a minimum of functions kept 'on' will still leak current. This leakage current adds to the minimum current necessary to maintain microcontroller operation, enough to wake up the system when called upon, often specified as Sleep current. Sleep current consumption generally has a positive correlation with temperature, consuming energy 24/7 whether the device is measuring and transmitting or not.

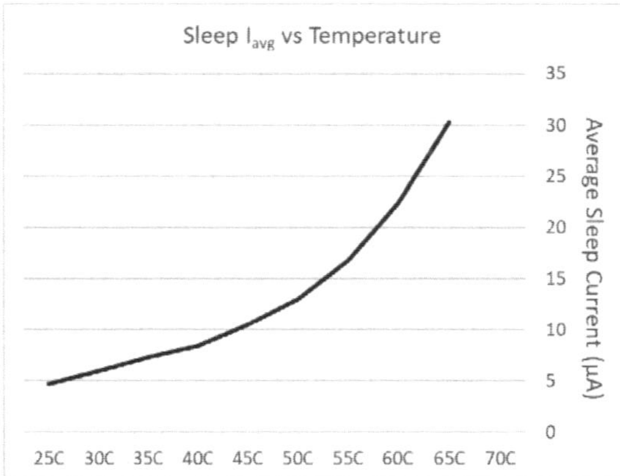

Figure 4. Average Sleep current of a wireless vibration sensor equipped with BLE 5.0, measured over temperature.

Energy Storage Self-discharge:

The impact that 24/7 circuit leakage has on battery life applies to Electro-chemical energy storage device *self-discharge* as well. Chemical ions will migrate between electrodes even when the battery is not in use.

Self-discharge specifications are not always published by the vendors, but the literature on self-leakage for various energy storage technologies suggests a qualitative comparison can be made. Typical self-discharge rates for various energy storage chemistries range from 30% monthly (Nickel based chemistries), 1 – 5% for lithium based batteries, to less than 1% for rechargeable energy storage technologies such as Supercap (EDLC), Hybrid Supercap and Lithium Titonate Oxide batteries. A qualitative summery of self-discharge rates for various energy storage technologies is shown in Table 1.

Table 1. A qualitative comparison of self-discharge rates for energy storage chemistries.

Charge Storage Type	Chemistry	Short form	Typical Monthly Self-Discharge rates
Alkaline	ZnMnO2	AFC	-
Nickel Cadmium	Ni-Cd	NiCd	---
Nickel Metal Hydride	Ni-MH	NiMH	---
Lithium Ion	Li-ion	LIB	-
Lithium Polymer	Li-Po	LPO	-
Lithium Manganese Dioxide	Li-MnO2	LMO	+
Lithium Nickel Manganese Cobalt Oxide	LiNiMnCoO2	LiCB	
Lithium Iron Phosphate	LiFePO4	LFP	-
Lithium Thionyl Chloride	Li-SoCL2	LTC	++
Lithium Titanate Oxide	Li4Ti5O2	LTO	++
Hybrid Super Capacitor		LIC	+
Electric Double Layer Capacitor		EDLC	+

Measurement Interval:

One of the few controls provided to the end user with potential for direct impact on the battery life is the measurement interval. As noted, Sleep current will dominate the power budget unless the frequency of measurement and transmission becomes high enough that the higher average transmit energy starts to impact the percentage of overall energy consumed.

Figure 5 shows a plot of both Sleep current and the Active mode consumed by waveform data measurement and transmission as a function of the measurement interval, collected with a wireless vibration sensor equipped with BLE 5.0. Note that the relative ratio of Active transmission to Sleep energy consumption is low and doesn't reach parity until a significant number of measurements in a 24 hour period are called for by the user.

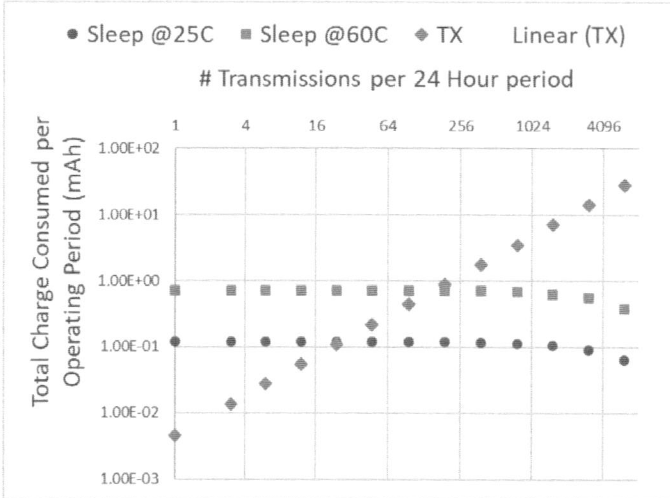

Figure 5. Total charge consumed during Sleep mode at both 25C and 60C, as well as Transmit power as a function of the number of measurement and transmission cycles in a 24 hour period.

Battery Life Specifications:

It should be clear by now that estimating battery life for a wireless sensor is an inprecise science. A selection of battery life guidance from several wireless vibration sensor datasheets is shown in Figure 6. With a few exceptions, it would be hard to construct a battery replacement maintenance schedule based on these specifications. Coming from a product development engineering background, I can sympathize with the product manager. It takes a significant characterization effort that includes analyzing large amounts of data to determine such a specification with any confidence.

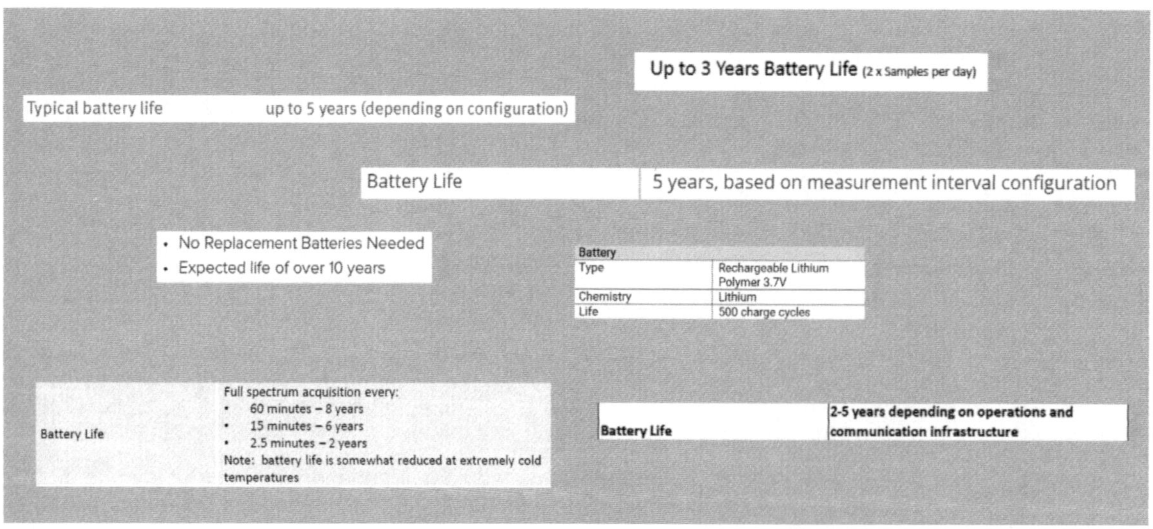

Figure 6. A sample collection of battery life specifications selected from wireless vibration sensor datasheets.

A reasonable estimate of energy consumption can be made by adding up the contributing components of each operational cycle (in this example- Active, Sleep, Advertise). These components can then be summed in a way consistent with the application of the sensor.

For example, average charge consumed during the Active cycle may be counted just once in a 24 hour cycle for the duration of measurement and transmission, leaving the rest of the day for Sleep mode to share with Advertise pulses. This total average power can then be compared to a battery capacity specification for an expected average temperature and load current.

$$Total\ charge\ per\ period = t_{active} * I_{avg} + t_{sleep} * I_{avg} + t_{advertise} * I_{avg} \qquad (1)$$

This exercise requires measuring and analyzing the current consumption on a time basis under the operating conditions expected (ex. over temperature and transmission distance). Lab or field measurements can be performed with specialized test equipment that replaces the battery with a constant voltage source to power the sensor operation while measuring the current consumed while operating. Multiple sensors should be tested to account for manufacturing variability and identify and anomalies.

Like buying a new car, the only way to really validate the mileage cited on the sticker is to drive it off the lot and calculate it yourself. Depending on your driving habits and normal daily use, your mileage may vary.

VIATC TRAINING CONFERENCE & EXPO
COVINGTON, KENTUCKY | AUGUST 7-9, 2024

MODAL ANALYSIS OF A VERTICAL TURBINE PUMP USING MOTION MAGNIFICATION AND TRADITIONAL METHODS

Andrew Lerche, Senior Staff Engineer
Maki Onari, President
Mechanical Solutions, Inc.
11 Apollo Drive
Whippany, NJ 07981
(001) 973-326-9920
Andy.Lerche@mechsol.com

Mr. Lerche has more than 20 years of professional experience in the areas of design, testing, analysis (fluid and structural), and troubleshooting of mechanical equipment and machinery such as pumps and turbomachinery. He is experienced in the areas of turbomachinery, high pressure fluid control equipment, and shipboard equipment. He has extensive experience with flow-induced vibration of turbomachinery blades, as well as shock and vibration testing and analysis of fixed mechanical and electrical equipment. Mr. Lerche has authored and presented several technical papers in the area of turbomachinery and holds three patents related to industrial compressors and valves. He holds a B.S. and M.S. in Mechanical Engineering from the University of Texas at Austin and the University of Texas at San Antonio, respectively.

Mr. Onari is the President and Director of Turbomachinery Testing at Mechanical Solutions, Inc. (MSI), in Whippany, New Jersey. He is responsible for field vibration testing involving Operating Deflection Shape (ODS) and Experimental Modal Analysis (EMA) testing. His career spans more than 25 years primarily working with rotating equipment analysis and troubleshooting in the petrochemical, refinery, and power generation industries. Before joining MSI, Mr. Onari was a Rotating Equipment Engineer in PDVSA-Venezuela responsible for the predictive maintenance of one of the largest petrochemical complexes in Latin America. Mr. Onari received his B.S. degree (Mechanical Engineering, 1996) from Zulia University in Venezuela. Mr. Onari is co-author of the pump vibration chapter, McGraw-Hill Pump Handbook. He is a member of ASME and the ISO TC108/S2 Standards Committee for Machinery Vibration as well as the Hydraulic Institute Vibration Committee.

ABSTRACT:

This paper will evaluate the use of motion magnification video technology to perform modal analysis on a vertical turbine pump and compare the results to the traditional modal impact test method. Results of tests, along with the advantages and disadvantages of each method will be discussed.

KEY WORDS: Experimental Modal Analysis, Machinery Diagnostics, Motion Magnification Video, Rotating Machinery, Video Amplification

Introduction:

Over the years, vibration signature data has been traditionally used in determining whether or not a machine is operating properly, to perform a diagnosis of vibration problems, or if reliability issues have been experienced (e.g., crack failures due to fatigue, or premature failure of internal components such as bearings, seals, shafts, ancillary components, etc.). An important part of such an investigation is the determination of natural frequencies and their mode shapes, in a test known as Experimental Modal Analysis (EMA).

When the EMA test is being conducted, the structural system is impacted with a modally tuned hammer. This is typically a specially designed hammer that measures the impact force with an integrated force sensor (load cell). The hammer is also designed so that its dynamic properties do not affect the measurement. The impact of the hammer excites a wide range of natural frequencies in the system, and the corresponding response is measured. When this method is performed over multiple locations on a system, either by roving the hammer, or by roving the sensor measuring the response, a large database is built up of amplitude vs frequency and phase angle. Each data point can then be mapped to a representation of the structure model. The results can then be amplified and filtered at specific frequencies so that the mode shape of the structure can be visualized for a specific natural frequency. This method has been used for years to better understand the dynamic characteristics of a structural system. Currently, there are multiple variations of this test method.

A newer method has recently been developed, that utilizes a high-speed camera vision-based motion magnification system that measures the response, and displays the mode shape for specific frequencies. This method utilizes existing motion magnification and/or video amplification tools and is discussed in greater depth by Marscher and Lerche [1]. This paper evaluates such methods against traditional methods when applied to a vertical turbine pump (VTP).

Background:

Vertical turbine pumps are used across multiple industries and applications such as water and wastewater, power generation, and oil and gas business sectors. Since vertical turbine pumps typically have a large length-to-diameter ratio, with an exciter (i.e., electric induction motor) usually mounted to the top of them, they can often have vibration issues. Typically, these vibration issues are due to a structural natural frequency in, or near, the operating range of the pump. The problem becomes even more common when the pumps are driven by variable speed drives (VFD), increasing the excitation forcing function range of frequency that could potentially excite a structural natural frequency in the pump.

When troubleshooting a vibration issue on a VTP it is common to perform an EMA test to understand if a natural frequency is causing the elevated vibration. Often, the vibration analyst will impact the top of the motor or the top of the discharge head to excite the above-ground structure of the system, and then measure the response to determine natural frequencies. Since VTPs are typically designed to have the discharge nozzle from the above-ground discharge head, the frequency response is typically not axis-symmetric. For this reason, the natural frequencies are determined in two directions. One of these directions is perpendicular to the discharge piping, and the other direction is parallel or in line with the discharge piping.

Testing:

Testing was performed on a 45,000 gallon per minute (GPM), 1,000 HP vertical turbine pump. The pump was variable speed from approximately 350 rpm to 595 rpm (5.8 Hz to 9.9 Hz). EMA tests were performed using a 12 lbm instrumented hammer. The pump was impacted near the top of the motor in the direction perpendicular to the discharge nozzle/piping, and parallel to the discharge pipe.

For this particular case, the goal was to determine if there was a structural natural frequency within the operating speed range of the pump, and the mode shapes were not as much of a concern. For this reason, full modal characterization to determine natural frequencies and mode shapes using a modal hammer was not performed.

Results – Traditional EMA

The traditional method of impact testing resulted in the Frequency Response Function (FRF), which is a transfer function of the input to output. The first fundamental natural frequency of the motor mounted on the discharge head, which is also called the reed frequency, is marginally lower (7.5 Hz) in the direction perpendicular to the discharge pipe, than in the direction parallel (7.8 Hz) to the discharge pipe. This is due to the additional stiffness gained from the discharge piping.

Results – Modal Analysis using Motion Magnified Video (MMV) Based Method

The MMV modal method is new and unique, in that it combines a traditional method of impacting a structure with a modally tuned hammer to excite structural natural frequencies, but instead of using accelerometers, it utilizes the MMV to measure and analyze the frequency response. This explanation is somewhat simplified, but can be further examined in Ref. [1], by Marscher and Lerche.

The results of the testing, besides just a natural frequency, are also magnified mode shapes. Since this method measures the response at all the locations axially along the pump's structure simultaneously, it is much quicker than roving an accelerometer to obtain different amplitudes, as the traditional method.

Figure 1 below shows a still frame image of the motion magnified mode shape at 7.5 Hz, with the impact and resulting motion in the direction perpendicular to the discharge piping. Figure 2 shows a still image of the motion magnified mode shape at 7.8 Hz, with the impact and resulting motion in the direction parallel to the discharge piping. These natural frequencies were the same as those determined through the traditional method using an accelerometer.

Figure 1: The MMV modal process applied to a vertical turbine pump and motor, Determining "reed" natural frequency and mode shape perpendicular to discharge head at 7.5 Hz

Figure 2: The EMA MMV modal process applied to a vertical turbine pump and motor, Determining "reed" natural frequency and mode shape parallel to discharge head at 7.8 Hz

Result Comparison

Figure 3 and Figure 4 compare the traditional EMA against the MMV-based modal method in the perpendicular direction, and parallel direction, respectively. Since the MMV modal test method is based on displacement, it has different units than the accelerometer-based method, which utilizes g's/lb_f. The frequencies from both methods correlated very well.

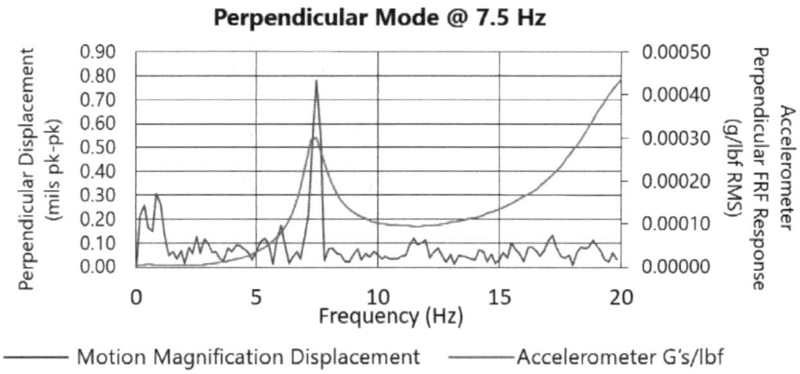

Figure 3: Comparison of traditional accelerometer-based EMA to MMV-based modal method in the perpendicular direction

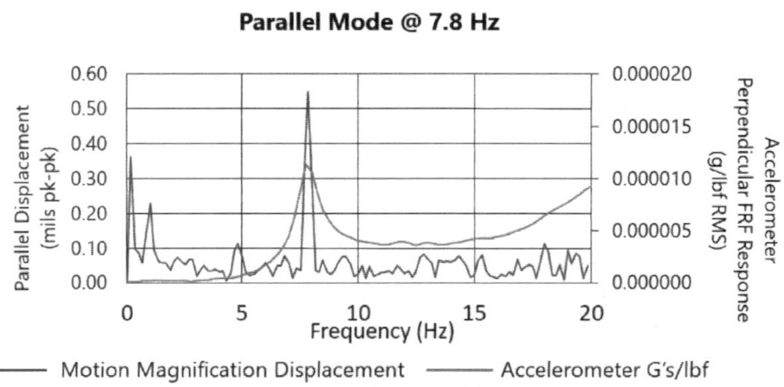

Figure 4: Comparison of traditional accelerometer-based EMA to MMV-based modal method in the parallel direction

Conclusion:

This testing demonstrated that the MMV-based modal testing method showed good correlation to the traditional accelerometer-based EMA test when evaluating the natural frequency. However, other aspects such as mode shape and other modal parameters were not compared due to the lack of test data available. In practice, the MMV-modal technique is very powerful in that it quickly evaluates the mode shapes and frequencies, whereas the traditional method would take more time and effort to determine the mode shapes.

However, the new technology is not without its limitations. Similar to MMV test vibration data, MMV modal analysis requires a system that is capable to detect and measure very small displacements. For this example of a large VTP, the response from the impact of a 12 lbm modal hammer was very small, however, the system and technique that the authors utilized were able to measure accurate results. In other words, the MMV modal method works only when the response from the impact pulse is large enough to be measured by the video vibration amplification system.

Thus, practically, the user must be aware of the capabilities of the video amplification system they are using so that they will obtain measurable data from the impact test (both in amplitude and frequency). This comes from practice and experience in conducting modal testing.

REFERENCES:

[1] Marscher, W., and Lerche, A., Motion Magnified Video Applied to Vibration Modal Testing, Proceedings of the Vibration Institute Annual Training Conference, Niagara Falls, NY, 2023

[2] Marscher, W., Modal Testing of Rotating Machinery While it is Operating, Sound and Vibration, Vol 33, pages 263-269, 1999

BIBLIOGRAPHY:

Marscher, W., and Lerche, A., Using Video to Magnify and Quantify Vibrating Motion, Proceedings of the Vibration Institute Annual Training Conference, Lexington, KY, 2019

Marscher, W., Lerche, A., Frolov, S., What Drives Accuracy for Video Vibration? Proceedings of the Vibration Institute Annual Training Conference, Niagara Falls, NY, 2023

Marscher, W.D., and Onari, M.M. (2018), Video Motion Magnification vs. Operating Deflection Shapes for Machinery Diagnosis, MFPT2018, Virginia Beach, May

VIATC TRAINING CONFERENCE & EXPO
COVINGTON, KENTUCKY | AUGUST 7-9, 2024

TROUBLESHOOTING HIGH VIBRATION ON A STEAM TURBINE GENERATOR USING A NEW VIDEO ALIGNMENT METHOD AND TRADITIONAL VIBRATION METHODS

Andrew Lerche, Senior Staff Engineer & Business Development
Maki Onari, President
Timothy Lebo, Senior Staff Engineer & Business Development
Mechanical Solutions, Inc.

11 Apollo Drive
Whippany, NJ 07981
(001) 973-326-9920
Andy.Lerche@mechsol.com

Mr. Lerche has more than 20 years of professional experience in the areas of design, testing, analysis (fluid and structural), and troubleshooting of mechanical equipment and machinery such as pumps and turbomachinery. He is experienced in turbomachinery, high-pressure fluid control equipment, and shipboard equipment. He has extensive experience with flow-induced vibration of turbomachinery blades, shock and vibration testing, and analysis of fixed mechanical and electrical equipment. Mr. Lerche has authored and presented several technical papers in the area of turbomachinery and holds three patents related to industrial compressors and valves. He holds a B.S. and M.S. in Mechanical Engineering from the University of Texas at Austin and the University of Texas at San Antonio, respectively.

Mr. Onari is the President and Director of Turbomachinery Testing at Mechanical Solutions, Inc. (MSI), in Whippany, New Jersey. He is responsible for field vibration testing involving Operating Deflection Shape (ODS) and Experimental Modal Analysis (EMA) testing. His career spans more than 25 years primarily working with rotating equipment analysis and troubleshooting in the petrochemical, refinery, and power generation industries. Before joining MSI, Mr. Onari was a Rotating Equipment Engineer in PDVSA-Venezuela responsible for the predictive maintenance of one of the largest petrochemical complexes in Latin America. Mr. Onari received his B.S. degree (Mechanical Engineering, 1996) from Zulia University in Venezuela. Mr. Onari is co-author of the pump vibration chapter, McGraw-Hill Pump Handbook. He is a member of ASME and the ISO TC108/S2 Standards Committee for Machinery Vibration as well as the Hydraulic Institute Vibration Committee.

Mr. Lebo is an Engineer and Business Development Manager for Mechanical Solutions. He received his BS in Chemical Engineering from Rutgers University and has worked in a number of different roles in the Oil & Gas, Refinery, Chemical, Power, Marine, and Water and Wastewater industries including Operations, Design, Logistics, Applications Engineering, and Sales over the past 15 years. At MSI, his focus is on working with industry partners to eliminate vibration problems in rotating equipment preferably before they even happen.

ABSTRACT:
This paper investigates the root cause of a high vibration and rotor rub issue on a steam turbine generator unit, that had a load-sensitive vibration issue. An advanced video-based motion tracking and measurement system was used to help characterize alignment and system movements from offline to running. Operating Deflection Shape (ODS) testing was also utilized to better understand the vibration under a particular loading condition. The two methods revealed the mysterious cause of the vibration issue.

KEY WORDS: Machinery Alignment, Machinery Diagnostics, Measurement Accuracy, Motion Magnification, Rotating Machinery, Turbine Vibration.

INTRODUCTION:
Many factors can contribute to machinery vibration. In turbines, these factors can include, but are not limited to: rotor imbalance, misalignment, aerodynamic forces, installation issues such as soft foot, or structural and lateral resonances issues. Imbalance and misalignment between the driver/driven machines are the most common mechanical causes, for which vibration responses can be observed at 1x & 2x running speeds. Many traditional tools and methods such as accelerometers, proximity probes, operating deflection shape (ODS) testing, thermocouples, dial indicators, string potentiometers, Essinger bars, and laser alignment tools can be used to investigate turbine movement and vibration. Also, newer video-based methods can be used to measure, visualize, and analyze vibration such as motion-magnified video or video vibration amplification tools. Other new video-based tools can be used to track the overall motion over longer periods to evaluate the thermal growth and alignment of machinery. These tools and methods will be discussed in the context of troubleshooting a vibration issue on a steam turbine generator unit.

BACKGROUND:
A steam turbine was installed approximately 25 years ago. The unit started to experience high vibration levels after a preventative maintenance overhaul in the previous year. The turbine was a 226 MW General Electric design. When the unit returned to service it experienced a rubbing phenomenon at bearing 1 (T1) and bearing 3 (T3) as depicted in the schematic on Figure 1, along with high steady-state vibration. The rub at T1 was related to differential expansion. Differential expansion occurs because, from a cold condition, the rotor is expected to grow at a higher rate than the casing as the system heats up.

During the fall overhaul, the mid-standard (T2 foundation) was suspected of not sliding axially as it was designed. As a result, the T2 foundation was reworked. Also, the lower halves of the high pressure (HP) and intermediate pressure (IP) turbine casings were removed, and two rows of IP blades were replaced.

In the following spring, a corrective overhaul was performed on the unit. During this outage it was discovered that significant rubs were present in: the lower half of the IP/low pressure (LP) packing (see Figure 1), HP rotor packings (significant axial rubs), the upper half of the HP turbine. A high-speed balance of the rotor was performed, and the cold alignment was corrected. It was also discovered that the front standard foundation (T1 location) had a significant tilt towards the front standard. The foundation was corrected with an epoxy repair.

More than a month after the unit was returned to service, the bearing 1 vibration increased with each load cycle. Vibration increased from 3.5 mils pk-pk to 4.8 mils pk-pk. For reference, the vibration alarm was set at 4 mils pk-pk, and the trip level at 7 mils pk-pk. The bearings have 8 mils of diametral clearance. The dominant vibration was observed to be at 1x rpm. This vibration was load-sensitive with high vibration at lower loads and low vibration at higher loads.

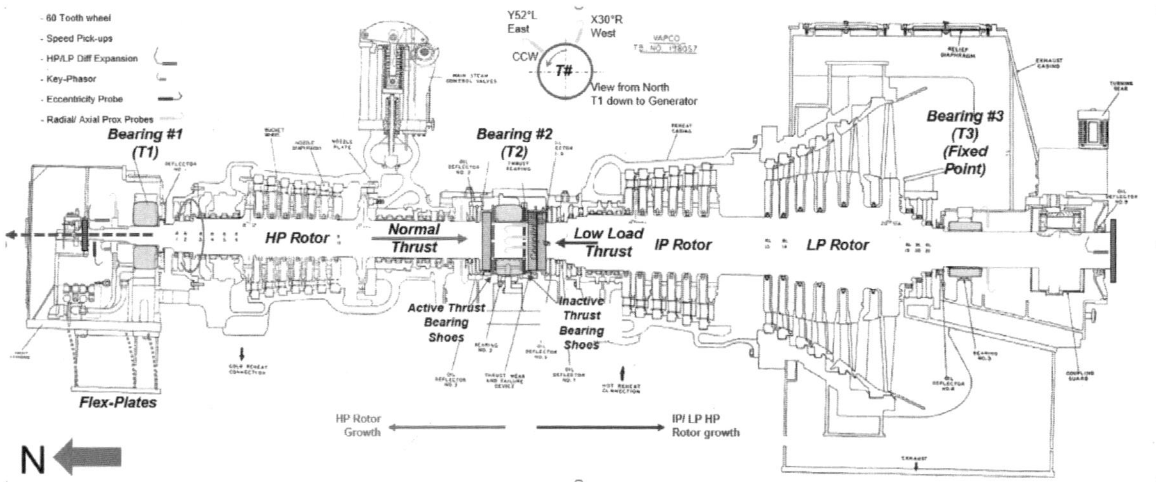

Figure 1: Steam turbine general info

VIBRATION TROUBLESHOOTING

In order to identify the root cause of the vibration, several different diagnostic tests were performed while the turbine was started from cold, and then operated at steady load, at variable loads, until it reached its full load. These tests included vibration monitoring of the casing, rotor, and support structure, tracking the thermal growth of the casing and support structure using a video-based motion system, experimental modal analysis (EMA) or impact testing, motion magnified video (MMV) testing during steady load, and operating deflection shape (ODS) testing of the entire unit.

The main objective of performing these tests was to characterize how the system was moving and vibrating at different temperatures and different loads to uncover the root cause of the vibration and rub events. This data, when taken simultaneously, showed how the structure, casing, and rotor, were moving relative to each other.

General Instrumentation

The following proximity probe data was utilized in the troubleshooting process to understand the rotor vibration and movement:

- Radial X and Y at each bearing
- Axial probe
- HP/LP differential expansion probes
- Eccentricity probes

In addition to the proximity probes, accelerometers were used on the casing and foundation at the following locations:

- Each of the 5 bearing locations (3 axes)
- Front standard supporting flex plates (3 axes)

Other parameters were also tracked to help in the diagnostics:

- Generator load
- Rotor speed
- Bearing temperatures

The above are point measurements at specific locations that provide significant insight into how one particular location of the system is responding over a period of time. However, there were a few other methods that were utilized during the testing that captured how the entire system was moving relative to each other, both from a vibration point of view and also from a thermal growth point of view. These methods included EMA, ODS, MMV testing, and video motion tracking. ODS testing and MMV have been covered in previous literature, and thus are not going to be covered here in great detail.

Experimental Modal Analysis

EMA is an important tool to identify the natural frequencies and mode shapes of a system. This method utilizes an impact hammer that excites natural frequencies in a system, while the frequency response is measured, often with accelerometers. This method is repeated over different locations, where either the hammer or accelerometer can be "roved" to different key locations. If the hammer impacts one location, the impacts can be repeated at the same spot while the accelerometers are moved to a different location of the system. This frequency response data is recorded versus amplitude and phase angle. The data can be exported onto a 3-D computer model so that the mode shapes can be animated. Besides the natural frequencies and mode shapes, modal damping characteristics can also be determined (when an instrumented impact hammer is used, and the impact force is measured).

For the case of this steam turbine system, since ODS testing was going to be performed, the focus of the EMA was to obtain the structural natural frequencies at a few particular locations, such as the natural frequencies of the flex plates supporting the front standard as well as the main bearing housings. The resulting data from the EMA was frequency response functions (FRF), which are a transfer function of the input to output signals. An example of an FRF from the EMA of the T1 bearing in the inline direction is shown in Figure 2. This FRF plot showed several natural frequencies, but there was one natural frequency at 62 Hz, which was within 3% of the operating speed of the unit.

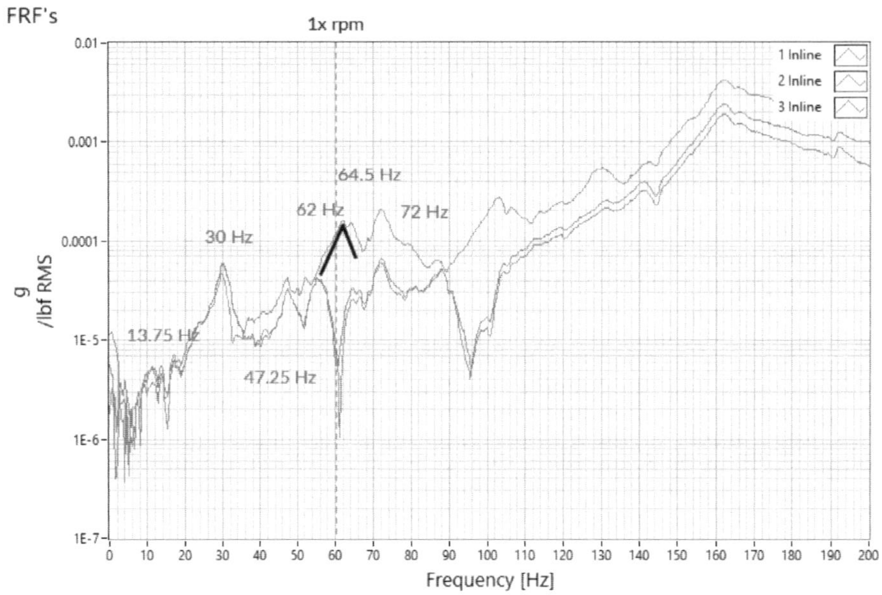

Figure 2: T1 bearing housing FRF plot showing a natural frequency at 62 Hz (3% separation margin from 1x operating speed) in the inline or axial direction.

Operating Deflection Shape Testing

ODS testing involves gathering extensive vibration data (amplitude and phase) over the entire system, and then visualizing the data in a computer animated 3D model. Vibration at individual frequencies can be animated, with motion highly amplified to better understand how the system is moving. This visualization method provides insight into how the system is moving, what is not moving, excessive flexibility, separation between parts, and facilitates identifying the potential root causes. Onari and Boyadjis discuss using ODS, along with other methods such as finite element analysis (FEA) to resolve vibration issues [1].

In addition to accelerometer data, proximity probe data can be incorporated into the ODS computer model to see also how the rotor is moving relative to the casing of the unit. Figure 3 shows an image of the ODS animation taken on the steam turbine. This method typically takes several hours of on-site measurements of the entire equipment assembly, depending on the size of the system, and then several hours/days in office to build the model, process the vibration database, and assign the vibration data to each point/sub-structure, before it can become animated using interpolation equations.

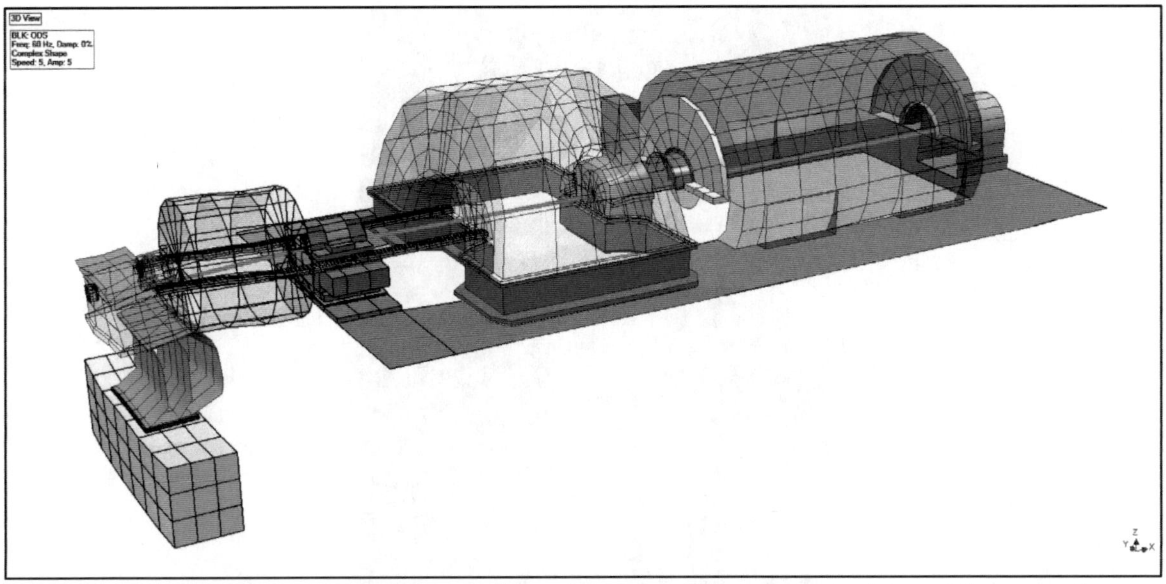

Figure 3: Still frame image of an ODS animation of the steam turbine at 1x operating speed. The animated video showed excessive pitching (axially) at T1 bearing.

Motion Magnified Video Testing

Similar to ODS, motion magnified video also displays exaggerated, slowed-down motion of the system vibration. This method utilizes a high speed camera to capture a few seconds of video of the equipment system during operation. The video can then be processed immediately to result in an animated vibration mode shape at any frequency. More on this method has also been covered in recent literature by various authors [2-9]. Unlike ODS, rotordynamic data from the proximity probes cannot be incorporated into the animated motion magnified video. This test tool method is extremely fast and powerful. A sample of a still frame image from the MMV of the front standard is shown in Figure 4 below. The actual MMV of the still image in this figure depicts the front standard rocking from front to back, along with axial movement.

Figure 4: Still frame image from motion magnified video of the front standard of turbine at 1x operating speed.

Video Movement Tracking
High resolution videography was also used for tracking overall movement of the whole system while it was heated up and powered up to full load. This test tool method is newer than the other methods mentioned above, so it will be discussed in greater detail below. Marscher, Lerche, and Piotrowski discussed this method along with other traditional methods for measuring offline to running alignment and movement [10]. They summarized that the video-based tracking method is advantageous over traditional methods, such as laser-based, proximity probe-based, or survey transient-based methods, in that it is faster and simpler to implement. Additionally, video-based methods record all the pixels within the field of view of the cameras, allowing for measurement of all points simultaneously on the system. Other methods only take measurements where the tool is located at a certain point in time and mounting arrangements can also be affected by thermal growth which reduces the accuracy of the measurement. Additionally, a video of the movement can be magnified and played back, if needed, to see how the system is moving as a whole, and relative to other points on the system.

For testing the steam turbine, several different camera angles were used to simultaneously capture the DC motion during the transient startup and gradual loading. Figure 5 below shows the location of the cameras for recording the data. Since high resolution cameras are used, this method does require the object of interest to be well-lit with consistent lighting. In this case, the testing was performed at night so that sun light through the windows did not provide inconsistent lighting.

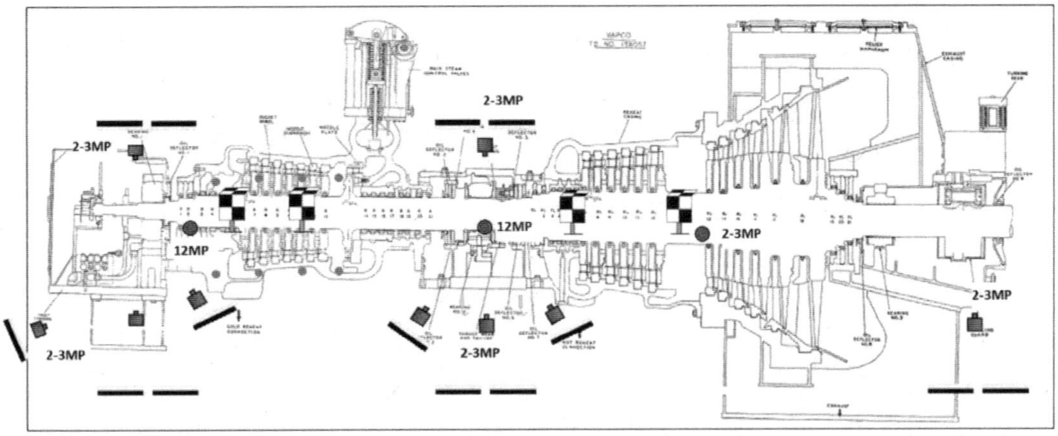

Figure 5: Camera and lighting layout for motion tracking of a steam turbine during the transient start up and loading (red circles and squares indicate camera locations, black/yellow rectangles indicate lighting locations).

RESULTS:

This section captures the key results of the above testing that was performed. However, due to the amount of data that was gathered, only key data is shown.

Overall Vibration

The overall vibration points measured by accelerometers were mostly within reasonable limits, with the exception of the vibration at the T1 bearing cap and flex plate in the axial direction. The T1 Bearing cap location was measured to be 0.46 in/s Pk at 65 MW (lower load). The vibration amplitude at the outboard flex plates were indicating 0.89 in/s Pk and 1.45 in/s Pk in the axial direction.

The vibration at the T1 bearing, where vibration was the highest, is shown in the overall vibration trend plot in Figure 6 below. The plot also shows how the vibration increased when the load decreased down to about 65 MW.

Radial proximity probe data (Figures 7 and 8) indicated that the centerline at the T1 bearing moved downward from full load to low load, to approximately 2 mils eccentric with respect to the initial cold conditions. At this position the orbit plot turned more elliptical with the X (East) vibration at T1 being much greater than the Y (west) vibration.

The axial proximity probe (Figure 9) indicated that as the load increased, the rotor was pushed towards the south, away from the axial probes, to the active side of the thrust bearing, with a total end play of about 12 mils. This was expected.

The still image of the motion tracking video shown in Figure 10, from the perspective of the overhead cameras, indicated that the T1 Bearing East side tracking point (green) grew 220 mils more than the west side (yellow), indicating uneven growth, which was not expected. Also, Figure 11 shows similar trends from the cameras on the East and West sides of the mid-standard.

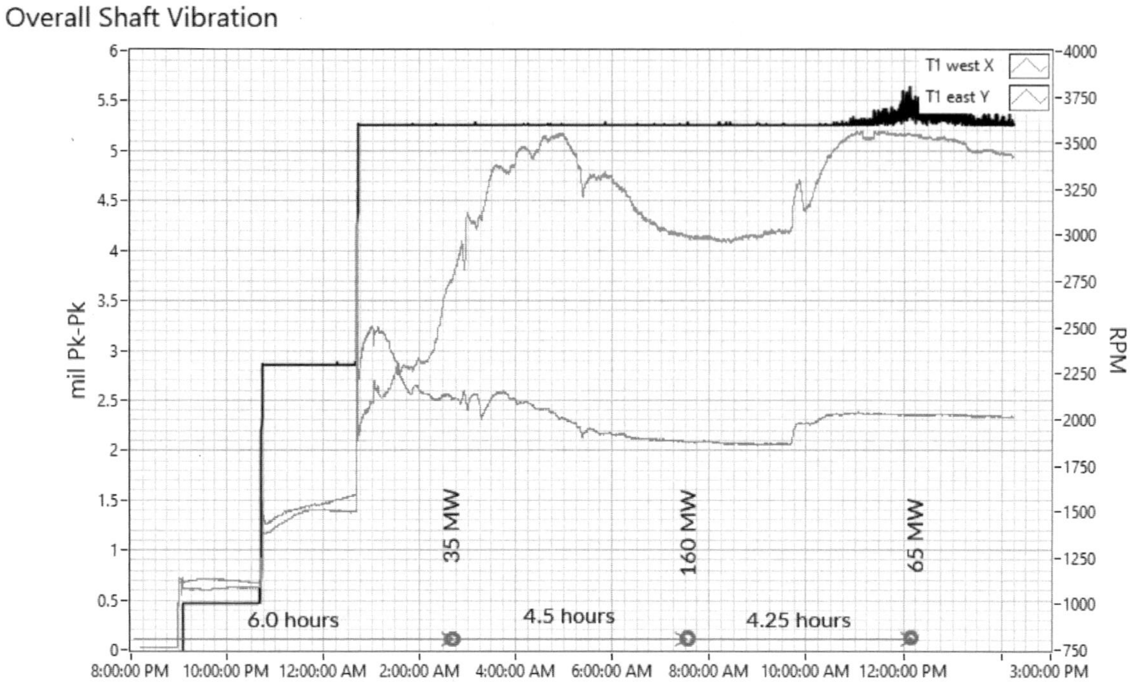

Figure 6: Overall vibration plot with varying load at the T1 bearing. The vibration appears to have the larges amplitude around 65 MW.

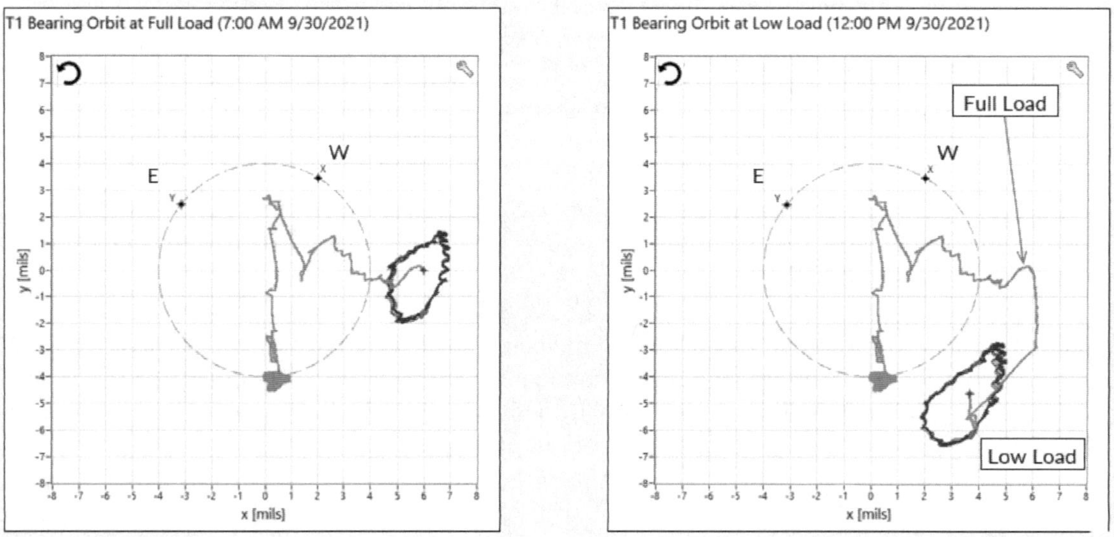

Figure7: Orbit plots of the T1 Bearing location at full load (left), and low load (right). At the low load condition, the orbit plot becomes more elliptical with greater X (West) vibration.

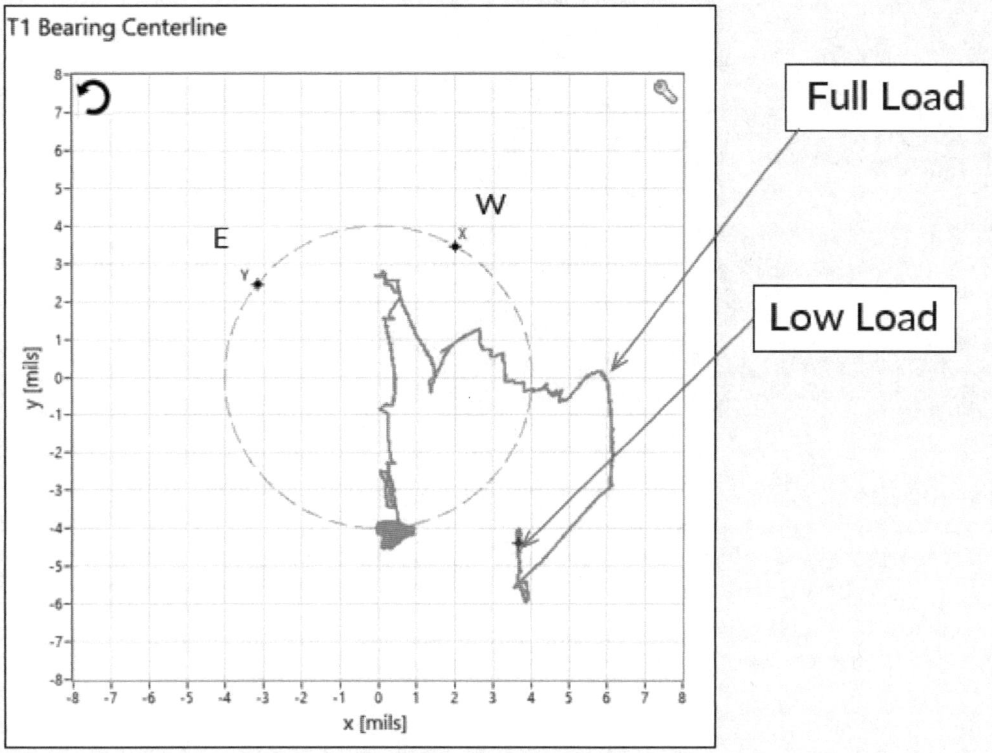

Figure 8: Shaft centerline plots of the T1 bearing location. There is a shift of downward at low load, and the plot indicates that the HP rotor is 2mils eccentric with respect to the initial cold conditions.

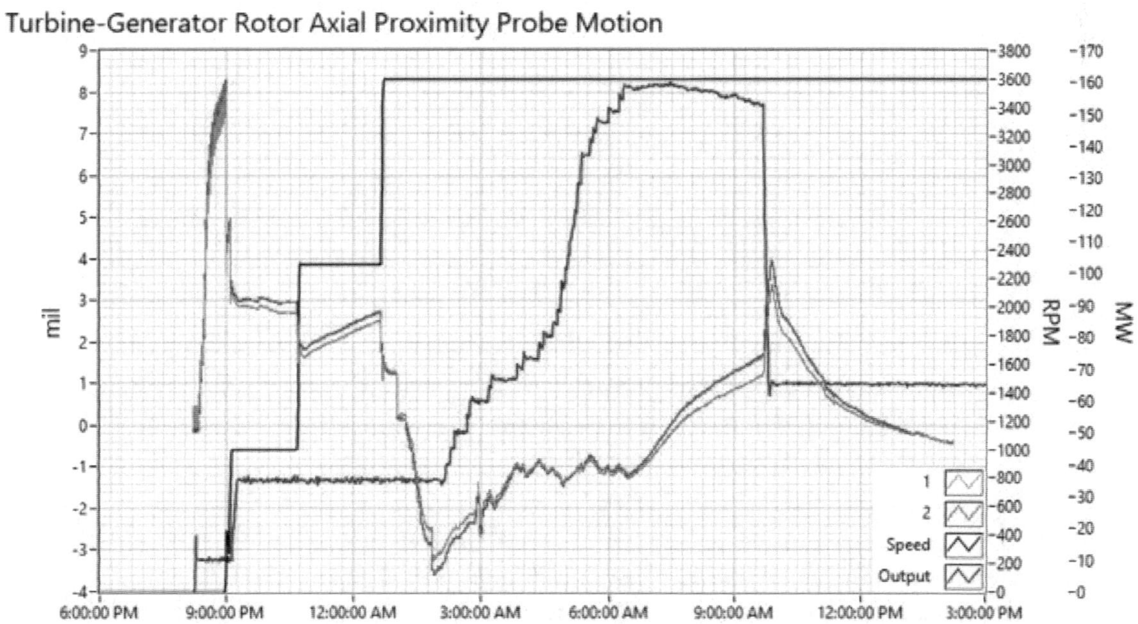

Figure 9. Turbine-generator rotor axial position during changing load. As the load increased, the rotor is pushed towards the south, away from the axial probes, to the active side of the thrust bearing. Total end play was 12 mils

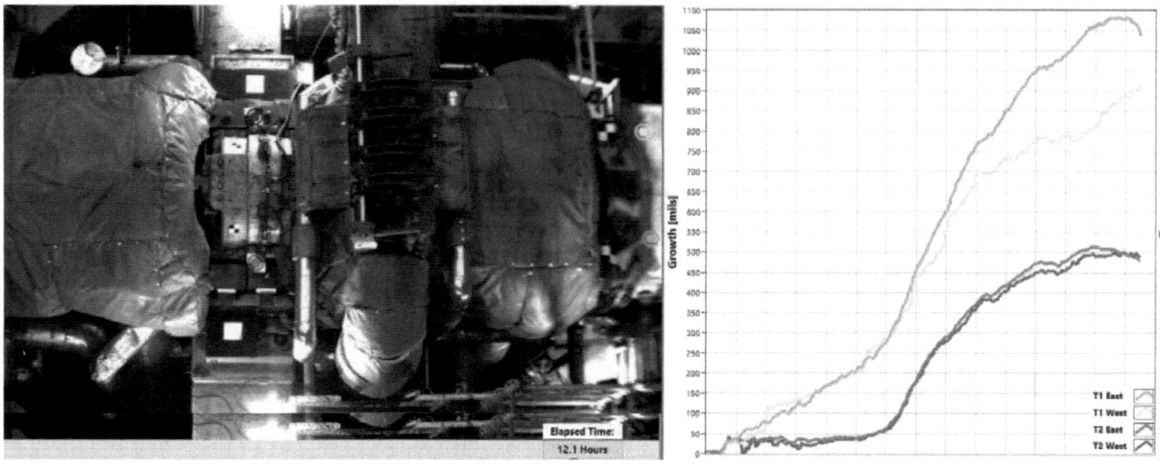

Figure 10. Still image of motion tracking video from overhead cameras with resulting plot of displacement of T1 and T2 bearings over 12-hour period. The T1 Bearing East tracking point (green) grew more than 220 mils more than the T1 Bearing West tracking point (yellow), indicating uneven growth, which is not expected.

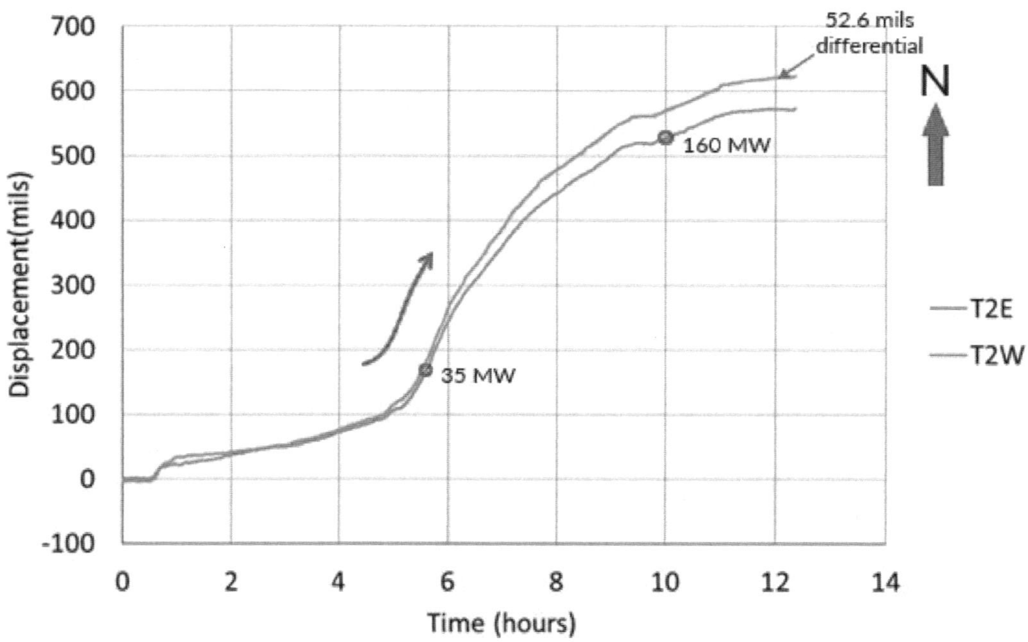

Figure 11. Thermal growth at T2 location (taken from East and West Cameras) shows that the west side grew 52.6 mils more than the east side after the system reached full load operation.

CONCLUSIONS:

Before coming to a conclusion about the cause of the high vibration, it is important to note how the turbine was designed, and how it is supposed to grow when the unit is starting or shutting down when the load is changing.

The casing is fixed at the T3 location, while the thrust bearing is at the T2 location. Thus, when the system heats up the casing grows from the T3 location. Also, the thrust direction on the high-pressure rotor is toward the thrust bearing at T2 (see Figure 1).

If the casing is not growing properly at T2 and T1, it creates a condition where the casing binds up with limited movement. It is likely that the twisted position of the mid-standard (T2) and front standard (T1) forced the operation of the T1 bearing in the lower right quadrant of the orbit as shown in Figure 8 above.

The limited HP casing growth, due to binding, could cause bending in the rotor, since the rotor still grows as it should. It should be noted that each of the casings are positioned by keys similar to those shown in Figure 12. If these keys do not allow the casing to grow smoothly, the binding that was observed in the testing could occur. These keys require periodic lubrication.

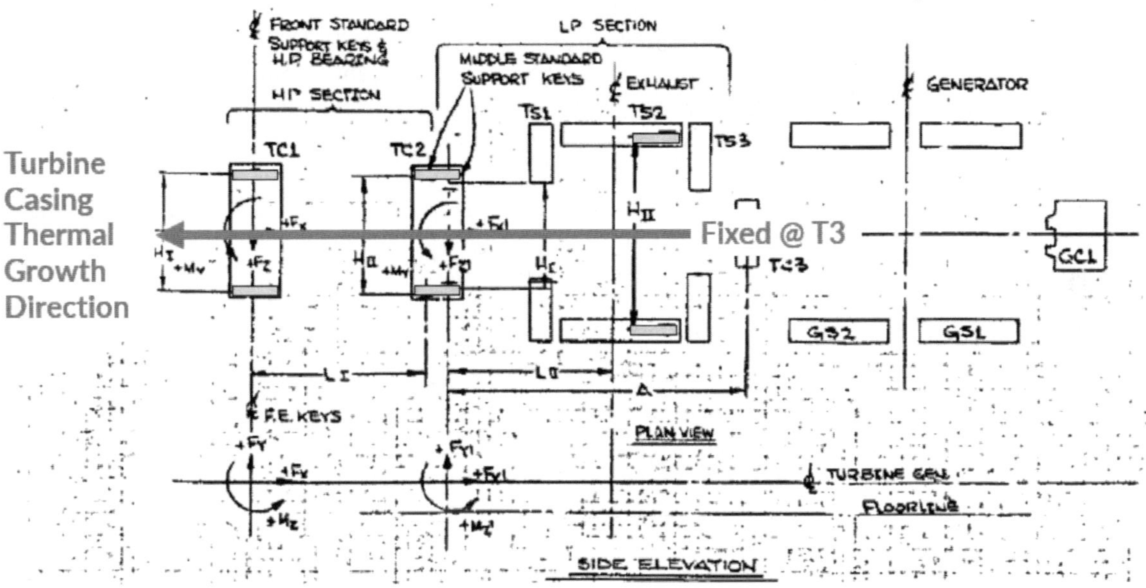

Figure 12. Configuration of positional guide keys on the turbine.

The EMA and ODS testing indicated a structural natural frequency at the front standard on the flex plates near 1x operating speed. It was also observed that this natural frequency did not change significantly with load. As the turbine casing grew, the stiffness of the flex plates increased due to the axial pre-load. If the turbine was allowed to grow as designed, the stiffening effect would have increased the separation margin of the natural frequency and the synchronous speed, thus reducing the overall vibration at high load. Therefore, at low load operation, this natural frequency was in resonance increasing the overall vibration. Therefore, the combination of the natural frequency of the front standard mounted on the flex plates while the casing was binding up contributed to the excessive vibration that was observed.

In subsequent testing after maintenance, including lubrication of the main accessible guide keys on T1 and T2, the differential growth from the East and West sides of T1 and T2 was eliminated, and vibration was reduced down to acceptable levels.

REFERENCES:

[1] Onari, M., Boyadjis, P., "*Solving Structural Vibration Problems Using Operating Deflection Shape and Finite Element Analysis*", Proceedings of the 25th International Pump Users Symposium, Turbomachinery Laboratory, Texas A&M University, College Station, Texas, 2009.

[2] Chen, J.G., et al, "*Modal Identification of Simple Structures with High-Speed Video Using Motion Magnification*", J. of Sounds & Vibration, vol. 345, 9 June 2015

[3] Liu et al (2005), "*Motion magnification*", ACM Transactions of Graphics (TOG) – Proc. of ACM SIGGRAPH, TOG Homepage, v. 24, Issue 3, July, 2005.

[4] Marscher, W., and Lerche, A., "*Motion Magnified Video Applied to Vibration Modal Testing*", Proceedings of the Vibration Institute Annual Training Conference, Niagara Falls, NY, 2023.

[5] Marscher, W., and Lerche, A., "*Using Video to Magnify and Quantify Vibrating Motion*", Proceedings of the Vibration Institute Annual Training Conference, Lexington, KY, 2019.

[6] Marscher, W., Lerche, A., Frolov, S., "*What Drives Accuracy for Video Vibration?*", Proceedings of the Machinery Failure Prevention Technology (MFPT) Conference, Niagara Falls, NY, 2023.

[7] Timoner, SJ, and Freeman, DM, "*Multi-image gradient-based algorithms for motion estimation*", Optical Engineering 40.9, pp. 2003-2016, 2001.

[8] Wu, H-Y et al (2012), "*Eulerian video magnification for revealing subtle changes in the world*", ACM Trans. of Graphics (Proc. SIGGRAPH 2012), v. 31, August 2012.

[9] Marscher, W.D., and Onari, M.M. (2018), "*Video Motion Magnification vs. Operating Deflection Shapes for Machinery Diagnosis*", MFPT2018, Virginia Beach, May 2018.

[10] Piotrowski, J., Marscher, W., Lerche, A., "*Characterizing Off -Line to Running Movement Using High-Resolution Imagery*", Proceedings of the Vibration Institute Annual Training Conference, Niagara Falls, NY, 2023.

BIBLIOGRAPHY:

ANSI/ASA S2.75-217, Shaft Alignment Methodology

The Importance of Motor Shaft Alignment, US Dept. of Energy, Motor Systems Tip Sheet #4, No. 2012

VIATC TRAINING CONFERENCE & EXPO
COVINGTON, KENTUCKY | AUGUST 7-9, 2024

Online Oil Health Monitoring for Land Fill Gas Engines

Jeremy Sheldon
Director Asset Management and Analytics, Poseidon Systems, LLC
jsheldon@poseidonsys.com

Jeremy Sheldon is Director Asset Management and Analytics at Poseidon Systems, LLC. He has over twenty years of experience in the Condition Based Maintenance and Prognostic Health Management domain with a strong background in signal processing, data analysis, health monitoring system implementation, testing, vibration diagnostics. In his current role, Jeremy manages Poseidon's condition monitoring team while simultaneously contributing to technical and programmatic aspects of several programs.

ABSTRACT:

Online oil sensing technologies offer an opportunity to improve how gas-powered engines are operated and reduce operation costs. Engines powered by recovered land fill gas (LFG) for instance can suffer increased oil degradation due to the fuel's properties. These engines require frequent lab analysis to monitor the effects of the fuel on the oil. The effort needed to maintain appropriate lubrication for this class of engines can be reduced with online sensors. Within this paper the authors discuss sensor technologies that are applicable to LFG powered engines via an actual in field case study.

KEY WORDS: Condition monitoring, debris, engine, generator, impedance, land fill gas, oil condition, online, real time, sensors, wear

INTRODUCTION:

Online oil sensing technologies offer an opportunity to improve how gas-powered engines are being operated and reduce operation costs. Engines powered by recovered land fill gas for instance can suffer increased oil degradation due to the fuel's properties. These engines require frequent lab analysis to monitor the effects of the fuel on the oil. In addition, many of these engines operate in remote areas making the oil sampling expensive and time consuming. Utilizing advanced sensing technology focused on oil condition can lead to higher uptime through lubrication health tracking and event detection, enabling informed, timely preventative maintenance. These benefits lead to lower maintenance costs as operators can reduce failures and take corrective action earlier, a critical advancement in sensing technology was required.

Advances in sensor design have enabled the required deeper understanding of the oil properties needed by a comprehensive oil condition monitoring program necessary to replace the need for periodic oil sampling. These types of online lubrication condition monitoring sensors allow operators to know the current health of their oil, trend degradation, project fluid remaining useful life (RUL), and significantly reduce overall oil consumption as part of a truly condition-based fluid change program.

SENSOR OVERVIEW:

Implementing the proper condition monitoring system can be a complicated process. Multiple aspects must be considered, foremost is the application and prevalent failure modes. While there are many different types of sensors that could be used to monitor the health of an engine, the authors focused on two. Specific to the land fill gas application, oil condition is of paramount importance due to the low fuel quality compared to traditional fuels. Secondary mechanical damage to the engine due to the contamination of the fuel leading to detrimental byproducts in the oil, is also common. Therefore, the authors identified mechanical wear debris and oil condition sensors as the proper online real time sensors for field trial on two land fill gas powered generators.

Mechanical Wear Debris

Online wear debris sensors provide the unique capability of constantly monitoring the levels of ferrous and nonferrous wear debris particles within a mechanical system. Accelerating wear generation rates in particle counts can be easily recognized and acted upon before a critical gearbox failure occurs. The utility of online wear debris monitoring as a means of detecting mechanical component failures and continually assessing their severity makes them an attractive monitoring option in many applications, especially oil-lubricated gearboxes.

Various online debris monitoring technologies are available. Inductively Coupled Plasma Atomic Emission Spectroscopy (ICP-AES) or analytical ferrography methods are commonly used in laboratory settings [1]. Optical based particle counters are common in labs but can also be found in certain field applications. Magnetic-based chip collectors and detectors are typically used in gearbox monitoring applications [2]. Another technology used for debris sensing is based on an inductive sensor created by one or more coil windings around a non-conductive fluid passage (sensing bore). The coil is excited with a high-frequency oscillating voltage and the resulting magnetic field couples with any metallic particle present within the field. The particle's passage results in a measurable impedance change in the sensing coils providing information about the particle's size, type (ferrous or nonferrous), and speed. Unlike many other sensing technologies, the output of wear debris is a direct measurement of mechanical damage. In addition, the results are easy to interpret with little expertise needed to analyze and understand. The authors selected an inductive coil-based debris sensor for this case study.

Oil Condition

Oil sampling analysis is the backbone of most oil reliability programs, including a key aspect of ensuring proper operation of LFG powered systems. From early analysis technology used by oil labs to the more modern miniaturized benchtop equipment, manual oil sampling is critical to most rotating machine preventative maintenance. The latest advancements are new in-line, real-time oil quality monitoring sensors that deliver many of the benefits of manual oil sampling in an online monitoring capability.

Typically, online sensors are electrochemical impedance, dielectric, conductivity, or permittivity measurement devices detecting degradation of the oil. Some methodologies can detect degradation of not only overall quality, but estimate percent soot, varnish, total base number, relative humidity, additive depletion, etc. Regardless of sensing method, effective oil condition sensors must detect most, if not all, key oil events and have an output that trends with the life of the oil while the asset is in operation. While these sensors cannot duplicate lab analysis results, they can provide the necessary insight to make preventative maintenance decisions before damage occurs. Online oil quality sensors are key enablers to shift oil sampling from periodic to condition-based sampling and remote monitoring.

Various technologies provide the ability to measure multiple properties of a lubricant. As mentioned above, two of the more widely used methods for online oil monitoring are electrochemical impedance spectroscopy (EIS) and measurement of the dielectric constant or ratio (simply referred to as dielectric) of the oil/fluid, see Figure 1. Dielectric is a measurement of how well a material conducts electric current compared to how the same field would conduct in a perfect vacuum [3]. The relative simplicity of the sensing technology is well suited for online applications. As oil degrades contaminates and chemical changes affect the measured dielectric of the oil. Dielectric may increase or decrease depending on the change to the oil. Typically, for many important health changes, for example depletion of additives, the oil dielectric will decrease. The developed sensor incorporates a dielectric measurement for tracking oil condition.

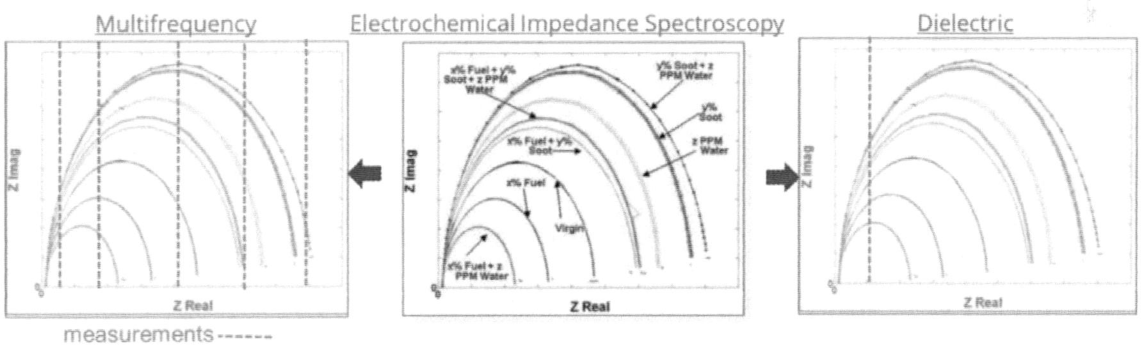

Figure 1: Illustrative Responses from Impedance Based Oil Condition Sensors

An EIS based sensor was selected for this case study. The application of EIS to the characterization of lubricating oil provides the ability to measure multiple properties of a lubricant using a pair of simple electrodes immersed in the fluid of interest. The technique does not require any consumables or exotic materials, it does not require any moving parts, and it is applicable to a wide range of fluid types. Accordingly, it is well suited to in-situ oil monitoring applications.

CASE STUDY:

The following sections detail an actual in field application of multiple oil sensors on two different landfill gas power generation systems. The main objective for installation was to detect costly engine failures due to premature failure of engine's plain bearings. Both wear debris and oil condition sensors were installed in the engine lubrication system to detect signs of premature engine wear and deterioration in oil condition that might be a pre-cursor to engine damage respectively. A secondary objective was to determine the correlation of the online sensors and the traditional manual lab-based oil analysis. Bio-derived fuels often cause unique degradation of the engine oils that are typically monitored by frequent, potentially as frequent as every 10 days, oil sampling. If the online oil sensors could reduce or eliminate the frequency of sampling, operating costs could be reduced.

The sensor systems were installed on two generator engines, per illustration in Figure 2. Generator 1 was expected to be completely healthy and free of abnormal conditions based on repair history. Generator 2 experienced regular issues relating to fuel (gas) quality. Three sensors were installed on each generator's engine, a metallic wear debris sensor, an impedance-based oil condition sensor, and a tuning fork-based viscometer. A small amount of engine lubrication oil is pulled from and returned to the engine in a bypass type installation as required. Data from the sensors is collected by a local data logger and transmitted to remote monitoring system for analysis and alarming.

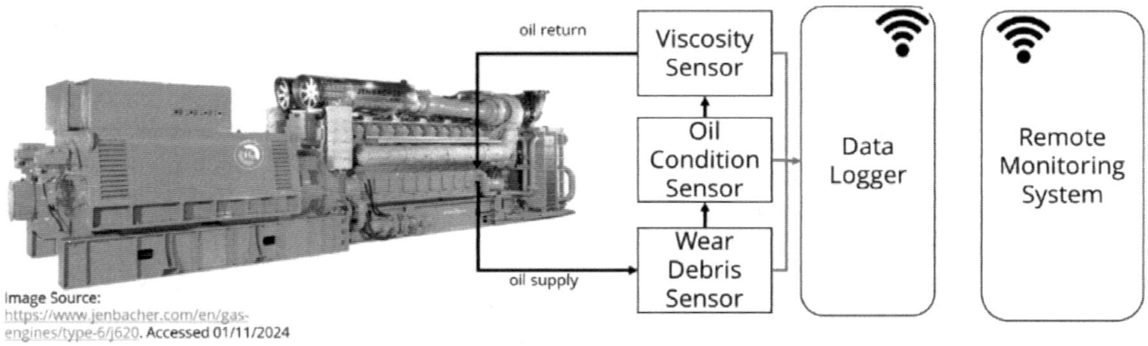

Figure 2: Installation Diagram

The first interesting finding occurred soon after installation when elevated levels of wear debris was detected in Engine 1. As shown in Figure 3, Engine 1 had eight times the total mass (75 mg) of ferrous debris compared to Engine 2 (9 mg). The site was notified of potential issue and a borescope inspection was performed on Engine 1. Relatively minor damage was found on one cylinder as shown in Figure 4, a proactive repair was preformed to replace the piston rings and liner. It was concluded that this early detection had alerted the company to an issue that would ultimately have resulted in serious engine damage. For Engine 2 wear debris was generated, but not in significant quantities to indicate abnormal condition.

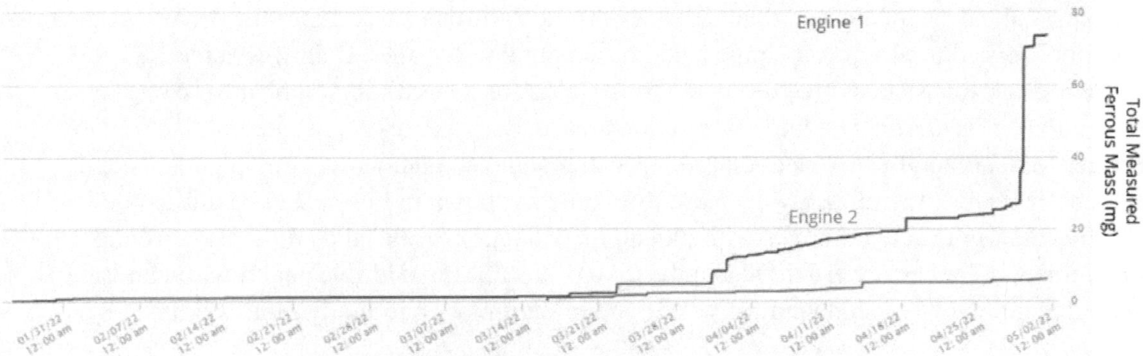

Figure 3: Total Measured Wear Debris Mass

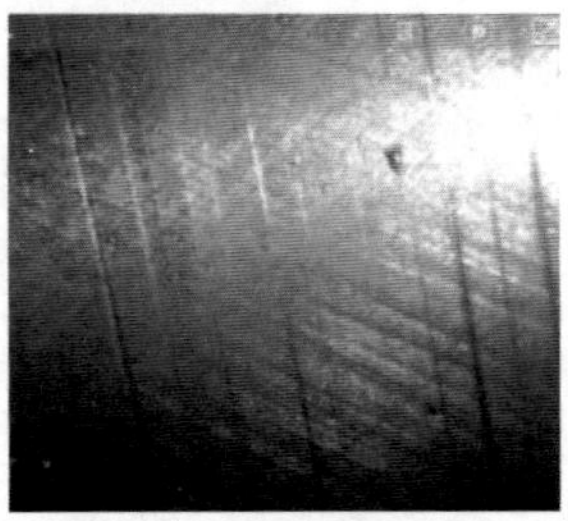

Figure 4: Confirmed Engine 1 Damage

An overall oil health indicator for both engines is shown in Figure 5. The oil health indicator ranges from one perfectly healthy oil to zero completely depleted or contaminated oil. The indicator can go negative to indicate poor oil health. From this comparison it is clear the oil condition of Engine 2 degrades faster than Engine 1 as evident by the quickly decreasing value seen over the two oil drain intervals. Engine 1 has a more typical trend in the oil condition, while Engine 2 oil clearly degrades very rapidly.

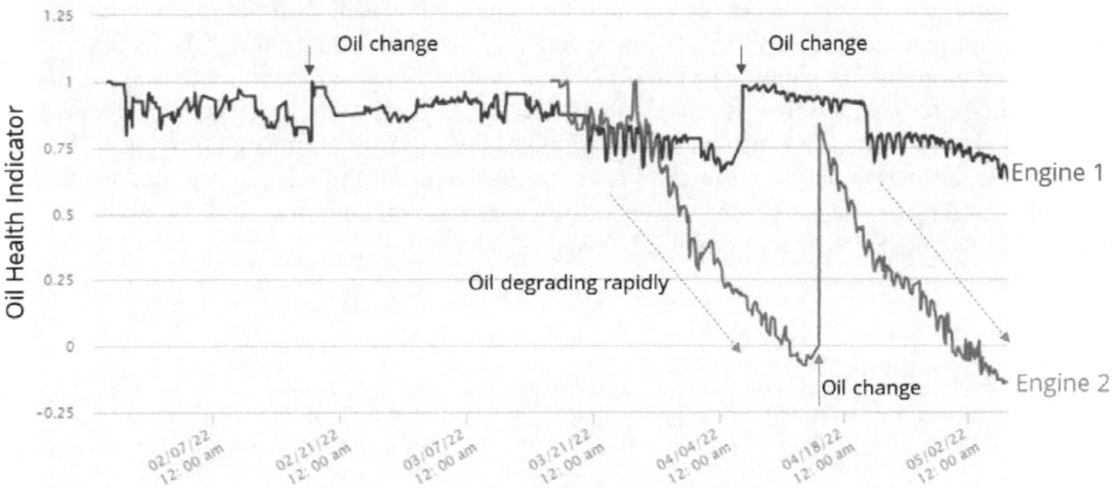

Figure 5: Oil Health Indicator Trends

Comparing the directly measured impedances provides further insight into the drivers of the oil condition. A medium frequency impedance-based parameter called Bulk Resistance is shown in Figure 6. Bulk Resistance often offers the most value for determining oil health. In engine applications, it is driven by additive depletion and oxidation. Bulk Resistance is also sensitive to changes to polar additives, oxidation products, and some contaminants. The impedance will rise after an oil change, then linearly decrease over time as shown in Figure 6. The initial rise in value occurs as the new oil mixes with old, additives agglomerate, and additives surround contaminants. The downward trend is indicative of decreasing additive health with abnormal accelerating downward slope indicative of contamination or abnormally rapid aging.

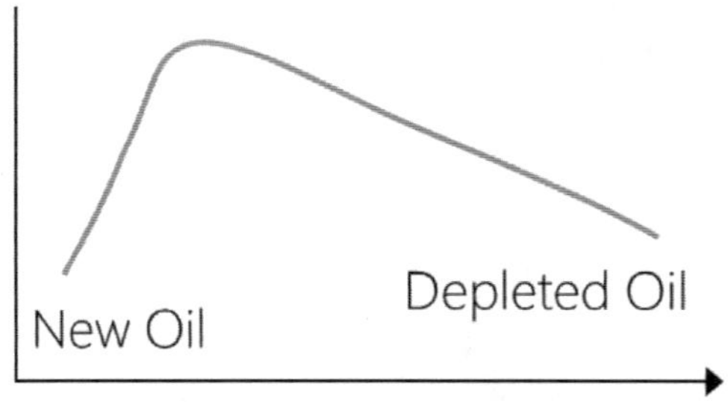

Figure 6: Characteristic Bulk Resistance Response

The measured Bulk Resistance from both engines is shown in Figure 7. From Figure 7, each engine had three oil changes during the interval shown. Note Engine 2's system was installed after Engine 1, hence the lack of data at the beginning of the trend.

During the second and third oil drain intervals of Engine 1, the Bulk Resistance parameter began decreasing more rapidly than during the first oil drain interval. The more rapid decrease indicates the oil was also more quickly degrading. Given the application and data pattern, there are two probable causes for the more rapid oil decline. First, since this is a land fill gas powered engine, there is potential worsening gas quality resulted in more rapid acidification of the oil, thus more rapidly degrading oil condition. Second, the presence of the increased wear debris, as described above, caused more rapid oxidation of the oil since metal particles can function as an oxidation catalyst in oils. From a condition monitoring perspective, the critical point is that the steepening trend during this second oil interval was highlighted to the end user therefore allowing appropriate investigation before more serious problems occurred. In this instance, the mechanical damage detected by the debris sensor was a contributor to the oil condition and is further described below compared to traditional lab-based oil sampling.

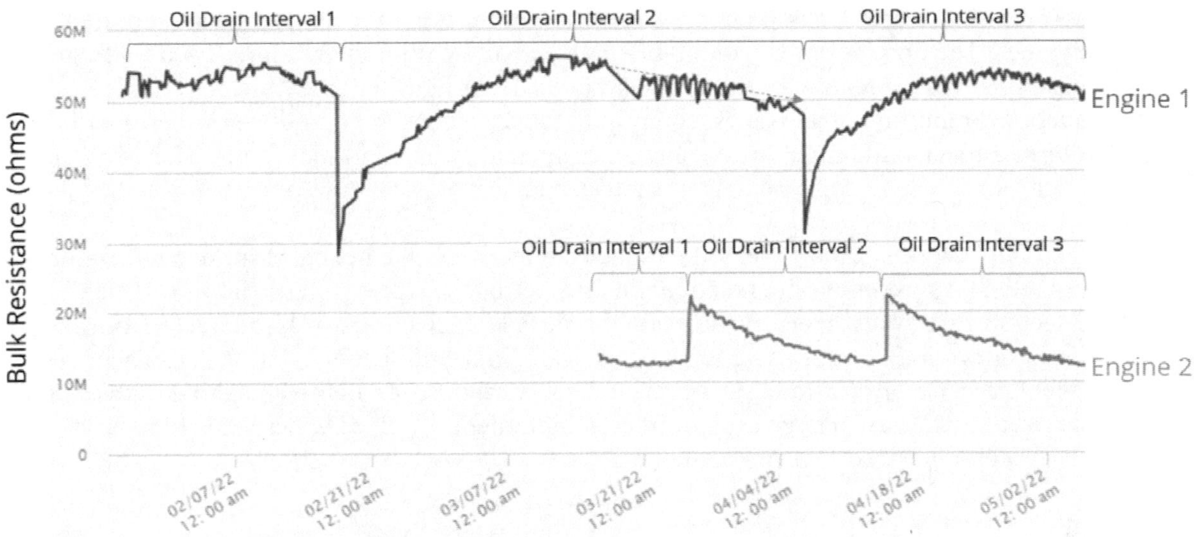

Figure 7: Bulk Resistance Sensed Parameter Trends

Engine 2's oil condition was clearly much different, as shown in Figure 7 and Figure 8. The oil health indicator and Bulk Resistance both decrease extremely rapidly. Such an extremely rapid decrease indicates short oil life and the flattening of the parameters curve at the end of the period is suggest completely depleted oil condition. After consulting with the customer, the oil sensor response is accurate as Engine 2's fuel is of low quality resulting in much higher acidification of the oil.

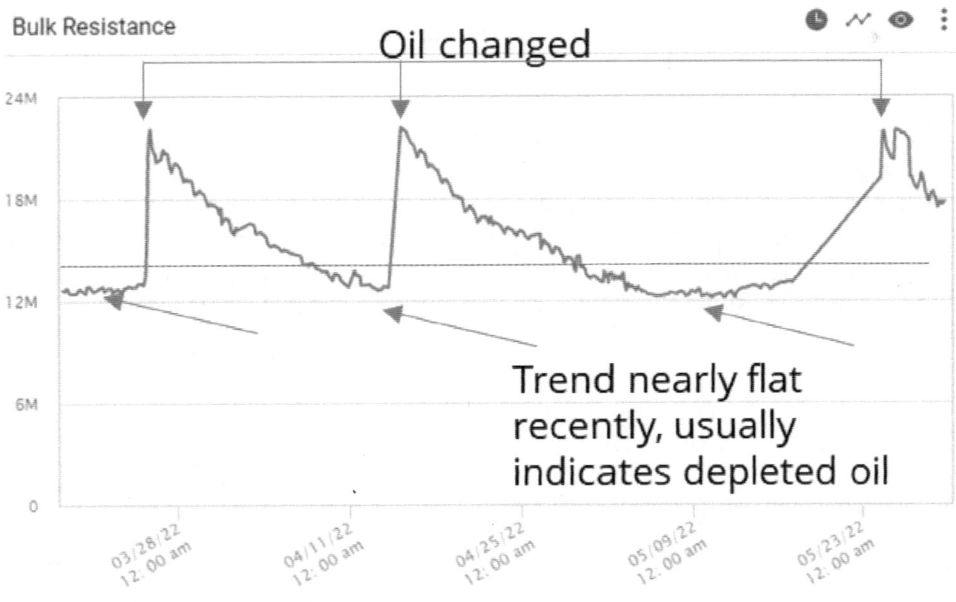

Figure 8: Bulk Resistance Sensed Parameter Trend, Engine 2

Additional analysis was performed for Engine 1 in which the sensor data was compared to oil lab results. Sensor parameters were compared to the various parameters measured in the lab analysis.

Analysis was particularly focused on any correlation of the real time data to the lab reported silicon values. The presence of silicon in the oil is one of the primary measures used to trigger an oil change due to the potential for the silicon to cause wear in the engine. Silicon dioxide is a byproduct of combustion of siloxanes, commonly found in consumer products and present in land-fill gas. Silicon build-up in the engine is detrimental to its performance [4]. Therefore, a deeper analysis of the sensed parameters versus lab results was conducted.

Interfacial Impedance is another sensed parameter sensitive to the health of surface protection additives including presence of contamination on electrode surfaces. Interfacial Impedance is a good detector of free water contamination on the surface of electrodes. The characteristic behavior is shown in Figure 9, a rise in value as new surface protection additives displaced old oil and contaminants on electrode surface then downward trend as additive health decreases. For example, water will cause a large drop in Interfacial Impedance since water tends to remove, or certainly react with, the surface protection layer.

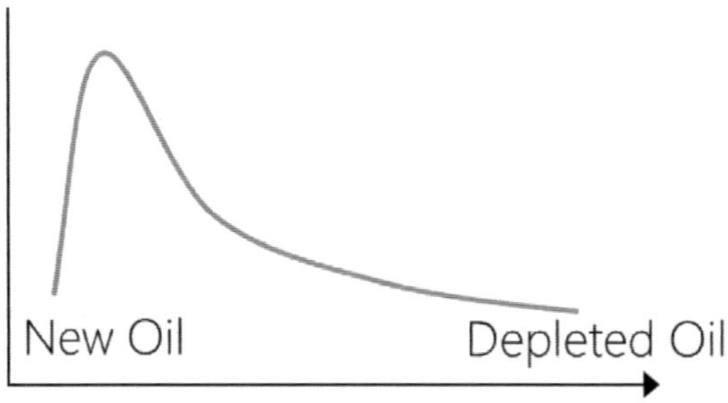

Figure 9: Characteristic Interfacial Impedance Response

Figure 10 shows the daily average Interfacial Impedance and the lab measurements of silicon in ppm for Engine 1. There is a correlation in the sensor parameter and the lab results. The Interfacial Impedance parameter is inversely related to the silicon (note the inverted left axis). Since there is a clear correlation, the sensor based Interfacial parameter can be used to infer the amount of silicon present in the oil ignoring other effects in the oil since they are less critical to the operation of these types of engines. Using an inline sensor, a generator operator can monitor in real time a parameter sensitive to the critical aspects of their engine thus removing unnecessary oil sampling.

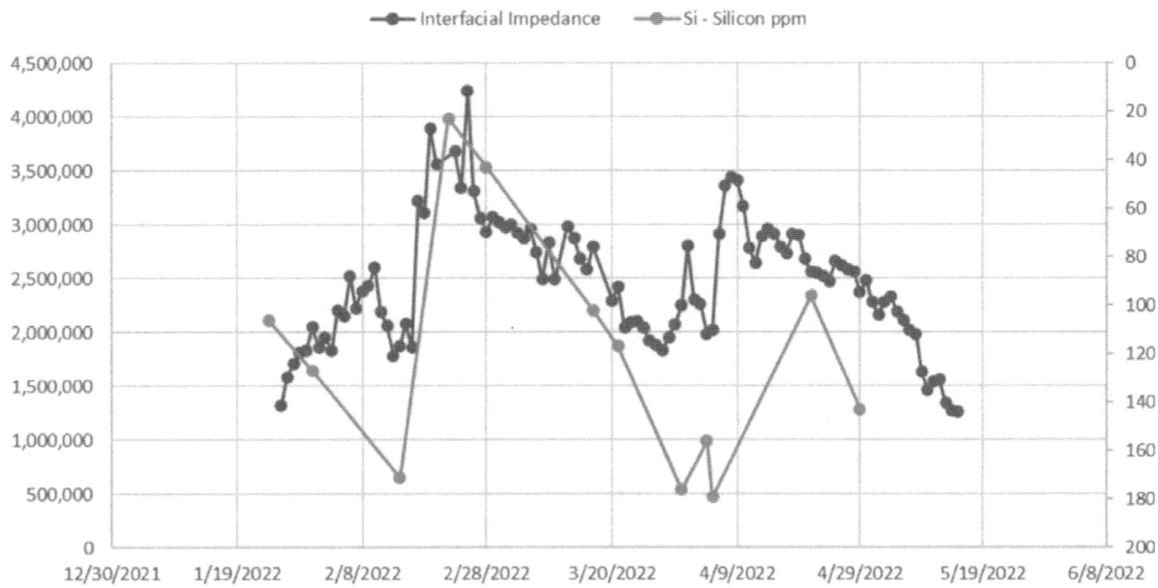

Figure 10: Interfacial Impedance Compared to Lab Based Silicon Measurement, Engine 1

SUMMARY:

In conclusion, online monitoring can be a valuable tool for protecting equipment in challenging applications especially when specific sensors can be used to address specific needs. Land fill gas powered engines offer unique challenges to their operators, including increased importance of proper maintenance, in particular oil sampling. By detecting and diagnosing oil problems early, it is possible to minimize the risk of engine failure, extend the lifespan of the equipment, improve machine reliability and uptime, reduce maintenance costs, and reduce plant safety risks. Too often this means frequent manual oil samples, that for remote sites can be challenging. The data presented herein proves online sensors are a viable and effective alternative. It was shown that tested sensors could detect, trend, and be used to identify the key issues facing LFG engines offering end users alternative solutions increase their uptime.

During the sensor trial period, careful attention was paid to the correlation between the sensor outputs and specific conditions important to the application. Implementing online oil condition sensors in LFG engines allows operators to reduce oil analysis frequency while providing constant monitoring for unexpected conditions that may require fast response. The proven sensor correlation with condemning criteria provides the foundational confidence to enable a transition from time-based sampling to condition based sampling, and eventually possibility of condition-based oil changes.

REFERENCES:

1. Booser, E. Richard, "*Tribology Data Handbook: An Excellent Friction, Lubrication, and Wear Resource*," CRC Press, Boca Raton, FL 1997, pg 900-907.

2. "*Engineered Sensors Tedeco Products Lubrication System Diagnostics Sensors Magnetic Chip Collectors*," http://www.tedecoindustrial.com/mag.htm.
3. https://www.machinerylubrication.com/Read/28778/dielectric-instruments-oil-analysis.
4. US Department of Energy, "*LFG Energy Project Handbook*" January 2024, pg 1-4.

VIATC TRAINING CONFERENCE & EXPO
COVINGTON, KENTUCKY | AUGUST 7-9, 2024

Evaluation of Energy Losses Due to Degradation and Defects in Electric Machinery with Electrical Signature Analysis, vibration Analysis, and Motion Amplification

Howard W Penrose, Ph.D., CMRP, CEM
President, MotorDoc LLC
hpenrose@motordocllc.com

Howard W Penrose, Ph.D., CMRP, CEM is the president of MotorDoc LLC, a veteran owned small business, chair of wind, solar and energy storage standards for American Clean Power, a past chair of the Society for Maintenance and Reliability Professionals and Certified Maintenance and Reliability Professional, a member of the Association of Energy Engineers and a Certified Energy Manager, represents the USA for international electric machinery standards with CIGRE, serves on the IEEE Power Engineering Society's materials subcommittee and Dielectrics and Electrical Insulation Society, is a past chair of Chicago Section IEEE, and a past senior research engineer for the University of Illinois' Energy Resources Center where he also served as an Adjunct Professor of Industrial Engineering. He served on the US DOE challenge committees related to electric machines, compressed air, pumps and related technologies, and won 5 UAW-GM People Make Quality Happen Awards for reliability, maintenance and energy programs at General Motors. His developments include advanced ESA testing systems, ML and AI, and electric machine and controls design for the GM hybrid Tahoe, Volt and John Deere 644 and 944 hybrid construction tractors, high temperature and low atmosphere electric machines, and the LVAD heart pump. He is a life member of SMRP and MENSA, certified in ML and AI through University of Michigan, Stanford University, IBM, AWS, and digital transformation through the Darden School of Business, a graduate of Kennedy-Western University, and presently attending University of Colorado at Boulder in advanced data science.

ABSTRACT: Condition-Based Maintenance (CBM) technologies, including Electrical Signature Analysis (ESA), vibration analysis, and motion amplification, offer powerful means to detect and quantify energy losses and associated emissions stemming from component degradation and defects in electric machinery. This paper delves into the methodologies for calculating these losses using the mentioned technologies and explores their limitations through direct comparisons. Case studies and research focusing on bearings, misalignment and unbalance conditions will illustrate practical applications. The insights gained from this analysis can be directly linked to achieving corporate energy efficiency and emissions reduction goals.

KEY WORDS: Vibration Analysis, Motion Amplification, Electrical Signature Analysis, Energy, Emissions, Reliability, VA to Energy

INTRODUCTION

In the quest for operational excellence and sustainability in industrial operations, understanding and mitigating energy losses in electric machinery is paramount. As industries strive to align with stringent energy efficiency and emissions targets, the role of advanced diagnostic technologies has become crucial. This paper explores three pivotal CBM technologies: Electrical Signature Analysis (ESA), Vibration Analysis (VA), and Motion Amplification (MA), highlighting their utility in identifying energy losses attributable to machinery degradation, defects, and shortcomings. The study involves direct relationships of such measurements across machines in a variety of conditions including field studies, shop bench inspections, and laboratory testing. The purpose is to provide mechanical measurement technology methodologies to relate the detected signatures to a reasonable value in equipment energy losses. As the selected ESA technology provides a direct loss measurement by defect as measured by the electric machine airgap, the developed algorithms will be compared from the vibration data obtained by the same instrument and MA data.

The need for understanding the losses associated with each defect or deficit frequency is the relation to additional energy and emissions. The results can be presented in a number of ways and can be tied to several corporate KPI and corporate sustainability goals including Energy Use Intensity, zero emissions, and Energy Star. For the purposes of this paper, we will convert to watts and kilowatt-hours and emissions in terms of metric tonnes (MT) of CO2 per Megawatt-hour and the general USA value of 0.709 MT/MWh for usage reduction. Local data can be obtained through US EPA's eGrid and AVERT programs, as well as through several calculators available on the US EPA's website.

The technologies used in this study are defined, for our purposes, as follows:

- ESA: utilizes voltage and current to identify power quality issues, electric machinery, and powertrain degradation and defects through spectral analysis. For power analysis, the ESA device must be chosen that provides a watt/kilowatt spectral analysis that identifies individual defects. For this study, it is understood that the kilowatt spectral losses are the combined energy losses for the specific defect including motion, heat, noise, and other parasitic losses at that frequency. ESA will provide the baseline for energy consumption by defect.
- VA: utilizes converted motion to identify equipment defects at the point of data collection and defects through spectral analysis. The primary direct energy type measured with some effort is kinetic energy.

- MA: the method of converting video and each video image pixel into a sensor to identify motion and motion frequencies that appear similar to vibration. As a version of Operation Deflection Shapes (ODS) the actual video provides a direct visual on the structure associated with equipment.

Losses associated with defects appear in a number of forms. In the case of VA and MA, kinetic energy can easily be converted to energy loss, but will not directly identify noise, heat, and other parasitic losses. Due to some of the variations, we will explore 1X RPM initially to understand the differences between direct measurements and indirect measurements. Some of the data required to review VA and MA data, at least the following data is required:

- Mass (m in kilograms) of the vibrating system: the mass of the part of the machine involved in the vibration.
- Frequency of Vibration (f): the frequency at which the system vibrates.
- Deflection or Amplitude (A): the maximum displacement of the vibrating system from its rest position.
- Velocity (v): the speed of vibration in meters/second (m/s).
- Dampening Coefficient (c): a parameter that quantifies the damping force relative to the velocity of the vibrating object.
- Stiffness (S): the stiffness of the equipment and base as well as any natural frequencies associated with the system.
- Operational Time (t) or number of cycles: for continuous operation, how long a sample of the system operates, or the number of cycles over which energy dissipation is considered.

REVIEW OF SMALL MACHINE ON VFD TEST BENCH:

To review the concepts associated with the study, a small 0.37 kW (0.5 hp), 1700 RPM, 230 Volt, 1.8 Amp electric motor was instrumented up using a .75hp variable frequency drive to a 230 volt single phase power supply. The motor running speed was limited to 50Hz due to the quality of 3-phase voltage output due to the single phase supply. The selected table is designed for power saws and relatively stable and placed in a way to be level by visible bubble on a concrete floor. The test bench (Figure 1) was used for initial testing with a focus on 1X RPM, which would be related to balance and structure, which are also easily viewed by motion amplification.

Figure 1: Bench test setup with ESA and VA setup.

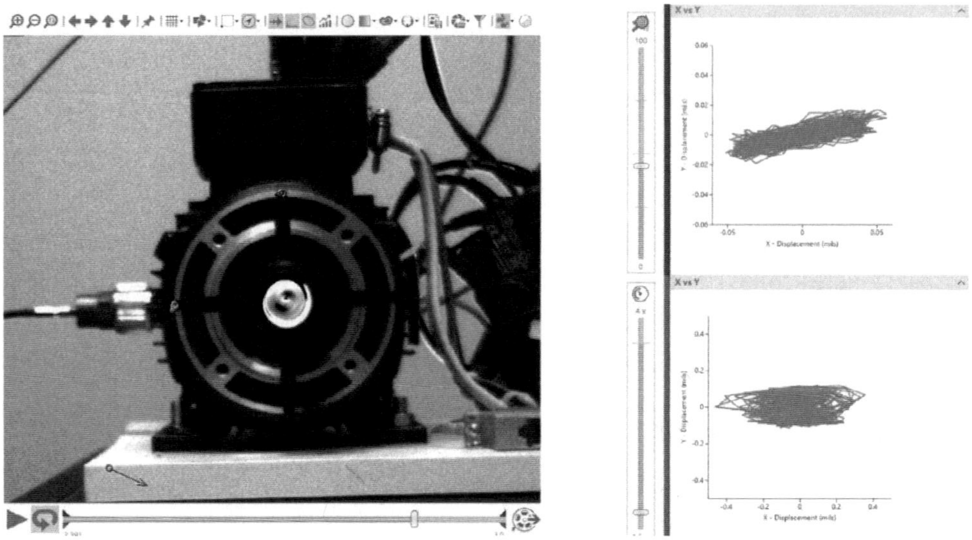

Figure 2: Motion Amplification of motor and related base movement.

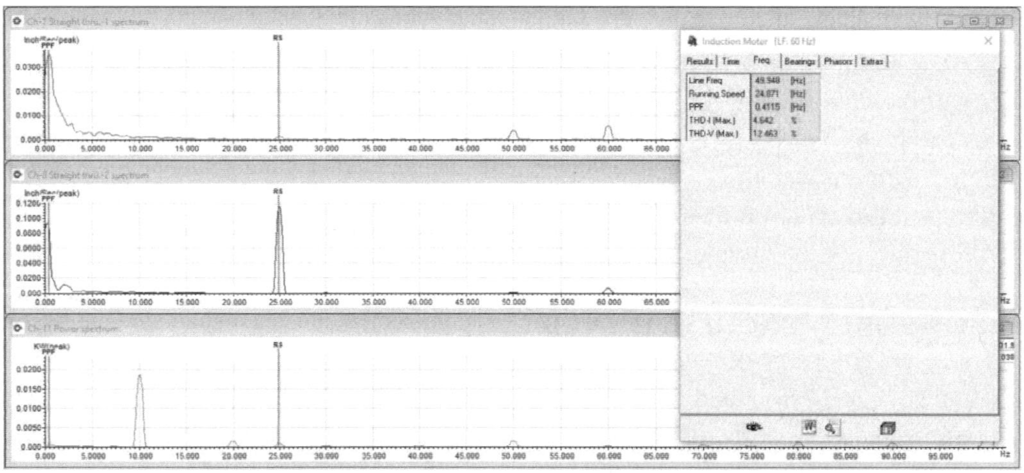

Figure 3: Direct comparison of VA in/sec and watts loss.

The initial measurements in the horizontal (X) and vertical (Y) are the same at both the drive end (DE) and opposite drive end (ODE) in vertical and horizontal positions. The mass for the assembly is 12 kg, the vertical reading is 0.002 in/sec and the horizontal reading is 0.116 in/sec at 24.871 Hz while the 1X energy loss is 1 watt. Motion amplification identified horizontal motion of 0.820 mils and vertical of 0.235 mils in the base and 0.004 in/sec horizontal and 0.001 in/sec vertical at the motor.

It should be noted that there is a discrepancy between the vibration analyzer and motion amplification. In addition to measurement and angle errors, there is a difference between vibration measured with accelerometers and video (pixels). Motion Amplification is a video-based technique that utilizes high speed cameras to visualize and measure displacement and movement of structures under operational conditions. It amplifies the visible motion to detect and measure displacements that are otherwise invisible to the eye. It directly measures displacement or movement over time, making it highly sensitive to actual motion observed in mils p-p. Elsewise, accelerometers are devices that measure acceleration forces directly. Accelerometers are inherently sensitive to acceleration which allows direct measurement of velocity. Errors can be introduced due to a number of noise-related inputs. Motion amplification may more effectively visualize global motion affecting the entire motor or its components coherently, whereas accelerometers can capture differential or localized motion that may not significantly alter the motor's overall position or orientation but still contributes to the vibrational energy. This will tend to show more dynamically in cases where a motor base is loose or a structure's stiffness is poor.

This means that the conditions of the base that would affect our readings, such as stiffness or dampening, are more readily determined by MA and velocity for conversion to energy consumption with VA. ESA provides more detailed and overall energy measurement with the following steps, which provide direct kinetic energy (KE) energy consumption and indirect additional losses. This requires a series of steps, which we will later convert to a single algorithm:

1. Convert velocity to meters per second: m/s = in/sec * 0.0254. In this case, the result from the accelerometer would be vertical 5.08×10^{-5} m/s and horizontal 2.95×10^{-3} m/s, respectively.
2. Calculate w: $w = 2\varpi f = 2\varpi * 24.871$ Hz = 156.2 Hz
3. Calculate kinetic energy (KE): $KE = 0.5m*(wA)^2 = 0.5 * 12kg * (152.2\ Hz * A)^2$. With $A = 5.08 \times 10^{-5}$ m/s the result is: 4.68×10^{-4} joules. With $A = 2.95 \times 10^{-3}$ m/s, the result is 1.21 joules.
4. Power, in watts, is watt = joules/second, which would allow us to add each value for each bearing as the loads would be evenly distributed. This would result in 2.4 watt-s, which at first appears to be 2.4 times the direct measured data. This may be a direct result of the horizontal movement found in motion amplification. We can then convert to watt-hours by dividing by 3600 (seconds to hours). In this case, 1 watt in ESA converts directly to 1 watt-hour and 2.4 watt-s converts to 6.7×10^{-4} watt-hours.

The result at a single bearing, even when doubled, is short of the 1 watt-hour expected. These values go along with the stiffness and dampening effects from the machine. It is noted, in this case, that the closest value is the horizontal, although, with the MA data that shows as the highest value of p-p displacement. The addition of the operation on a VFD at no load would also have some impact, which is an observation that will require additional study beyond the scope of this project.

LABORATORY TESTING 25 HORSEPOWER MACHINE:

In order to more closely replicate the conditions found in the field versus the laboratory, the tests were performed on dynamometers with motors that had 1X vibration and were aligned to the dynamometer loads. The motor was an 18.5 kW (25 hp), 1765 RPM, 460 Vac, electric motor weighing 142.7 kg and 29.825 Hz 1X, operating at a light load across the line (Figure 4).

Figure 4: 25 hp energy evaluation configuration on dynamometer.

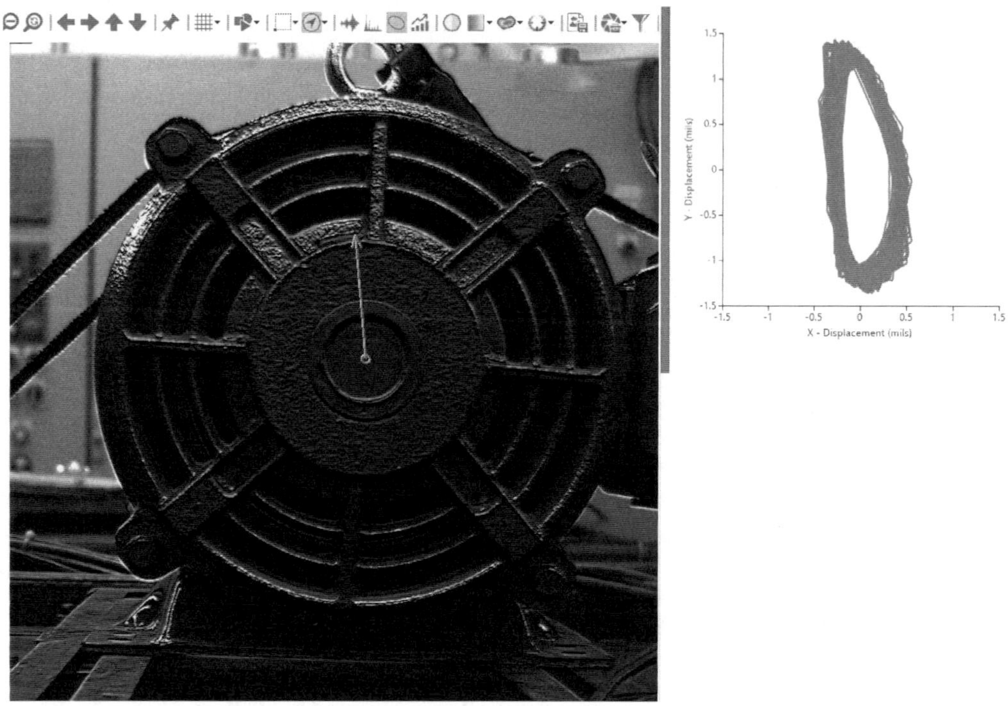

Figure 5: 25 hp dynamometer orbital evaluation.

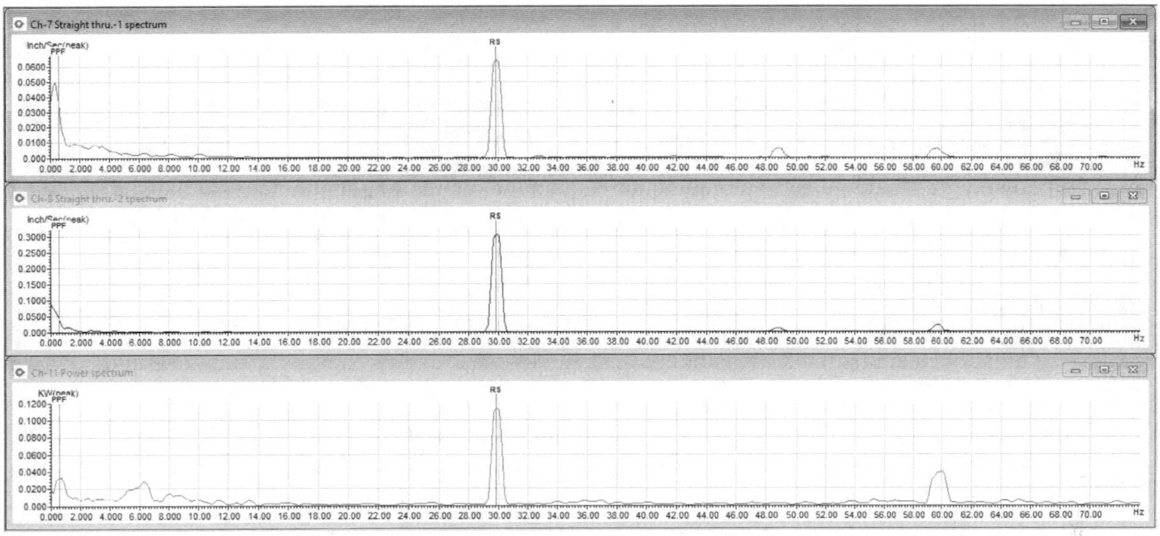

Figure 6: 25 hp motor loaded horizontal readings with ESA and VA.

Table 1: Data from the 25 hp electric motor vibration in inches per second.

	Speed	Watts	Weight	kW in	kW out	DE	ODE
No Load - H	29.955/ 1797.3	62	142.7	7.6	6.86	0.186	0.345
Loaded – H	29.825/ 1789.5	116	142.7	13.7	12.22	0.065	0.310
Loaded - V	29.825/ 1789.5	114	142.7	13.7	12.22	0.591	0.901

The MA data, while the equipment was loaded, was DE V = 0.2 in/sec; ODE V = 0.25 in/sec; ODE H = 0.06 in/sec; and ODE V = 0.25. The vertical base deflection was 1.67 mil p-p. This indicates that the base stiffness and dampening was poor in the vertical direction. Excluding stiffness and dampening, convert to KE using the complete algorithm (Equation 1).

$$KE = 0.5 * m * (2\pi f * A)^2 \quad \text{Equation 1}$$

Table 2: the losses calculated unloaded and partially loaded vibration in meters per second.

	Speed	Watts	Weight	DE	ODE	Watt-s DE	Watt-s ODE
No Load - H	29.955/ 1797.3	62	142.7	0.0047	0.0088	56.4	193.9
Loaded – H	29.825/ 1789.5	116	142.7	0.0017	0.0079	6.82	155.2
Loaded - V	29.825/ 1789.5	114	142.7	0.0150	0.0229	564.0	1311.0

The conversion from watt-s to watt-h (divide by 3600) results in Table 3.

Table 3: converted losses

	Speed	Watts	Weight	Watt-s DE	Watt-s ODE	Watt-h DE	Watt-h ODE
No Load - H	29.955/ 1797.3	62	142.7	56.4	193.9	0.016	0.054
Loaded – H	29.825/ 1789.5	116	142.7	6.82	155.2	0.002	0.043
Loaded - V	29.825/ 1789.5	114	142.7	564.0	1311.0	0.157	0.364

As with the initial study, the values do not result in the losses associated with the total losses in the system. The watts (measured) are the direct measured Watt-h from ESA at 1X RPM.

DISCUSSION OF DISCREPANCIES:

Through a review of the experimental setups and related movement, it becomes clear that the stiffness and dampening has an impact. In this case, either the horizontal relationship, the vertical relationship, or both. For instance, in Figure 2 the orbital plot provides detail that the highest freedom of movement is horizontal for this system, and in Figure 5 the highest freedom of movement is vertical.

Freedom of movement exposes a consideration for where components are in relation to their natural frequency. An evaluation of a second dynamometer setup with horizontal freedom of movement was evaluated to verify these assumptions.

Additionally, losses such as sound, heat, and energy other than vibration are not accounted for. The key is to determine the best approach to determine a closer value to the actual losses.

LABORATORY TESTING 50 HORSEPOWER MACHINE:

In order to evaluate the freedom of movement concerns the test was performed on a 50 horsepower motor on a dynamometer with horizontal freedom of movement. The motor was 37.3 kW (50 hp), 1775 RPM, 460 Vac, electric motor weighing 305.7 kW and 29.784 Hz 1X running speed, operating at 86% load across the line (Figure 6).

Figure 7: 50 hp electric motor on dynamometer with horizontal freedom of movement.

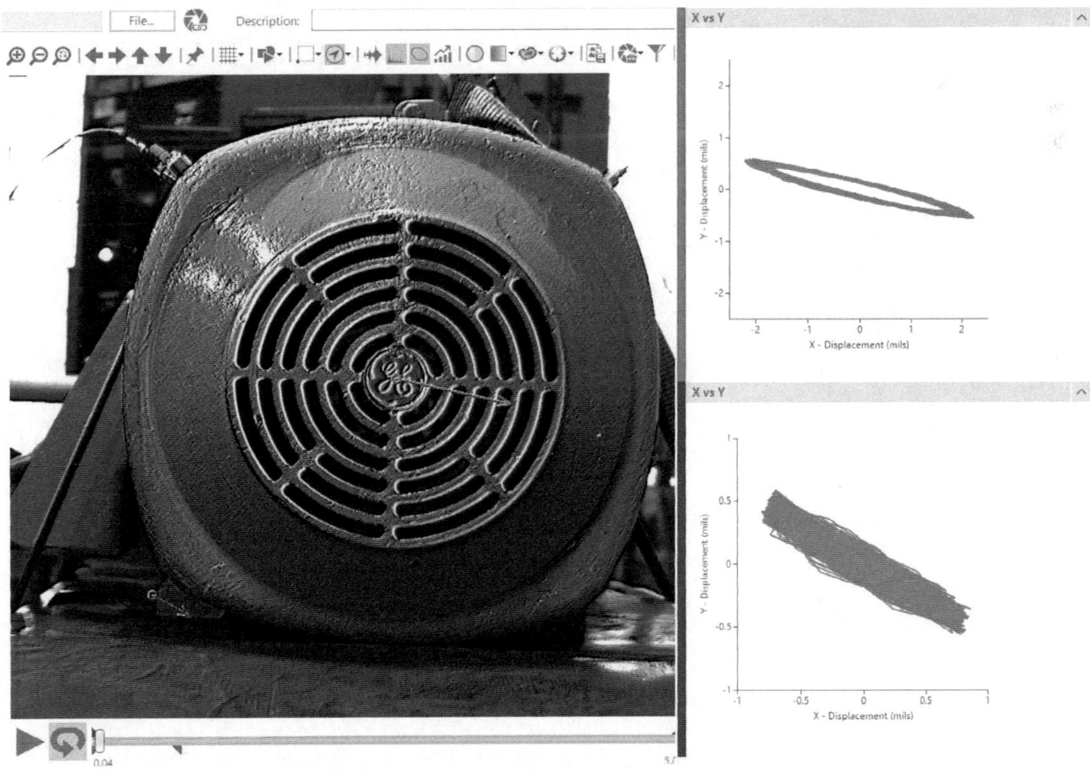

Figure 8: Orbital analysis of 50 hp motor with Motion Amplification.

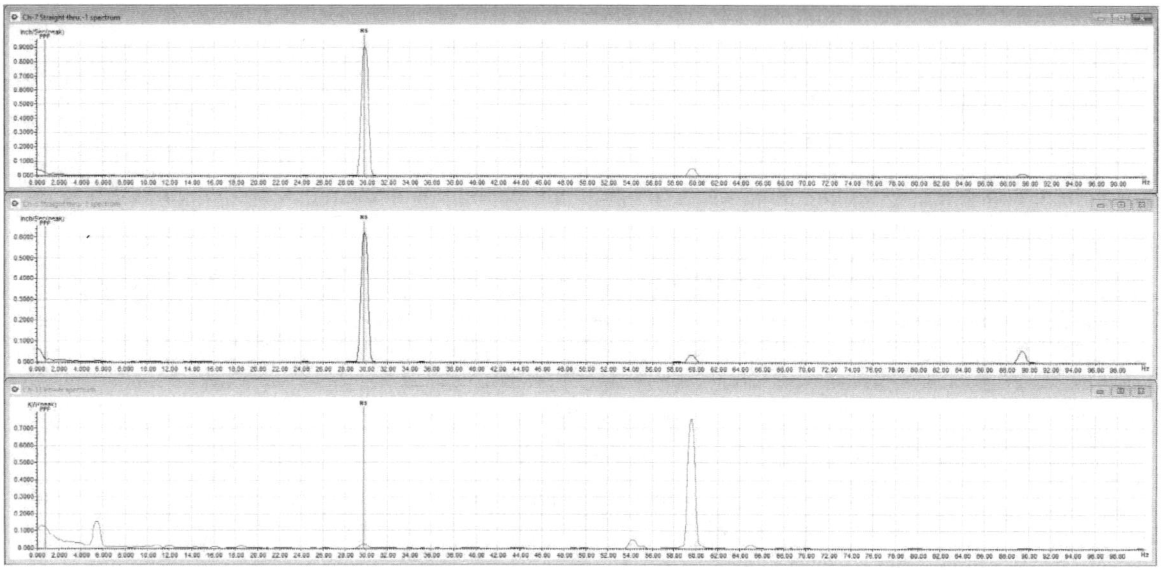

Figure 9: ESA and vibration data in inches per second and kilowatts.

Figure 8 shows the freedom of movement to approximately horizontal. Channel 7 data from Figure 9 is horizontal on the drive end and Channel 8 is in the same direction as the freedom of movement as shown in Figure 8. The amplitudes in this position are 0.9 in/sec drive end, 0.623 in/sec opposite drive end, and 24 watts in Channel 11 at running speed. Motion amplification shows as 0.4 in/sec vertical and 0.1 horizontal with 5 mil p-p horizontal.

Table 4: Data from the 50 hp electric motor, vibration in inches per second.

	Speed	Watts	Weight	kW in	kW out	DE	ODE
No Load - H	29.968 Hz/ 1798.1 RPM	14	305.7	15.2	12.76	0.924	0.784
Loaded – H	29.784 Hz/ 1787 RPM	24	305.7	32.1	28.04	0.9	0.623
Loaded - V	29.971 Hz/ 1787.5 RPM	19	305.7	31.8	27.82	0.246	0.663

Table 5: The losses calculated unloaded and partially loaded on 50hp motor, vibration in meters per second.

	Speed	Watts	Weight	DE	ODE	Watt-s DE	Watt-s ODE
No Load - H	29.968 Hz/ 1798.1 RPM	14	305.7	0.0235	0.0199	2989.8	2143.9
Loaded – H	29.784 Hz/ 1787 RPM	24	305.7	0.0229	0.0158	2804.3	1335.0
Loaded - V	29.971 Hz/ 1787.5 RPM	19	305.7	0.0062	0.0168	208.1	1528.3

Table 6: converted losses

	Speed	Watts	Weight	Watt-s DE	Watt-s ODE	Watt-h DE	Watt-h ODE
No Load - H	29.968 Hz/ 1798.1 RPM	14	305.7	2989.8	2143.9	0.830	0.596
Loaded – H	29.784 Hz/ 1787 RPM	24	305.7	2804.3	1335.0	0.779	0.371
Loaded - V	29.971 Hz/ 1787.5 RPM	19	305.7	208.1	1528.3	0.058	0.425

As in the first study, with the high horizontal value, the result is still far less than the measured value even when all four positions are added. In this case, the total would be 1.633 Watt-hours, which is still well below the average of 22 Watt-hours directly measured.

FIELD CASE STUDY 655 HP COMPRESSOR MOTOR:

The following involves a properly installed 655 horsepower, 1789 RPM, 460 Vac air compressor motor overloaded to 180%. The 1X vibration at 1789 RPM/29.822 Hz is 0.007 in/sec drive end and 0.018 in/sec opposite drive end with 447 watts load and weighing approximately 2675 lbs with a D-flange and foot mounted. The motor and compressor (Figure 10) includes rigid mount to the compressor and the compressor is on a vibration isolation mounting.

Figure 10: 488kW/655hp 1789 RPM 460 Vac electric motor weighing 2675 kg.

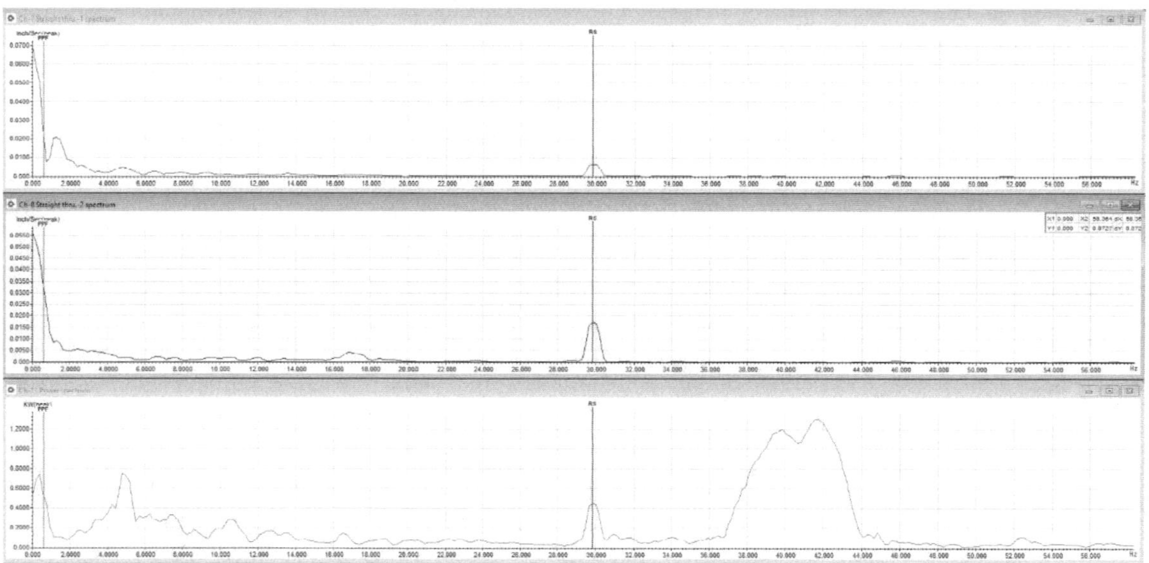

Figure 11: As found data on the machine including vibration and kW.

As noted earlier in the paper, the in/sec must first be converted to meters per second. In this case the drive end changes from 0.007 in/sec to 0.000178 m/sec and the ODE changes from 0.018 in/sec to 0.000457 m/sec.

We then convert the data based on Equation 1, which results in the following:

- Drive End: 1.486 Watt-s, which converts to 4.12×10^{-4} Watt-h.
- Opposite Drive End: 9.806 Watt-s, which converts to 2.72×10^{-3} Watt-h.

The total KE related losses come out to double each value and add them together which results in 6.26×10^{-3} Watt-h. A review of the relationship results in a curve shown in Figure 12 based upon a power formula shown in Equation 2.

$$Multiplier = a * X^b \quad \text{Equation 2}$$

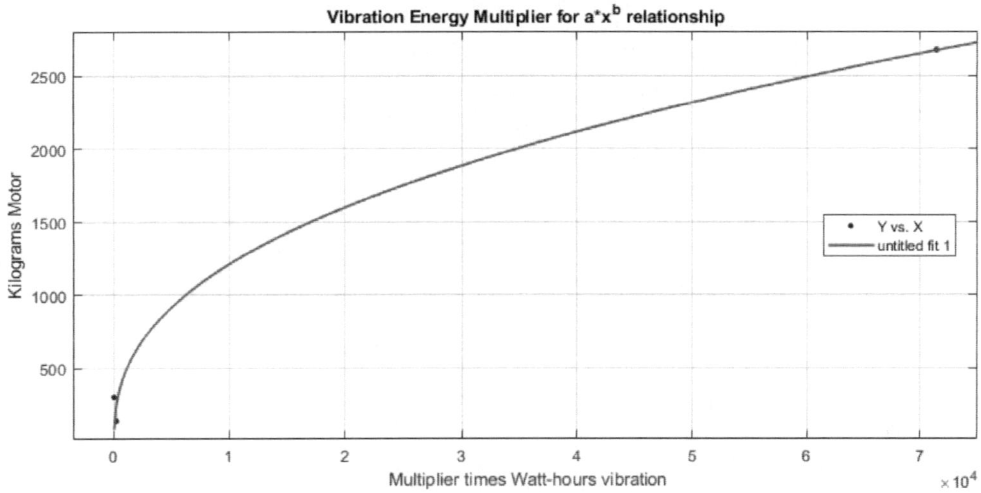

Figure 12: Multiplier based upon equipment weight from study to provide estimate.

CASE STUDY VERTICAL MOTOR ON VFD:

A 40 hp, 3520 RPM, 460 volt, vertical motor weighing 113 kg on a variable frequency drive was tested with a 1X RPM at 47.019 Hz (2821.1 RPM) at a 48.85 Hz operating frequency. The1X RPM had a measured loss of 3 Watts with vibration at 0.057 in/sec shaft end and 0.116 in/sec top end.

Figure 13: 40 hp 2-pole electric motor supplied by a VFD prior to testing.

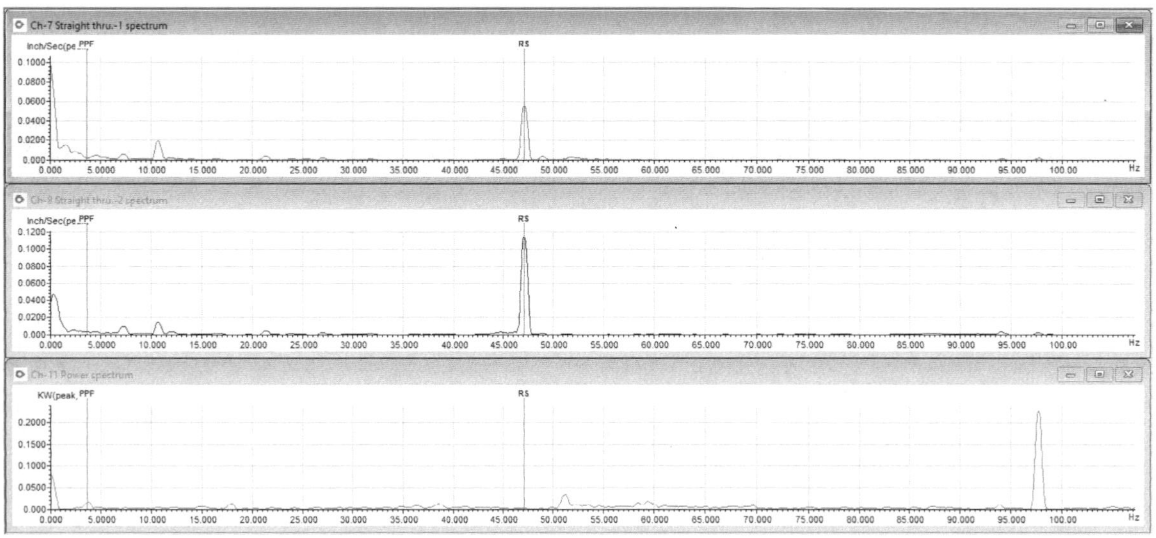

Figure 14: Vibration and energy loss data at 1X RPM on 40 hp motor and VFD.

Using the calculations above, the lower end of the motor, which is stiffer, has a loss of 0.00287 Watt-h at the lower end and 0.0119 Watt-h at the upper end. This results in a total of 0.0295 Watt-h based on KE. We can then apply the multiplier from Figure 12, which is found in Figure 15 with a value of ~110 times the found value. This would result in a total of 3.25 Watt-h, which is close to the measured value.

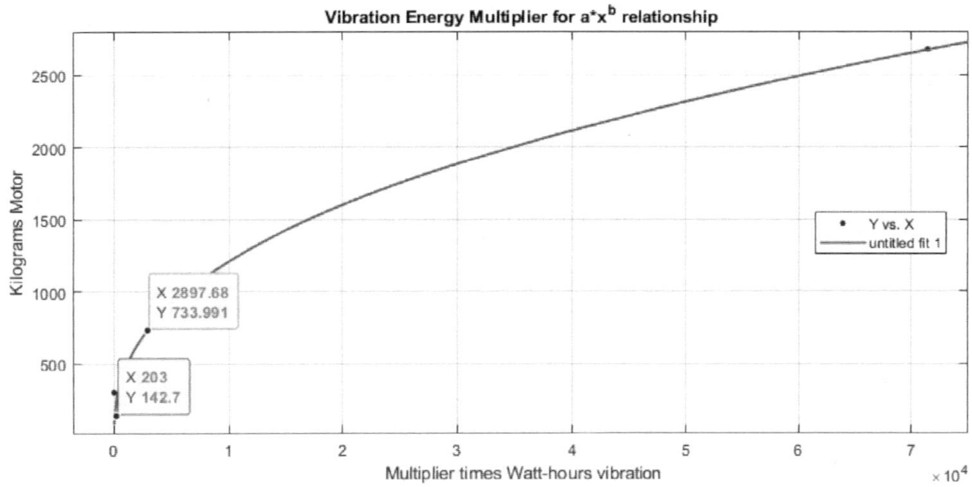

Figure 15: Approximate multiplier for curve.

CONCLUSION:

A method for developing a curve and algorithm for calculating the loss by frequency value for vibration as compared to the kilowatt analysis from ESA can be readily developed. As the values are variable by load in vibration and the watt values are also dependent upon load in ESA, the end result is that the weight of the machine may be used to develop the curve for evaluation. The algorithm for the analysis is as follows and requires the following information:

1. Electric motor weight.
2. Horizontal and vertical velocity readings per bearing.
3. Frequency of the peak(s) in question.

Equation 3 is the algorithm that assumes the data is collected in in/sec and requires a conversion to meters per second. The algorithm needs to be adjusted if the data is obtained in mm/sec.

$$Wh = X * ([0.5 * M * [\{2\pi f_a A_a\}^2 + \{2\pi f_b A_b\}^2 + \cdots + \{2\pi f_n A_n\}^2]]/3600) \qquad \text{Equation 3}$$

Where X is from the power curve, M is the mass of the motor in kg, f is the frequency in question, A is the velocity in meters per second.

Separate curves should be developed for motion amplification and may need to be developed for other vibration data collection systems. There are also a few considerations that may impact data accuracy including equipment stiffness, calibration, dampening, and other variables. The data should be considered an estimate in watt-hours.

This data may be converted over to GHG emissions by using the US DOE and US EPA eGrid values for emission reduction. The losses would be converted over to kWh/year or MWh/year and then multiplied by the eGrid value. For instance, the national average is 0.709 metric tonnes of CO2 per MWh and there are 8,760 hours per year. If we take the compressor example above and assume that it operates 24/7, then the value would be 0.447 kW * 8760 = 3,916 kWh/yr, which would be 3.916 MWh. If we then multiply that by 0.709 then we end up with 2.8 tonnes CO2/year for the defect, which can be assumed to change over time.

BIBLIOGRAPHY:

P. Janphuang, R. A. Lockhart, D. Isarakorn, S. Henein, D. Briand and N. F. de Rooij, "Harvesting Energy From a Rotating Gear Using an AFM-Like MEMS Piezoelectric Frequency Up-Converting Energy Harvester," in Journal of Microelectromechanical Systems, vol. 24, no. 3, pp. 742-754, June 2015, doi: 10.1109/JMEMS.2014.2349794.

A. Ermolaev, A. Plekhov, D. Titov and Y. Vagapov, "Vibration damping in a motor drive shaft system operating under active power flow oscillation," 2018 IEEE Conference of Russian Young Researchers in Electrical and Electronic Engineering (EIConRus), Moscow and St. Petersburg, Russia, 2018, pp. 1723-1727, doi: 10.1109/EIConRus.2018.8317437.

VIATC TRAINING CONFERENCE & EXPO
COVINGTON, KENTUCKY | AUGUST 7-9, 2024

THE IMPACT OF ACCELEROMETER MOUNTING TECHNIQUES ON THE ACCURACY OF VIBRATION DATA

Peter Eitnier
Senior Applications Engineer, Wilcoxon Sensing Technologies
peter.eitnier@wilcoxon.com

Peter Eitnier is a Senior Applications Engineer at Wilcoxon Sensing Technologies in Frederick, Maryland. He holds an ISO CAT II certification as a vibration analyst and a degree in Mechanical Engineering from the University of Maryland. Since joining Wilcoxon in 2012, he has specialized in providing technical expertise to customers in a variety of applications.

ABSTRACT:
The sensor mounting method can significantly inhibit the frequency performance of an accelerometer. This paper will offer guidance for vibration practitioners in evaluating and determining the ideal mounting location, method and installation guidelines for permanently mounted accelerometers. For route-based data collection, where the mount can be especially impactful, best practices will be shared for achieving consistent and complete vibration data. Vibration sensor mounting in specialty applications will also be covered, including low-clearance locations, triaxial measurements, electrical isolation, temporary mounting, and fin mounting on electrical motors.

This paper will include both theoretical limitations and demonstrative vibration data taken in the field to indicate applications in which the vibration signals of interest can be readily obtained with various mounting options, as well as which mounting methods are critical to the success of high-frequency condition monitoring applications.

KEY WORDS: accelerometer mounting, high-frequency vibration monitoring, permanently installed accelerometers, route-based data collection

ACCELEROMETER PLACEMENT:
The physical mounting location of the sensor is the first decision that impacts the validity of vibration data. In general, the sensor should be mounted as close as possible to the bearings. The shaft turning speed is the driving force of both the machine itself and the vibration signal, and it transmits vibration directly through the bearings into the machine case. Therefore, the closer the sensor is mounted to the bearings, the better it will detect the vibration for analysis.

On most machinery, vibration sensors are mounted in the horizontal direction with respect to the shaft because most heavy machinery is bolted to the floor or a cement fixture at several points. Therefore, movement in the vertical direction – shaking up and down –is minimized. The machine is more likely to move side-to-side, so horizontal sensor placement will measure the primary axis of movement. For vertically mounted machines, radial and axial measurements are most valuable. Figure 1 demonstrates the ideal locations and orientation of accelerometers on a motor and shaft driving production machinery such as a pump, compressor, blower, fan or agitator.

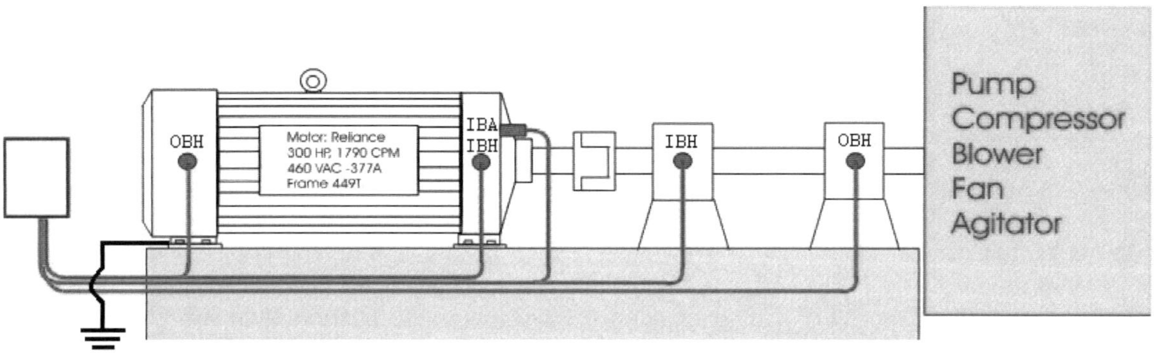

Figure 1: Ideal mounting locations and axes.

It is not always possible to mount in the preferred location and axis. The type of machine, spatial availability and budget are among myriad practical factors that can prevent “ideal” mounting. When an alternate location must be found, avoid mounting accelerometers on thin sections, guards, cantilevers, and antinodes (vibration-free areas).

Ultimately, repeatability is critical. Even when the sensor cannot be mounted directly on the bearings or in a perfectly horizontal orientation, repeatable measurements will indicate changes in the condition of a machine over time.

TRADITIONAL MOUNTING METHOD GUIDANCE:
Traditional guidance when mounting vibration sensors is to maximize surface contact between the mounting base and the machine to optimize signal transmission to the sensor, ultimately improving the accuracy of a vibration monitoring program. The ideal mounting method is a sensor stud or screw directly threaded into a drilled, tapped hole on the machinery.

When machinery and environmental factors prevent direct stud mounting, there are many other options, as shown in Figure 2. Not every mounting method is suitable for every application due to high frequency limitations, as demonstrated in Figure 3. Examining the tradeoffs ensures that the mounting method is suitable to achieve the vibration monitoring goals.

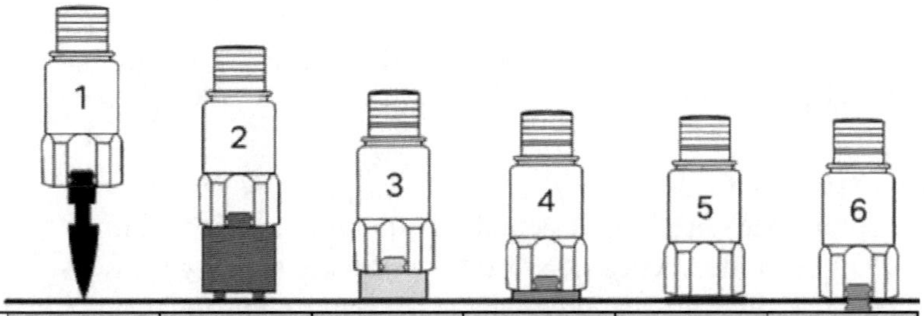

	Probe tip	Curved surface magnet	Flat magnet	Cement mounting pad	Adhesive	Stud
Frequency range (Hz)	500 - 1,000	2,000 - 5,000	5,000 - 7,000	10,000 - 15,000	10,000 - 15,000	Sensor max
	Acceptable		Good	Best		

Figure 2: Common accelerometer mounting options.

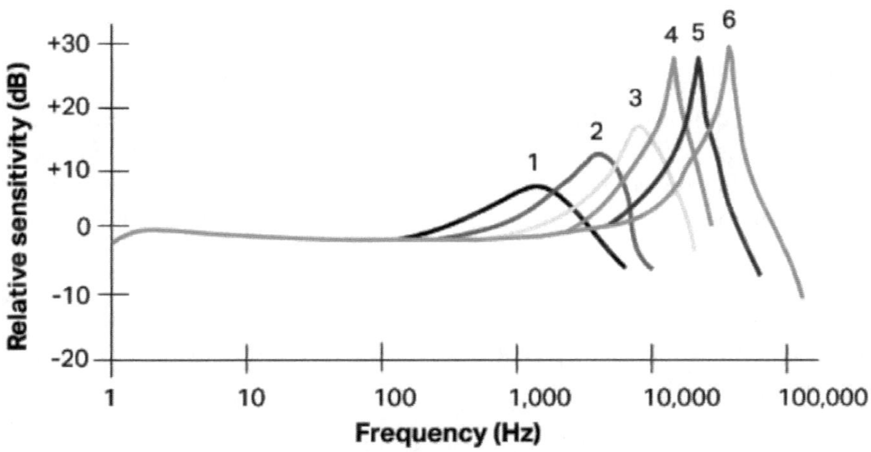

Figure 3: High frequency limitations of respective mounting methods.

PERMANENT INSTALLATIONS:
Threaded stud mounting, via a tapped hole drilled directly into the machinery structure, provides an electrical and mechanical connection between the sensor and machine. With this mounting method, the limitation on the measured frequency range is the sensor performance itself. When mounting a high frequency IEPE sensor that measures to 25 kHz, only direct stud mounting ensures vibration coupling that achieves the upper limits of the sensor specification.

Effective vibration coupling requires direct, rigid contact between the sensor and the machine. Achieving flush mounting requires careful surface preparation, beginning with a flat spot-faced surface that is 1.1 times larger than the diameter of the sensor housing. Debris such as dirt or burrs between the sensor and the machine must be eliminated. The centerline of the tapped hole must be perpendicular within 1° to the mounting surface to ensure no gaps are present between the base of the sensor and the structure. A thin coating of silicone grease between the sensor base and the spot-faced surface will increase the mounting stiffness and enhance the frequency response. A threadlock glue may be applied to the tapped holes in both the structure and the sensor for additional strength.

When the structure to be monitored cannot be drilled, cementing pads provide rigidity approaching that of stud mounts. With cementing pads, the maximum frequency response in the 10-15 kHz range is suitable for most general-purpose monitoring applications.

WALK-AROUND DATA COLLECTION:

Permanently installed cementing pads are ideal to create a quick connection point for accelerometers on data collection routes. They provide the requisite rigidity to measure frequencies up to 10-15 kHz as well as ensuring consistent measurement locations for walk-around applications. Changes in the measurement location can result in changes in the data that do not reflect changes in the machine condition.

As with stud mounting, a spot-faced surface that is 1.1 times larger than the diameter of the cementing pad is required. Allowing the machine tool to groove and abrade the surface prepares it for adhesive bonding. The flat side is bonded to the machine with an appropriately rigid adhesive, like cement. The opposite side contains a stud or tapped hole for mounting the sensor.

Higher frequencies in this range may not be attained with epoxy mounting. Epoxy provides a secure attachment without extensive machining but can reduce the operational frequency range since the adhesive acts like a shock absorber, known as damping.

While less ideal, magnetic bases are most commonly used for walk-around data collection because they allow the sensor to be quickly and easily moved from point to point. However, magnetic bases produce a significantly different response at higher frequencies compared to stud and cementing pad measurements. Poorer contact between the sensor and the machine, combined with the additional mass of the magnet, results in a significantly reduced frequency range and a lower resonant frequency.

Flat magnets are recommended for flat surfaces. Coupling fluids, such as silicone grease or oil, greatly improve measurements with flat bottom magnets and should be used whenever possible to measure frequencies as high as 5-7 kHz.

On curved machine surfaces, 2-pole curved magnets provide the best connection. The sensor performance is diminished to a max frequency response of 5 kHz and may not measure reliably over 2 kHz.

For both flat and curved magnets, the machine surface must be magnetically attractive and free of debris. Painted surfaces are acceptable, but should be clean and well maintained, not flaking or chipping.

For repeatable measurements and accurate trending, marking the measurement location or using magnetic mounting targets ensures readings are taken at the same place every time.

When the frequency of measurement is less than 200 Hz, probe tips (also called stingers) can be used to take readings on hard-to-reach areas or surfaces that are not conducive to mounting. In general, this method is not recommended, but sometimes it is necessary. In those situations, probe tips longer than six inches should be avoided. Drilling a countersink into the machine surface can increase repeatability and decrease tip chatter.

SPECIALTY CONSIDERATIONS

LOW-PROFILE INSTALLATIONS:

Where machines are located close to separate pieces of machinery, have protective shielding, or have tight tolerances for sensor mounting, side-exit accelerometers (Figure 4) provide rigid stud mounting in monitoring locations with low overhead clearance. With stud or screw mounting in this scenario, the only limitation on the frequency range is the sensor performance.

Figure 4: Side exit accelerometer

TRIAXIAL MEASUREMENTS:

For permanent installations, triaxial sensors should be stud or screw mounted, when possible, to achieve the full high-frequency performance of the sensor. Because the frequency response of most triaxial sensors is in the 2 - 10 kHz range, cementing pads can also be used effectively.

For walk-around data collection, magnet mounts are appropriate for triaxial accelerometers with a high frequency response in the 2 - 5 kHz range.

The limited frequency response of triaxial accelerometers may not be suitable for all applications, especially low frequency and high frequency monitoring. To achieve a wider frequency response, triaxial mounting cubes (Figure 5) can be installed with three single-axis, full performance sensors. However, the mounting cube introduces mass and will reduce frequency response of single-axis sensors depending on their style and weight.

Figure 5: Triaxial mounting cube

ELECTRICAL ISOLATION:

Isolator mounting bases (Figure 6) protect against high voltage, static electricity build-up, ESD shocks, and grounding issues such as poor ground bonding, ground loops or differing ground potentials. Like direct stud mounting, they are mounted on a prepared machine surface via tapped hole. All contaminants must be removed from the surface and isolation pad to avoid unintended conduction. With a cement or rigid adhesive between the machine surface and base, a frequency range up to 10-15 kHz can be achieved.

Figure 6: Isolator mounting base

TEMPORARY INSTALLATIONS:

For temporary installations, an oil or adhesive can be used to achieve a frequency range of 10-15 kHz. Suitable materials include silicone grease and beeswax. Ground/floor vibration analysis, calibration testing or other non-intensive vibration monitoring applications could benefit from this method.

ELECTRIC MOTORS:

On large electric motors, fin mounts (Figure 7) are typically used on non-drive locations where cooling fan shrouds prevent monitoring on the end belt. The mounts are designed to be epoxied or welded between cooling fins of the motor. With this mounting method, the frequency range limitation is 5-10 kHz.

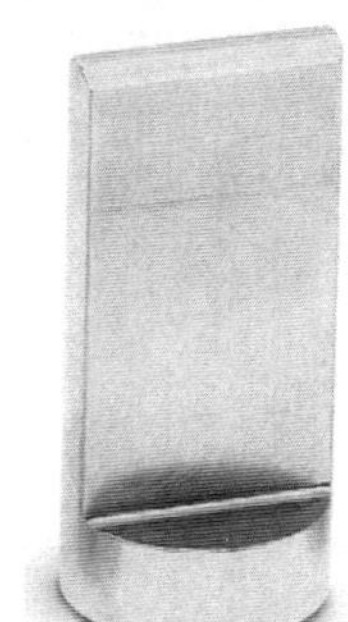

Figure 7: Fin mount

FIELD VERIFICATION OF TRADITIONAL MOUNTING METHODS

METHODOLOGY:

The Cascades' Bear Island project in Virgina manufactures recycled lightweight linerboard. In collaboration with Vibration Analyst Nicholas Allen, vibration data was collected using multiple mounting options on a refining train employed in the production of containerboard (Figure 8).

Vibration data was taken via the following methods: stud mounting, a cementing pad, a flat magnet, and a two-pole magnet.

Vibration data was taken on the gearbox input shaft in the horizontal axis. Vibration was also measured on the refiner with a horizontally mounted accelerometer. Two other vertically mounted motors used in the papermaking process were analyzed radially, generating a range of machinery inputs for which to draw conclusions.

Machinery vibration was measured using an accelerometer with a ±3dB frequency response of 0.4 Hz - 12,000 Hz. The handheld analyzer has an upper frequency response of 10 kHz and performed the velocity integrations. To achieve comparable results, the same accelerometer and analyzer were used across mounting methods and machinery.

Figure 8: Refining train. (Left) Full refining process: input, gearbox, refiner. (Right) Closeup of the refiner.

GEARBOX INPUT SHAFT MEASUREMENT COMPARISONS:

As seen in Figure 9, acceleration measurements in the horizontal axis of the gearbox input shaft show clear peaks that are repeated across the various mounting methods at roughly 1, 1.5 and 2 kHz. These are indicative of primary run speed harmonics that can be tied specifically to developing faults, a sign that the lower-level harmonics can be confidently detected across these various mounting methods.

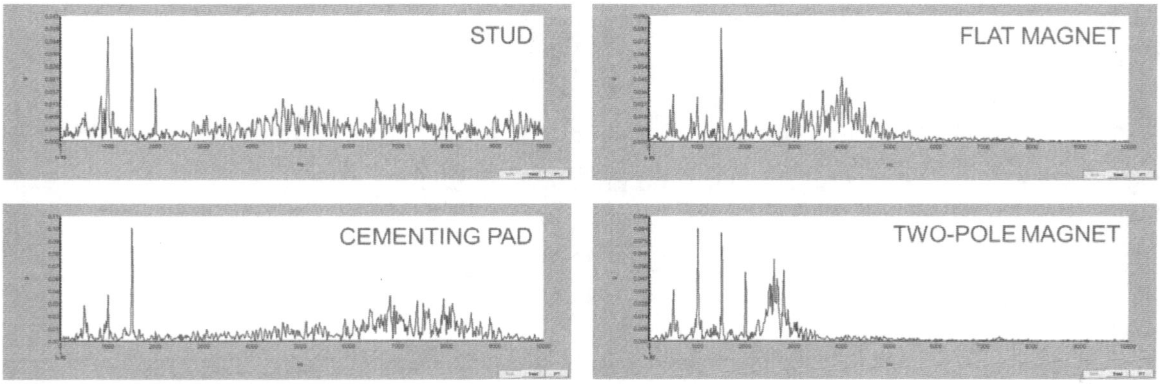

Figure 9: Gearbox input shaft horizontal acceleration comparisons.

As seen in Table 1, scaling increases across each mounting plot, showing the influence of the mounting method resonance on the vibration amplitude. The cementing pad measurements are not quite in line as expected, likely due to improper pad sealing because the surface could not be appropriately spot faced. Ideal surface preparation techniques, which should be used for all vibration monitoring applications, specifically those that are permanent, would help minimize any potential signal effects by the mounting accessory itself. Finally, high-frequency data is lost because the resonant frequency decreases in proportion to the rigidity of the mounting method.

Table 1: Gearbox input shaft horizontal acceleration comparisons.

Mounting method	Stud mount	Cementing pad	Flat magnet	Curved magnet
Scaling	0.043 g	0.11 g	0.09 g	0.094 g
Resonance	not detected	7-8 kHz	4 kHz	2.5-3 kHz
	← Rigidity			

As seen in Figure 10, velocity measurements did not result in the loss of high-frequency data due to the limited velocity bandwidth of 1 kHz. The primary run speed of ~20 Hz (1190 RPM) is visible on each of the four plots. With stud mounting, the harmonics are much clearer due to minimal high-frequency amplification. The cementing pad method is seen amplifying the 500 Hz signal, which could again be due to the improper surface finish. The flat magnet and two-pole magnet greatly increase amplification, as well as other secondary peaks (125, 175 and 425 Hz), due to their having lesser surface contact and stiffness. Aside from these specific frequencies being affected, minimal change in scaling is detected across the mounting methods, indicating little resonance impact below 1 kHz.

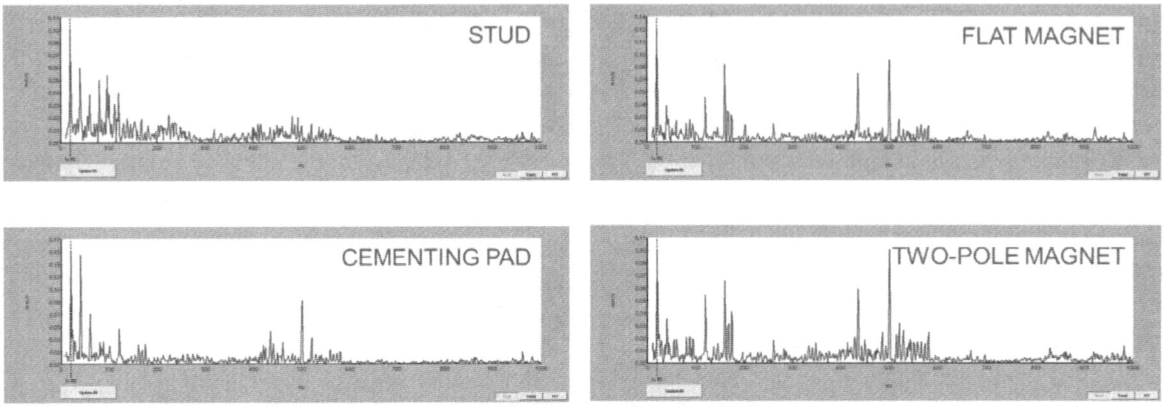

Figure 10: Gearbox input shaft horizontal velocity comparisons.

REFINER MEASUREMENT COMPARISONS:

Figure 11 shows the acceleration measurements taken on the refiner, in the horizontal axis.

- Stud mounting shows clear peaks, with a primary signal at 1325 Hz, ~3X the primary run speed of 431 Hz, and little high-frequency data.
- Cement pad mounting shows smaller peaks, though scaling is slightly higher, with a high peak around 4.5 kHz that is likely resonance of the pad itself. This is again due to imperfections in the mounting technique. Due to time constraints, the mounting surface was not as carefully prepared, in terms of removing the paint, machining a spot-face, and clearing debris for the cementing pad mounts. Because of this, and the epoxy being quickly carried from mounting location to mounting location, ideal mounting did not occur for each cementing pad location. This is primarily represented in the aforementioned resonance and some offset in overall amplitude, but otherwise did not adversely impact the data.
- Flat magnet mounting measures the 1325 Hz peak at a similar amplitude (~0.06 g), but clear resonances around 2.75 and 4.25 kHz distort the remaining data. No high frequency content is attained.
- Two-pole magnet mounting shows little data aside from a resonance at ~2.25 kHz and 0.6 g that washes out all other resolution.

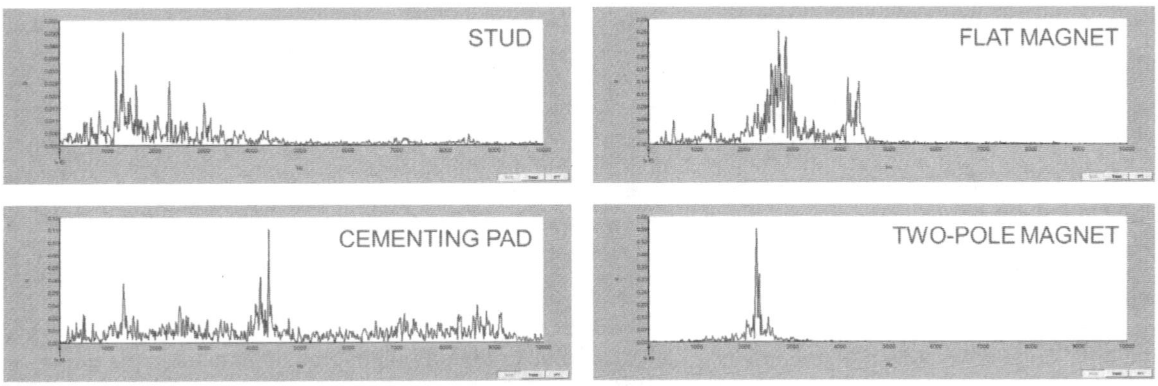

Figure 11: Refiner horizontal acceleration comparisons.

BROKE SCREEN RUN MOTOR MEASUREMENT COMPARISONS:

Radial velocity measurements taken on the main broke screen run motor are shown in Figure 12. The primary run speed of ~20 Hz (1186 RPM) is clearly visible on all four plots, with more sideband influence present on the non-stud-mounting options. The 100 Hz and ~716 Hz peaks are also clearly visible on all four plots. As the rigidity of the mounting method decreases, the ~225 Hz peak wanes in influence. This could be due to increasing scaling of the plots, which results in the resonance inducing larger vibration amplitudes.

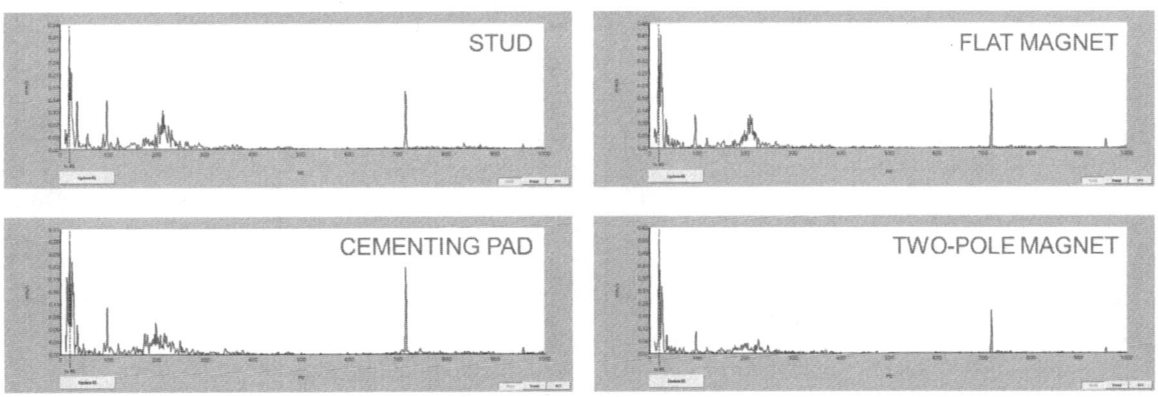

Figure 12: Main broke screen run motor radial velocity comparisons.

Figure 13 shows the radial velocity measurements taken on the trim broke screen run motor. Consistent peaks are visible between 1-2 kHz for all four mounting methods. Differing amplitudes are indicative of resonance-inducing vibration amplitude increases. Poor cementing pad fixturing contributes to elevated vibration levels visible in that measurement. Stud mounting captured data between 5-6 kHz that could be an early warning bearing failure, data which is lost in all other mounting methods.

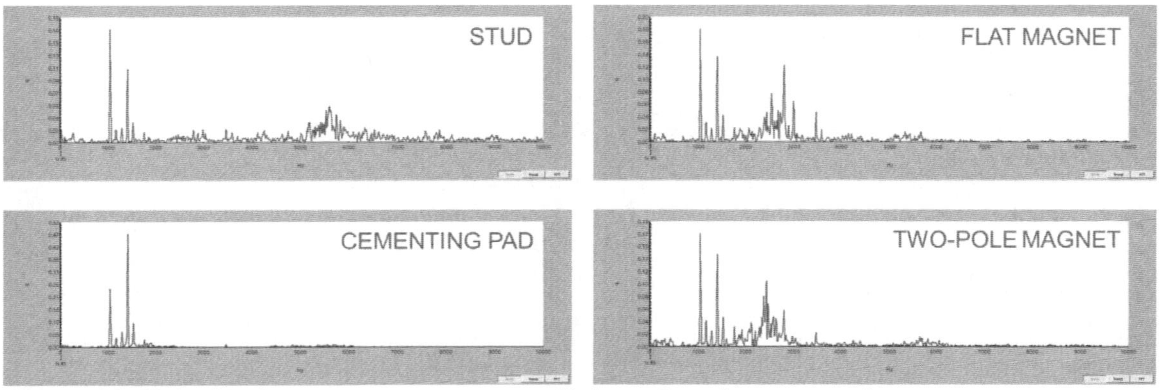

Figure 13: Trim broke screen run motor radial acceleration comparisons.

CONCLUSIONS:

Based on the field trials, stud mounting provides clear acceleration data up to and beyond 10 kHz. In this trial, the upper frequency response of 10 kHz was a limitation of the data collector. There is no indication of an upper limit between the vibration analyzer used and the sensor's max performance, but it is a consideration that needs to be taken when performing field measurements with various handheld devices.

Based on field trials, the following mounting method observations were made in relation to traditional guidance:

- Cementing pad measurements indicated a resonance around 7-8 kHz, but best-practice mounting techniques and guidance need to be used to achieve this. Best-case scenario results could accomplish traditional guideline values.
- Measurements with a flat magnet did not result in useful data above 4-5 kHz.
- Two-pole magnet measurements were rarely informative up to 3 kHz.

Additional trials using various sensors, analyzers, and mounting fixtures would be required to revise the guidance, but as always, real-life applications and techniques will determine the viability of your data.

Velocity readings up to 1 kHz did not vary greatly across mounting methods, and while there were differences in resolution and amplitudes, large signal influences remained consistent. Thus, velocity measurements using all four mounting techniques support the notion that for simple, walkaround vibration data collection applications, magnetic mounts are suitable for basic measurements and identifying primary machine faults.

Table 2: Field trials comparison to traditional mounting guidance.

Mounting method	Traditional guidance	Field trial upper limit
Stud mounting, acceleration	sensor max	> 10 kHz
Cementing pad, acceleration	10-15 kHz	7-8 kHz
Flat magnet, acceleration	5-7 kHz	4-5 kHz
Two-pole magnet, acceleration	2-5 kHz	< 3 kHz
All methods, velocity		1 kHz

AKNOWLEDGEMENTS:
Nicholas Allen, Vibration Analyst Technician & Millwright, Cascades, Bear Island Project, Virginia

BIBLIOGRAPHY:
www.wilcoxon.com